Lecture Notes in Earth Sciences 43

Norbert Clauer   Sambhu Chaudhuri (Eds.)

Isotopic Signatures and Sedimentary Records

With 151 Figures and 20 Tables

Springer-Verlag Berlin Heidelberg GmbH

Editors

Norbert Clauer
Centre de Géochimie de la Surface
1, rue Blessig, 67084 Strasbourg, France

Sambhu Chaudhuri
Kansas State University
Manhattan, KS 66506, USA

"For all Lecture Notes in Earth Sciences published till now please see final page of the book"

ISBN 978-3-540-55828-6 ISBN 978-3-540-47294-0 (eBook)
DOI 10.1007/978-3-540-47294-0

Originally published by Springer-Verlag Berlin Heidelberg New York in 1992

Typesetting: Camera ready by author

32/3140-543210 - Printed on acid-free paper

PREFACE

The emergence of new information from drilling in deep-sea and coastal areas and the surfacing of the plate tectonics theory probably had the greatest impacts in recent decades on the highly accelerated growth of knowledge regarding the evolution of sediments and sedimentary rocks. Studies in recent years have also provided new insights on global sedimentary processes, and isotopic tools in many ways have enhanced our knowledge and have provided even an unexpected added dimension to the mechanisms of some specific processes. Many different uses of isotopic tools in studies of sedimentary processes can be found in the literature, but the information is highly scattered in the vast field of sedimentology. The disseminated state of existing isotopic knowledge on sedimentary systems has undoubtedly deprived many practitioners in the field to fully appreciate the benefits and limitations, and even the apparent confusion, concerning the use of isotopic tools. We have endeavored here to bring together discussions on some major sedimentary systems in the sedimentary cycle and to analyze them according to isotopic evidence. To accomplish such a task required contributions from many individuals. We were fortunate to have friends who accepted to share our goals. We most sincerely thank all the contributors to this book and deeply appreciate their patience and fortitude despite our undue demands on them to reach our objectives.

A collective effort like this one could not be accomplished within any predetermined date, because the contributors had many other professional goals and objectives which they also had to meet. Consequently, some contributors submitted their articles earlier than others. This may be reflected in the omissions of some very recent works, but this should not be viewed as either the lack of awarness or the negligence on the part of the authors.

The news of the untimely death of our colleague and friend Douglas G. Brookins, Jr., an early explorer in the field of sedimentary isotope geochemistry, reached us in the midst of our work on this volume. Besides his unflagging enthusiasm in the works of others, he helped many to overcome the barriers of prejudice in their attempts to create a solid foundation for a professional career. Doug, consider this book a token of our appreciation for leaving us a legacy to build a truly multicultural professional world.

Sambhu Chaudhuri

Norbert Clauer

Strasbourg, January 1992

CONTENTS

INTRODUCTION TO "ISOTOPIC SIGNATURES AND SEDIMENTARY RECORDS"

Urey's (1947) suggestion, that fractionations of oxygen and carbon isotopes between water and shells of different organisms may be used to determine paleotemperatures of ocean waters, marked the dawn of isotope geochemical research on sedimentary deposits. Soon after, Wickman's (1948) proposal of using strontium isotopic data as a means of defining ages for marine chemical sedimentary rocks opened another new frontier of isotopic research on sedimentary rocks. Wickman suggested that as the ^{87}Sr content in the crustal rocks has increased with time due to the decay of ^{87}Rb to ^{87}Sr, the ^{87}Sr content of sea water deriving its Sr from weathering of crustal rocks must have also increased with time. His suggestion implied that the Sr isotopic compositions of those Rb-poor but Sr-rich marine chemical sedimentary rocks, which formed at different times, should reflect the time-dependent increase in the $^{87}Sr/^{86}Sr$ ratios of the ocean waters. Although Wickman's predictions on the magnitude and the trend of the isotopic changes proved to be wrong some twenty years later, the works that followed have led to the development of a presently existing best-estimate Sr isotope-age curve which is being increasingly used for stratigraphic correlations and understanding of the processes of global crustal evolutions.

Since the works of Urey and Wickman, isotopic approaches to studies of geologic phenomena have diversified to new dimensions to produce a large body of knowledge giving a clear identity of a research field that may be called Isotopic Sedimentology, which is the study of evolutions of organic and inorganic sediments, sedimentary rocks, waters and atmosphere from a perspective of isotopic analyses. Isotopic solutions have been a major force in recent advances to our understanding of such broad questions in the history of sedimentary cycles as:

-Ages of Sedimentary Rocks,

-Physical and Chemical Controls on Changes of Sediments to Sedimentary Rocks,

-Cycling of Sediments in the Continental Crustal Evolution,

-Tectonic and Climatic Controls on Fluxes of Chemical Elements from Different Sources into Oceans,

-Variations in Oceanic and Global Temperatures in the Past,

-Chemical Equilibria of the Present and Past Oceans,

-Evolution of Basinal Waters and their Dissolved Constituents,

-Source of Ore Fluids and Timing of Mineralization of Strata-Bound Ore Deposits,

and others. Increasing numbers of isotopic tools are coming into use providing new evidence for the resolution of these different problems and giving fresh perspectives on the dimensions of these problems. But promises from the use of isotopic tools cannot be fully realized unless the materials to be analyzed are also thoroughly characterized by relevant field, mineralogical, chemical, morphological studies.

A major aspect of the sedimentary cycle involves movements of materials of the lithosphere through many different reservoirs largely under the influence of forces of the hydrosphere, the biosphere and the atmosphere. The processes and products in this dynamic evolution are diverse and complex, and are too numerous to be covered in a single volume of texts. The limitation of space has forced us to be selective in the contents illustrating how isotopic analyses can shed much light on our understanding of these processes and products of the sedimentary cycles.

This book has endeavored to demonstrate for advanced undergraduate and beginning graduate students and interested professional geologists the values of some currently popular methods of isotope analyses in understanding a part of the sedimentary cycle which is governed largely by the interactions between the lithosphere and the hydrosphere. The topics covered here have not been meant to ignore the importance of interactions between any of these two spheres and the biosphere or the atmosphere, as influences of the biosphere and the atmosphere in many instances are inextricable from processes governed by interactions between the lithosphere and the hydrosphere. The illustrations covered in this book are by no means all encompassing. For example, little or no reference has been made to the use of Pb and Ca isotopes in many of the discussions. Very limited work in these fields is the reason for the lack of coverage in the book, but the coming years will undoubtedly see changes in the scopes of isotopic research on sedimentary materials, as technical innovations for some yet untested methods are already in progress.

As our intent has been to focus on the application of isotopes in problems related to sedimentary systems, the basis of the isotopic principles used in such studies has not been discussed or developed. Several books have discussed rigorously and elegently on the fundamental principles of different isotopic methods (e.g; Faure, 1986; Valley et al., 1986; Kyser, 1987). We will, therefore, mention briefly only those isotopic principles which have been dealt in this book.

STABLE ISOTOPE COMPOSITIONS

The composition of the stable isotopes of an element in a sample is an expression, which is now followed by almost every laboratory, of the ratio of the abundance of a heavy to that of a light isotope. For example, the hydrogen-isotope composition is taken as the D/H ratio. Similarly, the oxygen-isotope composition is the $^{18}O/^{16}O$ ratio, the carbon-isotope composition is the $^{13}C/^{12}C$ ratio, and the sulfur isotope composition is the $^{34}S/^{32}S$ ratio. As the absolute abundance ratio of the heavy isotope is extremely small relative to that of the light isotope, for which the precision of the ratio measurement is not extremely high, any small difference in the particular isotope compositions among different samples is best described relative to individual differences with some accepted standard substance. This practice of expressing the isotope composition of a substance relative to a standard substance is now an accepted mode in all investigations of stable isotope analyses. The specific isotope composition of a substance is expressed by its delta (δ) value given in per mil, which is stated as:

$$\delta = ((R_{sample} / R_{standard}) - 1) \times 1000,$$

where R is the ratio of the heavy to the light isotopes.

Differences in the physical and chemical properties among the isotopes can lead to varied degrees of partitioning of isotopes among different substances. The isotope ratio between two substances A and B is defined as the isotope fractionation factor α, which is expressed as:

$$\alpha_{A\text{-}B} = R_A/R_B$$

Since the R_A and the R_B are also expressed in terms of their δ-values, the fractionation factor $\alpha_{A\text{-}B}$ between the two substances is approximately related to the difference between their delta values by the following expression:

$$1000\ln\alpha_{A\text{-}B} = \delta_A - \delta_B$$

Thermodynamic equilibrium factor and kinetic effect are major influences that cause isotope fractionation between two substances. At temperatures in excess of 1000°K the following relationship between fractionation factor and temperature exists:

$$\ln\alpha \simeq 1/T^2$$

and consequently:

$$1000\ln\alpha_{A\text{-}B} = C/T^2 + D \text{ or } \delta_A - \delta_B = C/T^2 + D$$

where C and D are constants and T is the temperature in degrees Kelvin. The above expression may be used to determine the temperature of the equilibrium crystallization of a solid phase provided all other parameters for the equation are known and the isotopic composition of the solid phase was not altered since the equilibrium crystallization.

RADIOGENIC ISOTOPE COMPOSITIONS

The radiogenic isotope composition of an element in a substance is described by the ratio of the abundance of the radiogenic isotope to that of a non-radiogenic (or stable) isotope. For a high precision of the abundance ratio, a non-radiogenic isotope with an abundance not very greatly different from that of the radiogenic isotope, is selected and mass compared to other non-radiogenic isotopes to minimize the effect of analytical mass fractionation. Thus for Ar, the isotopic composition is expressed by $^{40}Ar/^{36}Ar$, for Sr by $^{87}Sr/^{86}Sr$, for Nd by $^{143}Nd/^{144}Nd$, for Pb by $^{206}Pb/^{204}Pb$, $^{207}Pb/^{204}Pb$, and $^{208}Pb/^{204}Pb$, and so forth.

The abundances of any radiogenic isotope of an element in sedimentary materials are varied, due to variations in the sources of the materials or in the chemical environments of their formation and in the growth of the radiogenic isotopes from time-dependent natural decay of the radioactive isotopes. The decay controlled growth of a radiogenic isotope is expressed by the well known equation for the decay of the radioactive isotope. The equation for the decay of a radioactive isotope is given as:

$$N = N_0 \, e^{-\lambda t}$$

where N is the number of the radioactive parent isotope present in time t, N_0 is the number of the radioactive isotope present initially, and λ is the decay constant. But this equation cannot be used directly in most instances to calculate the time for geologic materials, as we will never know the value for N_0. Hence the above equation is modified to express in such a way that we can calculate the value for t when all other parameters are known. The modified expression is given as:

$$D = N_0 - N = Ne^{\lambda t} - N = N\,(e^{\lambda t} - 1)$$

where D is the number of the radiogenic daughter isotope and all other parameters are the same as in the equation above. Both sides of the equation for the growth of radiogenic isotope may be divided by a constant factor, or in this case by a non-radiogenic or stable isotope of the same element as that of the radiogenic isotope, without changing the mathematical relationship. The discussion which follows gives expressions for Ar, Sr and Nd isotopic compositions of natural materials.

The K-Ar Method

The argon isotopes in natural materials consist essentially of ^{36}Ar, ^{38}Ar, and ^{40}Ar. Of these three, the ^{40}Ar is produced from the branched decay (electron capture) of ^{40}K which also decays to produce ^{40}Ca with beta emission. The decay constants for ^{40}K are: λ_{ec} = 0.581 x 10^{-10} yr^{-1}, and λ_{total} = 5.543 x 10^{-10} yr^{-1}. The radiogenic growth of ^{40}Ar is expressed as:

$$^{40}Ar_{rad} = (\lambda_{ec} / \lambda_{total})\ ^{40}K\ (e^{\lambda t} - 1).$$

The K-Ar age is calculated from the ratio of the radiogenic ^{40}Ar to ^{40}K, after subtracting any atmospheric Ar component. For the calculated age to be geologically meaningful, the sample should not have contained any extraneous ^{40}Ar and must have remained closed to both argon and potassium since the time of its formation.

The Rb-Sr Method

In natural materials, strontium consists of four isotopes: ^{84}Sr, ^{86}Sr, ^{87}Sr, and ^{88}Sr. The isotopic composition of strontium in natural materials is described by the $^{87}Sr/^{86}Sr$ ratio. As ^{87}Rb, one of the two isotopes of rubidium in natural materials and the other being ^{85}Rb, decays with beta emission to produce ^{87}Sr, the $^{87}Sr/^{86}Sr$ ratios of natural materials are varied. The decay constant of ^{87}Rb is considered to be 1.42 x 10^{-11} yr^{-1}. Following the common mode of expression for the growth of radiogenic isotopic composition of an element, the growth of the radiogenic Sr is given as:

$$(^{87}Sr/^{86}Sr) = (^{87}Rb/^{86}Sr)\ (e^{\lambda t} - 1)$$

But for an age determination of any natural material which is always likely to contain some amount of strontium at the time of its formation, we must know their initial strontium isotopic composition, in which case the equation for the strontium isotopic composition can be written as:

$$(^{87}Sr/^{86}Sr)_{total} = (^{87}Sr/^{86}Sr_{initial} + {}^{87}Rb/^{86}Sr)\ (e^{\lambda t} - 1)$$

In the absence of knowledge of the initial isotopic composition of any mass of analyzed natural material, the age can be calculated only when different components or minerals of the same material have sufficiently different $^{87}Rb/^{86}Sr$ (or Rb/Sr) ratios, but had the same initial $^{87}Sr/^{86}Sr$ isotopic ratio. The age can be determined graphically by plotting the analyzed data in the co-ordinates of $^{87}Sr/^{86}Sr$ as the Y-axis and $^{87}Rb/^{86}Sr$ as the X-axis. A least-square fit of the data gives the best slope of the line from which the age can be calculated (Fig. 1). The line is known as an isochron and the method of calculating an age by such a method is commonly called isochron method.

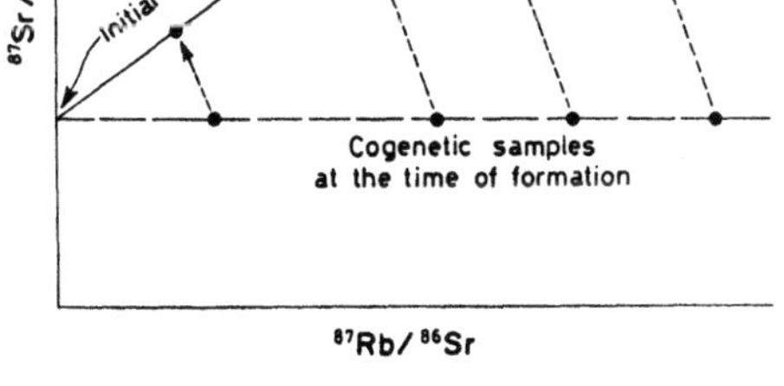

Figure 1: Sketch of a Rb-Sr isochron diagram

The Sm-Nd Method

Samarium and neodymium each has seven isotopes in natural materials. ^{147}Sm decays by alpha emission to ^{143}Nd with a decay constant, λ , of 6.54×10^{-12} yr^{-1}. The isotopic composition of neodymium in a geologic material is expressed by the $^{143}Nd/^{144}Nd$ ratio, and the growth of the Nd isotopic ratio from the decay of ^{147}Sm through time is expressed by:

$$(^{143}Nd/^{144}Nd)_{total} = (^{143}Nd/^{144}Nd)_{initial} + (^{147}Sm/^{144}Nd)\ (e^{\lambda t} - 1)$$

Both samarium and neodymium belong to the group of rare-earth elements, and their geochemical behaviors in natural crustal systems are quite similar. Consequently, the Sm/Nd ratios did not fractionate significantly in crustal materials.

The neodymium isotopic composition of the "bulk" earth has been commonly assumed to have evolved in a uniform reservoir whose Sm/Nd ratio is equal to that of chondritic meteorites, and this uniform reservoir is known as the "Chondritic Uniform Reservoir" (CHUR). The present-day ratios of $^{143}Nd/^{144}Nd$ and $^{147}Sm/^{144}Nd$ for the CHUR are considered to be 0.512638 and 0.1967, respectively, using a $^{146}Nd/^{144}Nd$ ratio of 0.7219 for fractionation correction. The differences in Nd isotopic compositions can also be expressed in terms of ϵ-values, which are deviations in parts per 10,000 from the CHUR evolution. The $_{Nd}$ expression for a sample relative to the CHUR today is as follows:

$$\epsilon_{Nd}(0) = ((^{143}Nd/^{144}Nd)_{sample}/(^{143}Nd/^{144}Nd)_{CHUR} - 1) \times 10^4$$

Positive $_{Nd}$ values point to mantle-type sources for the samples with Sm/Nd ratios higher than that of CHUR and negative $_{Nd}$ values indicate that the samples originated from crustal sources with Sm/Nd ratios lower than that of CHUR.

The Sm-Nd isotopic data on sediments can be used to calculate a "model" age for the provenance of the sediments. This "model" age expresses the time when the neodymium in the sediment had the same isotopic ratio as that of the reference reservoir, which could be the depleted mantle reservoir or the CHUR. Therefore, this model age or so-called crustal residence age is an estimation of how long the Nd of a rock resided in the crust. Model ages are calculated by:

$$t = 1/\lambda \ln [(D_{sample} - D_{CHUR}) / (P_{sample} - P_{CHUR}) + 1]$$

where D refers to the $^{143}Nd/^{144}Nd$ ratio, P refers to the $^{147}Sm/^{144}Nd$ ratio and subscripts sample and CHUR to the analyzed sample and the CHUR reservoir, respectively.

The U-Th-Pb Method

Naturally occurring materials have three isotopes of uranium: ^{238}U, ^{235}U, and ^{234}U and one isotope of thorium, ^{232}Th, which are all radioactive. Each with a long chain of decays, ^{238}U, ^{235}U, and ^{232}Th are transformed to stable isotopes of ^{206}Pb, ^{207}Pb, and ^{208}Pb, respectively. ^{234}U is an intermediate daughter product in the decay scheme of ^{238}U. The time-dependent growths of lead isotopes from radioactive decays of the uranium and thorium isotopes are expressed as:

$$(^{206}Pb/^{204}Pb) = (^{206}Pb/^{204}Pb)_{initial} + (^{238}U/^{204}Pb)\ (e^{xt}-1),$$
$$(^{207}Pb/^{204}Pb) = (^{207}Pb/^{204}Pb)_{initial} + (^{235}U/^{204}Pb)\ (e^{yt}-1),$$
$$(^{208}Pb/^{204}Pb) = (^{208}Pb/^{204}Pb)_{initial} + (^{232}Th/^{204}Pb)\ (e^{zt}-1),$$

where x, y, and z are decay constants of ^{238}U ($1.55125 \times 10^{-10} yr^{-1}$), ^{235}U ($9.8485 \times 10^{-10} yr^{-1}$), and ^{232}Th ($4.9475 \times 10^{-10} yr^{-1}$), respectively.

The equations for the growths of ^{207}Pb and ^{206}Pb may be combined to express the ^{207}Pb-^{206}Pb equation:

$$(^{207}Pb/^{206}Pb)_{radiogenic} = (^{235}U/^{238}U)\ (e^{yt}-1)/(e^{zt}-1)$$

where the ratio of $^{235}U/^{238}U$ is a constant with the value of 1/137.88 for terrestrial materials with normal U isotopic composition.

The dating of geologic materials by individual U-Pb isotope

method gives highly discordant ages, probably as a result of differential loss of radiogenic lead during the geologic history of the materials. A reliable age may be obtained by the use of the combined ^{207}Pb-^{206}Pb data.

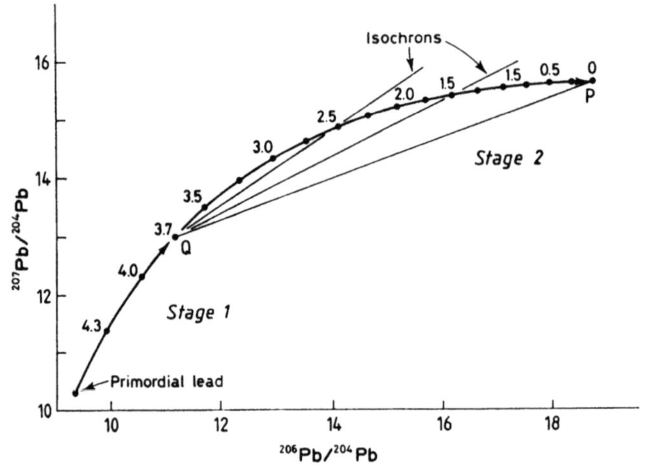

Figure 2: Two-stage evolution of Pb isotopes (Stacey and Kramers, 1975). Stage 1 involves evolution from 4.57 to 3.7 Ga in a reservoir with $^{238}U/^{204}Pb$ of 7.192. The evolution then continues from 3.7 Ga, Q, to the Present, P, in a reservoir with $^{238}U/^{204}Pb$ changed to 9.735 by chemical differentiation. The straight lines between 3.7 Ga and the Present are isochrons (Adopted from Faure, 1986)

Ore minerals, whose lead has been separated from uranium and thorium at the time of their formation so that their isotopic composition has been conserved, may be dated by one of several models with single or multiple stage evolution which has been proposed for the common lead. A single stage model assumes that the isotopic compositions of common lead in all samples evolved in different environments with different U/Pb and Th/Pb ratios from the same primordial isotopic value at 4.55 x 10^9 yr until the lead is separated from these different sources to form lead ore minerals. The lead isotopic compositions of different samples of ore deposits would then define an isochron in the $^{207}Pb/^{206}Pb$ and $^{206}Pb/^{204}Pb$ coordinates, and an age can be calculated from slope of the line. A frequently used multi-stage model is one of two stage evolution model (Stacey and Kramers, 1975) which assumes that the lead evolved first from primordial isotopic value between 4.55 and 3.7 Ga in a reservoir with $^{238}U/^{204}Pb$ of 7.19 and $^{232}Th/^{204}Pb$ of 32.21, and then as a result of differentiation at 3.7 Ga, in an reservoir with $^{238}U/^{204}Pb$ ratio of 9.735 and $^{232}Th/^{204}Pb$ ratio of 36.937 until the leads were separated from these sources. An isochron defined by the $^{207}Pb/^{204}Pb$ and $^{206}Pb/^{204}Pb$ data relate to the time elapsed since the lead in a sample was isolated from second reservoir (Fig. 2).

ACKNOWLEDGEMENTS

We sincerely thank Peter Stille (Centre de Géochimie de la Surface, Strasbourg) for his helpful comments.

REFERENCES

Faure G. (1986) Principles of Isotope Geology. Second edition. J. Wiley & Sons, New York, N.Y., 589 p.

Kyser T.K. (ed.) (1987) Stable isotope geochemistry of low temperature processes. Miner. Soc. Canada, Short Course Handbook, 13, 452 p.

Stacey J.S. and Kramers J.D. (1975) Approximation of terrestrial lead isotope evolution by a two-stage model. Earth Plan. Sci. Lett., 26, 207-221.

Urey H.C. (1947) The thermodynamic properties of isotopic substances. Jour. Chem. Soc., 1947, 562-581.

Valley J.W., Taylor H.P.Jr. and O'Neil J.R. (eds.) (1986) Stable isotopes in high temperature geological processes. Miner. Soc. Amer., Rev. in Miner., 16, 570p.

Wickman F.E. (1948) Isotope ratios - A clue to the age of certain marine sediments. Jour. Geol., 56, 61-66.

Sam CHAUDHURI Norbert CLAUER

DEPOSITIONAL AND DIAGENETIC HISTORY OF LIMESTONES: STABLE AND RADIOGENIC ISOTOPES

Ján VEIZER
Institut für Geologie, Ruhr Universität
4630 Bochum, Federal Republic of Germany
and
Derry/Rust Research Group, Ottawa-Carleton
Geoscience Centre, University of Ottawa
Ottawa, Canada K1N 6N5

Carbonates represent one of the dominant rock types in the sedimentary shell of this planet. They contain a wealth of information about the nature of sedimentary environments, their temporal evolution and their post-depositional history. Carbonates are also commonly repositories of mineral wealth, including metals, hydrocarbons, and water, and they are also the major resource for production of cement aggregate and building stones. Information contained in these rocks, if properly deciphered, goes beyond what we usually perceive as geology <u>sensu stricto</u>. Carbonates reflect the evolution of the ocean-atmosphere system over geologic time and permit to decipher its nature and variability prior to the advent of man as a geological agent. Environmental questions are emerging as perhaps the most important political issue of our times and they may dominate the forthcoming research agenda. If properly understood and applied, the isotope techniques discussed briefly in the subsequent section can serve as powerful tools for study of problems with scientific as well as economic and environmental overtones.

ISOTOPE SYSTEMATICS APPLIED TO CARBONATE ROCKS

Carbonate sediments and carbonate rocks contain a plethora of isotopic tracers that can be utilized either for dating (absolute or relative) or for tracing of processes involved in their formation and post-depositional stabilization.

The isotope systematics for dating purposes include the ^{14}C and the so called U-series disequilibrium methods, all utilized for dating of Holocene sediments up to several hundred thousand years old. These geochronometers are not the subject of the present chapter, but their overview is available in Faure (1986) and Bernat and Church (1989). In addition, the last few years witnessed an effort by several laboratories (e.g. Jahn et al., 1990) to utilize the U-Pb geochronometer for absolute dating of carbonates. This geochronometer has a potential for dating of old, particularly Precambrian, carbonate rocks, because their ages are often so poorly constrained that, even with large uncertainties, the technique may provide useful data.

The present chapter will concentrate instead on techniques that are applicable to tracing of processes involved in production of carbonate components and in their subsequent inversion into limestones. Such tracers include the isotopic ratios of $^{18}O/^{16}O$, $^{13}C/^{12}C$ and $^{87}Sr/^{86}Sr$. In specific situations, these tracers can be utilized also for "Isotopic Stratigraphy", that is for correlation and dating purposes.

INCORPORATION OF ISOTOPES INTO CARBONATE MINERALS

The bulk of modern carbonate sediments originate in the marine realm. This has been even more the case for ancient limestones, because lake sediments have a low preservation potential due to the ease of their erosional dispersal. I shall therefore concentrate on isotopic properties of marine carbonates. A good review of isotopic properties for non-marine environments is available in Buchardt and Fritz (1980) and Talbot (1990).

Carbonate components formed in sea water are orthorhombic aragonite (A) and rhombohedral calcite, the latter further subdivided into low-Mg calcite (LMC) with <4 mole % $MgCO_3$ (Chave, 1954, a,b), intermediate-Mg calcite (iHMC) with 4-12 mole % $MgCO_3$ and high-Mg calcite sensu stricto (hHMC) with 12-28 mole % $MgCO_3$ (Milliman, 1974, p. 267). The molar Mg/Ca ratio for present-day sea water is 5:1 and inorganically precipitated phases should be either A or HMC with about 7 mole % $MgCO_3$ (Fig. 3.1 in Morse and Mackenzie, 1990). LMC in marine environments is known mostly from intracellularly precipitated shells of some organisms, such as pelagic foraminiferans, that manage to retard the diffusion of Mg^{2+} through cellular tissue during shell secretion. Other organisms have shells composed of various combinations of A, HMC and LMC. On average, the shallow marine carbonate sediments, containing inorganically precipitated phases and skeletal debris, are composed of variable mixtures of HMC and A. As a generalization, the Mg content of HMC correlates with temperature and thus declines with latitude and/or depth (Mackenzie et al., 1983). Deep-sea oozes, on the other hand, consist almost exclusively of pelagic skeletal material (e.g. coccoliths and foraminiferans) and are predominantly LMC. Thus, primary minerals which compose the sediment become exposed to the influence of meteoric waters and/or are subjected to higher pressures and temperatures during burial, rendering the original mineralogical assemblages unstable and converting them into a monomineralic limestone (LMC) or dolostone (Land, this volume).

The incorporation of O and C isotopes into carbonate minerals is summarized well in O'Neil (1986). It is governed by the so called fractionation factor:

$$R_S = \alpha_{S\text{-}W} R_W$$

where R is the ratio of relative abundances of isotopes, in this instance $^{18}O/^{16}O$ and $^{13}C/^{12}C$, the subscript S represents the solid carbonate phase, and W stands for water (and its solutes, such as carbon species). Since α-values are close to unity and their variations are mostly in the third decimal place, the difference in α_{s-w} is better expressed in whole units as per mil (0/0$_0$). Thus,

$$\Delta_{s-w} = [(R_s/R_w) - 1] \times 10^3 = (\alpha_{s-w} - 1) \times 10^3.$$

For reasons connected with the design of mass spectrometers, it is easier to compare the measured isotope ratio of an unknown sample R_s to a standard with known R, and to express the measured difference as

$$\delta_s = [(R_s/R_{standard}) - 1] \times 10^3$$

Similarly, δ_w can be expressed as a difference relative to the same standard. The fractionation factor can be obtained from the measured δ's (Hoefs, 1980) as

$$_{s-w} = \delta_s - \delta_w \cong \Delta_{s-w} \simeq 10^3 \times \ln \;_{s-w}.$$

For oxygen and carbon isotope studies in carbonate rocks, the standard usually utilized is the PDB (<u>Bellemnitella americana</u> from the Cretaceous Peedee Formation of South Carolina; McCrea, 1950). This standard is no longer available and most laboratories utilize secondary standards calibrated relative to PDB. For oxygen, the results are frequently reported also relative to the SMOW standard (Standard Mean Ocean Water; Craig 1961). The conversion equations between V-SMOW (Vienna SMOW) and PDB are (Coplen et al., 1983):

$$\delta^{18}O_{V-SMOW} = 1.03091 \times \delta^{18}O_{PDB} + 30.91$$

and

$$\delta^{18}O_{PDB} = 0.97002 \times \delta^{18}O_{V-SMOW} - 29.98.$$

The equilibrium isotopic fractionation factor is inversely proportional to temperature and the relationship takes the form:

$$\ln\alpha = AT^{-2} + BT^{-1} + C$$

where A, B, and C are constants determined experimentally and T is absolute temperature in Kelvins. At surficial to high temperatures, the term BT^{-1} becomes negligible and the fractionation factor α is proportional to T^{-2} (Bigeleisen and Mayer, 1947).

Carbon Isotopes

The temperature dependence of α for C isotopes is relatively small. Gaseous CO_2 that diffuses into sea water, dissociates, and can finally precipitate as a $CaCO_3$ mineral, with fractionation involved at each step. The temperature dependencies of fractionation factors between CO_2 (gas) and the CO_2 (aq) (α_1), HCO^{3-} (α_2), CO_3^{2-} (α_3) and calcite (α_4), respectively, are as follows (Emrich et al., 1970; Deines et al., 1974):

$$1000\ln\alpha_1 = 0.0063\ (10^6\ T^{-2}) - 0.91,$$
$$1000\ln\alpha_2 = 1.099\ (10^6\ T^{-2}) - 4.54,$$
$$1000\ln\alpha_3 = 0.87\ (10^6\ T^{-2}) - 3.4, \text{ and}$$
$$1000\ln\alpha_4 = 1.194\ (10^6\ T^{-2}) - 3.63.$$

The overall variations in $\delta^{13}C$ of natural calcites induced by temperature are usually less than the scatter caused by other factors (about 0.04 per mil per 1°C). Theoretically, the $CaCO_3$ should be enriched by about 1-3 per mil in ^{13}C relative to the total dissolved C, because bicarbonate is the dominant C species in natural waters. In addition, aragonite should be somewhat enriched in ^{13}C relative to calcite. At 25°C, the theoretical calculations yield about 0.9 per mil enhancement, whereas the observed experimental enrichment has been about 1.8 per mil (Rubinson and Clayton, 1969).

Oxygen Isotopes and Paleothermometry

In contrast to C, O isotope fractionation is strongly temperature dependent. As a rule of thumb, the $\delta^{18}O$ become depleted by about 1 per mil for every 4°C at near-surficial temperatures. The actual α is, however, somewhat dependent on the mineralogical and chemical form of the $CaCO_3$ phase. The pertinent equation for calcite (O'Neil et al., 1969) is:

$$1000 \ln \alpha_{C-W} = 2.78\ (10^6 T^{-2}) - 3.39.$$

Aragonite should be enriched, up to 1 per mil, in ^{18}O relative to calcite, and for Mg-calcites the enrichment is about 0.06 per mil, for every mole per cent of $MgCO_3$ (Tarutani et al., 1969). The above equation can be modified to permit a temperature estimate in °C (Epstein et al., 1953; Craig, 1965; Hays and Grossman, 1991).

$$T(^\circ C) = 15.7 - 4.36\ (\delta^{18}O_c - \delta^{18}O_W) + 0.12\ (\delta^{18}O_C - \delta^{18}O_W)^2,$$

where $\delta^{18}O_c$ and $\delta^{18}O_W$ are in PDB and SMOW, respectively (Anderson and Arthur, 1983). This is the basis of oxygen isotope paleothermometry and a grid of combinations of T, $\delta^{18}O_c$ and $\delta^{18}O_W$ is given in figure 1. For aragonite, the equation is slightly different (Grossman and Ku, 1986), but has a similar slope.

$$T(^\circ C) = 19.0 - 3.52\ (\delta^{18}O_A - \delta^{18}O_W) + 0.03\ (\delta^{18}O_A - \delta^{18}O_W)^2.$$

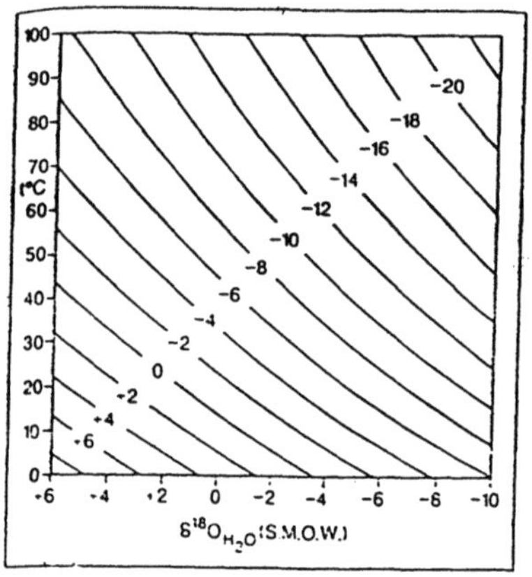

Figure 1: A grid of equilibrium relationships between $\delta^{18}O_W$ (SMOW), temperature (°C) and the $\delta^{18}O_{calcite}$, the latter expressed in per mil PDB on diagonal lines. Modified from Tucker and Wright (1990)

The equations contain three unknowns, so that one needs to measure the $\delta^{18}O$ of a mineral <u>and</u> water to utilize this paleothermometer. In most cases, the ambient water is no longer available. For marine minerals, the SMOW value of 0 per mil is usually substituted, but this is not without problems as will be discussed later. Alternatively, it is theoretically possible to utilize a cogenetic mineral phases in a manner similar to that of high temperature studies (see Faure, 1986; O'Neil, 1986; Valley, 1986). In this case, the $\delta^{18}O$ of ambient water is, by definition, the same for all minerals and this enables calculation of the formation temperature. Unfortunately, for most mineral phases in question - carbonates, phosphates, cherts - the $\Delta^{18}O$ (mineral A-B) changes little with temperature and this renders such pairs inapplicable for paleotemperature studies in sedimentary systems. In addition, although the minerals coexist frequently, their cogenetic nature is almost always questionable.

Disequilibrium Phenomena

In nature, observed isotopic ratios often deviate from calculated equilibrium values because of kinetic factors. Of these, perhaps the rate of precipitation of solid phases is the most important variable. Usually the faster the rate, the closer is the α to unity (e.g. Turner, 1982). The situation is still more complex if biomineralization is involved. Secretion of carbonate skeletons may proceed under equilibrium conditions or result in non-equilibrium fractionation of isotopes (O, C or both), known as biogenic fractionations or vital effects (Figs. 2 and 6). Such effects may be a consequence of biochemical processes that influence the kinetics of shell formation or, particularly for C, result from incorporation of metabolic CO_2 into $CaCO_3$ (Land, 1969; McConnaughey, 1990a,b). When this deviation is systematic, as for some foraminiferan species, it can be corrected for. A detailed review of these phenomena is available in Williams et al. (1988). In other cases, the degree - but not necessarily the sign - of disequilibrium is random. It is also possible, and by no means exceptional, to see different parts of the same skeleton having different isotopic composition.

The above discussion is not intended to discourage the potential user, only to point out that the rigorous theory has to serve only as a framework for practical applications. Nature obeys these rules to the extent of statistical approximations.

<u>Strontium Isotopes</u>

In contrast to O and C, the isotopes of Sr, ^{87}Sr and ^{86}Sr, are incorporated into carbonate minerals with no measurable isotope fractionation, because their mass difference is only 1.1%. The $^{87}Sr/^{86}Sr$ values are usually reported in absolute ratios of 0.70XXX and can now be reliably measured into the fifth decimal place. This usage is a consequence of the design of mass spectrometers. The O

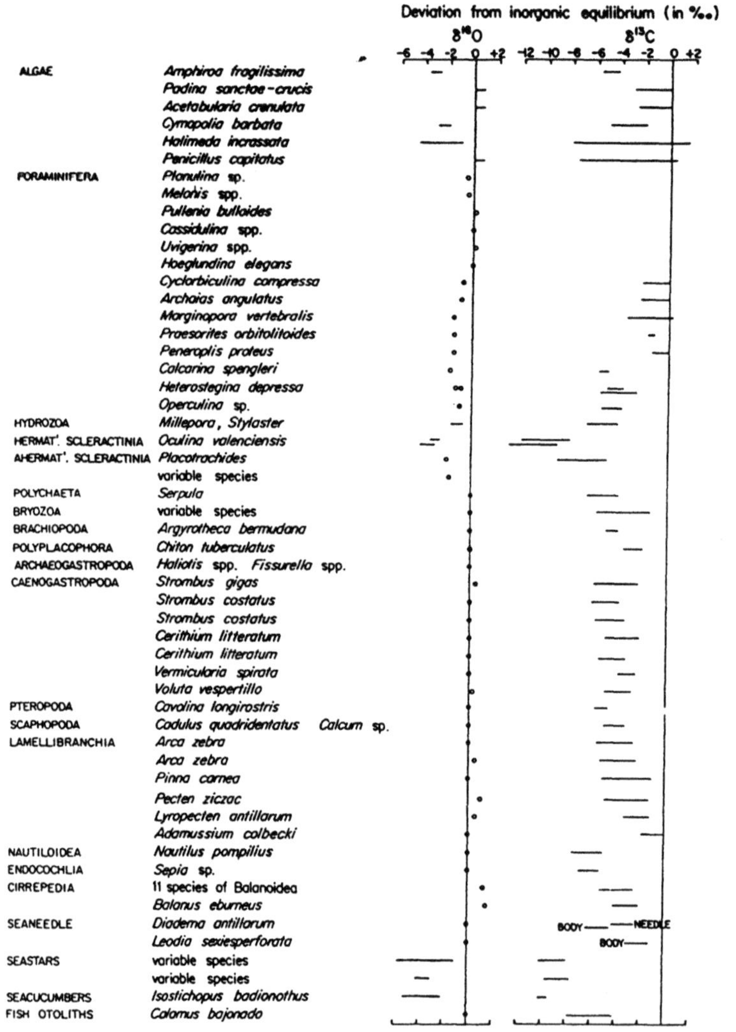

Figure 2: $\delta^{18}O$ and $\delta^{13}C$ departures from equilibrium values due to vital effect. Modified from Wefer (1983) and reproduced from Veizer (1983a)

and C isotopes are measured on CO_2 gas that is introduced into the source of ionization through a capillary. The passage through this capillary is accompanied by kinetic fractionation effects. For this reason, the measurement of an unknown sample must be immediately followed by a standard CO_2 gas with known isotopic composition that is introduced into the source via an analogous capillary. Isotopic values, if reported as δ (sample minus standard), selfcorrect for such kinetic factors. Sr samples, on the other hand, are introduced into the source of the mass spectrometer as solids, but the results - even among leading laboratories - differ in the fifth decimal place. They must be therefore normalized to a somewhat arbitrary value recommended for a given standard. Usually it is the Eimer and Amend $SrCO_3$ with an accepted $^{87}Sr/^{86}Sr$ ratio of 0.70800. Other standards include the NBS Sr carbonate (SRM 987), the USGS Tridachna shell (EN-1), and sea water. The results on carbonate samples are increasingly often reported in a way similar to that of the stable isotopes:

$$\delta^{87}Sr = [\ (^{87}Sr/^{86}Sr_{sample}/^{87}Sr/^{86}Sr_{sea\ water}) - 1] \times 10^5.$$

Since the $^{87}Sr/^{86}Sr$ of ocean sea water is uniform within analytical capabilities, the $\delta^{87}Sr$, reported relative to measurements of sea water (or EN-1) by the same laboratory, selfcorrects for interlaboratory differences. Further background to Sr isotope systematics is available in Faure and Powell (1972) and Faure (1986).

ISOTOPE CHARACTERISTICS OF NATURAL WATERS

Carbonate minerals, in one way or another, reflect the variability inherent in the composition of their parent solutions. It is therefore necessary to understand the isotopic patterns observed in natural waters and the dynamics that is involved in creation, maintenance, evolution, and dispersal of such patterns. The situation is much simpler for Sr than for O and C. The $^{87}Sr/^{86}Sr$ distribution in natural waters will therefore serve as a starting point of discussion.

Strontium

Present-day sea water contains about 8 μg/g Sr with a uniform isotopic composition. As already pointed out, the actual value varies somewhat between laboratories. For example, MOBIL (Burke et al., 1982), Cambridge (Elderfield, 1986), USGS (Ludwig et al., 1988), UCLA (DePaolo, 1986), Florida (Hodell et al., 1990) and our Bochum laboratory reported the following values: 0.70907, 0.70924, 0.70925, 0.70923, 0.70917 and 0.70914, respectively. These minute differences are not solely of academic interest because, as will be shown later, they are important for the so called "high resolution isotope stratigraphy". For future work, it is therefore advisable to report the Sr isotopic ratios in ^{87}Sr units.

In contrast to sea water, the Sr concentrations and isotopic compositions of rivers and lakes are highly variable. Nonetheless, most rivers and lakes have about two orders of magnitude lower Sr concentrations and they are enriched in radiogenic ^{87}Sr (Palmer and Edmond, 1989). The global average river flux has $^{87}Sr/^{86}Sr$ of 0.711, with Sr concentrations of 60 pg/g.

The residence time of Sr in the ocean is perhaps 2-4 million years (Holland, 1984; Hodell et al., 1990). Whatever the correct estimate, the input of extraneous Sr into the ocean is much slower than the oceanic mixing rate of some thousand years. This is the same reason why, at any given time, the open ocean water has a uniform Sr isotopic ratio. This ratio is a conservative property for even the most marginal marine environments, because the entire Sr endowment in any reasonable mixture of dilute riverine and salty sea water derives dominantly from sea water (cf. Veizer, 1989).

Oxygen

The variation in O isotopic composition of sea water is illustrated in figure 3. In the open ocean, the spread is at least 1.5 per mil. This range can even be exceeded in marginal seas and lagoons. High evaporation rates cause ^{18}O enrichment, while hyposaline and brackish situations, mostly caused by riverine influx, result in ^{18}O depletion. Usually the slope is approximately 0.1 per mil $\delta^{18}O$ for each 1 per mil salinity change between 0 and about 80 per mil (Ferronsky and Brezgunov, 1989; Anati and Gat, 1989).

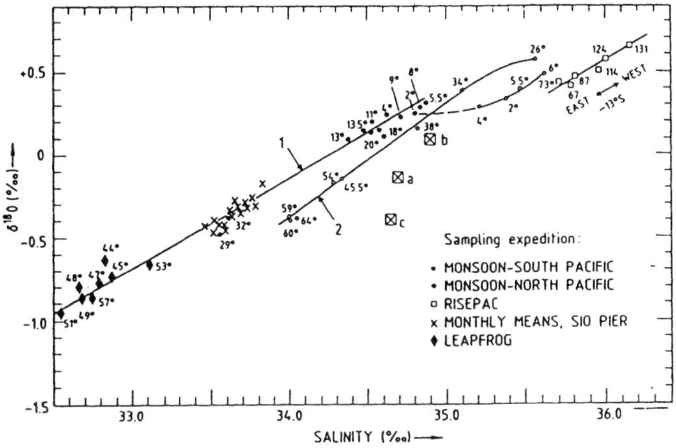

Figure 3: Variation of O isotopic composition in sea water. Modified from Ferronsky and Brezgunov (1989) and based on the data of Craig and Gordon (1965)

For low-latitude surficial waters of the open oceans, the loci of the bulk of carbonate production today, the $\delta^{18}O$ variations are usually less than those illustrated in figure 3 and they correlate positively with the net evaporation/precipitation (and runoff) balance in a given region. The ^{18}O depletion of deep waters, and of the circum-Antarctic surface waters (Fig. 3), is due to the dilution by cold and thus dense glacial melt waters that - for the reasons discussed below - are very depleted in ^{18}O. A complete melting of the present polar ice caps would decrease the $\delta^{18}O$ of the entire ocean by some 0.8 to 1.3 per mil (Savin and Yeh, 1981). This may have been the situation in the geological past.

The O isotopic composition of <u>meteoric water</u> is controlled by the process of evaporation/precipitation. As sea water evaporates at low-latitudes, the vapor is depleted in ^{18}O. This is because the $\alpha_{vapor-water}$ is inversely proportional to temperature. Subsequently, as the vapour diffuses across the surficial diffusion layer, additional kinetic fractionation occurs. The combined kinetic enrichment factor $\Delta\epsilon$ (Gonfiantini, 1986) can be expressed as:

$$\Delta\epsilon^{18}O \text{ (per mil)} = 14.2^{(1-h)}$$

where h is the relative humidity normalized to the surface temperature of the liquid. Cooling and condensation of the water vapor results in a reverse equilibrium fractionation. In this case, it is the heavy ^{18}O that is preferentially incorporated into condensing droplets. As the clouds move to higher, and colder, altitudes and latitudes, they and the precipitation become progressively more depleted in ^{18}O (Fig. 4).

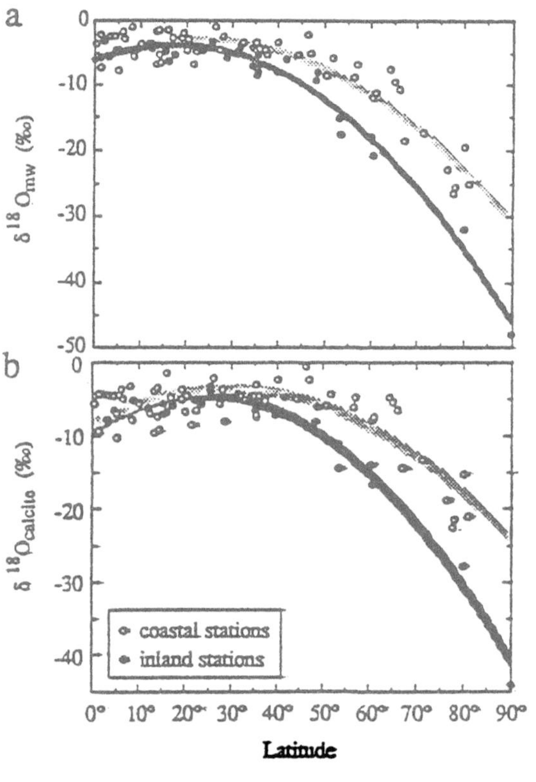

Figure 4: Mean $\delta^{18}O$ values of atmospheric precipitation (a) and equilibrium $\delta^{18}O$ values of calcite (b) vs. latitude for coastal and island stations. Reproduced from Hays and Grossman (1991). The $\delta^{18}O$ is in SMOW and PDB for water and calcite, respectively

This process, known as Rayleigh distillation, can be expressed as:

$$R_t = R_o f^{(x-1)}$$

where R_t is the $\delta^{18}O$ at time t, R_0 is the initial $\delta^{18}O$, f is the fraction of vapor remaining in the cloud, and x is the fractionation factor. The depletion gradient is somewhat steeper over the continents than over the oceans (Yurtsever, 1975; Gat, 1980; Hays and Grossman, 1991).

The net result of the above dynamics is that most, except some highly evaporated, meteoric waters are depleted in ^{18}O relative to sea water. Yet, in the low-latitudes that are the loci of carbonate production, the depletion is moderate, usually within 6 per mil (Fig. 4). Calcites precipitating in equilibrium with such meteoric waters should therefore have $\delta^{18}O_{PDB}$ of about -5 ± 2 per mil.

<u>Carbon</u>

The two dominant exogenic reservoirs of C are carbonate rocks and organic matter in sediments. They are linked in the carbon cycle via atmospheric CO_2 and the C species dissolved in the hydrosphere. The $\delta^{13}C$ for the total dissolved C (TDC) in <u>sea water</u> is about +1 ± 0.5 per mil (PDB), with surficial waters generally heavier and deep waters lighter than this average (Kroopnick, 1980; Tan, 1989). Atmospheric CO_2 in equilibrium with TDC of marine surface water has $\delta^{13}C$ of about -7 per mil. CO_2 is preferentially utilized by photosynthetic plants for production of organic C cau-

$$6CO_2 + 6H_2O \dashrightarrow C_6H_{12}O_6 + 6O_2,$$

sing further depletion in ^{13}C. Most land plants utilize the so called C_3, or Calvin pathway that results in tissue with $\delta^{13}C_{org}$ of about -25 to -30 per mil. In contrast, aquatic plants and tropical grasses with the C_4 (Hatch-Slack on Kranz) pathway have $\delta^{13}C$ of some -10 to -15 per mil. A third group that combines these two pathways, the CAM plants (algae and lichens), has intermediate $\delta^{13}C$ values. In detail, the nature of the discussed variations is far more complex (Deines, 1980; Sackett, 1989) and depends on the type of organic compounds involved. For our purposes, however, it is only essential to realize that C_{org} is strongly depleted in ^{13}C. This organic matter, which is very labile, is easily oxidized into CO_2 that inherits the ^{13}C-depleted signal.

Deep oceanic circulation during glacial times, such as the Quaternary, is dominated by production of cold, salty and dense waters at high latitudes. As these O_2-rich waters sink and migrate, their aging is accompanied by an increase in TDC and PO_4^{2-} and a depletion in dissolved O_2 (Fig. 5); all a consequence of oxidation of the "raining" planktonic matter from the surficial water layer.

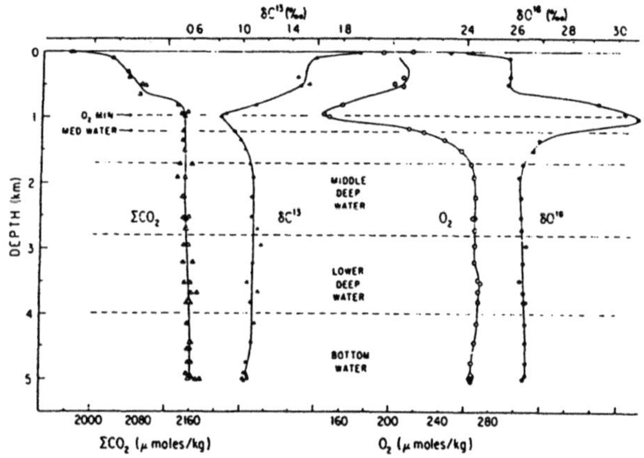

Figure 5: Changes in $\delta^{13}C$ of TDS (CO_2) and $\delta^{18}O$ of dissolved O_2 with depth for the North Atlantic Ocean. The $\delta^{13}C$ minimum at shallow depth is associated with oxygen minimum. The large $\delta^{18}O$ increase in the oxygen minimum zone is due to preferential consumption of ^{16}O by bacteria (Kroopnick and Craig, 1976). $\delta^{18}O$ in SMOW, $\delta^{13}C$ in PDB. From Kroopnick et al. (1972)

The addition of this "organic" CO_2 is the reason for the ^{13}C depletion of TDC in the deeper water column. However, the dynamics of the oceans may have been considerably different during times of warm, presumably non-glacial, climatic intervals, such as the Cretaceous. In this case, the dense waters may have originated not from cooling and ice formation, but from large degree of evaporation. Such waters would be O_2 poor and, as a result, the oceans may tend towards a greater degree of anoxia, retarding the oxidation of "raining" C_{org}. The oceans, however, are characterized not only by vertical, but also by horizontal gradients in $\delta^{13}C$ of TDC. In locations with carbonate deposition, these are typically within +2 $\pm$ 0.5 per mil. The lighter values are characteristic of upwelling regions and the heavier ones of mid-latitudes; a result of a net balance between organic productivity, rates of upwelling and sinking of waters, CO_2 invasion and evasion, and temperature (Kroopnick et al., 1977; Tan, 1989).

The $\delta^{13}C$ of TDC in meteoric waters is highly variable. It is determined (Anderson and Arthur, 1983) by the relative contribution to the overall TDC of C from sources such as: (1) degradation of C_{org} ($\delta^{13}C$ about -25 per mil), (2) dissolution of carbonates (+2 per mil), (3) diffusion from the air (-7 per mil) that results in TDC of +1 to +2 per mil, and (4) the extent of ^{12}C withdrawal into organic matter. The resulting $\delta^{13}C_{TDC}$ may range from positive values (in carbonate buffered water masses) to highly negative ones. In most natural situations, however, the TDC of meteoric waters is depleted in ^{13}C relative to sea water (Buchardt and Fritz, 1980).

RECENT MARINE CARBONATES

The components of Recent marine carbonates include skeletal parts, chemical and/or biochemical precipitates and early marin e cements. Low-latitude shallow water assemblages, called Chlorozoan are dominated by aragonitic corals and green algae, frequently accompanied by marine precipitates (whitings, ooids, peloids) and having ubiquitous early marine cements of HMC and A mineralogy. In contrast, carbonates of temperate zones, and of deeper slope settings in tropical regions, the Foramol assemblage, contain mostly skeletal remains with Mg-calcitic mineralogy. The dominant contributors are molluscs, foraminifera and bryozoans (Lees and Buller, 1972). Early marine cements in this assemblage are rare and of Mg-calcitic composition. Finally, the off-shore assemblage of pelagic organisms, such as calcitic foraminifera and coccoliths as well as aragonitic pteropods accumulate above their respective carbonate compensation depths to produce deep sea oozes on the ocean floor. Rare cementation is mostly by Mg-calcite cements that have been discovered on some seamounts and guyots in semi-enclosed basins. A more detailed description of the subject is available in

Milliman (1984) and James and Choquette (1990a). The wide spread in O and C isotopic compositions (Fig. 6) of Recent marine carbonates reflects this complexity and the relative proportions of constituent components.

As already pointed out (Fig. 2), certain groups of organisms (some molluscs, foraminifera) secrete their shells in approximate isotopic equilibrium with the ambient water, while others (e.g. articulate algae, scleractinian corals, echinoderms) produce strongly ^{18}O and/or ^{13}C depleted skeletons. The wide spread of isotopic values in O/C space (Fig. 6a; Morrison and Brand, 1986) is a direct reflection of this state of affairs. Bulk typical sediments (Fig. 6c) show less scatter than for individual components (Fig. 6a) due to the averaging effect of their mixing. Temperate assemblages and deep sea oozes have usually ^{18}O enriched signal due to colder ambient water temperatures.

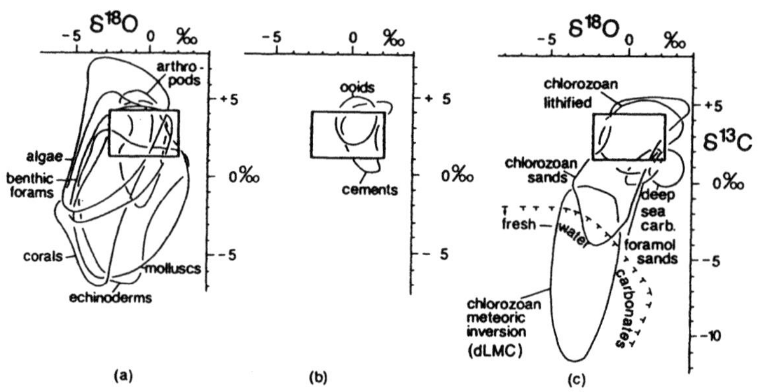

Figure 6: Generalized distribution of $\delta^{18}O$ and $\delta^{13}C$ in carbonate components and bulk sediments of Quaternary age. Box represents an approximate equilibrium range for $CaCO_3$ precipitated from shallow marine water. Domains of isotopic distribution in (a) marine skeletons, (b) precipitates, and (c) bulk sediments. Based on the data summarized by Land (1989), Milliman (1974) and James and Choquette (1990a and b)

The precipitates, such as ooids and early marine cements, tend to fall much closer to expected equilibrium values (Fig. 6b). In detail, however, they may be characterized by slight ^{13}C and/or ^{18}O enrichment (Milliman, 1974) and, if so, biochemically mediated processes may be involved in their precipitation.

ISOTOPIC COMPOSITION OF ANCIENT SEA WATER

The discussion of the dynamics of modern ocean and of superimposed inorganic and biological factors is reflected in the complex isotopic patterns of Recent marine carbonate sediments. While modern isotopic patterns may serve as a starting point for discussion of ancient carbonates, the task is not that simple, because the isotopic composition of sea water has evolved in the course of geologic history.

Strontium Isotopes

The $^{87}Sr/^{86}Sr$ ratio evolved from mantle-like non-radiogenic values of about 0.703 in the Archean to its present day value. The dominant control of Sr isotopic composition in sea water is exercised by two fluxes, the riverine input ($^{87}Sr/^{86}Sr$ of 0.711 at present) and the circulation of sea water through hydrothermal systems in the young oceanic crust ($\leq$0.703). The first order trend therefore reflects the transition from an ocean dominated by hydrothermal "mantle" buffering into an ocean dominated by riverine flux (Veizer, 1989). Due to waning and waxing of these two fluxes, the second and higher order oscillations are superimposed on the first order trend and they have been resolved for the Phanerozoic (Fig. 7b). This "curve" (Burke et al., 1982) approximates the $^{87}Sr/^{86}Sr$ evolution for the past 500 million years, but is better considered to be a band rather than a line. The width of the band decreases with decreasing age and it is particularly steep and well defined for the Cenozoic (Fig. 7c). This part of the curve is therefore frequently utilized for dating of unknown samples, by measuring their $^{87}Sr/^{86}Sr$ ratio and comparing it with the sea water curve. The approach is known as "high resolution isotope stratigraphy" (e.g. Elderfield, 1986). Theoretically, it is also be possible to utilize other steep slopes in the Phanerozoic curve for similar purposes. The mitigating factor is the width of the band, which cannot be constrained any better than the available dating precision for the

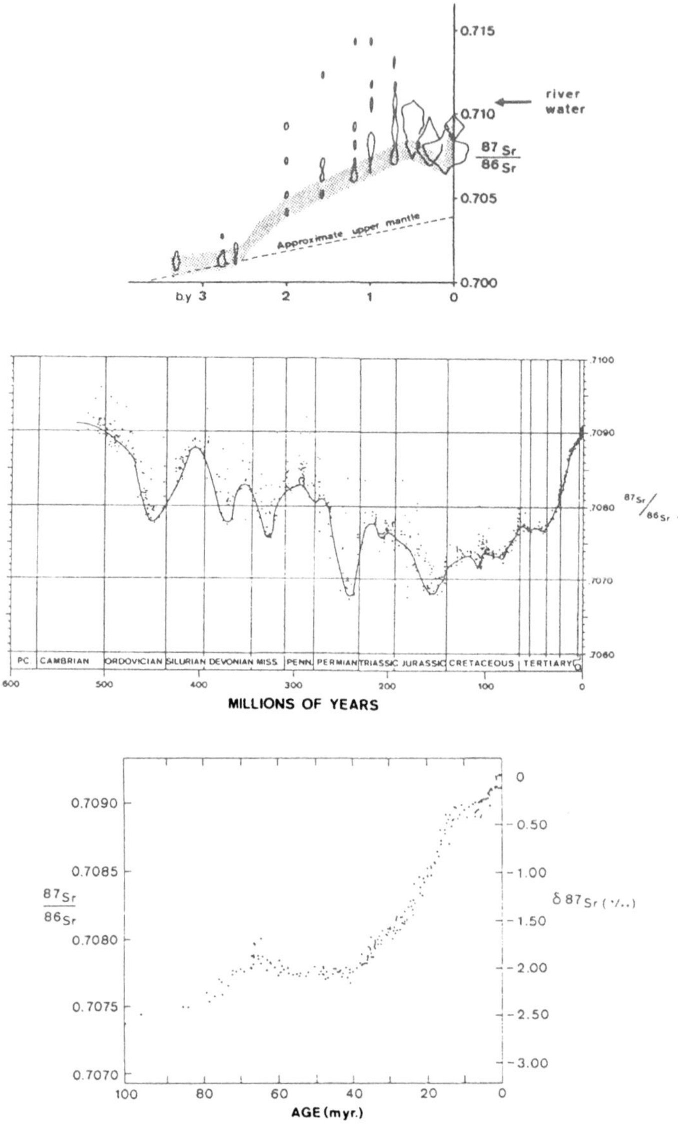

Figure 7: Sr isotopic evolution of sea water during geologic history. (a) First order trend on 10^9 years time scale, (b) Phanerozoic variations, and (c) details of variations for the Cenozoic. Reproduced in modified form from Veizer (1989), Burke et al. (1982) and Elderfield (1986)

samples that serve as calibration for the curve. Even if we could constrain all factors and exclude all post-depositional phenomena that alter the original signal, there will be a limit on the attainable resolution. It is apparent, that the time signal represents a hierarchy of wiggles and with each advance in resolution a set of higher order wiggles emerges. It may be that the structure of the curve approximates the so called fractal geometry (Mandelbrot, 1983). The secular curve will thus always represent a band, albeit of progressively smaller width with improving resolution. Whatever the arguments, the starting point of discussion for studies of ancient carbonates must be the $^{87}Sr/^{86}Sr$ of the contemporaneous ocean and not its modern value.

Carbon Isotopes

The $\delta^{13}C$ of past oceans varied mostly within a 0 ± 3.0 per mil band since at least 3.5 billion years ago (Schidlowski et al., 1983). As in the case of Sr isotopes, superimposed on this first order trend are higher order wiggles. The generalized Phanerozoic variations are given in figure 8. These ocillations are probably caused by the relative proportions of C sequestered into carbonates and organic matter, respectively. At times of large production and storage of C_{org} in sediments, ^{12}C is depleted in the marine reservoir and the dissolved bicarbonate becomes isotopically heavy (Veizer et al., 1980). Alternatively, the ^{12}C depleted C can be stored in the deep water body of a stratified ocean, causing the shallow water layer to be ^{13}C enriched (Hoffman et al, 1990). Whatever the cause, local phenomena (e.g. evaporation, photosynthetic drawdown) or vital fractionation effects are superimposed as a scatter on this general trend. As already discussed, oceanic circulation

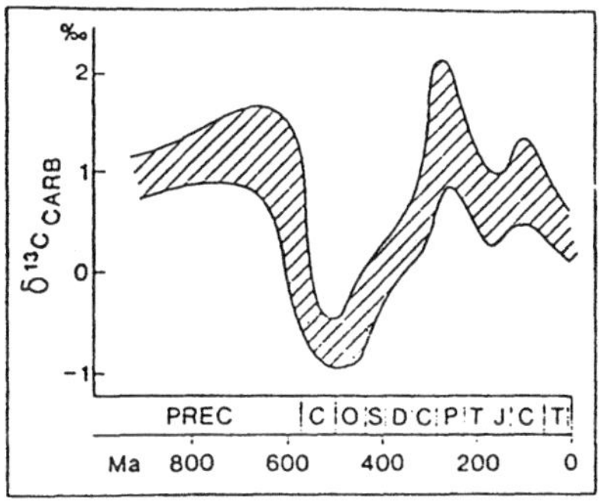

Figure 8: $\delta^{13}C$ variations in Phanerozoic carbonates. Reproduced from Tucker and Wright (1990) and based on the data summarized in Veizer et al. (1980)

patterns result in $\delta^{13}C$ variations with depth (Fig. 5). This gradient is recorded by pelagic and benthic organisms, respectively. In particular the foraminiferans (Fig. 9) have been utilized to map such gradients over geological past with a goal of deciphering the ancient oceanic circulation patterns (Berger, 1986). Similar to Sr isotopes, secular oscillations in $\delta^{13}C$ may too be utilized for "isotope stratigraphy" (Williams et al., 1988).

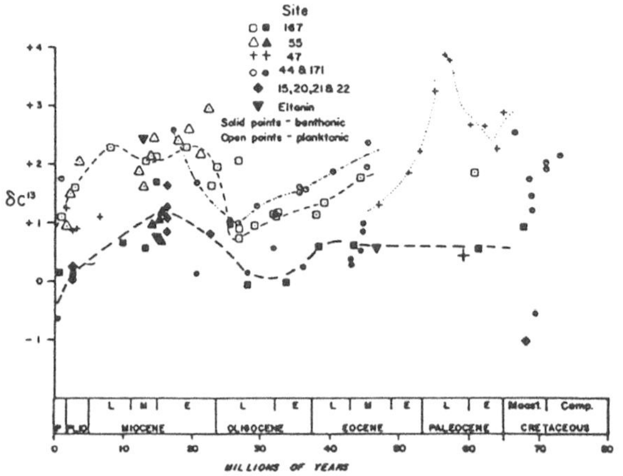

Figure 9: $\delta^{13}C$ values of Tertiary planktonic and benthonic foraminiferans from several DSDP sites. Reproduced from Savin and Yeh (1981)

Oxygen Isotopes

The story of the O isotopic evolution of sea water is by far the most controversial. The existing data over time scales of 10^8 to 10^9 years show a diminishing ^{18}O content for progressively older (bio)chemical sediments (Veizer and Hoefs, 1976). The pattern for the Paleozoic is given in figure 10. While the data as such are not disputed, their interpretation is contentious. Four alternatives have been proposed: (a) recrystallization in the presence of meteoric waters and/or at higher temperatures due to burial of the sediments, (b) the ancient, in particular the pre-Carboniferous, oceans have been considerably warmer, (c) sea water ^{18}O changed with time, and (d) the early oceans were strongly stratified, with deep waters formed by extensive evaporation at low latitudes. This is the "non-glacial" or "greenhouse" circulation pattern discussed previously that may have resulted in impoverishment of the surface layer in

^{18}O. My personal preference is for the alternative (c), perhaps complemented by the other factors, but the interpretation of this trend is still open to discussion. The pros and cons are summarized, for example, in Veizer et al. (1986).

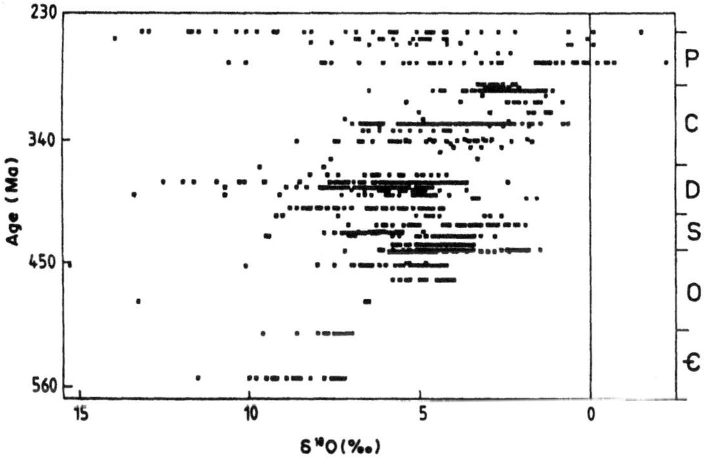

Figure 10: $\delta^{18}O$ variations in Paleozoic brachiopods. This summary represents 1428 samples and is reproduced from Wadleigh and Veizer (1992), who also list the sources of data

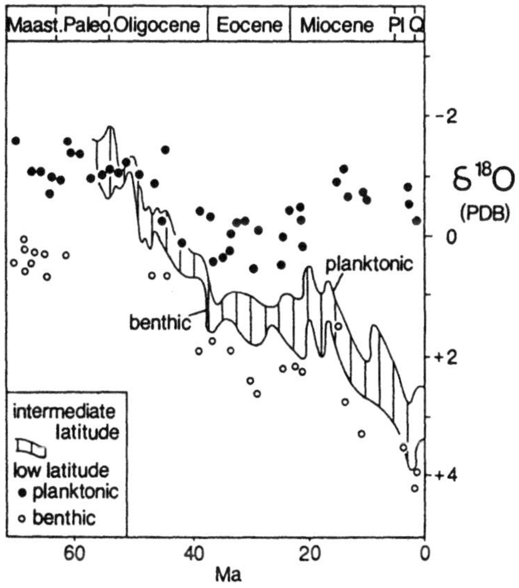

Figure 11: Oxygen isotopic composition of planktonic and benthonic foraminiferans. Modified from Berger (1979) and based on the data of Shackelton and Kennett (1975) and Douglas and Savin (1975)

Whatever the case for the Paleozoic, the Mesozoic and Tertiary $\delta^{18}O$ of sea water fluctuated in the vicinity of SMOW value, yet the results, particularly for the foraminiferans, indicate overall cooling since the Mesozoic (Fig. 11). At intermediate latitudes this is recorded by both the planktonic species at the surface and the benthic ones at depth. Note, however, that at low latitudes it is mostly the benthic organisms that follow this trend, with $\delta^{18}O_{benthic-planktic}$ reaching about 15°C (4 per mil) in the Quaternary. This increasing temperature/depth gradient at low latitudes results from initiation and expansion of the "glacial" type of ocean circulation, with production of deep water masses increasingly from dense cold waters from high latitudes.

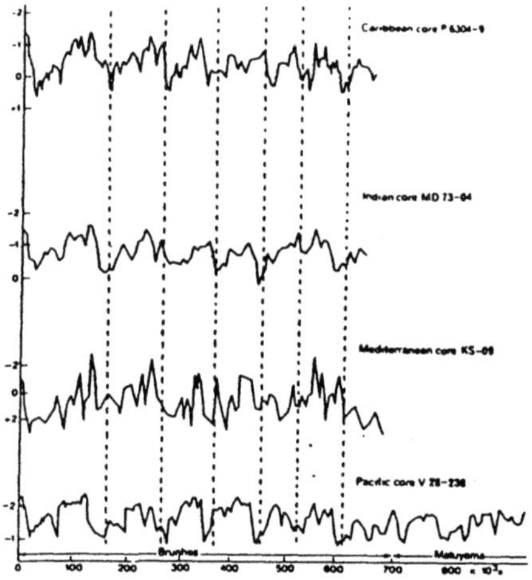

Figure 12: Oxygen isotopic compositions of foraminiferal tests in four cores from the Caribbean Sea, the Indian Ocean, the Mediterranean Sea and the Pacific Ocean. The $\delta^{18}O$ in PDB. Reproduced from Odin et al. (1982), who list sources of data

Superimposed on the overall cooling trend are higher level oscillations, culminating in the zig-zag pattern of $\delta^{18}O$ in Quaternary planktonic foraminiferans (Fig. 12; Emiliani, 1955; Shackleton and Opdyke, 1973; Williams et al., 1988). These fluctuations record glacial/interglacial cycles. However, the ^{18}O depletions are due to both, higher temperatures and melting of ^{16}O-rich polar ice caps. The relative proportions of these two causes at a specific time are difficult to disentangle. It is widely believed that these, the so called Milankovitch cycles, are driven by orbital parameters of the earth (Imbrie and Imbrie, 1980). This is the standard explanation based on the concepts of linear dynamics. Alternatively, one is dealing with a non-linear dynamic system and the wiggly time series is a fractal structure that represents only a one dimensional projection from an attractor embedded in a four dimensional space (Nicolis and Nicolis, 1984). In plain language, at least four parameters (including the orbital ones) would be required to explain the observed pattern. As in the case of Sr and C, the secular $\delta^{18}O$ curve can be utilized as a chronostratigraphic tool and for correlation purposes in post-Paleozoic sediments.

DIAGENETIC OVERPRINT

Diagenetic modification of freshly deposited sediments begins on the sea floor with breakdown (mechanical and biological) of the carbonate components and their binding by cements. This process of rock generation from a loose sediment continues via two major subsequent pathways, designated here as meteoric and burial diagenesis (Bathurst 1975; James and Choquette, 1990a,b; Choquette and James, 1990; Hesse, 1990). These two pathways converge with time and some overlaps are possible even in the early stages of diagenesis.

Meteoric Pathway

Carbonate sediments deposited originally in shallow marine environments are, as a result of tectonics, sediment progradation or eustasy, exposed to the influence of meteoric waters. Such waters are frequently charged with CO_2 and thus highly corrosive. The original unstable marine mineralogical assemblage (aragonite and Mg-calcite) transforms into a monomineralic low-Mg calcitic limestone via sequential dissolution-reprecipitation. The dissolution-reprecipitation process is driven by the degree of undersaturation that decreases from Mg-calcite with more than 12 mole % $MgCO_3$, to aragonite, to Mg-calcite with 12-4% $MgCO_3$, to low Mg-calcite. Usually the process begins with dissolution of the most soluble phase accompanied by its reprecipitation as diagenetic low-Mg calcite (dLMC). When this phase is obliterated the next most soluble phase maintains supersaturation with respect to dLMC, etc. But the differences in solubility, or the so called diagenetic potential, are not controlled solely by mineralogy and chemistry of the mineral phases, but also by factors that influence kinetics. For example, fine-grained sediments with large surface area and easy access from fluids (high porosity) dissolve faster than coarse grains of the same mineralogy and chemistry. This dissolution-reprecipitation process proceeds in a meteoric flow regime that is comprised of two components, regional flow and diffusion. The latter is relatively slow, of the order of $10^2 m/10^6 a$, whereas the former can vary considerably. At high flow rates the system is water-dominated. With high rates of dissolution, the resulting product becomes mineralogically, chemically and isotopically rather monotonous, albeit with possible gradients down-flow. At slow flow-rates, local dissolution-reprecipitation gradients, maintained by diffusion and local physical parameters, are not entirely dissipated. This is a rock-dominated system, where the precursor phases play a considerable role in controlling the chemistry and isotopic composition of waters in the microenvironments and thus of the locally precipitated successor dLMC. Depending on the relative rates of flow vs. diffusion and on the nature of the confining solid framework, all permutations are possible. But in most natural cases we deal with a partially rock dominated system, characterized by intermediate water/rock ratios. Note, however, that the same physical system may result in vastly different water/rock ratios for different chemical or isotopic constituents involved (see Banner and Hanson, 1990 and Land, this book). In the case of limestones, this will particularly be the situation for C and O isotopes. While C in the waters of the microenvironment originates mostly from the precursor carbonate phases, the proportion of O from this source - as opposed from the intervening water - is negligible. Thus the same physical system may yield an apparent low water/rock ratio for C and a high one for O isotopes.

For Sr isotopes, the $^{87}Sr/^{86}Sr$ in the successor phase reflects directly the proportion of Sr derived from the water and the precursor, respectively. As a rule, meteoric waters - relative to sea water - are mostly ^{87}Sr enriched. Thus the overall shift in diagenetic product is toward higher $^{87}Sr/^{86}Sr$ ratio. Exceptions do exist in environments dominated by volcanic rocks or by young volcaniclastic sediments. In this case a decrease in $^{87}Sr/^{86}Sr$ is possible. Regardless, for intermediate to high water/rock systems with high rates of recrystallization the entire microenvironment is usually swamped by Sr derived from precursor. This is because the partition coefficient $D^{Sr}_{calcite}$ is much less than unity and each recrystallization step therefore partitions more Sr into solution (cf, Veizer, 1983 a,b). In this case, the Sr isotopic composition of the precursor is sequestered into the successor phase with little isotopic shift.

The situation is more complex for C and O isotopes. Apart from recrystallization, the other important diagenetic process is the oxidation of organic matter. With burial, the sediment passes successively (Hesse, 1990) through the zones of (1) oxidation by dissolved O_2, (2) nitrate reduction, (3) sulfate reduction, (4) carbonate reduction, (5) fermentation, and (6) thermocatalytic decarbonation. The processes in the first five zones are mediated by assemblages of specific bacteria to temperatures of about 75°C. The abiogenic zone (6) extends to about 150°C, and at higher temperatures equilibria between graphite, CO_2 and CH_4 dominate. The processes (1) to (3) produce CO_2 and bicarbonate that are released to the pore space. The C in these species is thus derived from oxidation of C_{org} and is depleted in ^{13}C, with $\delta^{13}C$ usually around -25 per mil. The passage of sediment into the zone (4), particularly at the trailing edge of sulfate depletion, leads to production of methane. CH_4 is isotopically very light (-80 $\pm$ 20 per mil) and the residual C_{org}, that is subsequently oxidized to CO_2, becomes thus exceptionally heavy ($\delta^{13}C$ in excess of +10 per mil). A return to ^{13}C depleted values again characterizes zone (6).

The meteoric diagenetic environment usually encompasses the zones (1), (2) and, in some instances, the early stages of zone (3). From the point of view of hydrology, this environment can be divided into the vadose zone (infiltration and percolation of water through air filled pores) and the phreatic zone below a permanent water table (see James and Choquette, 1990b). At depth, the phreatic zone changes into mixing zone and finally into the deep phreatic or marine phreatic zone, the latter two usually containing dense waters with salinity in excess of 10 ppt. Some cementation and mineral neomorphism occurs in the vadose zone, but the bulk of diagenetic processes probably takes place in the phreatic zone. The usual isotopic signature of these processes will therefore be a light $\delta^{13}C$ in the products, the degree of ^{13}C depletion being proportional to the relative contribution of C from the precursor carbonate and from the oxidation of organic matter. The vadose zone in particular, because it includes the roots of vegetation, may be up to two orders of magnitude enriched in organically derived CO_2.

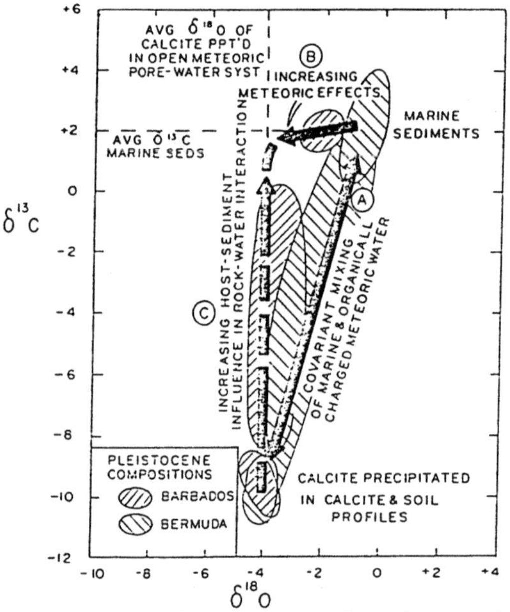

Figure 13: Diagram showing isotopic composition and trends in metastable carbonates where mineral-controlled alteration occurs in a meteoric system capped by a caliche-soil profile. The end members of trend (A) are sea water and ^{13}C depleted pore waters: diagenetic calcites precipitated in a marine-meteoric mixing zone might have compositions along that trend. If no caliche-soil profile existed, alteration of marine sediments might follow the trend shown by the top arrow (B). The long vertical trend (C) applies to a rock-water interaction beneath a caliche-soil profile in a shallow vadose zone, where $\delta^{13}C$ becomes heavier upward as "rock" CO_3^{2-} overpowers soil CO_2 while $\delta^{18}O$ remains the same. From James and Choquette (1990b)

The isotopic composition of O in meteoric water is, relative to sea water, also depleted in ^{18}O. This depletion is not very severe at low latitudes, with displacements mostly less than 5 per mil. Recall that dissolution of carbonate precursors plays only a marginal role in O budget of coastal aquifers. The outcome results in trends in O/C isotope space that resemble mixing lines and/or the so called inverted "J" or "L" trajectories (Figs. 13 and 6c; Lohmann, 1988). The construction of these patterns is based on a suite of isotopic measurements of bulk rocks or, a much better practice, their constituent components. If bulk rocks are considered, the ^{13}C depletion patterns are ubiquitous in Quaternary sequences (e.g. Gross, 1964), but surprisingly rare in their ancient counterparts (Hudson, 1977). The patterns emerge only after the rocks are decomposed into their constituent grains and cement generations. This is somewhat an enigma. Brought to its logical conclusion, it would suggest that organically derived CO_2 played only a subordinate role in diagenetic transformation of ancient carbonate sediments into limestones. Perhaps, as pointed out by Land (1989), the answer lies in large Quaternary sea level fluctuations of about 100 m, due to glaciation. Sudden drops in sea level during that period exposed large tracts of carbonates to prolonged influence of vadose diagenesis. In contrast, the ancient counterparts reflect mostly the diagenetic inprint of the phreatic zone, with $\delta^{13}C$ buffered by the stabilizing carbonate phases.

Burial Pathway

This type of diagenetic stabilization is best documented in deep sea environments and in continually subsiding shallow water basins. The entrapped pore waters are of marine origin and in equilibrium with the assemblage of carbonate minerals. This is particularly the case for the predominantly low-Mg calcitic deep sea oozes. The conversion of sediment into limestone in this case is achieved not by a chemical gradient but through loading initiated by deposition of younger sediments. This results in pressure and temperature rise that, in turn, increases the solubility of the solid phases. Their dissolution-reprecipitation proceeds sequentially in a manner similar to that described for the meteoric pathway (Choquette and James, 1990).

In burial diagenesis, in contrast to the meteoric pathway,

the flow (advection) is confined to squeezing of pore waters upwards into the overlying water column. This is a slow process, particularly for sites with low sedimentation rate (loading). As a consequence, diffusion through the pore water column dominates the transport of solutes. It is therefore the rate of dissolution-reprecipitation of solid phases, relative to the rates of diffusion, that controls the magnitude of gradients in the pore water column. The recrystallization front, as it migrates upwards with continuous loading, causes dissolution of the original carbonate components and their reprecipitation as cements into the pore space, lithifying the sediment and reducing its porosity. Chemical species entering the solution at this front migrate downgradient in both directions, upwards toward the sediment/water interface and downward into the sediment. This may cause "smoothing" of the original isotope signal, as can be illustrated using Sr isotopes (Fig. 14). The $^{87}Sr/^{86}Sr$ of sea water evolved by about 0.0015 in the course of the Cenozoic. At present, the recrystallization front, or the ooze/chalk transition, is in about 20 million year-old sediments (Fig. 14). The Sr with isotopic composition of 0.7085 that is solubilized at this front is being incorporated into cements, thus lowering the bulk $^{87}Sr/^{86}Sr$ in the overlying younger sediments and increasing it in the underlying older sediments. This smoothing process diminishes the magnitude of the preserved temporal signal, particularly for higher order oscillations.

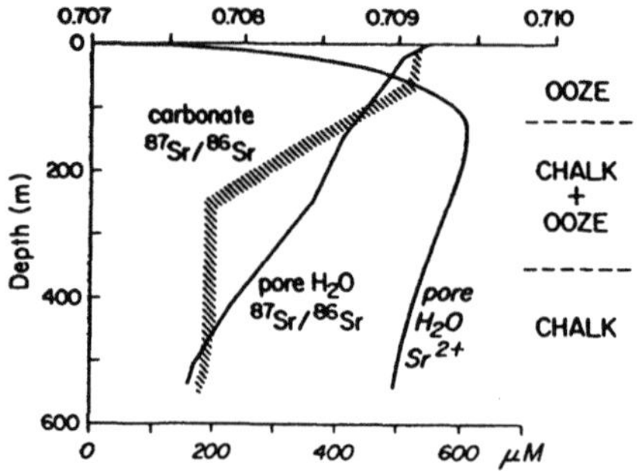

Figure 14: Dissolved Sr^{+2} and $^{87}Sr/^{86}Sr$ at DSDP site 528 as functions of depth. Note the intersection of pore water and carbonate $^{87}Sr/^{86}Sr$ profiles at ooze/chalk transition that marks the recrystallization front. Reproduced from Veizer (1989) and based on the work of Gieskes et al. (1986)

A similar smoothing overprint operates also for C and O isotopes, but here the overall picture is more complex. As for meteoric diagenesis, the ^{13}C depleted C may originate from oxidation of organic matter but, at greater depths, methane production can supply C enriched in ^{13}C. Such progression, from ^{13}C depleted to ^{13}C enriched processes is often observed in diagenetic carbonate concretions that may begin with marine $\delta^{13}C$ in the center, followed by ^{13}C depleted layers, and finally by an outer rim with ^{13}C-rich carbonate. This isotopic pattern thus records the passage of the sediment through oxidation zones (1) to (4) detailed earlier.

Theoretically, the ^{18}O, being of sea water origin, should not change appreciably with burial, whether in the solid phases or in the pore waters. Yet, with increasing depth the deep-sea sediments - and often also the water - show ^{18}O depletion by several per mil (Lawrence, 1989). It is believed that the major reason for this ^{18}O depletion is a low-temperature exchange of O either with volcaniclastic sediments in the underlying sedimentary pile or with the oceanic crust. Secular evolution of sea water (e.g. absence of polar caps) or other phenomena may have been contributory factors. The shift towards ^{18}O depletion in the solid phases is mostly due to increase in temperature. However, in contrast to pore waters, the bulk of this shift is achieved at the time of deep burial.

The carbonate sediment, regardless whether following the meteoric or the burial pathway, eventually reaches the <u>deep burial</u> environment. The boundaries can be set at cessation of most bacterial processes at about 75°C (or 2 km depth), but other authors would propose an earlier onset. The major process occurring at these depth is pressure-solution that results in dissolution of a portion of the already stabilized low-Mg calcitic limestone and reprecipitation of this material as late, mostly ferroan, blocky calcite cements that occlude the residual porosity (Bathurst, 1975). The deep phreatic waters that are involved in this process are mostly saline, with $\delta^{18}O$ heavier than sea water and with endowment of ^{87}Sr from interactions with silicate rocks in the sedimentary pile. Plagioclase, in particular, is subject to high dissolution rates. In terms of C isotopes, the system is mostly rock buffered, but introduction of light $\delta^{13}C$ from thermocatalytic decarboxylation may locally be important. The net result is forma-

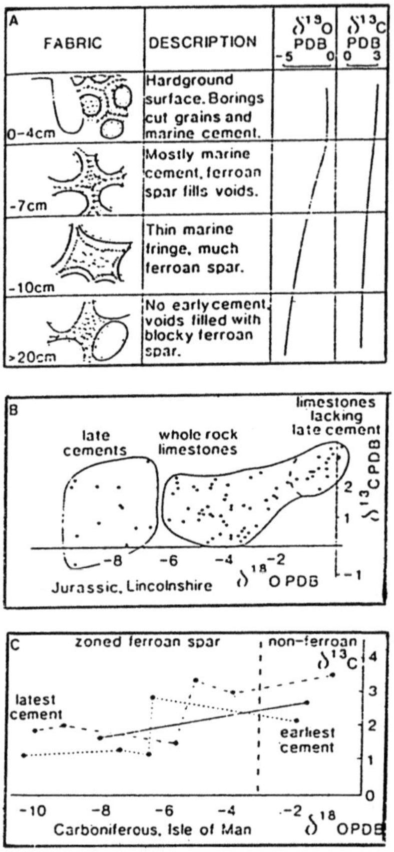

Figure 15: Stable isotopes, marine and burial cementation. (A) Diagram illustrating the changes in cement type and stable isotope signatures from a hardground surface with borings and abundant marine cements down to a grainstone with only burial calcite spar, 20 cm below. Middle Jurassic, Lincolnshire Limestone, England. (B) Isotopic data from the Jurassic Lincolnshire Limestone showing trend to more negative ^{18}O values with increasing content of late spar. (C) Isotopic signatures of early to late calcite cement phases in Carboniferous limestone from Isle of Man, UK. Later cements are more depleted in ^{18}O and more ferroan than early cements. Reproduced from Tucker and Wright (1990) and the sources listed therein.

tion of cements with a signal that is "marine" in δ^{13}C, radiogenic in ^{87}Sr/^{86}Sr and depleted in ^{18}O (Fig. 15). The O isotope shift reflects mostly the increase in temperature that counteracts, and usually overrides, any shift toward ^{18}O-enrichment that may have been imprinted on deep phreatic waters during burial. It is essential to understand the fundamentals of these post-depositional processes if we are to interpret the origin of an isotopic signal preserved in the ancient rock record.

CONCLUDING REMARKS

This description of the carbonate system, as it relates specifically to limestones, intentionally emphasizes principles rather than case histories. The reasons for such approach are twofold. Firstly, the allotted space would not have enabled a sufficiently representative review of the plethora of existing and potential applications. The reading list appended to this chapter may serve as an introduction into the subject. Secondly, the understanding of principles is a prerequisite for appreciation of case histories and for theoretical formulation of problems tractable by isotopic techniques. These techniques are unusually powerful if properly applied, but also relatively costly. Commercial rates for a combined O and C isotopic measurement are about 40$ and for a high quality Sr isotopic measurement 250$. They have to be applied therefore judiciously. As with every other technique, the isotopic approach has its advantages and weaknesses and it should be applied only with the former in mind.

REFERENCES

Anati D.A. and Gat J.R. (1989) Restricted marine basins and marginal sea environments. In: Handbook of Environmental Isotope Geochemistry, 3, Fritz P. and Fontes J.C. (eds.), Elsevier, Amsterdam, 29-73.

Anderson T.F. and Arthur M.A. (1983) Stable isotopes of oxygen and carbon and their application to sedimentologic and paleoenvironmental problems. In: Stable Isotopes in Sedimentary Geology, Soc. Econ. Paleont. Miner., Short Course, 10, Arthur, M.A., Anderson T.F., Kaplan I.R., Veizer J., and Land L. (eds.), I/1-I/151.

Banner J.L. and Hanson G.N. (1990) Calculations of simultaneous isotopic and trace element variations during water-rock interaction with application to carbonate diagenesis. Geochim. Cosmochim. Acta, 54, 3123-3138.

Bathurst R.C.G. (1975) Carbonate Sediments and their Diagenesis. Elsevier, Amsterdam, 658 p.

Berger W.H. (1979) Stable isotopes in foraminifera, foraminiferal ecology and paleoecology. Soc. Econ. Paleont. Miner., Short Course 6, 156-198.
Berger W.H. (1988) Deep-sea carbonates: Reading the carbon-isotope signal. Geol. Rundsch., 75, 249-269.
Bernat M. and Church T.M. (1989) Uranium and thorium decay series in the modern marine environment. In: Handbook of Environmental Geochemistry, 3, Fritz P. and Fontes J.C. (eds.), Elsevier, Amsterdam, 357-383.
Bigeleisen J. and Mayer M.G. (1947) Calculation of equilibrium constants for isotopic exchange reactions. Jour. Chem. Phys., 15, 261-267.
Buchardt B. and Fritz P. (1980) Environmental isotopes as environmental and climatic indicators. In: Handbook of Environmental Geochemistry, 1, Fritz P and Fontes J.C. (eds.), Elsevier, Amsterdam, 473-504.
Burke W.H., Denison R.E., Hetherington E.A., Koepnick R.B., Nelson H.F. and Otto J.B. (1982) Variation of seawater $^{87}Sr/^{86}Sr$ throughout Phanerozoic time. Geology, 10, 516-519.
Chave K.E. (1954a) Aspects of the biochemistry of magnesium-1. Calcareous and marine organisms. Jour. Geol. 62, 266-283.
Chave K.E. (1954b) Aspects of biochemistry of magnesium-2. Calcareous sediments and rocks. Jour. Geol. 62, 587-599.
Coplen T.B., Kendall C. and Hopple J. (1983) Comparison of stable isotope reference samples. Nature, 302, 236-238.
Choquette P.W. and James N.P. (1990) Limestones - The burial diagenetic environment. In: Diagenesis, McIlreath I.A. and Morrow D.W. (eds.), Geosc. Canada Reprint Series, 4, 75-112.
Craig H. (1961) The measurements of oxygen isotope paleotemperatures. In: Stable Isotopes in Oceanographic Studies and Paleotemperatures, Tongiorgi E. (ed.), Spoleto 1965, CNR, Pisa, 1-24.
Craig H. and Gordon L. (1965) Deuterium and oxygen-18 variations in the ocean and marine atmosphere. In: Stable Isotopes in Oceanographic Studies and Paleotemperatures, Tongiorgi E. (ed.), Spoleto 1965, CNR, Pisa, 9-130.
Deines P. (1980) The isotopic composition of reduced organic carbon. In: Handbook of Environmental Isotope Geochemistry, 1, Fritz P. and Fontes J.C. (eds.), Elsevier, Amsterdam, 329-406.
Deines P., Langmuir D. and Harmon R.S. (1974) Stable carbon isotope ratios and the existence of a gas phase in the evolution of carbonate groundwaters. Geochim. Cosmochim. Acta, 39, 1147-1154.
DePaolo D.J. (1986) Detailed record of the Neogene Sr isotopic evolution of seawater from DSDP site 5908. Geology, 14, 103-106.
Dickson T. (1990) Carbonate mineralogy and chemistry. In: Carbonate Sedimentology, Tucker M.E. and Wright V.P. (eds.), Blackwell, 284-313.
Douglas R.G. and Savin S.M. (1975) Oxygen and carbon isotope analyses of Tertiary and Cretaceous microfossils from Shatsky Rise and other sites in the North Pacific Ocean. DSDP Intl. Reports, 32, 509-520.
Elderfield H. (1986) Strontium isotope stratigraphy. Palaeogeogr. Palaeoclimatol. Palaeoecol., 57, 71-90.
Emiliani C. (1955) Pleistocene temperatures. Jour. Geol., 63, 538-578.

Emrich K., Erhalt D.H. and Vogel J.C. (1970) Carbon isotope fractionation during the precipitation of calcium carbonate. Earth Plan. Sci. Lett., 8, 363-371.

Epstein S., Buchsbaum R., Lowenstam H.A. and Urey H.C. (1953) Revised carbonate-water isotopic temperature scale. Geol. Soc. Amer. Bull., 64, 1315-1326.

Faure G. (1986) Principles of Isotope Geology. J. Wiley & Sons, Chichester, 589 p.

Faure G. and Powell J.L. (1972) Strontium Isotope Geology. Springer Verlag, Heidelberg, 188 p.

Ferronsky V.I. and Brezgunov V.S. (1989) Stable isotopes and ocean mixing. In: Handbook of Environmental Isotope Geochemistry, 3, Fritz P. and Fontes J.C. (eds.), Elsevier, Amsterdam, 1-27.

Gat J.R. (1980) The isotopes of hydrogen and oxygen in precipitation. In: Handbook of Environmental Isotope Geochemistry, 1, Fritz P. and Fontes J.C. (eds.), Elsevier, Amsterdam, 21-47.

Gieskes J.M., Elderfield H., and Palmer M.R. (1986) Strontium and its isotopic composition in interstitial waters of marine carbonate sediments. Earth Plan. Sci. Lett. 77, 229-235.

Gonfiantini R. (1986) Environmental isotopes in lake studies. In: Handbook of Environmental Isotope Geochemistry, 2, Fritz P. and Fontes J.C. (eds.), Elsevier, Amsterdam, 113-168.

Gross M.G. (1964) Variations in the O^{18}/O^{16} and C^{13}/C^{12} ratios of diagenetically altered limestones in the Bermuda Islands. Jour. Geol., 72, 170-194.

Grossman E.L. and Ku T.L. (1986) Oxygen and carbon isotope fractionation in biogenic aragonite: temperature effects. Chem. Geol., 59, 59-74.

Hays P.D. and Grossman E.L. (1991) Oxygen isotopes in meteoric calcite cements as indicators of continental paleoclimate. Geology, 19, 441-444.

Hesse R. (1990) Early diagenetic pore water/sediment interaction: Modern offshore basins. In: Diagenesis, McIlreath I.A. and Morrow D.W. (eds.), Geosc. Canada Reprint Series-4, 277-316.

Hodell D.A., Mead G.A. and Mueller P.A. (1990) Variation in the strontium isotopic composition of seawater (8 Ma to present): Implications for chemical weathering rates and dissolved fluxes to the oceans. Chem. Geol. (Isot. Geosc. Sect.), 80, 291-307.

Hoefs J. (1980) Stable Isotope Geochemistry. Springer Verlag, Heidelberg, 208 p.

Hoffman A., Gruszczynski M. and Malkowski K. (1991) On the interrelationship between temporal trends in $\delta^{13}C$, $\delta^{18}O$ and $\delta^{34}S$ in the world ocean. Jour. Geol., 99, 355-370.

Holland H.D. (1984) The Chemical Evolution of the Atmosphere and Oceans. Princeton Univ. Press, 582 p.

Hudson J.D. (1977) Stable isotopes and limestone lithification. Jour. Geol. Soc. London, 133, 637-660.

Imbrie J. and Imbrie J.Z. (1990) Modeling the climatic response to orbital variations. Science, 207, 943-953.

Jahn B.-M., Bertrand-Sarfati J., Morin N. and Macé J. (1990) Direct dating of stromatolitic carbonates from the Schmidtsdrift Formation (Transvaal Dolomite), South Africa, with implications on the age of the Ventersdorp Supergroup. Geology, 18, 1211-1214.

James N.P. and Choquette P.W. (1990a) Limestones - The seafloor diagenetic environment. In: Diagenesis, McIlreath I.A. and

Morrow D.W. (eds.), Geosc. Canada Reprint Series, 4, 13-34.
James N.P. and Choquette P.W. (1990b) Limestones - The meteoric diagenetic environment. In: Diagenesis, McIlreath I.A. and Morrow D.W. (eds.), Geosc. Canada Reprint Series, 4, 35-73.
Kroopnick P. (1980) The distribution of ^{13}C in the Atlantic ocean. Earth Plan. Sci. Lett., 49, 469-484.
Kroopnick P. and Craig M. (1976) Oxygen isotope fractionation in dissolved oxygen in the deep sea. Earth Plan. Sci. Lett., 32, 375-388.
Kroopnick P., Weiss R.F. and Craig H. (1972) Total CO_2, ^{13}C and dissolved oxygen -^{18}O at GEOSECS II in the North Atlantic. Earth Plan. Sci. Lett., 16, 103-110.
Kroopnick P., Margolis S.V. and Wong C.S. (1977) ^{13}C variations in marine carbonate sediments as indicators of the CO_2 balance between the atmosphere and oceans. In: The Fate of Fossil Fuel CO_2 in the Oceans, Andersen N.R. and Malahoff A. (eds.), Plenum Press, 295-321.
Land L.S. (1989) The carbon and oxygen isotopic chemistry of surficial Holocene shallow marine carbonate sediments and Quarternary limestone and dolomite. In: Handbook of Environmental Isotope Geochemistry, 3, Fritz P. and Fontes J.C. (eds.), Elsevier, Amsterdam, 191-217.
Lawrence J.R. (1989) The stable isotope geochemistry of deep-sea pore water. In: Handbook of Environmental Isotope Geochemistry, 3, Fritz P. and Fontes J.C. (eds.), Elsevier, Amsterdam, 317-356.
Lees A. and Buller A.T. (1972) Modern temperate water and warm water shelf carbonate sediments contrasted. Mar. Geol., 19, 159-198.
Lohmann K.C. (1988) Geochemical patterns of meteoric diagenetic systems and their application to studies of paleokarst. In: Paleokarst, James N.P. and Choquette P.W. (eds.), Springer Verlag, Heidelberg, 58-80.
Ludwig K.R., Halley R.B., Simmons K.R. and Peterman Z.E. (1988) Strontium isotope stratigraphy of Enewetak Atoll. Geology, 16, 173-177.
Mackenzie F.T., Bischoff W.D., Bishop F.C., Loijens M., Schoonmaker J. and Wollast R. (1983) Magnesian calcites: low-temperature occurrence, solubility and solid-solution behavior. Rev. Miner., 11, 97-144.
Mandelbrot B.B. (1983) The Fractal Geometry of Nature. Freeman, San Francisco, 468 p.
McConnaughey T. (1990a) ^{13}C and ^{18}O isotopic disequilibrium in biological carbonates: 1. Patterns. Geochim. Cosmochim. Acta, 53, 151-162.
McConnaughey T. (1990b) ^{13}C and ^{18}O isotopic disequilibrium in biological carbonates: 2. In vitro simulation of kinetic isotope effects. Geochim. Cosmochim. Acta, 53, 163-171.
McCrea J.M. (1950) On the isotopic chemistry of carbonates and a paleotemperature scale. Jour. Chem. Phys., 18, 849-857.
Milliman J.D. (1974) Marine Carbonates. Springer Verlag, Heidelberg, 375 p.
Morrison J.O. and Brand U. (1986) Geochemistry of Recent marine invertebrates. Geosc. Canada 13, 237-254.
Morse J.W. and Mackenzie F.T. (1990) Geochemistry of Sedimentary Carbonates. Elsevier, Amsterdam, 707 p.
Nicolis C. and Nicolis G. (1984) Is there a climatic attractor? Nature, 311, 529-532.

Odin G.S., Renard M. and Vergnaud-Grazzini C. (1982) Geochemical events as a means of correlation. In: Numerical Dating in Stratigraphy, Odin G.S. (ed.), J. Wiley & Sons, Chichester, 37-71.

O'Neil J.R. (1986) Theoretical and experimental aspects of isotopic fractionation. Rev. Miner., 16, 1-40.

O'Neil J.R., Clayton R.N. and Mayeda T.K. (1969) Oxygen isotope fractionation in divalent metal carbonates. Jour. Chem. Phys., 51, 5547-5558.

Palmer M.R. and Edmond J.M. (1989) Strontium isotope budget of the modern ocean. Earth Plan. Sci. Lett., 92, 11-26.

Rubinson M. and Clayton R.N. (1969) Carbon-13 fractionation between aragonite and calcite. Geochim. Cosmochim. Acta, 33, 997-1002.

Sackett W.M. (1989) Stable carbon isotope studies on organic matter in the marine environment. In: Handbook of Environmental Isotope Geochemistry, 3, Fritz P. and Fontes J.C. (eds.)., Elsevier, Amsterdam, 139-169.

Savin S.M. and Yeh H.W. (1981) Stable isotopes in ocean sediments. In: The Sea, 7, Emiliani C. (ed.), Wiley-Interscience, 1521-1554.

Schidlowski M., Hayes J.M. and Kaplan I.R. (1983) Isotopic inferencies of ancient biochemistries: carbon, sulfur, hydrogen and nitrogen. In: Earth's Earliest Biosphere: Its Origin and Evolution, Schopf J.W. (ed.), Princeton Univ. Press, 149-186.

Shackleton N.J. and Kennett J.P. (1975) Paleotemperature history of the Cenozoic and the initiation of Antarctic glaciation: oxygen and carbon isotope analyses in DSDP sites 277, 279 and 281. DSDP Intl. Reports, 29, 743-755.

Shackleton N.J. and Opdyke N.D. (1973) Oxygen isotope and palaeomagnetic stratigraphy of equatorial Pacific core V-28-238: oxygen isotope temperatures and ice volumes on a 10^5 year and 10^6 year scale. Quat. Res., 3, 39-55.

Talbot M.R. (1990) A review of the palaeohydrological interpretation of carbon and oxygen isotopic ratios in primary lacustrine carbonates. Chem. Geol. (Isot. Geosc. Sect.), 80, 261-279.

Tan F.C. (1988) Stable carbon isotopes in dissolved inorganic carbon in marine and estuarine environment. In: Handbook of Environmental Isotope Geochemistry, 3, Fritz P. and Fontes J.C. (eds.), Elsevier, Amsterdam, 171-190.

Tarutani T., Clayton R.N. and Mayeda T.K. (1969) The effects of polymorphism and magnesium substitution on oxygen isotope fractionation between calcium carbonate and water. Geochim. Cosmochim. Acta, 36, 1237-1253.

Tucker M.E. and Wright V.P. (1990) Carbonate Sedimentology. Blackwell, Oxford,482 p.

Turner J.V. (1982) Kinetic fractionation of carbon-13 during calcium carbonate precipitation. Geochim. Cosmochim. Acta, 46, 1183-1191.

Valley J.W. (1986) Stable isotope geochemistry of metamorphic rocks. Rev. Miner., 16, 445-489.

Veizer J. (1983a) Trace elements and isotopes in sedimentary carbonates. Rev. Miner., 11, 265-299.

Veizer J. (1983b) Chemical diagenesis of carbonates: theory and application of trace element technique. In: Stable Isotopes in Sedimentary Geology, Arthur M.A., Anderson T.F., Kaplan I.R., Veizer J., and Land L.S. (eds.), Soc. Econ. Paleont.

Miner., Short Course, 10, III-1/III-100.
Veizer J. (1989) Strontium isotopes in seawater through time. Ann. Rev. Earth Plan. Sci., 17, 141-167.
Veizer J. and Hoefs J. (1976) The nature of O^{18}/O^{16} and C^{13}/C^{12} secular trends in sedimentary carbonate rocks. Geochim. Cosmochim. Acta, 40, 1387-1395.
Veizer J., Fritz P. and Jones B. (1986) Geochemistry of brachiopods: oxygen and carbon isotopic records of Paleozoic oceans. Geochim. Cosmochim. Acta, 50, 1679-1696.
Veizer J., Holser W.T. and Wilgus C.K. (1980) Correlation of $^{13}C/^{12}C$ and $^{34}S/^{32}S$ secular variations. Geochim. Cosmochim. Acta, 44, 579-587.
Wadleigh M.A. and Veizer J. (1992) $^{18}O/^{16}O$ and $^{13}C/^{12}C$ in Lower Paleozoic articulate brachiopods: isotopic composition of sea water. Geochim. Cosmochim. Acta, in press.
Wefer G. (1983) Die Verteilung stabiler Sauerstoff-und Kohlenstoff-Isotope in Kalkschalen mariner Organismen - Grundlage einer isotopischen Palokologie. Habilitationsschrift, Univ. Kiel, 151 p.
Williams D.F., Lerche I. and Full W. E. (1988) Isotope Chronostratigraphy: Theory and Methods. Academic Press, 345 p.
Yurstsever Y. (1975) Worldwide survey of stable isotopes in precipitation. Rep. Sect. Isotope Hydrol., IAEA Vienna, November 1975, 40 p.

THE DOLOMITE PROBLEM:
STABLE AND RADIOGENIC ISOTOPE CLUES

Lynton S. LAND
Department of Geological Sciences
The University of Texas at Austin
AUSTIN, Texas, 78712 (U.S.A.)

The "dolomite problem" is a classic geological enigma: we cannot explain, to the satisfaction of a great number of scientists, many rocks composed dominantly of dolomite. Although numerous Holocene examples, possibly analogous to some ancient deposits, have been described and are still being described, no analog exists to massive stratal dolomites common in the geologic past. Considering the brevity of the shallow water "platform-like" Holocene record, it is unlikely such analogs will ever be found. To compound the problem, dolomite has yet to be synthesized in the laboratory under earth-surface conditions. Practically, this means that we cannot test quantitative measures of dolomite behavior such as trace element distribution coefficients and isotopic fractionation factors.

Another complicating factor, which is also true of virtually all other chemically-generated sedimentary phases, is that the mineral which forms during or immediately following sedimentation is almost never preserved without chemical modification. Limestones provide an excellent parallel. Modern carbonate sediments are composed of the minerals aragonite and Mg-calcite, yet limestones are composed of calcite. The metastable "precursor" phases aragonite and Mg-calcite are transformed stepwise during diagenesis, according to Ostwald's rule, to a more stable configuration. Primary chemical signatures are either irrevocably lost, or modified to the point that we are not yet able to untangle the original from the secondary. Dolomite behaves similarly.

Most Holocene dolomite is Ca-rich, very fine-grained, and is characterized by numerous cystallographic imperfections (e.g. Wenk et al., 1983). Aragonite and Mg-calcite, the precursor phases to limestones, are replaced by calcite as limestones form, and early-formed dolomite is similarly replaced during diagenesis by more stable dolomite phases. Our goal in utilizing isotopic data to understand dolomite (or any other chemically precipitated sedimentary phase) should be not only to characterize the primary sediment, or characterize the water from which it formed, but to understand the history of the sediment, from its chemical inception through its subsequent modification. We should not be surprised or disappointed if we are unable to deduce conditions of genesis, which remains only one (elusive) goal of sedimentology. In the case of dolomite, isotopic data have proven more useful than many other kinds of observations in documenting the geologic history of various kinds of dolomite.

Dolomite, $CaMg(CO_3)_2$, contains major cations (Ca and Mg), minor cations (e.g., Sr, Fe, Mn, and Rare Earth Elements), major anions (C and O), and possibly minor anions like borate and sulfate, all of which can substitute into the crystal lattice (as opposed to being adsorbed, or trapped in fluid inclusions). The potential number of isotopic systems which bear on the mineral dolomite is clearly very large! Only a few of the potentially useful isotopic systems have been utilized, however. All dolomite crystals are precipitates whether they "replace" other phases or form primary sediment or cements. Solid state reactions are precluded at the temperatures characteristic of sedimentary basins. Because dolomite precipitation from aqueous solution involves vast amounts of oxygen, oxygen isotopes have proven most useful in elucidating dolomite geochemistry. Carbon isotopes, though less useful, have the capacity to document both redox conditions and secular changes in ocean chemistry. One of the most nagging problems of dolomite genesis has been the source of sufficient amounts of Mg to account for large volumes of dolomite.

Of the "cation-tracers", Sr isotopes are proving to be of great utility in constraining the Mg source. Each of these isotopic systems, for which considerable data exist, will be treated in turn. A final section is devoted to the remainder, and to speculations about useful paths for future research.

OXYGEN ISOTOPES

Like most oxygen-bearing minerals, we presume there exists a temperature-dependent equilibrium relationship between the oxygen isotopic composition of a growing dolomite crystal and the water from which it precipitates. We cannot currently precipitate dolomite in the laboratory under earth-surface conditions, and therefore the relationship we seek must be approximated. Two methods of approximation have evolved, and, fortunately, they are in reasonable agreement. The first determines the needed relationship at sufficiently high temperatures so the reaction takes place in the laboratory on a reasonable time scale, and extrapolates the results to earth-surface temperatures, and the second utilizes natural examples. Before investigating the results of these two approaches, however, we must understand how analytical methodology bears on our results.

Methodology

When Urey first recognized the potential utility of carbonate minerals to determine past earth surface temperatures, he also recognized the difficulty of simply extracting the oxygen isotope ratio from these minerals. At that time, fluorination techniques were not available, and it was quickly recognized that the decomposition reaction:

$$CaCO_3 \longrightarrow CaO + CO_2$$

yielded CO_2 (the preferred gas for analysis) whose isotopic composition depended on numerous uncontrollable variables (surface area, heating rate, crystal defect density, trace element composition, etc.). Fortunately, Urey's student, McCrea, recognized that anhydrous phosphoric acid could be used to release CO_2 reproducibly, albeit not quantitatively, from carbonates. But if we only analyze two of the O atoms in carbonate minerals with this technique, how do we relate the isotopic composition of the mineral (all three O atoms) to the isotopic composition of water (as a function of temperature)? The answer is that we must know the fractionation factor (known as the phosphoric acid fractionation factor) between the oxygen in the mineral (all three O atoms), and the oxygen in CO_2 released from the mineral (as a function of temperature). Just as each carbonate mineral has a unique mineral-water fractionation factor, each carbonate mineral has a unique phosphoric acid fractionation factor.

A further complication has developed because of the use of two different standards, SMOW (Standard Mean Ocean Water) and PDB (Peedee Belemnite). Urey chose a natural skeletal calcite for a standard (not wisely, in hindsight) and used the $^{18}O/^{16}O$ of the CO_2 released from that calcite in phosphoric acid for a reference. To complicate matters, he was also unable to determine the $^{18}O/^{16}O$ of H_2O. Instead, Urey recognized that the reaction between CO_2 and H_2O proceeds reasonably rapidly in the laboratory, and so if he added a small amount of CO_2 (sufficient for analysis) to a large amount of H_2O, and allowed equilibrium to be obtained, analysis of the $^{18}O/^{16}O$ in the CO_2 would be proportional to the $^{18}O/^{16}O$ in the H_2O. The proportionality is, of course, the fractionation factor. Urey's classic "paleotemperature equation" relates the $^{18}O/^{16}O$ in CO_2 evolved from calcite in anhydrous phosphoric acid at 25°C to the $^{18}O/^{16}O$ of CO_2 equilibrated with water at 25°C. In order to convert to the SMOW scale we need three fractionation factors (Fig. 1): (a) the relation between the $^{18}O/^{16}O$ of calcite and the $^{18}O/^{16}O$ of CO_2 evolved by reaction of calcite in anhydrous phosphoric acid at 25°C, (b) the relation between the $^{18}O/^{16}O$ in water and the $^{18}O/^{16}O$ in CO_2 in equilibrium with water at 25°C, and (c) the relation between the two CO_2 samples (1 and 2) when the calcite and water are in equilibrium with each other.

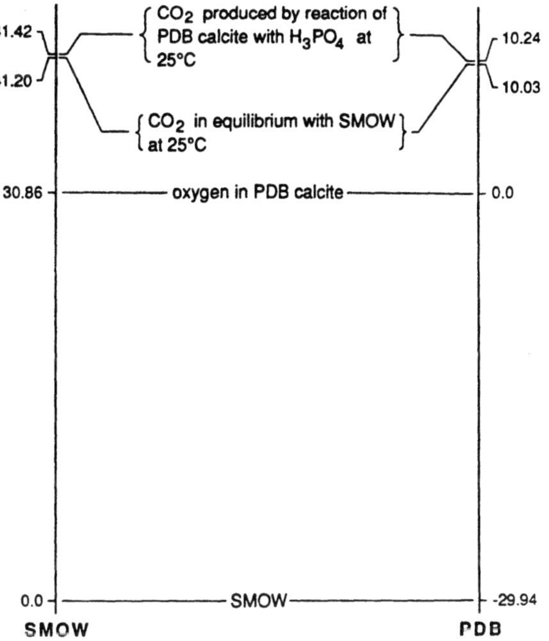

Figure 1: Graphical representation between SMOW and PDB scales (after Land, 1980, his figure 6)

If we are interested in minerals other than calcite (like dolomite), then all relations must be known for each mineral. Because different researchers may have used different methodology (e.g. different standards, different reaction temperature, etc.), we must be certain to understand what each investigator actually measures and reports, in making comparisons and in interpreting the analyses.

High Temperature Studies

Surprisingly, only five studies nave been made of dolomite water fractionation at elevated temperature (Fig. 2), and none in almost 15 years! None of these early studies characterized the microstructure of the dolomite which was synthesized, or examined the reaction products carefully with the electron microscope to document the petrographic nature of the precipitates. Nevertheless, we can be quite certain that dolomite is enriched in ^{18}O by 3 ± 3 per mil at 25°C relative to calcite precipitating under the same conditions of $\delta^{18}O$ and temperature. This is often expressed by the statement that the "delta" value (the difference in the δ-values between calcite and dolomite) is about 3 per mil.

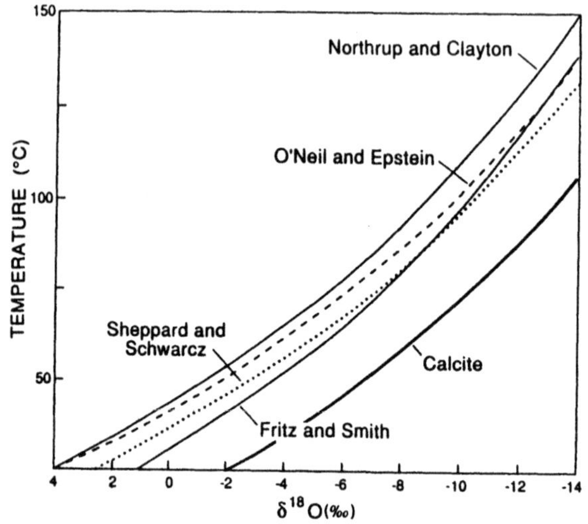

Figure 2: Four available equations for dolomite-water fractionation as a function of temperature for water having a $\delta^{18}O$ of 0 per mil (SMOW; after Land, 1983 his Fig. 4-1)

Natural Examples

New examples of modern dolomite are being found and described almost every year. Most examples are marine, ranging in salinity from hypersaline to normal marine, but many examples of dolomite forming in playas are also known. The problem in using modern examples is that the dolomite is rarely forming at the actual time of sampling, and therefore the $\delta^{18}O$ of the water and the temperature must be assumed for the recent past. Additionally, the dolomite often replaces precursor Ca-carbonate phases, and although Ca carbonate and dolomite coexist, they are not commonly cogenetic. These are difficult problems in the case of an environment like a playa, where conditions can be very ephemeral. But they are less difficult in the case of marine and marine hypersaline environments where sufficient sedimentological study has been accomplished to constrain past conditions. Table 1 lists modern environments where the $\delta^{18}O$ of water, $\delta^{18}O$ of dolomite, and temperature are all known. A Δ value is then calculated by subtracting from the analyzed dolomite value the value of calcite which should precipitate under the observed conditions of $\delta^{18}O$ water and temperature. Recognizing that ranges of values are characteristic of all environments which have been studied so far, the conclusion from the high temperature studies that $\Delta = 3 \pm 3$ per mil can be further constrained to 5 ± 2 per mil.

Table 1

Locality	$\delta^{18}O$ water	$\delta^{18}O$ dolomite	Temperature (°C)	Δ
Jamaica	+0.8	+2.0	26-29	4.2
(Mitchell et al., 1987)				
Baffin Bay	+2.0	+4.8	24-30	2.9
(Behrens and Land, 1972)				
Florida	+0.9	+2.8	30-32	5.6
(Carballo et al., 1987)				
Gulf of Elat	+11.0	+9.0	49-57	6.7
(Aharon et al., 1977)				
Persian Gulf	+3.9	+2.1	32-39	3.0
(Mackenzie, 1981)				
Deep Springs	-2.1	+2.5	15-25	6.0
(Clayton et al., 1968)				

$\delta^{18}O_{PDB}$ was converted to $\delta^{18}O_{SMOW}$ using the equation: $\delta^{18}O_{SMOW} = 1.03086 \times \delta^{18}O_{PDB} + 30.86$. $\delta^{18}O$ calcite was calculated using the equation from Friedman and O'Neil (1977): $10^3 \ln_{calcite-water} = 2.78 * 10^6\ T^{-2} - 3.39$ where T is in degrees Kelvin ($T^{o}C + 273.16$). Δ is $\delta^{18}O_{observed}$ dolomite - $\delta^{18}O_{calculated}$ calcite

To a first approximation, if the measured $\delta^{18}O$ of dolomite is made 3.8 per mil more negative (3 per mil because of the Δ value between calcite and dolomite, and 0.8 per mil because of different phosphoric acid fractionation factors between the two minerals), and the fractionation between calcite and water is used (see Table 1), then a satisfactory approximation for the equilibrium relations between dolomite and water, as a function of temperature, can be made. Future work may refine this approximation, probably upward, but it seems to be sufficiently accurate that useful conclusions can be reached. More experimental work, and more analyses of existing natural examples, are clearly needed. Figure 3 graphs the above relationship and presents, as a histogram at the top, a summary of dolomite values from the literature as of about 1979. Quite clearly, few described samples could have precipitated from seawater (or hypersaline seawater) at earth-surface conditions. Either most dolomite did not form under earth surface conditions, or dolomite is "reset" during diagenesis to more ^{18}O-depleted values. Considerable support can be advocated to support the latter supposition.

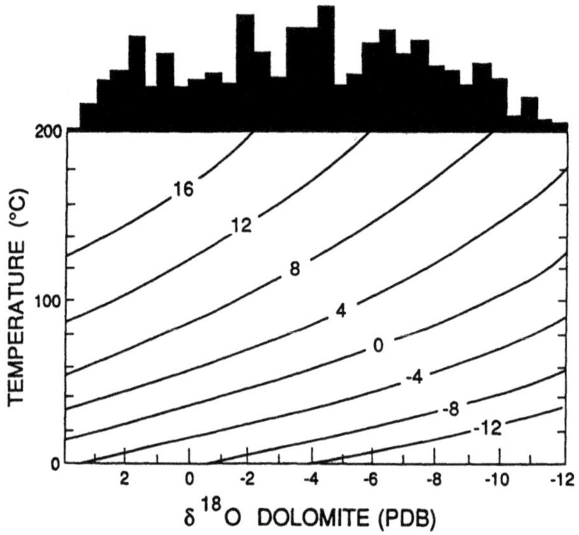

Figure 3: Graphical representation of the oxygen isotopic composition of dolomite (PDB scale), $\delta^{18}O$ water (SMOW scale), and temperature. At the top of the diagram is a summary of data obtained on sedimentary dolomite as of about 1979 (after Land, 1985, his Fig. 9)

Modern dolomite is trace element rich, and forms small crystals with numerous imperfections (e.g. Mitchell et al., 1987). Ancient dolomites are typically trace element depleted (or contain abundant Fe, an element rare in modern carbonates), and form large crystals characterized by simple, more nearly ideal micro-structure. Stepwise stabilization of metastable precursor phases is the most likely explanation for these observations, and is substantiated by oxygen (and Sr) isotopic analyses. More stable phases are typically depleted in ^{18}O (e.g Fritz and Jackson, 1972), and can be either enriched or depleted in ^{87}Sr (Gao, 1990), due to stabilization during burial at elevated temperature by solution-reprecipitation (replacement).

If we consider the amount of oxygen in a carbonate sediment relative to the amount of oxygen in the pore fluids, we can demonstrate that the ratio of those two "reservoirs" (the rock/water ratio) is the lowest of almost any isotopic or trace element system we can conceive. For example, using 40% porosity as a reasonable value for unconsolidated sediments (and typical modern carbonate sediments in seawater for subsequent calculations), the ratio of the number of O atoms in the minerals to the number of oxygen atoms in the water (the rock/water ratio) is 2.2. If replacement takes place (by dissolution-precipitation, since solid state processes can be ignored at earth surface temperatures), then the oxygen isotopic composition of the secondary mineral can easily be changed because subequal numbers of O atoms are present in both the minerals and the pore fluids. This suggests a "test" for unaltered minerals. If any trace component (including isotopes) having a rock/water ratio higher than oxygen has been altered, then the oxygen isotopic composition of the mineral must have been altered as well. It needs not to have changed from its original composition, by pure happenstance, but it must have been altered. As a corollary, if components with higher rock/water ratios than oxygen have not been altered, that does not prove that the oxygen isotopes are unaltered also. Because the rock/water ratio for oxygen is so small, a very small amount of replacement can affect the oxygen isotopic composition of a mineral without causing measurable changes in other components. Dolomite formed in a sabkha (say $\delta^{18}O$ = +4 per mil PDB) can be replaced (in meteoric lenses, for example) by a more stable phase, shifting the $\delta^{18}O$ to, say -3 per mil, without any measurable effect on the C or Sr isotopic composition (but with a large effect on the Sr concentration). For this reason, the use of oxygen isotopes along with other isotopic (and trace element) systems provides a potentially much more powerful tool than the use of oxygen isotopes alone.

CARBON ISOTOPES

Barring small effects due to the solubility of CO_2 in phosphoric acid, the $^{13}C/^{12}C$ ratio of the CO_2 released on digestion of carbonate minerals in phosphoric acid is presumed to be identical to the $^{13}C/^{12}C$ ratio in the mineral. Equilibrium reltionships between calcite, dissolved carbon (mostly HCO_3^- at the pH's typically encountered in water in contact with carbonate rocks), and CO_2 gas, are reasonably well known (Fig. 4). But unfo-

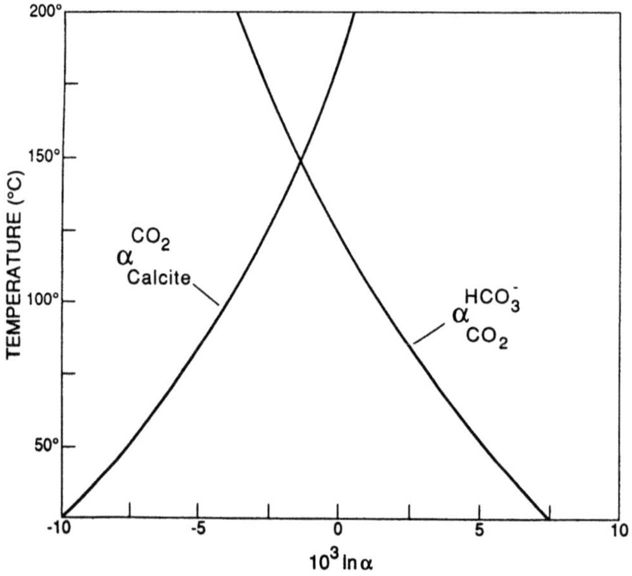

Figure 4: Fractionation between calcite and CO_2, and between bicarbonate and CO_2 as a function of temperature (after Friedman and O'Neil, 1977). Given an atmospheric CO_2 value of -7 per mil at 25°C , equilibrium bicarbonate should be about +0.5 per mil and calcite +3 per mil

rtunately, we also know that precipitation rate can affect the carbon isotopic composition of calcite (and presumably dolomite), something which has not been (and hopefully never will be!) demonstrated for oxygen. The situation is similar to the uptake of trace elements in that rapidly precipitating crystals simply sequester the component of interest ($CO_3^=$ in this case) from solution so rapidly that equilibrium processes, limited by the slowest reaction (in this case the isotopic exchange that takes place via the reaction:

$$H_2O + CO_2 \text{ ------> } H_2CO_3$$

cannot prevail. As a result (see Fig. 4), rapidly precipitated calcite (and presumably dolomite) can be depleted in ^{13}C relative to an equilibrium precipitate.

Assuming that water in contact with carbonate rocks contains about 200 mg HCO_3^-/l (3.3 millimoles C/l), then a carbonate sediment or rock (36.94 cm^3/mole) having 40% porosity has a rock/water ratio of about 12,000. Appreciable changes in the carbon isotopic composition of carbonate sediments during replacement by dolomite will, therefore, be very difficult to achieve. If approximately 1000 pore volumes of seawater are sufficient to completely dolomitize a sediment (e.g. Land, 1985), and if the carbon isotopic composition of the water were to differ from the sediment by 10 per mil (an extreme case!), then the dolomite which formed would still be no more than about 1 per mil different from the precursor sediment. Carbon isotopes, therefore, have been of less use than oxygen isotopes in interpeting the origin and diagenetic history of dolomite, and only few authors have systematically considered them.

Aerobic and anaerobic microbial processes during early sediment burial, aerobic oxidation during subareal exposure, and thermal maturation of kerogen during deep burial all release ^{13}C-depleted carbon from organic matter, which may be incoporated in carbonate phases. Dolomite depleted in ^{13}C relative to expected normal marine values is known from a variety of environments, including lithified supratidal crusts associated with subareal exposure (Carballo et al., 1987), generally poorly lithified deposits associated with extensive sulfate reduction of organic-rich sediments (Behrens and Land, 1972; Garrison et al., 1984), and late stage dolomite cements associated with hydrocarbon emplacement (Woronick and Land, 1985). Distinguishing these various environments in the rock record on the basis of carbon isotopes alone is impossible, but used in conjunction with sedimentary structures and textures, and in the context of sequence stratigraphy, carbon isotopes can be useful in confirming depositional environments. For example, an upward trend toward depleted ^{13}C values as one approaches an exposure surface would provide support for subareal exposure and the development of a soil cover (Allen and Matthews, 1982). In the absence of petrographic, stratigraphic, or paleogeographic evidence of exposure, however, depleted carbon values could be explained in many other ways.

Both extremely enriched and extremely depleted carbon isotopic values do, however, seem to be diagnostic of specific sedimentary environments and processes (see figure in companion paper of Holser). $\delta^{13}C$ values in carbonates in excess of about +6 per mil seem to require a process commonly known as "fermentation". Taking place microbially (and therefore at temperatures less than about 80°C) after elemental O_2 and oxygen from compounds like Fe-oxides, Mg-oxides, and dissolved sulfate have been consumed, $^{12}CO_2$ is preferentially utilized by bacteria as an O source, leaving behind a reservoir of CO_2 enriched in ^{13}C. Nodules are able to record evolving carbon diagenesis, beginning with oxidation and sulfate reduction at shallow burial depths (meters or tens of meters), which are characterized by depleted $\delta^{13}C$ values, and proceeding into the zone of methanogenesis (Irwin et al., 1977). At the other end of the spectrum, carbon isotopic values more depleted than about -28 per mil seem to be indicative of methane oxidation, since methane is the only natural C-containing compound known to contain such depleted carbon. Little dolomite has been documented from such environments.

With few exceptions, most dolomite is characterized by carbon isotopic values between about +6 and -2 per mil, values characteristic of marine limestone precursors. Although some data suggest dolomite should be enriched in ^{13}C relative to calcite by about 2 per mil (Sheppard and Schwarcz, 1970), little confirmatory evidence has emerged from the study of modern examples. Dolomite rocks associated with hypersaline environments are commonly relatively ^{13}C-enriched, in keeping with the inorganic (rather than skeletal) nature of the precursor sediments and the paucity of abundant organic matter in highly evaporitic environments. Absolute values should never be used as diagnostic of any particular environment without reference to the petrography of the rocks themselves, and their age must be kept in mind since secular variations in marine carbon of approximately 4 per mil have been determined and are well documented (Fisher and Arthur, 1977).

STRONTIUM ISOTOPES

Strontium is the most abundant trace element in carbonate sediments and most carbonate rocks. Reasons for the abundance of Sr include the fact that it is the ninth most abundant ion in seawater (not counting H and O), and it is preferentially incorporated in many skeletal and inorganic precipitates of aragonite up to, and even exceeding concentrations of 10,000 μg/g. Most modern carbonate sediments are deposited from marine water and contain aragonite as a primary phase. Even Holocene marine dolomite, whether of normal marine or hypersaline marine origin, contains about 800 μg/g Sr. Therefore, the reservoir of Sr in primary carbonates is quite large, about 1250 times the reservoir of Sr in the pore fluids in a primary marine sediment. But this is less than the rock/water ratio for carbon previously discussed by about a factor of ten, so Sr will be about ten times more sensitive than carbon to early diagenetic modification. The situation is very different during late diagenetic modification, however, when porosity is much lower (say about 10%). The rocks no longer contain Sr-rich phases like aragonite (typical limestones contain only about 200 μg/g Sr), and saline pore fluids can be quite Sr-rich (about 2000 μg/g). In this case the rock/water ratio is about 2.4, less than that of O (which is 13.2) under the same circumstances. So under subsurface conditions, if the rock is out of ionic equilibrium with its pore fluids, and dissolution-reprecipitation takes place, Sr isotope ratios can be more easily altered then O isotope ratios.

Carbonates contain too little Rb to be good candidates for Rb-Sr age dating, but Sr isotopes still can provide some extremely useful information. The reason for this is that we know that the $^{87}Sr/^{86}Sr$ of seawater has varied over a range more than one hundred times larger than our ability to measure that ratio (see companion paper of Holser in this volume). The driving forces for the secular change include changes in continental runoff and hydrothermal circulation through ocean ridges, both processes being driven by variations in plate dynamics. The Sr isotopic system represents a fascinating and productive way to study these processes. But our emphasis here is on understanding how to use the Sr isotopic system to understand dolomite paragenesis. Therefore, we want to use the secular variation in seawater without necessarily having to understand it. But first we must be certain that we understand how the $^{87}Sr/^{86}Sr$ of seawater has changed with time. How can we do this if the recorded ratio in the rocks has been modified by post-depositional changes? The answer is that we must always proceed on parallel paths of investigation. We must understand diagenesis, using as many tools as possible, in order to document both the primary conditions of deposition, and to quantify subsequent changes. Because the rock/water ratio is usually relatively large, and barring extensive burial modification (which would be evidenced by depleted $\delta^{18}O$ values) documenting the $^{87}Sr/^{86}Sr$ (and $^{13}C/^{12}C$) ratio of ocean water is much easier than documenting the $^{18}O/^{16}O$ of ancient oceans.

Heavy isotopes like Sr (we are concerned with masses 86, 87, and 88) do not fractionate in the same way as is observed with light isotopes like oxygen and carbon. There simply isn't sufficient mass difference for ^{86}Sr to fractionate from ^{88}Sr as processes like mineral precipitation take place. So the mineral (dolomite in this discussion) always bears the same isotope ratio as the fluid from which it precipitates. If dolomite formed in, say, mid-Ordovician time from coeval seawater, it might have a $^{87}Sr/^{86}Sr$ of 0.7085 (see seawater curve in Holser's companion paper). If the $^{87}Sr/^{86}Sr$ "age" of a dolomite agrees with its stratigraphic "age", then contemporaneous dolomitization by seawater is logical. Of course we must always be careful to examine other evidence as well. For example, a mid-Ordovician dolomite having a $^{87}Sr/^{86}Sr$ of 0.7085 could also have formed from mid-Silurian, early Devonian, or even early Miocene (see seawater curve in Holser's companion paper) seawater, or have been reset by groundwater which had flowed through strata of these ages. Given the stratigraphic constraints and the Sr isotopic data we would need to assess, on geological grounds, whether or not post-Ordovician seawater having a $^{87}Sr/^{86}Sr$ of 0.7085 was a possible agent of dolomitization.

Conversely, if we know from paleontological or other evidence that a certain dolomite replaced lower Cretaceous rocks, and we measured a $^{87}Sr/^{86}Sr$ of 0.7098 on the dolomite (again, see the seawater curve in Holser's companion paper), we could be quite certain that if the dolomite formed contemporaneously with deposition, it must have been subsequently modified by a fluid other than seawater. Alternatively, the dolomite did not form from seawater in the first place. Saller (1984) used similar reasoning to conclude that dolomite found in Eocene rocks of Enewetak Atoll was formed by circulation of Middle Miocene or younger seawater through the atoll. He measured a $^{87}Sr/^{86}Sr$ between 0.7086 and 0.7090 on the dolomite, eliminated all other possible reservoirs of Sr (easy to do in this Pacific Ocean setting), recognized that seawater is the only source of mobile Mg at the earth's surface, documented a reasonable way to circulate seawater through the buried sediments (by low temperature hydrothermal convection), and thus destroyed previous dogma (Berner, 1965).

Obviously, Sr isotopes can provide powerful constraints, but constraints which depend on accurate knowledge of variations in the $^{87}Sr/^{86}Sr$ ratio of seawater, and the assumption that the world ocean always had a uniform Sr isotopic composition. The latter assumption is certainly true today (based on actual measurements), and supported by knowledge that the residence time for Sr in the ocean (the time it would take for all the Sr in the ocean to be replaced by continental discharge, hydrothermal circulation, and diffusion of Sr out of sediments undergoing burial diagenesis) is relatively short, on the order of 4 million years. But the world ocean today is very well connected because of the continent distribution and is being very actively circulated on a time scale of only a few thousand years because of our present "ice-house" climate. In the past, neither of these requirements may have been met, and although variations in $^{87}Sr/^{86}Sr$ between coeval ocean basins or within a single stratified ocean basin has not yet been demonstrated, we must consider that possibility.

If we can demonstrate that a given dolomite is characterized by an $^{87}Sr/^{86}Sr$ ratio inconsistent with coeval dolomitization by seawater, then what information is provided? The first hypothesis we should test is that dolomitization was caused by seawater subsequent to deposition of the rocks (as Saller concluded). Because it provides such a ready source of Mg, seawater should always be the solution of "first choice" for dolomitization. Because of the complex nature of the secular variaton of seawater $^{87}Sr/^{86}Sr$, the Sr isotopic composition of dolomite could be either higher or lower than the composition of coeval seawater. If geologic constraints eliminate such an hypothesis (or the $^{87}Sr/^{86}Sr$ of the dolomite falls outside the range of seawater variation), then we must conclude that the dolomite has been "reset", or it did not form from seawater in the first place. Using our conclusions from the oxygen isotopic composition of dolomite (Fig. 3), replacement of metastable dolomite by ^{18}O-depleted, more stable phases, is most easily caused by meteoric water, or by increased temperature during burial (Gao, 1990). In both cases, sources of ^{87}Sr from Rb-bearing silicates tend to elevate the $^{87}Sr/^{86}Sr$ ratio. We know that modern rivers and modern formation waters exhibit wide ranges in $^{87}Sr/^{86}Sr$ (see companion papers by Chaudhuri and Clauer, this volume), so once we are sure that the measured Sr isotope ratios do not record coeval or younger seawater, interpretation of absolute values will be very difficult.

Although absolute values may be hard to interpret, spatial variation, as in the case of the other isotopic systems, may be extremely useful. If rock chemistry is changed by import of allochthonous material (as opposed to autochthonous "cannibilization" by stylolitization), then an ancient flow system must have existed. Systematic changes in isotope ratios (and trace element concentrations) can define the direction of flow assumed by diagenetic fluids and "point to" the source of the diagenetic fluids themselves. For example, suppose a dolomite formation displayed typical "marine" isotope ratios in one area, depleted ^{13}C, and ^{18}O, and more radiogenic $^{87}Sr/^{86}Sr$ in another area, and systematic intermediate values in between. A possible interpretation might be that the formation once behaved as an aquifer, and hot, ^{13}C-depleted, radiogenic basinal fluids invaded the aquifer and "reset" an early generation of dolomite. Quantitative modelling of the different isotopic systems (accompanied by other geochemical and petrographic measurements) could constrain not only the direction of fluid flow, but the composition and volume of fluid involved. Although these types of calculations are in their infancy (e.g. Land, 1980, his Appendix III; Banner et al., 1988, their Fig. 14), they hold great promise.

OTHER ISOTOPIC SYSTEMS

The oxygen, carbon, and strontium isotopic systems just discussed are only three of the potentially useful isotopic systems that have been extensively investigated. Continued improvement in mass spectrometers and in the chemistry involved in sample preparation suggests that many more isotopic systems will be utilized in the future. Ca and Mg are both major elements which have natural stable isotopes which may prove to be of use. Measurable natural fractionation of Mg may or may not occur between ^{24}Mg and ^{26}Mg. It has been shown that natural fractionation as much as 2.5 per mil can affect the $^{40}Ca/^{44}Ca$ ratio (Russell et al., 1978), but whether or not useful conclusions will result from the study of dolomite remains to be demonstrated. Sulfur and boron isotopes also have potential application if substitution for carbonate in the dolomite lattice takes place, or if sufficient amounts of either element are incorporated in fluid inclusions.

Recently, a few analyses of dolomite involving the Sm-Nd system have been accomplished. Rare earth elements are vanishingly soluble in water so that rock/water ratios are extreme, typically around 10^6 (Banner et al., 1988). The Mississippian dolomites studied by Banner et al. apparently record the Ce-depleted signature of oxidizing seawater, and Nd isotopic analyses can be interpreted (not unequivocally, however) as being controlled by late Paleozoic seawater. Considerably more data will be needed, however, before application of rare earth isotopic systems to dolomite is fully understood.

CONCLUSION

At what time(s) in the history of a sediment did dolomite form? What kinds of solutions were involved? Were conditions (e.g. temperature, ocean composition) at the earth's surface different than they are today? Isotopic analyses are just one tool which can be used to constrain answers to these important questions. But isotopic data alone are almost never unequivocal, and a single isotopic system rarely useful. Isotopic data must be obtained within a sedimentologic and petrographic framework, and used in conjunction with elemental analyses and quantitative models. Only in this way can these powerful tools be wisely used.

REFERENCES

Aharon P., Kolodny Y. and Sass E. (1977) Recent hot brine dolomitization in the "solar lake", Gulf of Elat. Isotopic, chemical, and mineralogical study. Jour. Geol., 85, 27-48.

Allen J.R. and Matthews R.K. (1982) Isotope signatures associated with early meteoric diagenesis. Sedimentology, 29, 797-817.

Banner J. L., Hanson G. N. and Meyers W. J. (1988) Rare earth element and Nd isotopic variations in regionally extensive dolomites from the Burlington-Keokuk formation (Mississippian): Implications for REE mobility during carbonate diagenesis. Jour. Sed. Petrol., 58, 415-432.

Behrens E. W. and Land L. S. (1972) Subtidal Holocene dolomite, Baffin Bay, Texas. Jour. Sed. Petrol., 42, 155-161.

Berner R. A. (1965) Dolomitization of Mid-Pacific atolls. Science, 147, 1297-1299.

Burke W. H., Denison R. E., Hetherington E. A., Koepnick R. B., Nelson H. F. and Otto J. B. (1982) Variation of sea-water $^{87}Sr/^{86}Sr$ throughout Phanerozoic time. Geology, 10, 516-519.

Carballo J. D., Land L. S. and Miser D. E. (1987) Holocene dolomitization of supratidal sediments by active tidal pumping, Sugarloaf Key, Florida. Jour. Sed. Petrol., 57, 153-165.

Clayton R. N., Jones B. F. and Berner R. A. (1968) Isotope studies of dolomite formation under sedimentary conditions. Geochim. Cosmochim. Acta, 32, 415-432.

Fisher A. G. and Arthur M. A. (1977) Secular variations in the pelagic realm. In: Deep-water carbonate environments, Cook H.E., and Enos P. (eds.), Soc. Econ. Paleontol. and Mineral., Spec. Pub., 25, 19-50.
Friedman I. and O'Neil J. R. (1977) Compilation of stable isotope fractionation factors of geological interest. In: Data of Geochemistry, 6th edition, United States Geol. Survey Prof. Paper, 440KK.
Fritz P. and Jackson S. A. (1972) Geochemical and isotopic characteristics of Middle Devonian dolomites from Pine Point, Northern Canada. Proc. 24th Intern. Geol. Congress, Montreal, Section 6, 230-243.
Fritz P. and Smith D. C. W. (1974) The isotopic composition of secondary dolomites. Geochim. Cosmochim. Acta, 34,1161-1173.
Gao G. (1990) Geochemical and isotopic constraints on the diagenetic history of a massive stratal, late Cambrian (Royer) dolomite, Lower Arbuckle Group, Slick Hills, SW Oklahoma, USA. Geochim. Cosmochim. Acta, 54, 1979-1989.
Garrison R. E., Kastner M. and Zenger D. H. (1984) Dolomites of the Monterey Formation and other organic-rich units. Pacific Soc. Econ. Paleontol. and Mineral., 215p.
Hazma M. S. and Broecker W. S. (1974) Surface effect on the isotopic fractionation between CO_2 and some carbonate minerals. Geochim. Cosmochim. Acta, 38, 669-681.
Irwin H., Curtis C. and Coleman M. (1973) Isotopic evidence for source of diagenetic carbonates formed during burial of organic-rich sediments. Nature, 269, 209-213.
Land L. S. (1980) The isotopic and trace element geochemistry of dolomite: the state of the art. In: Concepts and Models of dolomitization, Zenger D.H., Dunham J.B. and Ethington R.L. (eds.), Soc. Econ. Paleontol. and Mineral., Spec. Pub., 28, 87-110.
Land L. S. (1983) The application of stable isotopes to studies of the origin of dolomite and to problems of diagenesis of clastic sediments. In: Stable Isotopes in Sedimentary Geology, Arthur M. A. (ed.), Soc. Econ. Paleontol. and Mineral., Short Course, 10, 4-1 to 4-22.
Land L. S. (1985) The origin of massive dolomite. Jour. Geol. Educ., 33, 112-125.
McKenzie J. A. (1981) Holocene dolomitization of calcium carbonate sediments from the coastal sabkhas of Abu Dhabi: A stable isotope study. Jour. Geol., 89, 185-198.
Mitchell J. T., Land L. S. and Miser D. E. (1987) Modern marine dolomite cement in a north Jamaican fringing reef. Geology, 15, 557-560.
Northrup D. A. and Clayton R. N. (1966) Oxygen-isotope fractionation in systems containing dolomite. Jour. Geol., 28, 259-320.
O'Neil J. R. and Epstein S. (1966) Oxygen isotope fractionation in the system dolomite-calcite-carbon dioxide. Science, 152, 198-201.
Russell W. A., Papanastassiou D. A. and Tombrello T. A. (1978) Calcium isotope fractionation on the Earth and other solar system materials. Geochim. Cosmochim. Acta, 42, 1075-1090.
Saller A. H. (1984) Petrologic and geochemical constraints on the origin of subsurface dolomite: An example of dolomitization by normal seawater, Enewetak Atoll. Geology, 12, 217-220.

Sheppard S. M. F. and Schwarcz H. P. (1970) Fractionation of carbon and oxygen isotopes and magnesium between metamorphic calcite and dolomite. Contrib. Mineral. and Petrol., 26, 161-198.

Wenk H. R., Barber D. J. and Reeder R. J. (1983) Microstructures in carbonates. In: Carbonates: Mineralogy and Chemistry, Reeder R.J. (ed.), Mineral. Soc. Amer., Rev. in Mineral., 11, 301-367.

Woronick R. E. and Land L. S. (1985) Late burial diagenesis, Lower Cretaceous Pearsall and Lower Glen Rose Formations, South Texas. In: Carbonate Cements, Schneidermann N., and Harris P.M. (eds.), Soc. Econ. Paleontol. and Mineral., Spec. Pub., 36, 265-275.

ISOTOPE SIGNATURES IN PHOSPHATE DEPOSITS: FORMATION AND DIAGENETIC HISTORY

Yehoshua KOLODNY

and

Boaz LUZ

Department of Geology, The Institute of Earth Sciences
The Hebrew University of Jerusalem
Jerusalem, Israel.

Most sedimentary rocks contain less than 1% of P. The formation of rocks that are highly enriched in P_2O_5, and in which apatite is the main mineralogical constituent, requires special conditions, during both deposition and diagenesis. Our interest in phosphate rocks is not only academic. Phosphorus is an essential nutrient of both plants and animals; hence, repeated growth of crops on limited land, demands the artificial addition of P to the soil. Application of phosphate fertilizers became an essential part of modern agricultural techniques. The combined scientific and economic importance of phosphate rocks made them an object of extensive geological, mineralogical and geochemical research for over a century. Research has intensified after World War II, and peaked in the seventies and eighties. The results of these studies are summarized in several thorough reviews and books (Cook, 1976; Cook and McElhinny, 1979; Bentor, 1980; Kolodny, 1981; Sheldon, 1981; Baturin, 1982). The present chapter is a summary of the contribution which has been made by the application of isotopic methods to the study of phosphate rich rocks, as well as an attempt to outline those isotopic methods that may be applied in the future to that purpose. The central theme of this chapter will be phosphorites, but at several points attention will be drawn to the study of phosphatic biogenic components of non-phosphatic sedimentary rocks: biogenic apatite of fish (ichthyoliths), bones and teeth of other vertebrates, conodonts, brachiopods.

The main mineral in phosphate rich rocks is apatite, more specifically *francolite* - the carbonate-fluoride variety (CFA) of apatite. Both, the structure and the chemical composition of francolite have been the subject of extensive research and dispute. The extremely small size of francolite crystallites combines with a very extensive range of substitutions in the lattice. The key substitution seems to be that of a distorted $(CO_3 \cdot F)^{-3}$ tetrahedron replacing a tetrahedral PO_4^{-3} ion (Altschuler et al., 1952; Borneman-Starynkevitch and Belov, 1940; Bacquet et al., 1980). To resolve both structural and charge imbalance, additional substitutions, such as F^-, OH^- for PO_4^{-3}, and Ca^{2+} for Na^+, have been proposed. The general formula that has been suggested for francolite is

$$Ca_{10-a-b}\ Na_a\ Mg_b\ (PO_4)_{6-x}\ (CO_3)_{x-y-z}\ (CO_3 \cdot F)_y\ (SO_4)_x\ F_2$$

where $x=y+a+2c$, and x varies from near zero to 1.5, $y=0.4x$ on the average and $b=0.4a$. (McClellan and Lehr ,1969; Nathan, 1984).

Marine rocks in which apatite is the main mineralogical constituent are referred to as *phosphorites*. Phosphorites are found in sediments of practically all geological ages from the Proterozoic (2,200 m. y. B.P.) to the present. The distribution of phosphorites in space and time does not seem to be random, however. Cook and McElhinny (1979) have shown that phosphorites are confined to specific, phosphogenic epochs, and form phosphogenic provinces, the most prominent of which are the Cambrian deposits of the Karatau (Kazakhstan) and the Georgina Basin (Australia), the Permian Phosphoria Formation and associated formations in the Western U.S., the Upper Cretaceous to Eocene Tethyan province of the Mediterranean, and the Plio-

Miocene deposits of the southeastern U.S. On the other hand, it has been noticed (Strakhov, 1970; Kolodny, 1980) that phosphorites occur in a characteristic lithologic association - most frequently accompanied by cherts and organic rich deposits. It is quite likely that the latter association, can be explained by the biolimiting nature of P and Si (Broecker, 1974, p.21), and by the connection between these elements and areas of high biological productivity in the ocean. Most recent theories, which aim at explaining the genesis of phosphorites, have made use of the strong relationship between organic productivity and P. The problem of explaining phosphorite formation may be reduced to the location of a break in the oceanic cycle of upwelling - high productivity - raining out of organic matter to the sediment - organic matter rich sediment - decomposition of the organic matter into interstitial water. Kazakov (1937) was the first to stress the importance of the marine cycle of P for understanding phosphorite formation. Practically all phosphate is extracted from sea-water by organisms in the photic zone. The P-containing, organism-produced particles are then oxidized while falling to the sea-floor. A sharp concentration gradient in dissolved phosphate is thus formed between surface water and the deep water masses. It was Kazakov's suggestion that phosphorites may form when "phosphate rich", cold, deep water rises to the surface in areas of upwelling; upon reaching the surface, such water warms up, loses CO_2, and therefore precipitates apatite.

Ever since Kazakov, theories of phosphorite genesis have been intimately tied with studies and speculations on marine upwelling. The main advocate of Kazakov's theory in its "chemical " form has been the phosphate group of the U.S. Geological Survey (McKelvey et al., 1953, 1967; Sheldon, 1964). The correlation between occurrences of phosphorite nodules on present day marine shelves and areas of upwelling, has been extended to equating upwelling and phosphorite formation. In the sixties and seventies, it became obvious that whereas upwelling is very likely a necessary requirement for many phosphorite deposits, it is certainly neither a sufficient nor a sole condition for their formation. The reasons for this realization were several: (a) most phosphorites occur among sedimentary sequences with clear imprints of rather shallow water deposition, not leaving much room for upwelling from great depth; (b) when Recent phosphorite formation was identified off the coasts of Namibia and Peru (Baturin, 1969, Veeh et al., 1973; Burnett and Veeh, 1977; see below), there were many indications that the apatite of these nodules is forming interstitially within the uppermost few centimeters of sediment, rather than by precipitation from the bottom water (Froelich et al., 1988); and (c) in laboratory studies, Mg has been shown to inhibit crystallization of apatite from sea-water. Thus Kazakov's "classical" theory of phosphorite precipitation from sea-water has been gradually replaced by a strongly altered version of it - his compatriot's G.N. Baturin's theory of phosphorite formation by ocean dynamics (Baturin, 1971a, b, 1982).

Baturin's theory envisages the formation of a phosphorite rock in two principal stages:

(1) *the formation of apatite pellets*, which form within the sediment, by growth from interstitial water, during a relatively high sea stand, by remobilization of organic matter phosphorus. The apatite pellets are at this stage sparsely distributed within the sediment;

(2) *the formation of a phosphorite rock*, which is principally a mechanical concentration stage. It is tied to a lower sea level stand, and is a stage of winnowing in which apatite is enriched to a major rock component. Compton et al. (1990) suggested that as long as 10 Ma could have separated the two stages which finally resulted in the enormous phosphorite deposits of the southeastern U.S.

Even in Baturin's version of phosphogenesis the connection between phosphorites and upwelling remains strong. Once such a relationship has been established, it may be used both as a means for explaining the origin of phosphorites, or in its inverse form - the distribution of phosphorites may be used for the tracing of past areas of upwelling. The latter approach has been applied by J. Parrish and her co-workers in recent years (Parrish, 1982; Parrish et al., 1982) for the reconstruction of upwelling zones in Paleozoic oceans. The relationship between phosphatic sediments and upwelling also encompasses the other bioproductive constituents of sedimentary sequences - chert and organic rich rocks. It is hence not surprising that phosphorites have been related to petroleum-producing source rocks as well (Waples, 1982). The occurrence of phosphorite deposits in areas where no evidence for simple, west-coast-type upwelling can be found is problematic in this context. Such are the young phosphorites of the southeast United States. In most proposed models, these "east-coast" type phosphorites are explained by invoking paleo-ocean dynamics that could result in large scale upwelling (Riggs, 1984; Compton et al., 1990).

Current ideas of phosphogenesis are based upon petrological observations. The two-stage model of Baturin is supported by the occurrence of two essentially different types of phosphatic rocks.

1. Phosphatic micronodules, nodules, pellets and lenses are common in brown or black shales or carbonates (Garrison et al., 1987; 1990). These rocks are laminated, organic rich, and usually contain large amounts of biogenic carbonate and/or silica. They bear distinct evidence of having been diagenetically formed in a reducing environment. The P_2O_5 content of such rocks does usually not exceed a few percent.

2. In contrast to these rocks, economic phosphorites are usually rocks which bear clear evidence of clastic reworking, winnowing and enrichment. Phosphorites were most probably formed in oxidizing conditions. They show mechanical sorting and traces of boring.

Most apatite in sedimentary rocks occurs as grains. Fine grained phosphorites, sometimes termed phospholutities were also described, but are by far less common. The majority of phosphatic grains are francolitic pellets. Those which are ovular, with no discernable internal structure, are usually referred to as "ovules", "ovulites", or just "pellets". When an internal concentric structure is observed in the pellet, the term oolite or "ooid" is usually applied. Another common type of apatitic grains are fossil fragments. These are common in practically all phosphorites, but sometimes they reach such predominance in the rock that it is referred to as a "bone bed " or "bone phosphorite". The cement or matrix in such rocks is usually calcitic, sometimes siliceous, and in the high grade phosphorites - francolitic. The association between phosphorites and apatitic skeletal fragments has been interpreted, by some investigators (e.g. Penrose, 1888; Gulbrandsen, 1969) as reflecting cause and effect. In the extreme case phosphorite formation was related to mass mortalities of fish (Brongersma-Sanders, 1957). Later studies influenced by Kazakov stressed the chemical aspects of phosphorite formation and viewed the apatitic pellets in phosphate rocks as principally inorganic formations. In recent years this tide has turned again: more and more cases of apatitic pellets and cements have been identified as biological precipitates. Thus, Williams and Reimers (1983) identified sulfur-oxidizing bacteria of the genus *Beggiatoa* in sediments off California, and from the Gulf of California. They suspect similar structures in phosphate rich sediments of the Miocene Monterey Formation of California. Bacterial structures were identified both in Cretaceous phosphorites of the Negev, (Israel) and in East Australian continental margin phosphorites (Soudry and Champetier, 1983; Soudry, 1987). A sequence of studies by Lucas and Prévôt (e.g. 1985) experimentally demonstrated bacterial mediation in replacement of Ca carbonate by apatite.

STABLE ISOTOPES AS RECORDERS FOR THE ORIGIN OF PHOSPHATES

The understanding of conditions under which phosphatic sediments form is a challenge of dual nature. On one hand, we aim a better understanding of the different stages of phosphogenesis, an apparently difficult and challenging task in itself. On the other hand, phosphorites and apatitic fossils are records of oceanic conditions. In this aspect the chemical (and isotopic) composition of a sedimentary apatite is not a reflection of a repeatedly recurring set of conditions which result in the formation of these rocks, but rather it reflects the ever changing oceanic environment. Table 1 is a preview of such an approach.

Table 1: Isotopes in apatite: what do they tell us?

Element	Measured Parameter	Genesis of phosphorites	About the world
Ca-sites			
Sr	$^{87}Sr/^{86}Sr$	Sr - Stratigraphy	
Nd	ε_{Nd}		Land-sea distribution
U	U(IV)/U(VI) $^{234}U/^{238}U$ $^{230}Th/^{234}U$ $^{231}Pa/^{235}U$	Age, Rate of growth, Redox, Submarine weathering	
PO_4 sites			
O	$\delta^{18}O_p$, $\delta^{18}O_c$	Temp. of formation, T of diagenesis	Paleoclimates
C	$\delta^{13}C$	Diagenetic mode	
S	$\delta^{34}S$	Diagenetic mode	

All stable and radioactive isotopes that are used as tracers or time measuring devices in the study of phosphatic deposits occur as trace constituents in the apatitic rocks. However, whereas the isotopic composition of O, C and S refer to constituents of CFA itself, being a part of its major structural components, several other isotopic tracers such as U, Sr and Nd are constituents that replace some major elements in the lattice. These latter elements are captured by or admitted to the lattice. The unique feature of sedimentary apatite is the ability to accommodate a large number of different trace elements in its lattice by coupled substitutions. Thus, as explained above, divalent carbonate may replace phosphate, the charge difference being balanced by a fluoride ion. Similarly, trivalent rare earths or tetravalent U may substitute for Ca^{2+} accompanied by a charge balancing substitution such as Na^+ for Ca^{2+} or CO_3^{2+} for F^-. Below, we shall treat the different tracers, element by element, and then we shall attempt to summarize our conclusions into an integrated picture of what isotopes can tell us about phosphorite formation and about the world in which they formed.

Oxygen Isotopes

Oxygen occurs in apatite as four oxyanions. The major site is in the phosphate ion, the other ones are in carbonate, sulfate and hydroxyl. Fortunately it is rather simple to isolate the O from the major two of these sites: carbonate in apatite can be separated by reaction of the mineral with phosphoric acid, and evolution of CO_2 (McCrea, 1950; Kolodny and Kaplan, 1970b). Oxygen in the phosphate can be analyzed by dissolving the mineral, precipitating the phosphate as ammonium phosphomolybdate and Mg-phosphate to ensure purity, and finally precipitating Bi-phosphate (Tudge, 1960; Longinelli, 1965; Kolodny et al., 1983; Karhu and Epstein, 1986). Throughout this procedure, O in the phosphate does not exchange isotopically with O in the aqueous solutions. Hence, the final $BiPO_4$ precipitate can be fluorinated and the resultant O analyzed, assuming that it adequately represents the isotopic composition of the original apatite. The recently developed procedure by Crowson et al. (1991) in which Ag_3PO_4 is used, is less time consuming, and yields identical, yet more precise results. The idea of using the isotopic composition of phosphate O as a geochemical indicator is almost as old as stable isotope geochemistry. When Urey and his co-workers (1951) presented the first paleotemperature estimates of the Upper Cretaceous, they noted its dependence on the knowledge of the isotopic composition of ancient sea water. Their suggestion for the elimination of this problem was the analysis of coexisting phosphate and carbonate in sediments. The fractionation of O between the two solid phases would then be independent of water composition. The first measurements were made by Longinelli (1965) using the procedure of Tudge (1960). Longinelli analyzed trace amounts of phosphate in carbonate shells, measured $\delta^{18}O$ of the environmental water in which they deposited their shells, and determined the growth temperature from $\delta^{18}O$ of the carbonate. The initial efforts of this pioneering research were marred by experimental difficulties, and it was not until eight years later

(Longinelli and Nuti, 1973), that the following phosphate - water temperature equation was proposed: (1)

$$t^{o}C \quad = \quad 111.4 - 4.3\ (\delta_p - \delta_w) \qquad (1)$$

where δ_p refers to the $\delta^{18}O$ of O in the phosphate and δ_w to that of the water, both on the SMOW scale (Fig. 1a). Kolodny et al. (1983) confirmed Eq.1 by analyzing bones and teeth of living fish. (Fig 1b).

A major advantage of the phosphate O paleothermometer is that isotope exchange between phosphate O and water is extremely slow in inorganic systems, but is very rapid in biochemical enzyme-catalyzed reactions (Kolodny et al., 1983). The sluggishness of inorganic exchange is best demonstrated by the validity of the laboratory method for apatite isotopic analysis: it involves several steps of dissolution and precipitation, some of them at elevated temperatures.

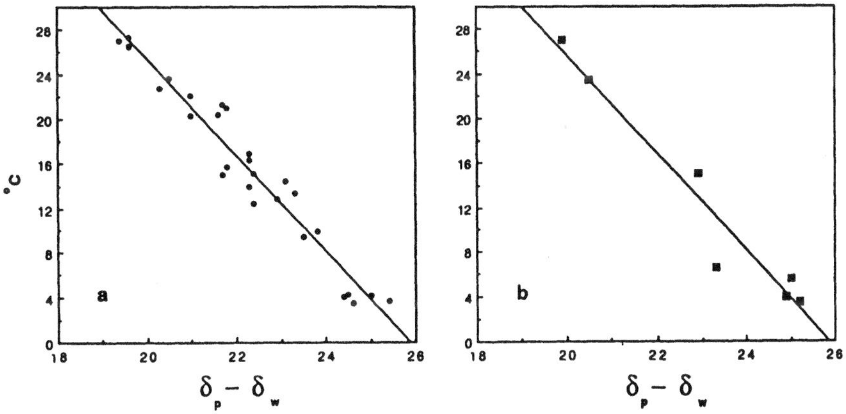

Figure 1: Temperature vs. (δ_p-δ_w) of mollusc shells. After Longinelli (1965) and Longinelli and Nuti (1973a). (b) Average temperature vs. (δ_p-δ_w) of fish bones and teeth. Data from Longinelli and Nuti (1973a) and Kolodny et al. (1983). Data points in which the temperature range is larger than 5°C, were rejected. After Luz and Kolodny (1989)

Throughout those steps the isotopic composition of the phosphate in apatite is preserved and remains unchanged in the laboratory precipitated $BiPO_4$, which is the precipitate one fluorinates in the procedure given by Tudge (1960). This sluggishness suggests that isotopic exchange between solid apatite and water at low temperatures should be extremely slow even on a geologic time scale. On the other hand, researchers of biochemical mechanisms have shown (e.g. Boyer at al., 1977) that isotopic exchange between water and dissolved phosphate is rapid when enzymatically catalyzed. Although the exact mechanism of phosphate biomineralization is not clear, similar reactions catalyzed by such enzymes as alkali-phosphatase, probably affect exchange of O between water and phosphate in bones, teeth, and apatitic shells. Kolodny et al. (1983) demonstrated the high rate of this exchange in an experiment with trout growing in a pond with constant temperature and δ_w. The trout were fed marine fish meal with much heavier O. Still, the isotopic composition of the trout bones is determined only by the temperature-dependent fractionation between bone-phosphate and water. Thus, biogenic apatites have the properties required for an ideal geochemical recorder: they respond sensitively during their formation to the aqueous environment, and they preserve very well their record after death of the organism.

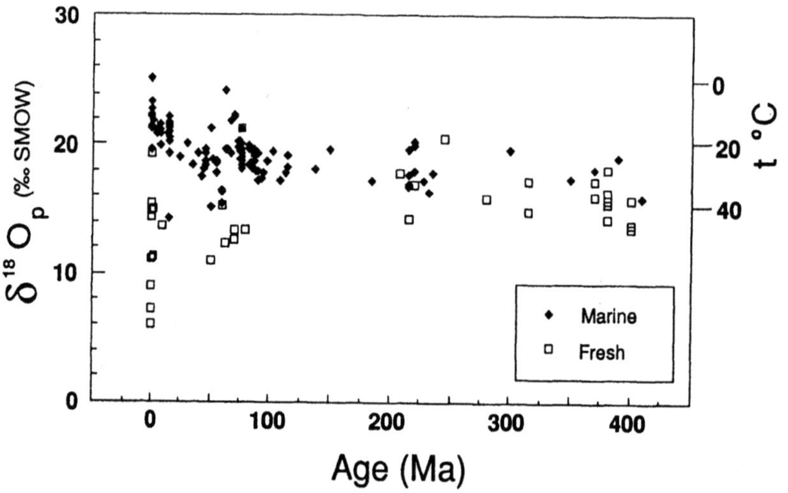

Figure 2: The secular change of $\delta^{18}O$ in phosphate of marine and fresh-water fish; Devonian to Recent. After Kolodny and Luz (1992)

The calibration of the phosphate thermometer enabled the application of this method to the fossil record. This application included the analysis of δ_p in conodonts of the North American

Paleozoic (Luz et al., 1984), of fossil fishes from the Cretaceous and Tertiary of Israel and from other Mediterranean countries (Kolodny and Raab, 1988). The latter analysis yields information about temperatures that prevailed during the Late Mesozoic-Early Cenozoic in the tropical parts of the Tethys. The thermal record is well correlated with the record obtained by carbonate O isotope thermometry of foraminifers from the deep Pacific (Douglas and Woodruf, 1981) and with the contemporaneous record for Northern Europe (Loewenstam, 1964). The deduced temperatures for the Cretaceous range between 20°C and 34°C, some 10° higher than Cretaceous surface water temperatures in Northern Europe. A survey of the isotopic record of phosphate in fish from their earliest appearance in the geological record to the Recent has recently been completed (Kolodny and Luz, 1992). Figure 2 is a plot of δ_p in fresh water and marine fishes in the last 400 million years. Throughout this time fresh water fishes had δ_p values which are lower than δ_p of contemporaneous marine fish. This is not only a demonstration of the very long term persistence of the hydrological cycle as we know it but also a reconfirmation of the essential lack of post depositional exchange in apatite O. No clear time trend in δ_p, and hence in temperature, is evident from figure 2.

Much effort has also been directed towards isotopic analysis of phosphate of non-skeletal origin. Shemesh et al. (1983) showed that ichthyoliths in phosphorites of the Campanian Mishash Formation have the same $\delta^{18}O$ as apatitic pellets from the same rocks. In a study of phosphorites from the Tethyan-Mediterranean Mesozoic province, Shemesh and Kolodny (1988) showed that δ_p of phosphorites were close to those of fish (Kolodny and Raab, 1988). Furthermore, a stratigraphic analysis of δ_p in the Mishash Formation of Israel shows that the major phosphorite-forming event is correlated with an increase in δ_p, which can be interpreted as an onset of an upwelling regime in the area. A similar conclusion, i.e. the onset of upwelling in the Tethys phosphogenic province, has been proposed by Grandjean et al. (1987, 1988) on the basis of REE pattern analysis.

Longinelli and Nuti (1968) noted a trend of decreasing δ_p in phosphorite whole rocks with increasing geological age. Similarly, Shemesh et al. (1983) have shown that whereas $\delta^{18}O$ in Recent phosphorites varies between 20 and 25 per mil, it decreases to 15-20 per mil in Mesozoic rocks, and drops further

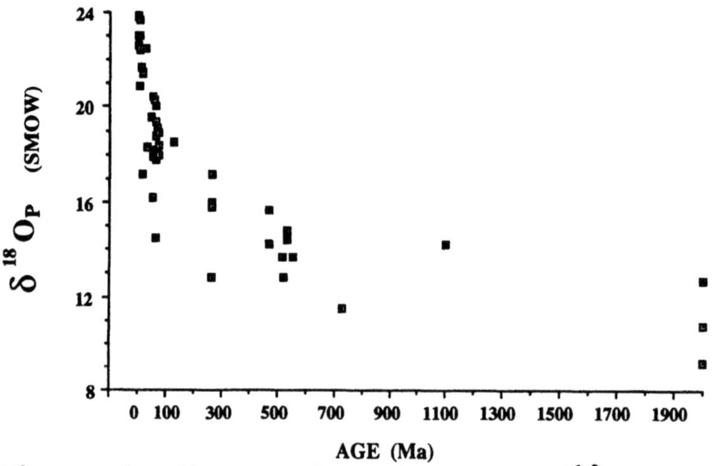

Figure 3: The secular change of $\delta^{18}O$ in phosphate of phosphorites. After Shemesh et al. (1988)

to about 10 per mil in the Archean (Fig. 3). A similar trend has been observed for $\delta^{18}O$ of carbonates, cherts and glauconites (Degens and Epstein, 1962; Knauth and Epstein, 1976; Keppens and O'Neil, 1984). Four alternative explanations have been offered for the isotopic time trends: (1) post-depositional isotopic exchange with fresh water, (2) changing $\delta^{18}O$ of sea-water with time, (3) changing temperature of deposition of phosphorite with time, and (4) high temperature alteration. Whereas the first explanation was favored for carbonates by Degens and Epstein (1962) and for phosphorites by Longinelli and Nuti (1968), the main proponent of an ocean with changing isotopic composition has been E.C. Perry and his co-workers (e.g. Chase and Perry, 1972). Shemesh et al. (1983) rejected the first hypothesis; they were strengthened in their belief in the resistance of phosphate to late isotopic exchange by evidence such as the narrow range of δ_p in the Mishash apatites compared to the broad range of $\delta^{18}O$ in coexisting carbonates and cherts (see also Vengosh et al., 1987). Hence they favored higher temperatures of deposition for ancient phosphorites. Such higher temperatures could be either a reflection of a global phenomenon, i.e. of warmer oceans in the geological past, or of a shift in the site of phosphorite formation in the ocean. A warmer ocean in the past 100 Ma has been suggested on the basis of $\delta^{18}O$ in carbonates (Savin, 1977); much debate has however been caused by the proposition of Knauth and Epstein (1976) that Precambrian oceans might have been as warm as 60-70°C (Veizer et al., 1986). The resolution of the dilemma of whether it was temperature or $\delta^{18}O$ of ocean water that caused the low δ_p in ancient rocks could be achieved by analyzing O in cogenetic phases. This is the approach taken by Karhu and Epstein (1986) who analyzed Precambrian and Phanerozoic cherts for both $\delta^{18}O$ of the silica in the chert and that of the very small amounts of phosphate in them. They concluded that: " The early Precambrian temperatures were as high as 80°C and decrease with time". A corollary of that conclusion was that: "The $\delta^{18}O$ value of the ocean has not varied more than a few per mil over the past $3x10^9$ years" (see also Perry, 1990, and Epstein and Karhu, 1990).

If chert and phosphate can be treated as two coexisting phases, then the two types of O in apatite (δ_p the $\delta^{18}O$ of PO_4^{3-}, and δ_c the $\delta^{18}O$ of CO_3^{2-}) may be even better candidates for such a pair. In such a treatment, δ_c of CFA-CO_2 is assumed to behave as in $CaCO_3$. Compare the phosphate paleothermometer (Eq. 1) to the carbonate paleothermometer (Eq. 2, after Epstein et al., 1953)

$$t = 16.5 - 4.3\ (\delta_c - \delta_w) + 0.14\ (\delta_c - \delta_w)^2 \qquad (2)$$

The striking similarity of the slope coefficient in the carbonate and phosphate paleothermometers was a disappointing observation for any attempt to estimate marine paleotemperatures independently of water isotopic composition. Karhu and Epstein (1986) pointed out that such parallelism is impossible over the entire temperature range, because at room temperature the $\delta^{18}O$ of marine calcite is about 10 per mil higher than apatite precipitated from the same water, whereas they must achieve zero fractionation at high temperature.

The establishment of a phosphate-water paleothermometer over a broad temperature range requires experimental equilibration of phosphate water pairs at different temperatures. Preliminary results (Shemesh et al., 1988), yielded the following equation:

$$1000\ln\alpha = 2.12\ (10^6/T^2) - 2.98 \qquad (3)$$

where $1000\ln\alpha \cong (\delta_p - \delta_w)$, and T is the temperature in °K. The slope here of 2.12 should be compared with that of 2.78 from the analogous equation for calcite (O'Neil et al., 1969):

$$1000\ln\alpha_{(c-w)} = 2.78\ (10^6/T^2) - 2.89 \qquad (4)$$

By subtracting (3) from (4), we obtain (5) which is a paleotemperature equation in $\delta^{18}O_p - \delta^{18}O_c$ space, independent of δ_w:

$$1000\ln\alpha_{(c-p)} \approx 0.66\ (10^6/T^2) + 0.09 \quad (5)$$

where again $1000\ln\alpha_{(x-y)}$ equals about $(\delta_x - \delta_y)$. Initially, Shemesh et al. (1983) were reluctant to use the $\delta_c - \delta_p$ pair, because the carbonate O appeared to be much more exchangeable than the phosphate one. Later however it was shown that δ_c and δ_p are very well linearly correlated (Fig. 4), so that if post depositional exchange is to be assumed, it must occur for both sites in a comparable (though not necessarily equal) degree. If δ_c and δ_p of figure 4 are equilibrium pairs, then both temperature and δ_w can be estimated. In figure 3 isotherms (with unity slope) and lines of constant $\delta^{18}O_w$ (with a slope of 0.76 = 2.12/2.78) are plotted as an overlay on the data points. Assuming isotopic equilibrium, temperature of formation and $\delta^{18}O$ can be estimated from each $\delta^{18}O_c - \delta^{18}O_p$ pair. Most points in the upper part of figure 4 yield temperatures and $\delta^{18}O_w$ well within the range of recorded earth surface values. However, most data points in the lower left part of the diagram indicate temperatures well above 100°C. The δ_w of water in equilibrium with these samples, had to be well above zero. Shemesh et al. (1988) concluded therefore that the isotopic composition of O in both apatite sites is a reflection of late diagenetic (to early metamorphic?) interactions between the apatite and fluids. It is also suggested that such interaction took place at a relatively low water/solid ratio in a semi closed system of pore waters. We can conclude by stating that although temperature may be recorded by phosphorites, it is not always the depositional temperature. Most phosphorites as well as phosphate bearing rocks may thus not tell us much about temperatures of ancient seas (Karhu and Epstein, 1986). Rather they tell us about conditions of burial. The latter conclusion is supported by the location of phosphorites from the Miocene Monterey Formation on figure 4 (Kastner et al., 1990).

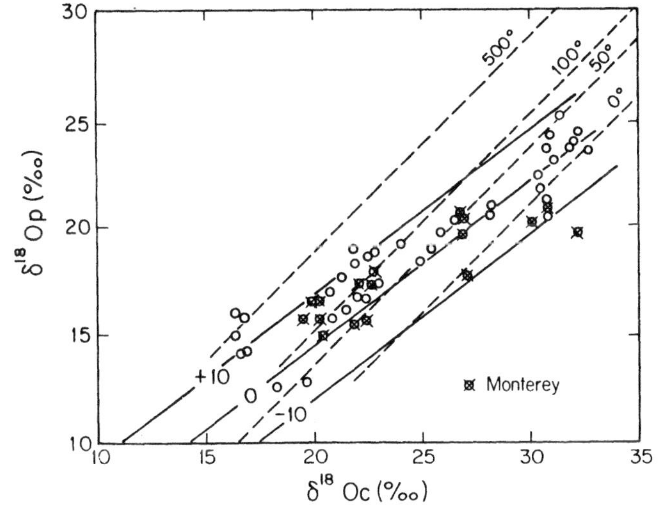

Figure 4: Relationship between $\delta^{18}O_p$ and $\delta^{18}O_c$ in Phanerozoic phosphorites. Iso-$\delta^{18}O_w$ lines and isotherms (=isoΔ_{c-p}) are overlaid upon the data points. Note the spread of Monterey data which cover the entire sample range. The high temperature data are all in the field of positive $\delta^{18}O_w$ values. After Shemesh et al. (1988)

The Monterey phosphorites were certainly **not** deposited in a very warm ocean, as there is good evidence that they were deposited from rather cold, upwelling water: δ_p of fishes from the Monterey yield temperatures of 16-21°C. Yet, the δ_p (and δ_c) values of the Monterey samples spread almost over the entire range of figure 4. Curiously the temperatures indicated by the O in the phosphate of Monterey phosphorites are very close to temperatures indicated by the isotopic analysis of dolomite concretions in the same beds. The concretions are clearly diagenetic, and the isotopic composition of their C bears indications of equilibration with methane (high $\delta^{13}C$ in carbonates). It is thus possible that the post-depositional interaction between phosphate and fluids was bacterially mediated, by the methanogenic bacteria.

The isotopic study of phosphatic sedimentary rocks should by no means include only the phosphatic phases. The availability of δ_p isotopic data opens the possibility of multi-phase isotopic analysis (Vengosh et al., 1987). Such studies should be especially useful as tracers of diagenesis. The Mishash Formation of Israel is a sequence of alternating cherts, porcellanites, chalks, limestones and phosphorites. A combination of chemically selective analytical methods with handpicking of textural components enables one to reconstruct a diagenetic sequence of events. Three isotopically concordant assemblages were identified: (a) a depositional assemblage formed in normal seawater at temperatures between 23°C and 30°C is reflected by benthic foraminifera (*Nodosaria*) skeletons ($\delta^{18}O_{SMOW}$ = +27.5 to 29.5 per mil) and apatitic bio-detritus (δ_p= +18 to +19.5 per mil), (b) an early diagenetic assemblage formed in equilibrium with a marine solution is recorded by calcite of micritic cement (+26 to +27.5 per mil), silica in opal-CT (porcellanite) (+31.5 to +33.3 per mil), silica as quartz in homogeneous cherts (+29 to +32 per mil), silica as quartz in "fragments" of chert breccias (+31 to +33 per mil), and some quartzose matrix in silicified phosphorites, and (c) a later diagenetic assemblage contains calcitic spar infilling of fossils (+20 to +26.6 per mil), silica in "matrix" of chert breccias (+21 to +33 per mil) and coarse quartz matrix in silicified phosphorites (+20 to +30 per mil). This last assemblage has been interpreted by Vengosh et al. (1987) as representing a late fresh water fluid.

One of the most striking constraints of isotopic paleothermometry is its limitation to reconstructing marine conditions. A necessary assumption of paleothermometry is the knowledge of the isotopic composition of paleo-seawater. In paleothermometry of non-Glacial periods, δ_w is assumed constant and equal to about -1 per mil. Such assumption can be justified by several approaches (Savin, 1977; see Holland, 1984, p. 241-251). In contrast, the isotopic composition of water in lakes, rivers, and groundwaters, all of which originate from meteoric water, is highly variable both in space and in time even in the present world (Dansgaard, 1964; Yurtsever and Gat, 1981). Thus, if continental, aquatic paleothermometry is to become operative, means for estimating δ_w of paleo-meteoric waters must be found. A possible way of solving the problem of the dependence of δ_p on two variables, temperature and δ_w, is fixing one of the two. In warm blooded animals the isotopic composition of O in bone phosphate should depend only on the isotopic composition of the body fluid, since the temperature of deposition is fixed (Longinelli 1984). Unlike in aquatic organisms (e.g. foraminifera, molluscs, fish), the isotopic composition of body fluids of mammals is not equal to δ_w of the drinking water. $\delta^{18}O$ of the body fluids (δ_{BW}) depends on the total flux of O through the body water reservoir (Lifson and McClintock, 1966; Luz et al., 1984). The major inputs are: drinking water, air O, metabolically produced O, and water in foods. Oxygen leaves the body by the following routes: as liquid water (urine, sweat), as water vapor, and as exhaled CO_2. When such a balance is performed the interesting result is that to a first (and good) approximation the slope of δ_{BW} versus δ_w does not depend on the isotopic composition of the different reservoirs (Eq. 6):

$$\delta_{BW} = \frac{F_1}{F_1+F_4+F_5}\,\delta_W + \frac{8(1-x)(F_1+F_4) - F_2\delta_{OX} + F_3\delta_F + 38F_5}{F_1+F_4+F_5} \qquad (6)$$

where F_1 is the flux of O in drinking water (and water in food) with an isotopic composition δ_w, F_2, with an isotopic composition δ_{Ox}, is the influx of atmospheric O and F_3, with δ_F, is the input of O from organic food compounds. Similarly, F_4 is the metabolic water formed within the body (hence possessing δ_{BW}); a fraction x of the water leaving the body is in liquid form and has a composition similar to body water, whereas a fraction (1-x) is in the form of water vapor and is hence depleted by 8 per mil with respect to body water. F_5 is the flux of respiratory CO_2 out of the body; its $\delta^{18}O$ being δ_{BW} +38 per mil.

The relationship between a mammal's drinking water and the composition of its body water is thus crudely dependent on the ratio between the amount of "drinking" to "drinking plus metabolism". Luz et al. (1984) verified the above relationship in a controlled experiment using laboratory rats. Only δ_w of the drinking water of the rats was modified, and a good linear fit between drinking water and body water was observed:

$$\delta_{BW} = 0.53\ \delta_w - 0.39 \quad (r = 0.99) \qquad (7)$$

the slope of eq. (7), 0.53, agrees remarkably well with the ratio of "drinking" to "metabolic" fluxes for rats as obtained from physiological literature. Eq. (6) is thus a guide for selecting optimal organisms for paleoenvironmental studies: animals which drink much and have a low metabolism should be the best recorders of environmental δ_w. As stated above, the isotopic composition of a mammal bone should be determined solely by δ_w, since body temperature is fixed. Moreover, according to Eq. (1), $(\delta_p - \delta_{BW})$ should not be far from 17.3 in most mammals, since their body temperatures are around 37°C. Indeed, in a controlled experiment, bones and teeth of rats were on the average 17.9 per mil heavier than their body water (see Fig. 5), close enough to the expected value (Luz and Kolodny, 1985). Although in principle it has been demonstrated that mammals can serve as good monitors of paleo-δ_w, several points remain unexplained and cause difficulties: (1) in some cases, a slope of unity or even larger has been observed for the δ_{BW} vs. δ_w relationship, suggesting that the model of Luz et al. (1984) might be an oversimplification (D'Angela and Longinelli, 1985); (2) Longinelli (1984) observed $(\delta_p - \delta_{BW})$ values quite different from the expected 17.3 per mil; he found 20.9 per mil in man and 21.4 per mil in pigs, and (3) in a study of well documented human teeth from locations over a broad range of latitudes and hence δ_w, Levinson et al. (1987) found a good linear correlation ($r = 0.93$) between the isotopic composition of local tap water and δ_p of the human teeth. The slope of the regression line was much lower (0.46) than that expected from physiological data (0.7). Mixing of water sources due to dietary habits of modern man has been suggested as an explanation. Luz et al. (1990) showed that when a significant part of intake water in

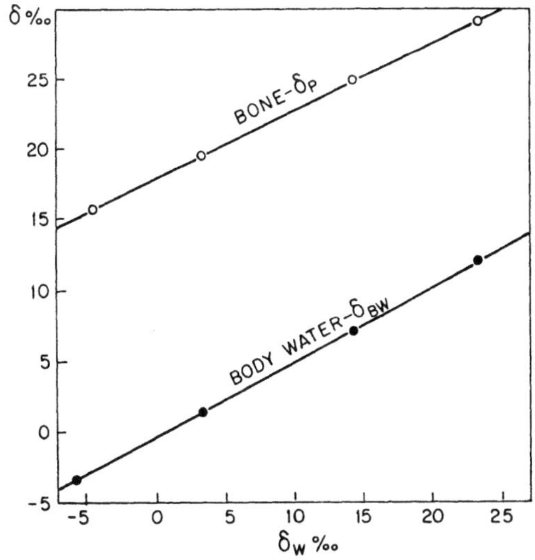

Figure 5: Plots of δ_{BW} ($\delta^{18}O$ of body water) and δ_p ($\delta^{18}O$ of bone phosphate) vs. δ_w ($\delta^{18}O$ of drinking water) of experimental rats. The separation between the two nearly parallel lines indicates equilibrium apatite precipitation at 37°C, and the slope of both lines is less than unity indicating metabolic activity (see text). After Luz and Kolodny (1985)

a mammal is obtained from leaves, relative humidity becomes a significant factor in determining the δ_w of the body fluid and hence of δ_p. Multiple regression of the isotopic composition of phosphate O of white tailed deer from North America yields a high correlation with local relative humidity and precipitation. The relative uniformity of δ_w over Australia, coupled with the variance of climate in this continent allowed L. Ayliffe and her co-workers (Ayliffe and Chivas, 1988, 1990) to isolate humidity as a variable which *via* its influence on leaf water isotopic composition is reflected in δ_p of Australian kangaroos. The isotopic composition of O in the carbonate of mammal bone apatite (δ_c) is probably also determined by the composition of body fluids and should hence also bear climatic information, albeit it is probably more susceptible to alteration. Land et al. (1980) demonstrated that δ_c of deer bone is correlated to local meteoric water δ_w values. More recently, Koch et al. (1989) have interpreted laminae in tusk and molar dentine of late Pleistocene mastodonts and mammoths as annual growth bands. They suggested that the variation in δ_c observed between these bands is a reflection of seasonality. The low $\delta^{18}O$ values representing winter growth.

Carbon Isotopes in Apatite Carbonate

We have shown above that $\delta^{18}O$ of apatite CO_2 can essentially be treated as representing a separate carbonate phase. Similarly $\delta^{13}C$ of CFA can be treated as a useful tracer of depositional and diagenetic conditions. The three following assumptions are necessary for a genetic interpretation of $\delta^{13}C$ in CFA.

1. There is little or no C isotope fractionation upon incorporation of dissolved CO_2 into the CFA lattice. This would mean that the isotopic composition of the CFA-CO_2 is representative of the bulk total of the CO_2 dissolved in the water from which the apatite formed. Although there are no experimental data for isotopic fractionation factors of carbonate ions between solutions and apatite, one can expect a rather small fractionation by analogy with inorganic Ca-carbonate precipitation.

2. An implicit assumption is that CO_2 enters the CFA lattice at the time of its formation. This is not as obvious as it may seem. Price and Calvert (1978) indicated that CO_2 content in phosphorites increases from 1.4% in poorly consolidated nodules to 4.2% CO_2 in concretionary and pelletal phosphorites, respectively. A similar trend was observed by Romankevich and Baturin (1972). Mineralogical and isotopic analysis of fossil fish bones suggests that not only must CO_2 be introduced *post mortem* into lattices of fish bones, but it is quite likely that the stable CFA phase in fish fossils is a result of a dissolution-reprecipitation process of an original hydroxyapatite phase.

3. If the isotopic analysis of CO_2 is to be used for reconstruction of depositional or diagenetic environments then one must assume that the CO_2 is well preserved in the lattice after its formation.

There are principally two different approaches to the role of CO_2 in the apatite lattice: the CO_2 content decreases with weathering and diagenesis from an initially uniform value of about 6% (McArthur, 1978; McClellan, 1980; McClellan and Saavedra, 1986; Jahnke, 1984), and the CO_2 content reflects a spectrum of formation conditions. Thus, Gulbrandsen (1970) attributed the variation in CO_2 content in Phosphoria Formation apatites to regional variations in the temperature of the Permian water column. Similarly, McArthur (1985) suggested that "the chemical composition of francolite reflects that of its formational environment". The relative constancy in Na, Sr, Mg, and SO_4 content are then a reflection of the constancy of composition of Phanerozoic seawater, whereas "the variable contents of CO_2 in unaltered francolite are a response to the pH of its formational environment". Romankevich and Baturin (1972), on the other hand, suggested that the carbonate content is controlled by its supply from degradation of organic matter in pore waters during apatite precipitation.

Isotopic analysis of CFA-CO_2 can be used to distinguish the role of these processes, as well as to elucidate several related problems. Benmore et al. (1983), McArthur et al.(1986) and Glenn and Arthur (1990) applied the approach of Goldhaber and Kaplan (1974), Irwin et al. (1977) and Froelich et al. (1979) to estimate the depth of formation of a carbonate beneath the sediment-water interface. The crux of this approach is the tracing of the relative amount of organic-matter-derived C in the C that is dissolved in the pore water. The organic-matter-derived C is labeled by low $\delta^{13}C$ (about -24 per mil) values. With increasing depth in a marine sediment, one observes a transition from the zone of oxidation of organic matter, through a suboxic zone, to an anoxic zone of (bacterial) sulfate reduction. The $\delta^{13}C$ of the total dissolved CO_2 decreases along this depth profile. When sulfate is exhausted at depth, organic matter is decomposed by methanogenic microorganisms, and part of the total dissolved CO_2 is reduced to form methane (which is highly depleted in ^{13}C). The remaining CO_2 is thus enriched in ^{13}C. McArthur et al. (1980) attempted to distinguish between "authigenic" apatites precipitated in anoxic sediments, and "diagenetic" apatite formed by limestone replacement. They proposed using ^{13}C as a discriminating indicator of mode of formation. According to McArthur et al. (1980) a strong depletion in $\delta^{13}C$

indicates void-filling apatite formation in organic rich sediments, whereas phosphatized carbonates retain the $\delta^{13}C$ value of their precursors, which lie around 0 per mil PDB. Such a simple relationship may hold only in rock dominated systems, where the amount of dissolved C species that are derived from oxidation of organic C is negligible. In systems with a high water/rock ratio, and a high organic matter content, such a distinction is no longer possible.

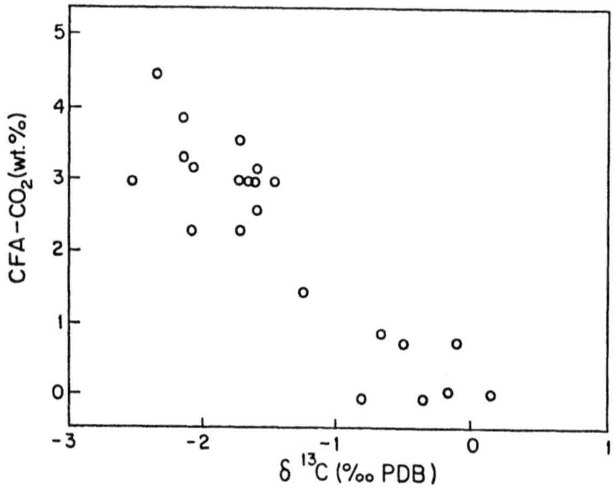

Figure 6: $\delta^{13}C$ vs. wt% CO_2 in CFA, in grains of young phosphorite from off Peru. Note the negative correlation. After Glenn et al. (1988)

Glenn and his co-workers (1988, 1990) compared $\delta^{13}C$ and CO_2 contents of phosphorite grains and crusts with $\delta^{13}C$ and concentration of total dissolved carbon, PO_4^{3-} concentration, and total alkalinity in pore waters of cores from the same area. They concluded that the amount of CO_2 in CFA depends mainly on the carbonate content of the pore water. It thus primarily reflects carbonate concentration (and thus carbonate alkalinity, but not pH), rather than the CO_3^{2-}/PO_4^{3-} ratio. The increased pore-water carbonate alkalinity is the result of microbial oxidation of organic matter. Increased carbonate substitution in CFA increases its solubility and limits the extent of its crystallite growth. Thus CFA with higher than about 6% CO_2 becomes unstable in typical pore waters. The $\delta^{13}C$ values indicate that most of the Peru-Chile phosphorite grains form between 5-10 cm below the sediment-water interface, in suboxic, perhaps anoxic, sediments.

The combination of the above processes results in a general negative correlation between $\delta^{13}C$ and the CO_3^{2-} content of CFA (see Fig. 6).

Since the turn of the century, the insular phosphorite deposits of Nauru, Ocean Island, Ebon Island and Makatea in the Pacific have yielded the main fertilizer demands of Australia and New Zealand. All these phosphorites are guano related. In an attempt to explain their formation, Aharon and Veeh (1984) measured δ_c and Ayliffe et al. (1991) determined δ_p in these apatites. They showed clear geochemical and petrological differences between the young phosphorites of the low islands and the >200 ka old deposits of the elevated, karstic islands. Whereas the former (δ_c = 32.4, δ_p = 21-23) are the result of interaction of seabird guano with slightly evaporated rain water, the latter (δ_c = 24.7, δ_p = 17-19) apparently formed by diagenetic reaction between ^{18}O depleted rain water, guano and karstic limestone.

Sulfate-Sulfur Isotopes

The presence of sulfate, probably substituting for PO_4^{3-} as a structural component within the CFA lattice, is today a well established fact (McConnell, 1973; Nathan, 1984). Similarly to the case of carbonate, the degree of isomorphous replacement of PO_4^{3-} by SO_4^{2-} depends on the sulfate content of the aqueous phase that was in equilibrium with the phosphate precipitate during its formation. (Bliskovskiy et al., 1977; McArthur, 1985). Hence, the S content and especially its isotopic composition ($\delta^{34}S$), become useful indicators of the depositional and diagenetic history of a phosphorite. The key observations here are the reduction of sulfate ions to hydrogen-sulfide by certain anaerobic bacteria which is associated with a strong fractionation (10-60 per mil) of S isotopes; the reduced phase being enriched in ^{32}S, while residual SO_4^{2-} in solution can be highly enriched in ^{34}S, the removal of sulfate into solid apatite which may, by analogy to evaporite sulfate precipitation, be assumed to have a rather small isotopic effect, and the sea-water sulfate $\delta^{34}S$ which has undergone significant secular variation in the Phanerozoic. The $\delta^{34}S$ of marine evaporites of different ages records these variations (Holser and Kaplan, 1966; Anderson and Arthur, 1983; Holser, this volume). Hence, the concentration and isotopic composition of sulfate sulfur can vary greatly in an organic-rich sediment, as a result of both $\delta^{34}S_{sw}$ and degree of reduction.

CFA-sulfur that formed in contact with seawater in an oxic environment, should thus have $\delta^{34}S$ values close to marine sulfate of contemporaneous seawater. Francolite that formed in the anoxic zone, where sulfate reduction was the controlling reaction, should contain sulfate enriched in ^{34}S. Generally, one should expect an association between negative $\delta^{13}C$ values and positive $\Delta^{34}S$ ($=\delta^{34}S_{CFA}-\delta^{34}S_{SW}$). Figure 7 demonstrates such a relationship. Data points with high $\delta^{13}C$ and $\Delta^{34}S$ near zero form a tight cluster in the oxic zone. The $\delta^{13}C$ is generally low and uniform in the anoxic zone, while (with one exception) $\Delta^{34}S$ varies from high to low positive values. This latter variation is related to the variable extent of reduction of sulfate.

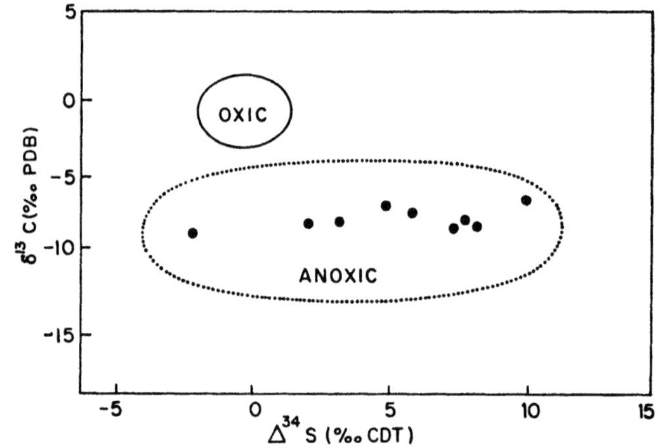

Figure 7: Variation of $\delta^{13}C$ and $\Delta^{34}S$ ($\delta^{34}S_{CFA}$ - $\delta^{34}S_{seawater}$) in oxic vs. anoxic sediments. After McArthur et al. (1986)

The first $\delta^{34}S$ measurements in apatite were performed by Bliskovskiy et al. (1977), followed by studies of Nathan and Nielsen (1980), Benmore et al. (1983), and McArthur et al. (1986). $\delta^{34}S$ of most young apatites (Miocene to Recent from Aghulas Bank, Blake Plateau, the California Borderland) are +20 to +23 per mil, and are indistinguishable from the modern seawater value. Most phosphorites ranging in age from the Precambrian to the Miocene have $\Delta^{34}S \geq 0$. Thus, Nathan and Nielsen (1980) drew the conclusion that whereas the phosphorites with $\Delta^{34}S \cong 0$ were apparently deposited from oxic ocean water, those with $\Delta^{34}S > 0$ formed in reducing environments. They supported the latter conclusion by the observation that gypsum associated with phosphorites is often isotopically very light, suggesting that it might have formed from aqueous sulfate released by the oxidation of biogenic Fe-sulfide. A decrease in sulfate concentration in interstitial water should thus be correlated with an increase in $\Delta^{34}S$. This, indeed, is the case for phosphorites of the Phosphoria Formation (Piper and Kolodny, 1987), where all $\delta^{34}S$ values are heavier than 14 per mil, (the value of Permian seawater), and negatively correlated with $\log(SO_4^{2-}/P_2O_5)$ supporting a "Rayleigh distillation" type reduction of sulfate to sulfide.

The simple relationship outlined above may be complicated at interfaces between reducing and oxidizing zones by the upward diffusion of ^{34}S depleted H_2S. In the absence of reactive Fe, this sulfide is oxidized to sulfate (in the suboxic or oxic zones), thus lowering $\delta^{34}S$ from its ambient value. An alternative explanation for the range of $\Delta^{34}S$ values between +6 and -2 is that these reflect different fractionations due to different rates of sulfate reduction (McArthur et al., 1986; Kaplan, 1983).

RADIOGENIC ISOTOPES AS RECORDERS FOR THE ORIGIN OF PHOSPHATES

Strontium and Neodymium Isotopes

The last decade has seen a sequence of impressive achievements in the application of isotopic ratios of Sr and Nd in sedimentary rocks for the understanding of the geological history of paleo-seas. The variations of both Sr and Nd isotopic ratios in seawater are the results of a combination of the relative amounts and the isotopic compositions of the sources that contribute these elements to the oceans. In this case, variations in the isotopic composition of the sources are the result of radioactive decay, not of mass-dependent fractionation. The concentrations of Sr and Nd in seawater are very different, reflecting their different balances of input/removal rates. Whereas the residence time of Sr in seawater is about 10^6 years (Turner and Whitfield, 1979), several orders of magnitude longer than oceanic mixing time, the residence time of Nd is only tens to hundreds of years (Piepgras et al., 1979), much shorter than the oceanic turnover time. The oceans are therefore well mixed with respect to Sr, but heterogeneous with respect to Nd. Whereas a single ocean-wide curve has been established for the variation of $^{87}Sr/^{86}Sr$ in seawater with age (Peterman et al., 1970; Burke et al., 1982; Koepnick et al., 1985), no such curve has been proposed, or should be expected for $^{143}Nd/^{144}Nd$ (expressed as ε_{Nd}).

Apatitic fossils are good indicators of $^{87}Sr/^{86}Sr$ in paleo-oceans, and probably the best available material for obtaining $^{143}Nd/^{144}Nd$ in paleo-seawater. Sr occurs in rather high concentrations (1100-2300 μg/g; Garavelli et al., 1979, Schmitz et al., 1991) in bones of living marine fish, and in comparable concentrations in apatitic fossils. Thus, it is reasonable that $^{87}Sr/^{86}Sr$ ratios in ichthyoliths reflect seawater values. Unlike Sr, Nd is apparently introduced into apatite during early diagenesis. The REE occur in extremely low concentrations in living fish bones, but reach rather high values relatively early after the organism's death (Arrhenius et al., 1957; Bernat, 1975; Shaw and Wasserburg, 1985). The concentration of REE in apatitic fossils is higher than in any other sedimentary mineral; it is 5-6 orders of magnitude higher than in seawater, and 2-5 orders of magnitude higher than in marine carbonates. Based on mass balance considerations, Grandjean et al. (1987) and Grandjean and Albarède (1989) have shown that REE must have been introduced into fish debris during very early diagenesis, via a carrier phase (probably biogenic particulates such as zooplankton fecal pellets and oxihydroxides) which transported REE into the near-bottom water.

The Sr isotopic composition of apatitic fossils is often in reasonably good agreement with the $^{87}Sr/^{86}Sr$ ratio in contemporaneous seawater. Such agreement has been demonstrated for conodonts, ichthyoliths and inarticulate brachiopods (Kovach, 1980; Shaw and Wasserburg, 1985; Staudigel et al., 1985; Keto and Jacobsen, 1987; Grandjean et al., 1987). Whereas Staudigel et al. (1985) concluded that about 70% of their analyses fall within the range of the seawater curve, the data spread of about half of the fish fragments analyzed by Grandjean et al.(1987, 1988) was larger than that of contemporaneous carbonates. Practically all of these deviations were towards higher than expected $^{87}Sr/^{86}Sr$ ratios. A similar spread has been recently observed by Schmitz and coauthors (1991), who attempted to distinguish between marine and fresh-water fossil fish by means of the Sr isotopic composit-

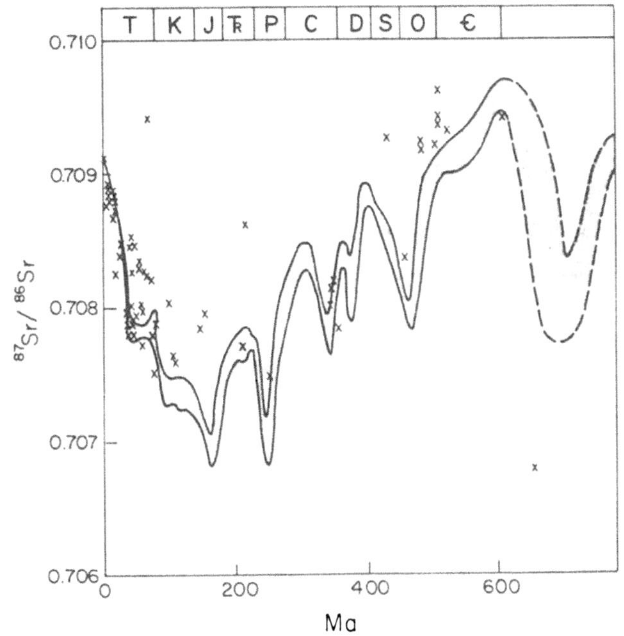

Figure 8: $^{87}Sr/^{86}Sr$ in phosphorites and fish debris plotted on the curve of the isotopic composition of Sr in seawater vs. time. Note the general agreement between phosphate and contemporaneous seawater; practically all deviations from the curve are towards more radiogenic values. Data from Shaw and Wasserburg (1985), Staudigel et al. (1985), and Grandjean et al. (1987, 1988)

ion. Most fossils that were presumed (on paleontological grounds) to be marine, plotted within error of the $^{87}Sr/^{86}Sr$ seawater curve. However, whereas living fresh-water fishes contain less than 100 µg/g Sr, the Sr content of the fossils was in the range of their marine counterparts. Hence, one cannot escape the conclusion that post-burial interaction between apatitic fossils and their surrounding fluids is common. Figure 8 is a summary of $^{87}Sr/^{86}Sr$ ratios in phosphorites and apatitic fossils. An attempt to date phosphogenesis by Sr isotopes has recently been been made (McArthur et al., 1990; Compton et al., 1990). This approach is supported by three observations: (1) $^{87}Sr/^{86}Sr$ ratios of modern phosphorites are within error of modern marine Sr, (2) during phosphatisation of carbonate the Sr isotopic composition of the newly formed apatite is not inherited from the carbonate but reflects seawater Sr, and (3) the $^{87}Sr/^{86}Sr$ ratio in shallow marine pore waters is indistinguishable from modern seawater. The age determinations for phosphorites from four different areas are in good agreement with geological reasoning and previous estimates. Compton and co-workers (1990) suggested on the basis of $^{87}Sr/^{86}Sr$ ratios in francolites and biogenic carbonates from Florida that phosphogenesis in the southeastern United States occurred in the Oligocene and early to middle Miocene. The apatites were reworked and weathered, *a la* Baturin during later Miocene regressions. These authors also suggested that whereas the depositional stage of apatite formation, which is basically a stage of deposition of organic rich sediments on the shelves, resulted in a global positive $\delta^{13}C$ excursion, the winnowing stage is the result of a regression, massive oxidation of organic carbon, and a return of $\delta^{13}C$ to its preexcursion values.

The reasonably good agreement between ichthyolith Sr isotopic values and the secular Sr-isotope curve should be considered in view of the demonstration (Nelson et al., 1986) that bones of prehistoric marine organisms (seals) acquire a "terrestrial" $^{87}Sr/^{86}Sr$ ratio after being exposed to freshwater diagenesis. Apparently this observation reflects an early dissolution-precipitation step in biogenic apatite diagenesis. The fossilized apatite ultimately acquires the chemical signatures of its very early diagenetic environment. If as in the "experiment" of Nelson et al. (1986) that environment is drastically different from the depositional environment, the preserved information may be misleading. However, if as in the great majority of cases, the early burial site is identical, or very similar to the environment of life, then $^{87}Sr/^{86}Sr$ (and perhaps $\delta^{18}O$ as well) reflect that environment. The influence of late alteration on the Sr isotopic composition of phosphorites was considered by McArthur et al. (1987). They found phosphorites of the Upper Varswater Formation (Cape Province, South Africa) to be enriched in radiogenic Sr as their alteration is intensified (reaching $^{87}Sr/^{86}Sr = 0.70929$ in the most weathered sample). The alteration can be detected from the change in Sr concentration: weathering reduces the Sr content of the phosphorite from a Sr/Ca ratio of 3.7×10^{-3} in unaltered phosphorites to less than half that value in the most altered ones. An independent criterion for the degree of alteration of apatites has recently been suggested by Shemesh (1990). He used the crystallinity index as determined by infrared spectroscopy (FT-IR) as an indicator to distinguish between pristine samples and those strongly affected by diagenesis and recrystallization. As the crystallinity index increases, both $\delta^{18}O_p$ and the Sr content decrease.

$^{143}Nd/^{144}Nd$ ratios were measured in conodonts (Wright et al., 1984; Shaw and Wasserburg, 1985; Keto and Jacobsen, 1987, 1988), ichthyoliths (Staudigel et al., 1985; Grandjean et al.,1987), phosphatic brachiopods (Keto and Jacobsen, 1987) as well as phosphorites (Shaw and Wasserburg, 1985; Keto and Jacobsen, 1988). Recent biogenic apatites from geographically separated localities bordering on a common sea, record approximately the same εNd. $^{143}Nd/^{144}Nd$ ratios in fish teeth separates from the Pacific and Atlantic oceans are similar to those ratios in Mn nodules, which in turn probably reflect some time-integrated values in the corresponding large water masses. The analysis of $^{143}Nd/^{144}Nd$ in apatitic fossils yields a rather consistent picture (Fig. 9). The present day oceanographic pattern seems to have existed for about the last 50 Ma. Throughout this time the Atlantic was characterized by relatively low $^{143}Nd/^{144}Nd$ (εNd between about -12 and -6) whereas the Pacific water masses had significantly higher ratios (εNd from -6 to 0). This difference

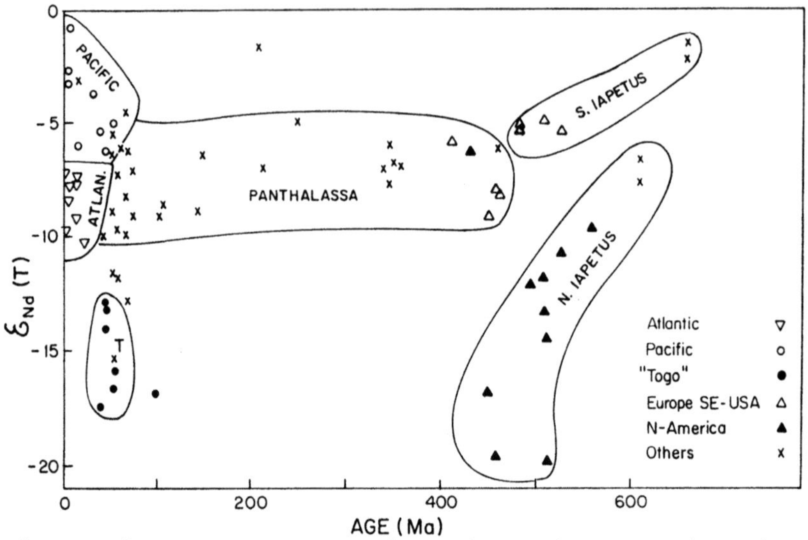

Figure 9: εNd(T) vs. time for Phanerozoic phosphatic fossils. Note two periods of geochemical provincialism: (a) between 0 and 50 Ma - the Pacific and Atlantic provinces (for data in field "T", see text), and (b) 400-650 Ma - the southern and northern Iapetus provinces (see text). Accordingly, data points from Pacific, Atlantic, North America, and SE-USA+Europe are marked. Data from Shaw and Wasserburg (1985), Staudigel et al. (1985), Grandjean et al. (1987, 1988), and Keto and Jacobsen (1987)

difference apparently reflects the dominance of old, LREE enriched crust as the source of Atlantic Nd, as opposed to relatively young crust as well as hydrothermal sources that supply Nd to the Pacific. Shaw and Wasserburg (1985) considered the convergence of εNd to a narrower spread prior to 100 Ma, as evidence for the existence of "Panthalassa", the uniform ocean which surrounded Pangea prior to its breakup. The studies of Keto and Jacobsen (1987, 1988) of $^{143}Nd/^{144}Nd$ in Phanerozoic apatitic fossils and phosphorites suggest that the picture of a single Paleozoic world ocean is an oversimplification. The Pacific/Panthalassa Ocean has been the largest Phanerozoic and Late Precambrian paleo-ocean (averaging 80% of world ocean area), but not the only one. Between the Middle Cambrian and the Middle Ordovician, geochemical provincialism shows up again. $\varepsilon Nd_{(T)}$ values in samples from Europe and the Southeastern United States (-5 to -9) differ sharply from those in the rest of North America (-10 to -20). The proposed interpretation is the existence of two separate oceans between North America and Baltica, with a barrier between them. Keto and Jacobsen refer to these as the Southern and Northern Iapetus Seas, respectively. Only with the onset of the Taconic Orogeny did this barrier apparently disappear, and a uniform ocean (in εNd terms) form.

Further insight into paleoceanographic connections between different oceanic basins of younger age is provided by the isotopic analysis of Nd by Grandjean et al. (1987). Thus it appears that the Middle Eocene was a critical period in establishing modern oceanic circulations. Prior to that time an isolated basin (with εNd of about -13.5) existed in the South Atlantic. Sediments exposed in Togo, Guinea and Angola were deposited in that basin (Fig. 9, field "T"). In some cases the ichthyoliths have $^{87}Sr/^{86}Sr$ ratios of contemporaneous seawater, but $^{143}Nd/^{144}Nd$ values of an isolated basin. This probably is a reflection of good mixing and connection with the world ocean on a time scale of the residence time of Sr, (about 4 Ma) but insignificant interaction with other basins on the scale of Nd residence time (about 100 a; Fig.9). It should be stressed that in many cases the analyzed fossils do not actually sample an open oceanic environment. They usually represent marginal seas and carbonate platforms with poorly known connections to the major ocean water masses.

Uranium Isotopes and its Daughter Nuclides

The significant enrichment of sedimentary apatite in U has been noted early in this century. R. J. Strutt (later Lord Rayleigh) was inspired by Ernest Rutherford to study the helium content of numerous uranium bearing rocks and minerals, aiming at establishing methods for calculating their ages. As part of this project Strutt discovered the exceptional radioactivity of fossil bones and phosphorites. Later it was shown that while igneous apatite is enriched in both U and Th, the radioactivity of sedimentary phosphates is due almost entirely to U alone (Altschuler et al., 1958). This discrimination is apparently related to the relative solubility of some uranyl ions, leading to U enrichment in seawater, compared to the extreme insolubility of Th compounds. The enrichment of phosphorites in U is so strong, that mapping the radioactivity of the sea floor by means of a deep-towed scintillometer proved an effective method for defining the distribution of phosphatic sediments in many parts of the oceans (Jones, 1989).

Phosphorite contains between 10 and 500 µg/g U. The average content is 120 µg/g, enriched by a factor of 30 over an average shale (Altschuler 1980). The study of the location of U in phosphatic sediments has been principally carried out by nuclear (though not necessarily isotopic) methods. In 1958 Altschuler et al. applied alpha-track autoradiography to demonstrate that practically all of the U in phosphorites is associated with the CFA. They further showed that much of the U in phosphorites is tetravalent, and proposed that U substitutes for Ca^{+2} in CFA as U^{+4} which has a similar ionic radius to that of Ca^{+2}. The structural location of hexavalent U in CFA is less clear. The uranyl $(UO_2)^{2+}$ ion is too large to fit structurally into any of the major lattice sites of CFA; hence its association with phosphorites has been explained by chemisorption of uranyl ions on growing apatite crystallites, or as a result of post-formational oxidation of primary U^{+4}.

The fission-track technique has also been successfully applied to the study of U distribution in phosphorites. Burnett and Veeh (1977), Kress and Veeh (1980), Kozlov (1970), Berzina et al. (1981) confirmed the early findings that U is CFA-related. A fission track study by Avital et al. (1983) of Israeli phosphorites enabled them to map concentrations of U in various textural components of a phosphorite. Optically anisotropic bone fragments have the lowest U concentration of all apatitic grains in the rock (80-150 µg/g); replaced isotropic bones are richer in U (120-260 µg/g), as are isotropic ovulites (140-300 µg/g). Non-apatitic grains and calcitic matrix and cement are U poor (usually less than 5 µg/g). A related study (Stein et al., 1982) dealt with the behavior of U in phosphorites during phosphate ore calcination. It has been found that during ore calcination at about 600°C, all the U is oxidized to the hexavalent state, and at about 900°C it migrates into U-rich phases that can be preferentially extracted with acid, without dissolving the entire rock. The significance of such a study is in providing background information for attempts to extract U as a side product during industrial treatment of phosphorite ores.

Extremely few radiometric age determinations have been carried out on pre-Pleistocene phosphatic rocks. Early attempts to apply fission track dating to phosphorites failed because the crystallites of CFA in phosphorites are too small to record tracks (D. Lal, personal communication). Soviet scientists were the first to attempt Pb isotope studies of Proterozoic phosphatic sediments (Tugarinov et al., 1963). These were followed by two age determinations of uraniferous phosphorites (Vaasjoki et al., 1980; Bell and Thorpe, 1986), in both of which good Pb-Pb isochrons were obtained. Kovach and Zartman (1981), reported a U-Th-Pb date on conodonts. They obtained a concordant age of about 325 Ma for conodonts of Upper Carboniferous age.

Uranium Series Dating of Young Phosphorites.

In contrast to the, so far, limited success in dating ancient phosphorites, isotope geochemistry was central to the intensive study of young, sea-floor phosphorites. The success in establishing their ages, rates and modes of growth, and the understanding of the environment of Recent and young phosphogenesis led to a better understanding of how phosphorites formed through time. The distribution of submarine phosphorites, especially along the western margins of continents, has been tacitly assumed by many workers, notably by Sheldon (1964) and McKelvey (1967), to represent sites of Recent phosphorite formation. This in turn was accepted as proof of Kazakov's theory of phosphogenesis. The development of U series disequilibrium dating techniques (See Faure, 1986, and Ivanovich and Harmon, 1982, for summaries and thorough treatment of methods), brought many sedimentary events in the range of tens of thousands to hundreds of thousands of years within the capability of precise dating. The original methods were based on high resolution alpha-spectrometry; the recent introduction of mass-spectrometric determination of $^{234}U/^{238}U$ ratios (Chen et al., 1986; Edwards et al., 1987), makes the prospects of the disequilibrium methods even more promising, though no mass-spectrometric measurements of phosphatic sediments have been reported so far.

In order to date a U-bearing phosphorite one must assume that: (1) U (but little if any Th) enters the apatite lattice at a time close to the mineral's formation, (2) this U is of marine origin with an initial $^{234}U/^{238}U$ activity ratio close to that of modern seawater, and (3) the apatite has remained a closed system with respect to U and Th isotopes. Kolodny and Kaplan (1970a) made these assumptions and attempted to date phosphorites from off California, Chatham Rise, and Aghulas Bank. None of these contained an excess of ^{234}U. In fact, many of the phosphorites showed $^{234}U/^{238}U$ ratios below unity, indicating that many phosphorites exposed now on the sea-floor are not forming at the present time, and are most likely even being leached. Thus it was concluded that none of the sea-floor phosphorites that have been studied by the end of the sixties, were Recent. When Recent phosphorite genesis was suspected in organic rich diatomaceous sediments beneath areas of intensive upwelling off the coasts of Namibia, Peru and Chile, it was by applying U-disequilibrium dating that their Recent age was confirmed (Baturin et al., 1972; Veeh et al., 1973; Veeh et al., 1974; Burnett and Veeh, 1977). It was found that these phosphorites contain U with a $^{234}U/^{238}U$ activity ratio very close to present seawater, and that although some ^{232}Th was found in the phosphorites, it could be assumed safely that none of it, and no ^{230}Th entered the apatite lattice.

The successful dating of young phosphorites began a sequence of radiochemical studies of phosphatic sediments, thanks to which our understanding of the modes of phosphogenesis increased greatly. These studies resulted in not only the location of further young phosphorites in zones of upwelling (off South Africa, Birch et al., 1983), the dating of very young insular (guano related) phosphorites on Ebon atoll in the Marshall Islands (Veeh and Burnett, 1978), but also the discovery of currently forming phosphorites off New South Wales, Australia (Kress and Veeh, 1980; O'Brien and Veeh, 1980; O'Brien et al.,

1986). This latter case was the first evidence for Holocene phosphogenesis not linked to strong upwelling. The Australian continental margin is an area of very limited seasonal upwelling and low biological productivity. This deposit as a possible analog of an "east coast" type phosphogenic province (McKelvey, 1967), eliminates the necessity to assume paleo-upwelling in areas where phosphorites are found with no indications of Recent upwelling. McArthur and co-workers (1988) dated phosphorites off western South Africa and concluded that phosphogenesis was episodic between the Eocene and 62 ka. Isotopic analysis of C and S suggested to those authors that phosphogenesis had occurred in oxic sediments and was probably unrelated to enhanced coastal upwelling.

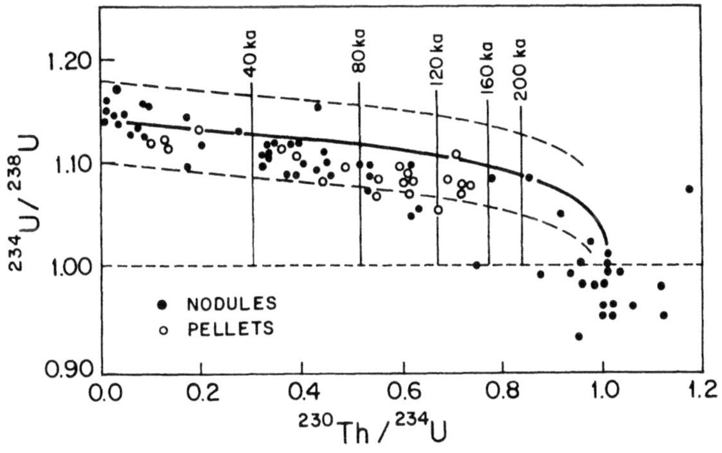

Figure 10: Ideal development in time or "Concordia (heavy curve) of $^{234}U/^{238}U$ activity ratios vs. $^{230}Th/^{234}U$ with analytical results of young pellets and nodules. Note the initial $^{234}U/^{238}U$ activity ratio of >1.14, and the occurrence of old nodules with $^{234}U/^{238}U$ <1.0. The two curves enclose a deviation of 2σ from the Concordia path. After Burnett et al. (1988), Burnett and Veeh (1977), and Veeh and Burnett (1982)

The U system offers several opportunities to test for concordance of radioactive dates. The decay of the two long-lived isotopes of U (^{235}U and ^{238}U) into a sequence of short lived daughters enables one to check concordance by comparing ages obtained by applying two different clocks. The graphical tool for such a check is the construction of "concordia" type diagrams, so that the location of data points within analytical error of the calculated concordia curves indicates system closure, and increases one's confidence in the obtained date. Figure 10 represents the evolution in time of $^{234}U/^{238}U$ and $^{230}Th/^{234}U$ for radioactive decay in a closed system with an initial $^{230}Th/^{234}U$ ratio of zero and a constant initial $^{234}U/^{238}U$. These are conditions approximated by many phosphorites. It can be seen from figure 10 that the initial $^{234}U/^{238}U$ activity ratio in phosphorites is slightly higher than that in seawater (1.14), in contrast to corals which show no such difference. Veeh and Burnett (1982), suggest that this is the result of phosphorite formation in anoxic, organic-rich sediments. Uranium in such environments is incorporated into CFA from interstitial waters, which may be enriched in ^{234}U by diffusion from the solid grains. A further conclusion from the plot is that again, unlike in the case of corals no complete system closure can be assumed for phosphorites. The spread of the data points can be best explained as the result of a radioactive decay of excess ^{234}U in a semi-closed system from which a small fraction of ^{234}U is continuously lost; it is this fraction which diffuses into the interstitial water and enriches it in ^{234}U. This semi-closed 'valve type' mechanism of isotopic fractionation was first proposed by Kolodny and Kaplan (1970a), and was later confirmed and elaborated by Burnett and Veeh (1977) and O'Brien and his co-workers (1980, 1987). We shall deal with it in more detail below.

^{230}Th is daughter of ^{238}U and ^{231}Pa of ^{235}U. Figure 11 is a $^{230}Th/^{234}U$-$^{231}Pa/^{235}U$ concordia plot. The solid curve represents concordance which is expected if the system remained closed to all relevant isotopes, and no initial ^{230}Th and ^{231}Pa were included in the apatite. It can be seen from figure 11 that several phosphorite samples from non-upwelling environments (New South Wales) or young samples yield concordant $^{230}Th/^{234}U$-$^{231}Pa/^{235}U$ ages (Veeh, 1982; Birch et al., 1983). On the other hand many of the phosphorite nodules from off Peru-Chile have an excess of ^{231}Pa, as compared to their $^{230}Th/^{234}U$ age (Roe et al.,1982). An excess of ^{231}Pa is probably the result of adsorption of both Th and Pa during the nodules exposure to seawater and pore-water at the sediment water interface. These bottom waters have a $^{230}Th/^{231}Pa$ activity ratio of less than the production ratio of 11.2, because ^{230}Th is preferentially scavenged by particulates in such a high productivity area; hence the apparent "excess" of ^{231}Pa. An other kind of openness is demonstrated by the pellets from the Peru shelf (Burnett et al., 1988). Their data points on figure 10 are predominantly below the concordia. Normally such tendency to plot below the concordia path would be explained by addition of U, preferential addition (relative to Pa) of Th, loss of Pa or a combination of these effects. Analyses of bones from limestone caves in South Australia (Ayliffe and Veeh, 1988) also plot below the concordia line. They were interpreted as representing episodic U assimilation, a conclusion supported both by fission track mapping and by $^{234}U/^{238}U$ analysis.

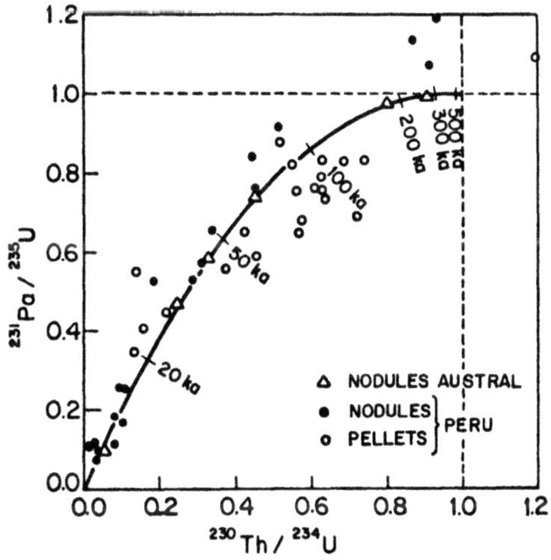

Figure 11: Ideal development in time or "Concordia" of activity ratios $^{230}Th/^{234}U$ with analytical results for young phosphatic pellets and nodules. Compare concordant ages of nodules from Australia to "excess Pa" in nodules from Peru, to "excess Th" in pellets from Peru. After Veeh (1982), Burnett and Roe (1983), and Burnett et al. (1988)

The geochronology of the Peru-Chile sea-floor phosphorites suggested to Burnett and Veeh (1977), that phosphorite formation was largely restricted to times of high sea level. This conclusion was not supported by the Eastern Australia data (Kress and Veeh, 1980; O'Brien and Veeh, 1980; O'Brien et al., 1986), where phosphogenesis seems to have been a more or less continuous process for the last 180,000 years. It is too early to tell whether the difference between the two groups is the result of inadequate sampling, or it is real, indicating different mechanisms of phosphorite formation in "west coast" and "east coast" type phosphorites.

Modes and Rates of Growth

The establishment of modes and rates of growth of phosphorites is essential to a better understanding of mechanisms of phosphogenesis. The discovery of young phosphorites has made this task possible. In addition it is now possible to compare rates of growth of pellets, nodules and crusts with the surrounding sediments. In their study, Burnett and his co-workers (1982, 1983) cut phosphorite nodules and crusts from off Peru into several slices, each of them a few millimeters thick, in a direction perpendicular to their suspected growth axis. The $^{230}Th/^{234}U$ age was then established for each slice. The top and bottom of the dredged nodules was inferred from an oxidized layer on the exposed surface. In a later study, Kim and Burnett (1986) succeeded in sampling from the Peruvian sea floor an undisturbed, oriented 20 mm thick, phosphatic crust, together with its overlying (14 cm) and underlying sediment. They concluded that whereas phosphate nodules located at the sediment water interface, grow *down* into the soft underlying mud, the direction of growth of the buried crust is *upward*.

Growth rates turn out to be slow relative to the surrounding sediment. The fastest growing nodule grew at a rate of 10 mm/1000 yrs.; the slowest - at a fraction of a mm per thousand years. It is therefore surprising that such nodules remain at the sediment-water interface in sediments that accumulate at a rate of about 1 mm per year. Even the fastest growing nodule should have been buried under about 2 meters of sediment. Burnett and his co-workers suggested that either burrowing activity by benthic organisms or intermittent periods of non-deposition could explain the interface location of the nodules. Burnett et al. (1988)

applied an analytically improved version of U-disequilibrium dating, combined with accelerator mass spectrometry (AMS) ^{14}C dating of structural CFA-CO_2 in Peruvian shelf phosphatic *pellets* (as opposed to nodules). Calculated rates of phosphate uptake are somewhat higher than rates for nodules. The pellets seem to form very quickly, probably within less than 10 years after growth is initiated.

Redox, Uranium Oxidation States and its Isotopic Composition

The effect of ^{234}U diffusion outward from the solid apatite into interstitial water upon the dating system of phosphorites has been mentioned above. Such diffusion is an expression of a fractionation of the U isotopes ($^{234}U/^{238}U$) between different oxidation states. Chalov and Merkulova (1966) found that the $^{234}U/^{238}U$ ratio in the tetravalent fraction of uranium oxides is always lower than that ratio in the hexavalent fraction. Kolodny and Kaplan (1970a) showed that upon chemical separation of U into its two oxidation states in apatite, the ratio of U isotopes varies as follows:

$$^{234}U/^{238}U(IV) \quad < \quad ^{234}U/^{238}U(T) \quad < \quad ^{234}U/^{238}U(VI)$$

where (IV), (T), and (VI), designate the tetravalent U, the total U, and the hexavalent U fractions respectively. The scheme which has been proposed in order to explain these differences is shown in figure 12. It is assumed that a fraction of the radiogenic ^{234}U in the tetravalent U in apatite is relocated due to alpha recoil, and then oxidized to hexavalent U. Hence there is a continuous leakage of ^{234}U from $U_{(IV)}$ to $U_{(VI)}$. If that fraction (termed "R") remains constant with time, then the $^{234}U/^{238}U_{(IV)}$ varies with time according to:

$$(^{234}U/^{238}U)_{IV} =$$

$$(1-R)\,[1-e^{(\lambda 238-\lambda 234)\,t}] + (^{234}U/^{238}U)_{IV0} \cdot e^{(\lambda 238-\lambda 234)\,t} \qquad (8)$$

where t = time since phosphorite formation and $(^{234}U/^{238}U)_{IV0}$ is the initial activity ratio at $t = 0$. When t becomes sufficiently large (> 1Ma), the right-hand side of Eq. (8) becomes equal to (1-R). Thus R can be estimated from:

$$R = 1 - (^{234}U/^{238}U)_{IV} \qquad (9)$$

by measuring the activity ratio of tetravalent U in sufficiently old phosphorites. In Kolodny and Kaplan's phosphorites this activity ratio averaged 0.7. They therefore estimated R as being 0.30, indicating that about 30% of the tetravalent ^{234}U atoms produced within the phosphorites are oxidized to the hexavalent state. Kolodny and Kaplan (1970a) considered the constant R to be the combined probability of a radiogenic ^{234}U atom in the tetravalent state to be displaced and then oxidized to hexavalent U. They showed that in order to produce the leakage of recoil $^{234}U_{(IV)}$ an oxidant is necessary, and that in sea-floor weathered, outer oxidized rims of phosphatic nodules, both $U_{(VI)}$ and $(^{234}U/^{238}U)_{IV}$ are lower than in their un-oxidized centers. Both the general model, and the estimate of the best value for "R" as 0.3, has since been confirmed by Burnett and Veeh (1977), O'Brien and Veeh (1980), and Veeh and Burnett (1982).

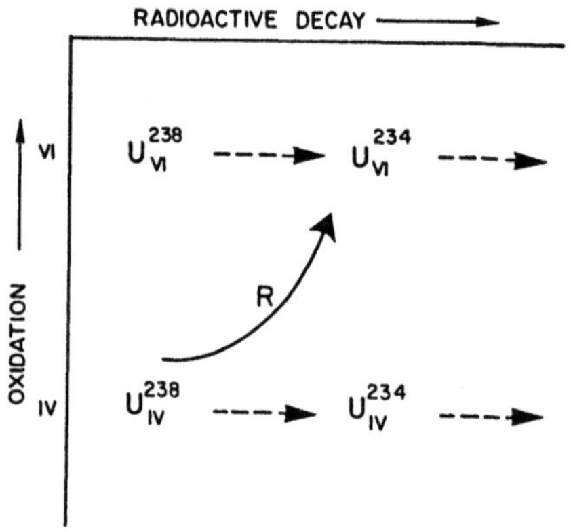

Figure 12: Diagrammatic representation of the partial oxidation of ^{234}U as a result of radioactive decay. After Kolodny and Kaplan (1970)

O'Brien et al. (1987) have applied a similar approach to the analysis of East Australian phosphorites. They demonstrate convincingly that these nodules underwent at least two distinct stages in their history: (1) an initial stage in which phosphate diffuses interstitially into a slowly growing layer of phosphorite and incorporates U only as $U_{(IV)}$ and (2) a subsequent exposure at the sediment water interface during which the layer is broken up into nodules, which are periodically reworked and exposed. Whereas in the first stage conditions might be mildly oxic or sub-oxic, clearly oxidizing conditions prevail during the weathering stage, as evidenced by a strong ferruginous impregnation of the phosphorites, oxidation of organic matter and oxidation of U to $U_{(VI)}$. Both the differences in $\%U_{(IV)}$ and "R" values of exposed and buried nodules are related to the length of time that the nodules have been exposed to oxic conditions at the sediment-water interface. Thus the isotopic analysis of U in its two oxidation states leads one to adopt a two-stage phosphogenic model, which is also required from numerous petrographic, geochemical and field evidences (Baturin, 1971 a,b; Kolodny, 1981): a suboxic (or anoxic) stage in which initial apatitic layers and concretions are formed, and an oxidizing winnowing stage in which apatite is mechanically enriched and concentrated.

SUMMARY

Figure 13 is a graphic summary of the different isotopic species which can be analyzed in apatite and of their structural location. Isotopic studies of phosphatic rocks (as of any other rock) are not a substitute for field, petrographic, elementary geochemical, and mineralogical studies. The isotopic analysis of apatites does however shed light on many aspects of their environment of deposition and diagenesis that cannot be understood without the application of such methods. The third column in table 1 summarizes these aspects.

Our understanding of phosphogenesis can be enormously increased by identifying the environments in which phosphorites are forming today, or where they were forming in the recent past, but ceased to form due to change of conditions. Studying radioactive disequilibrium in the U series, possibly in combination with ^{14}C analysis leads to the recognition of such environments. In addition to dating phosphatic sediments these studies also lead us to an understanding of the geometry and rate of phosphatic nodule growth, and to a more precise grasp of the alternating stages of apatite precipitation and winnowing.

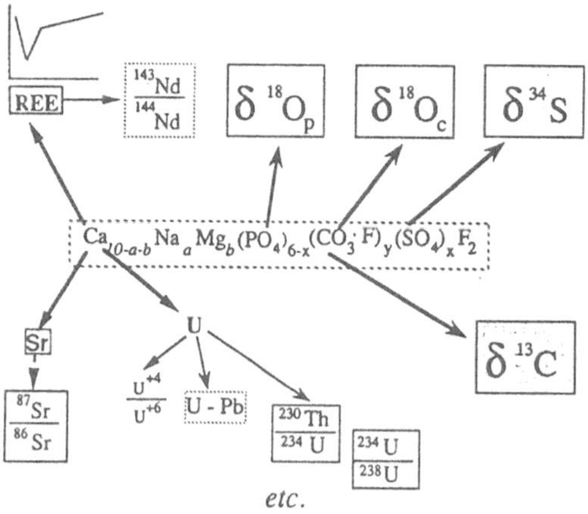

Figure 13: Scheme of location of the different isotopes in the carbonate fluor apatite structure. Note that the stable isotopes of the light elements are constituents of major anions in CFA, whereas the radiogenic isotopes of the heavy elements are trace substitutions for the Ca cation

It is still a matter of dispute whether the sites of Recent phosphorite deposition are indeed modern analogs of the large, economic phosphorite deposits of the geologic past (see Bentor, 1980). Nevertheless, data from phosphatic sediments forming off Peru-Chile, Namibia and Australia combined with the geological record lead to the conclusion that phosphorite pellets and nodules must be forming from interstitial solutions and that every phosphorite must be the end product of numerous cycles of deposition and reworking.

Phosphorites being essentially a diagenetic product, it is this environment that is monitored by both $\delta^{13}C$ in the structural CO_2 of apatite and $\delta^{34}S$ in sulfate. Both of these are strongly sensitive to redox conditions in the interstitial environment, and reflect such factors as sediment redox, degree and mode of organic matter oxidation, carbonate alkalinity, and degree of sulfate reduction. It appears that the site of CFA formation, a few centimeters below the sediment-water interface, is usually suboxic to anoxic. The broad range of $\delta^{13}C$ in apatite of different phosphorites supports the conclusion of Froelich et al. (1988), that CFA precipitation "is interface-linked, rather than driven by organic diagenesis of phosphorus deeper in the sediment". Froelich et al. (1988) have also suggested that the formation of CFA may be preceded by the precipitation of a metastable precursor, often referred to as amorphous Ca phosphate (ACP). The isotopic analysis of O in its two crystallographic sites (δ_p and δ_c) suggests that the formation of a phosphorite may involve dissolution-precipitation at considerably deeper burial conditions, resulting in isotopic re-equilibration.

The last column of table 1 summarizes that information which isotopic analysis of phosphorites (and phosphatic fossils) yields about the world in which they were deposited. $\delta^{18}O$ of phosphate oxygen is probably a very strong paleoclimatic indicator. In fact, for Paleozoic sediments it may be *the* paleoclimatic indicator. The paleoclimatic application of fossils of fresh-water and warm-blooded animals is only in its infancy. Whereas most isotopic studies of oxygen in sediments leave the temperature-water dilemma unsolved, the analysis of δ_p in mammal bones and teeth holds promise for separating these variables. $^{87}Sr/^{86}Sr$ and εNd values in phosphorites and phosphatic fossils carry information about the major paleoceanographic connections during those times. The distribution of land and sea, the opening and closure of land bridges and ocean passages have traditionally been subjects addressable mainly by faunistic analysis; geochemistry now extends its influence into these new territories.

ACKNOWLEDGEMENTS

We wish to express our sincere thanks to H.P. Schwarcz and H.F. Shaw who read different versions of the manuscript and improved both its style and substance.

REFERENCES

Aharon P. and Veeh H.H. (1984) Isotope studies of insular phosphates explain atoll phosphatization. Nature, 309, 614-617.

Altschuler Z.S. (1980) The geochemistry of trace elements in marine phosphorites. Part I: Characteristic abundances and enrichment, in: Bentor Y.K. (ed.), Soc. Econ. Paleont. Miner., Spec. Pub. 29, 19-30.

Altschuler Z.S., Cisney E.A. and Barlow I.H. (1952) X-Ray evidence of the nature of carbonate-apatite. Geol. Soc. Am. Bull., 63, 1230-1231 (Abstract).

Altschuler Z.S., Clarke R.S. and Young E.J. (1958): Geochemistry of Uranium in Apatite and Phosphorite. U.S. Geol. Surv., Prof. Paper, 314-D, 45-90.

Anderson T.F. and Arthur M.H. (1983) Stable isotopes of oxygen and carbon and their application to sedimentologic and paleoenvironmental problems In: Stable Isotopes in Sedimentary Geology, Arthur M.A., Anderson T.F., Kaplan I.R., Veizer J. and Land L.S. (eds.), Soc. Econ. Paleont. Mineral., Short Course, 10, 1-151.

Arrhenius G., Bramlette M.N. and Picciotto E. (1957) Localization of radioactive and stable heavy nuclides in ocean sediments. Nature, 180, 85-86.

Avital Y., Starinsky A. and Kolodny Y. (1983) Uranium geochemistry and fission track mapping of phosphorites, Zefa Field, Israel. Econ. Geol., 78, 121-131.

Ayliffe L. K. and Chivas A. R. (1988) Australian Macropods (Kangaroos and Wallabies): Palaeoenvironmental potential of bone phosphate oxygen isotopes. Chem. Geol., 70, 114.

Ayliffe L. K. and Chivas A. R. (1990) Oxygen isotope composition of the bone phosphate of Australian kangaroos: Potential as a palaeoenvironmental recorder. Geochim Cosmochim. Acta, 54, 2603-2610.

Ayliffe L. K. and Chivas A. R. (1991) Phosphate oxygen - isotope constraints on the origin of island phosphate deposits. Earth Plan. Sci. Lett. (in press)

Ayliffe L. K. and Veeh H H. (1988) Uranium - Series dating of speleothems and bones from Victoria Cave, Naracoorte, South Australia. Chem. Geol., 72, 211-234.

Bacquet G., Vo Quang T., Bonel G. and Vignoles M. (1980) Résonance paramagnétique du centre F dans les fluorapatites carbonatées de type B. J. Solid State Chem., 33, 189-195.

Baturin G.N. (1969) Authigenic phosphate concretions in recent sediments of the South West African shelf. Dokl. Akad. Nauk SSSR, 189, 1359-1362 (English Translation, pp. 227-230).

Baturin G.N. (1971a) Stages of phosphorite formation on the ocean floor. Nature, 232, 61-62.
Baturin G.N. (1971b) Formation of phosphate sediments and water dynamics. Oceanology, 11, 373-376.
Baturin G.N. (1982) Phosphorites on the sea floor: origin, composition and distribution. Developments in Sedimentology, 33, Elsevier, Amsterdam, 343 pp.
Baturin G.N., Merkulova K.I. and Chalov P.I. (1972) Radiometric evidence for recent formation of phosphatic nodules in marine shelf sediments. Mar. Geol., 13, M37-M41.
Bell R.T. and Thorphe R.I. (1986) Pb-Pb isochron age of uraniferous phosphorite at the base of the Menihek Formation, Labrador Trough, in: Current Research, B, Geol. Surv. Canada, 86-1B, 585-589.
Benmore R.A., Coleman M.L. and McArthur J.M. (1983) Origin of sedimentary francolite from its sulphur and carbon isotope composition. Nature, 302, 516-518.
Bentor Y.K. (ed.) (1980) Phosphorites - the unsolved problems. Soc. Econ. Paleont. Miner., Spec. Publ., 29, 3-18.
Bernat M. (1975) Les isotopes de l'uranium et du thorium et les terres rares dans l'environnement marin. Cah. ORSTOM, Sér. Géol., 7, 68-83.
Berzina I.G., Drushchits A.V., Karpova M.I., Maksimova I.G., Mel'nikov E.V. and Ozol A.A. (1978) Boron and Uranium distribution in concretionary phosphorites of the Aktyubinsk basin. Dok. Akad. Nauk, SSSR, 242, 159-161.
Birch G.F., Thomson J., McArthur J.M. and Burnett W.C. (1983) Pleistocene phosphorites off the west coast of South Africa. Nature, 302, 601-603.
Bliskovskiy V.Z., Grinenko V.A., Migdisov L.I. and Savina L.I. (1977) Sulfur isotopic composition of the minerals of phosphorites. Geochem. Int., 14, 148-155.
Borneman-Starynkevitch I.D. and Belov N.V. (1940) Isomorphic substitutions in carbonate-apatite. Dokl. Acad. Nauk, USSR, 26, 804-806.
Boyer P.D., deMeis L., Carvalho M.G.C. and Hackney D.D. (1977) Dynamic reversal of enzyme carboxyl group phosphorylation as the basis of the oxygen exchange catalyzed by sacroplasmic reticulum Adenosine Triphosphase. Biochem., 16, 136-140.
Broecker W.S. (1974) Chemical Oceanography. Harcourt Brace Jovanovich, New York, 214 pp.
Brongersma-Sanders M. (1957) Mass mortality in the sea. In: Treatise on Marine Ecology and Paleoecology, Hegpeth J.W. (ed.), 1, Ecology, Geol. Soc. Am. Mem., 67, 941-1010.
Burke W.H., Denison R.E., Hetherington E.A., Koepnick R.B., Nelson H.F. and Otto J.B. (1982) Variation of seawater $^{87}Sr/^{86}Sr$ throughout Phanerozoic time. Geology, 10, 516-519.
Burnett W.C. and Veeh H.H. (1977) Uranium-series disequilibrium studies in phosphorite nodules from the west coast of South America. Geochim. Cosmochim. Acta, 41, 755-764.
Burnett W.C., Beers M.J. and Roe K.K. (1982) Growth rates of phosphate nodules from the continental margin off Peru. Science, 215, 1615-1618.
Burnett W.C., Roe K.K. and Piper D.Z. (1983) Upwelling and phosphorite formation in the ocean. In: Coastal Upwelling - its Sedimentary Record, Part A: Responses of the Sedimentation Regime to Present Coastal Upwelling, Suess E. and Thiede J. (eds.), Plenum, New York, 377-397.

Burnett W.C. and Froelich P.N. (1988) Preface. In: The Origin of Marine Phosphorite. The Results of the R.V. *Robert D. Conrad* Cruise 23-06 to the Peru Shelf, Burnett W.C. and Froelicher P.N. (eds.), Mar. Geol., 80, III-VI.

Burnett W.C., Baker B., Chin P., McCabe W. and Ditchburn R. (1988) Uranium-series and AMS ^{14}C studies of modern phosphatic pellets from Peru shelf muds. In: The Origin of Marine Phosphorite. The Results of the R.V. *Robert D. Conrad* Cruise 23-06 to the Peru Shelf, Burnett W.C. and Froelich P.N. (eds.), Mar. Geol., 80, 215-230.

Chalov P.I. and Merkulova K.I. (1966) Comparative rate of oxidation of ^{234}U and ^{238}U atoms in certain minerals. Dokl. Akad. Nauk SSR, 167, 146-148.

Chase C.G. and Perry E.C. (1972) The Oceans: Growth and oxygen isotope evolution. Science, 177, 992-994.

Chen J.H., Edwards L.R. and Wasserburg G.J. (1986) ^{238}U, ^{234}U ^{232}Th in seawater. Earth Plan. Sci. Lett., 80, 241-251.

Clayton R.N., O'Neil J.R. and Mayeda T.K. (1972) Oxygen isotope exchange between quartz and water. Jour. Geophys. Res., 77, 3907-3915.

Compton J. S., Snyder S. W. and Hodell D. A. (1990) Phosphogenesis and weathering of shelf sediments from the southeastern United States: Implications for Miocene $\delta^{13}C$ excursions and global cooling. Geology, 18, 1227-1230.

Cook P.J. (1976) Sedimentary Phosphate Deposits. In: Handbook of Strata-Bound and Stratiform Ore Deposits, Wolf K.H. (ed.), 7, Elsevier, Amsterdam, 505-535.

Cook P.J. and McElhinny M.W. (1979) A reevaluation of the spatial and temporal distribution of sedimentary phosphate deposits in the light of plate tectonics. Econ. Geol., 74, 315-330.

Crowson R.A., Showers W.J., Wright E.K. and Hoering T.C. (1991) Preparation of phosphate samples for oxygen isotope analysis. Anal. Chem., 63, 2397-2400.

D'Angela D. and Longinelli A. (1990) Oxygen isotopes in living mammal's bone phosphate: Further results. Chem. Geol., 86, 75-82.

Dansgaard W. (1964) Stable isotopes in precipitation. Tellus, 16, 436-468.

Degens E.T. and Epstein S. (1962) Relationship between $^{18}O/^{16}O$ ratios in coexisting carbonates, cherts and diatomites. Amer. Ass. Petrol. Geol. Bull., 166, 534-542.

Douglas R.G. and Woodruff F. (1981) Deep sea benthic foraminifera. In: The Sea, Emiliani C. (ed.), 7, Interscience-Wiley, New York, N.Y., 1233-1327.

Edwards L. R., Chen J.H., Ku T.-L. and Wasserburg G.J. (1987) Precise timing of the last interglacial period from mass spectrometric determination of Thorium-230 in corals. Science, 236, 1547-1553.

Elderfields H. and Gieskes J.M. (1982) Sr isotopes in interstitial waters of marine sediments from Deep Sea Drilling Project cores. Nature, 300, 493-497.

Epstein S., Buchsbaum H.A., Lowenstam H.A. and Urey H.C. (1953) Revised carbonate-water isotopic temperature scale. Geol. Soc. Am. Bull., 64, 1315-1326.

Epstein S. and Karhu J. (1990) A reply to E. C. Perry's comments. Geochim. Cosmochim. Acta, 54, 1181-1184.

Faure G. (1986) Principles of Isotope Geology. 2nd edition, John Wiley & Sons, 589 pp.

Froelich P.N., Klinkhammer G.P., Bender M.L., Luetke N.A., Heath G.R., Cullen D., Dauphin P., Hammond D., Hartman B. and Mynard V. (1979) Early oxidation of organic matter in pelagic sediments of the eastern equatorial Atlantic: suboxic diagenesis. Geochim. Cosmochim. Acta, 43, 1075-1090.

Froelich P.N., Arthur M., Burnett W.C., Deakin M., Hensley V., Jahnke R., Kaul L., Kim K., Roe K., Soutar A. and Vathakanon C. (1988) Early diagenesis of organic matter in Peru continental margin sediments: Phosphorite precipitation. In: The Origin of Marine Phosphorite. The Results of the R.V. *Robert D. Conrad* Cruise 23-06 to the Peru Shelf, Burnett W.C. and Froelich P.N. (eds.), Mar. Geol., 80, 309-343.

Garavelli C.L., Melone N. and Nuovo G. (1979) Presence of whitlockite in the mineral substance of annealed fish-bones: character and possible significance. Oceanol. Acta, 2(4), 417-421.

Garrison R.E. Kastner M., and Reimers C.E. (1990) Miocene phosphogenesis in California. In: Phosphate Deposits of the world, vol.3. Neogene to Modern Phosphorites, Burnett W.C. and Riggs S.R. (eds.), Cambridge University Press, 285-299.

Garrison R.E., Kastner M. and Kolodny Y. (1987) Phosphorites and phosphatic rocks in the Monterey Formation and related Miocene units, Coastal California, in: Cenozoic Basin Development of Coastal California, Ingersol R.V. and Ernst W.G. (eds.), Rubey, VI, Prentice-Hall, 349-381.

Glenn C.R., (1988) Pore water, petrologic and stable carbon isotopic data bearing on the origin of Modern Peru margin phosphorites and associated authigenic phases. In: Phosphate Deposits of the world, vol.3. Neogene to Modern Phosphorites, Burnett W.C. and Riggs S.R. (eds.), Cambridge University Press,.46-61.

Glenn C.R., Arthur M.A., Yeh H.W. and Burnett W.C. (1988) Carbon isotopic composition and lattice-bound carbonate of Peru-Chile margin phosphorites. In: The Origin of Marine Phosphorite. The Results of the R.V. *Robert D. Conrad* Cruise 23-06 to the Peru Shelf, Burnett W.C. and Froelich P.N. (eds.), Mar. Geol., 80, 287-307.

Glenn C.R. and Arthur M.A. (1990) Anatomy and origin of a Cretaceous phosphorite-greensand giant, Egypt. Sedimentology, 37, 123-154.

Goldhaber M.B. and Kaplan R. (1974) The sulfur cycle. In: The Sea, 5, Marine Chemistry, Goldberg E.D. (ed.), Wiley-Interscience Publishers, New York, 569-655.

Grandjean P. and Albarède F. (1989) Ion probe measurements of rare earth elements in biogenic phosphates. Geochim. Cosmochim. Acta, 53, 3179-3183.

Grandjean P., Cappetta H., Michard A. and Albarède F. (1987) The assessment of REE patterns and $^{143}Nd/^{144}Nd$ ratios in fish remains. Earth Plan. Sci. Lett., 84, 181-196.

Grandjean P., Cappetta H., Albarède F. (1988) The REE and e_{Nd} of 40-70 Ma old fish debris from the West-African platform. Geophys. Res. Lett., 15, 389-392.

Gulbrandsen R.A. (1969): Physical and chemical factors in the formation of marine apatite. Econ. Geol., 64, 365-382.

Gulbrandsen R.A. (1970) Relation of Carbon Dioxide Content of Apatite of the Phosphoria Formation to Regional Facies. U.S. Geol. Surv. Prof. Paper, 700-B, B9-13.

Holser W.T. and Kaplan I.R. (1966) Isotope geochemistry of sedimentary sulfates. Chem. Geol., 1, 93-135.

Holland H.D. (1984) The Chemical Evolution of the Atmosphere and Oceans. Princeton University Press, Princeton, N.Y., 582 pp.

Irwin H., Curtis C. and Coleman M. (1977) Isotopic evidence for source of diagenetic carbonates formed during burial of organic rich sediments. Nature, 269, 209-213.

Jahnke R.A. (1984) The synthesis and solubility of carbonate fluorapatite. Amer. Jour. Sci., 284, 58-78.

Jones E. J. W. (1989) Radioactivity of the ocean floor and marine phosphorite deposits: observations with a new deep-towed scintillometer. Geophys. Res. Lett., 16, 123-126.

Karhu J. and Epstein S. (1986) The implication of the oxygen isotope records in coexisting cherts and phosphates. Geochim. Cosmochim. Acta, 50, 1745-1757.

Kastner M., Garrison R.E., Kolodny Y., Shemesh A. and Reimers C.E. (1990) Simultaneous changes of oxygen isotopes in PO_4^{-3} and CO_3^{-2} in apatite, with emphasis on the Monterey Formation California. In: Phosphate Deposits of the world, vol.3. Neogene to Modern Phosphorites, Burnett W.C. and Riggs S.R. (eds.), Cambridge University Press, 312-324.

Kazakov A.V. (1937) The phosphorite facies and the genesis of phosphorites. In: Geological Investigations of Agricultural Ores. Trans. Sci. Inst. Fertil. Insecto-fungicides, 142, 93-113.

Keppens E. and O'Neil J.R. (1984) Oxygen isotope variations in glauconies. Terra cognita, 5, 283.

Keto L.S. and Jacobsen S.B. (1987) Nd and Sr isotopic variations of Early Paleozoic oceans. Earth Plan. Sci. Lett., 84, 27-41.

Keto L.S. and Jacobsen S.B. (1987) Nd isotopic variations of Phanerozoic oceans. Earth Plan. Sci. Lett., 90, 395-410.

Kim K.H. and Burnett, W.C. (1986) Uranium-series growth history of a Quaternary phosphatic crust from the Peruvian continental margin. Chem. Geol., 58, 227-244.

Knauth L.P. and Epstein S. (1976) Hydrogen and oxygen isotope ratios in nodular and bedded cherts. Geochim. Cosmochim. Acta, 40, 1095-1108.

Koch P. L., Fisher D. C. and Dettman D. (1989) Oxygen isotope variation in the tusk of extinct proboscideans: A measure of season of death and seasonality. Geology, 17, 515-519.

Kolodny Y. (1980) Carbon isotopes and depositional environment of a high productivity sedimentary sequence - the case of the Mishash-Ghareb formation, Israel. Isr. Jour. Earth Sci., 29, 147-156.

Kolodny Y. (1981) Phosphorites. In: The Sea, 7, Emiliani C. (ed.), J. Wiley & Sons, New York, 981-1023.

Kolodny Y. and Kaplan I.R. (1970a) Uranium isotopes in sea floor phosphorites. Geochim. Cosmochim. Acta, 34, 3-24.

Kolodny Y. and Kaplan I.R. (1970b) Carbon and oxygen isotopes in apatite CO_2 and coexisting calcite from sedimentary phosphorite. Jour. Sedim. Petrol., 40, 954-959.

Kolodny Y. and Luz B. (1992) Oxygen isotopes in phosphates of fossil fish - Devonian to Recent. In: Stable Isotope Geochemistry: A Tribute to Samuel Epstein, Taylor H.P., O'Neil J.R. and Kaplan I.R. (eds.), The Geochem. Soc., Spec. Public., 3, 105-119.

Kolodny Y., Luz B. and Navon O. (1983) Oxygen isotope variations in phosphate of biogenic apatite. I: Fish bone apatite -

rechecking the rules of the game. Earth Plan. Sci. Lett., 64, 398-404.
Kolodny Y. and Raab M. (1988) Oxygen isotopes in phosphatic fish remains from Israel: Paleothermometry of tropical Cretaceous and Tertiary shelf waters. Palaeogeogr., Palaeoclimatol., Palaeoecol., 64, 59-67.
Kovach J. (1980) Variations in the Strontium isotopic composition of seawater during Paleozoic time determined by analysis of conodonts. Geol. Soc. Amer., Abst. with Progr., 12(7), 465.
Kovach J. and Zartman R.E. (1981) U-Th-Pb dating of conodonts. Geol.Soc. Amer., Abst. with Progr., 13(6), 285.
Kovach J. and Miller J.F. (1988) Eustatic sea-level changes near the Cambrian-Ordovician boundary: Data from Sr isotope analysis of biogenic apatites. Geol. Soc. Amer., Abstracts with Programs, 20, A393.
Kozlov A.A. (1975) On the distribution of Uranium in the phosphorites of the Russian Platform. Geochem. Intern., 136-138.
Kress A.G. and Veeh H.H. (1980) Geochemistry and radiometric ages of phosphatic nodules from the continental margin of northern New South Wales, Australia. Mar. Geol., 36, 143-157.
Koepnick R.B., Burke W.H., Denison R.E., Hetherington E.A., Nelson H.F., Otto J.B. and Waite L.E. (1985) Construction of the seawater $^{87}Sr/^{86}Sr$ curve for the Cenozoic and Cretaceous: supporting data. Chem. Geol., 58, 55-81.
Land L.S., Lundelius E. L., Jr. and Valastro S.,Jr. (1980) Isotopic ecology of deer bones. Palaeogeogr., Palaeoclimatol., Palaeoecol.,32, 143-151.
Levinson A.A., Luz B. and Kolodny Y. (1987) Variations in oxygen isotopic compositions of human teeth and urinary stones. Applied Geochem., 2, 367-371.
Lifson N. & McClintock R. (1966) Theory of use of the turnover rates of body water for measuring energy and material balance. Jour. Theoret. Biol., 12, 46-74.
Longinelli A. (1965) Oxygen isotopic composition of orthophosphate from shells of living marine organisms. Nature, 207, 716-719.
Longinelli A. (1984) Oxygen isotopes in mammal bone phosphate: A new tool for paleohydrological and paleoclimatological research? Geochim. Cosmochim. Acta, 48, 385-390.
Longinelli A. and Nuti S. (1968) Oxygen isotopic composition of phosphorites from marine formations. Earth Plan. Sci. Lett., 5, 13-16.
Longinelli A. and Nuti S. (1973a) Revised phosphate-water isotopic temperature scale. Earth Plan. Sci. Lett., 19, 373-376.
Longinelli A. and Nuti S. (1973b) Oxygen isotope measurements of phosphate from fish teeth and bones. Earth Plan. Sci. Lett., 20, 337-340.
Lowenstam H.A. (1964) Paleotemperatures of the Permian and Cretaceous Periods. In: Problems in Paleoclimatology, Nairn A.E.M. (ed.), Wiley Interscience, New York, N.Y., 227-248.
Lucas J. and Prévôt L. (1985) The synthesis of apatite by bacterial activity: mechanism. Sci. Géol. Mém., 77, 83-92.
Luz B., Cormie A. B.and Schwarcz H. P. (1990) Oxygen isotope variation in phosphate of deer bones. Geochim. Cosmochim. Acta, 54, 1723-1728.

Luz B., Kolodny Y.and Kovach J. (1984) Oxygen isotope variations in phosphate of biogenic apatites, III. Conodonts. Earth Plan. Sci. Lett., 69, 255-262.

Luz B., Kolodny Y.and Horowitz M. (1984) Fractionation of oxygen isotopes between mammalian bone phosphate and environmental drinking water. Geochim. Cosmochim. Acta, 48, 1689-1693.

Luz B.and Kolodny Y. (1985) Oxygen isotope variation in phosphate of biogenic apatites. IV. Mammal teeth and bones. Earth Plan. Sci. Lett., 75, 29-36.

Luz B.and Kolodny Y. (1989) Oxygen isotope variation in bone phosphate. Applied Geochem., 4, 317-323.

McArthur J.M. (1978) Systematic variations in the contents of Na, Sr, CO_3 and SO_4 in marine carbonate-fluorapatite and their relation to weathering. Chem. Geol., 21, 89-112.

McArthur J.M. (1985) Francolite geochemistry -- compositional controls during formation, diagenesis, metamorphism and weathering. Geochim. Cosmochim. Acta, 49, 23-35.

McArthur J.M., Coleman M.L. and Bremner J.M. (1980) Carbon and oxygen isotopic composition of structural carbonate in sedimentary francolite. Jour. Geol. Soc. London, 137, 669-673.

McArthur J.M., Benmore R.A., Coleman M.L., Soldi C., Yeh H. W. and O'Brien G.W. (1986) Stable isotopic characterisation of francolite formation. Earth Plan. Sci. Lett., 77, 20-34.

McArthur J.M., Hamilton P.J., Greensmith J.T., Boyce A.J., Fallick A.E., Birch G., Walsh J.N., Benmore R.A. and Coleman M.L. (1987) Phosphorite geochemistry: Isotopic evidence for meteoric alteration of francolite on a local scale. Chem. Geol., 65, 415-425.

McArthur J.M., Sahami A. R., Thirlwall M., Hamilton P.J.and Osborn A.O. (1990) Dating phosphogenesis with strontium isotopes. Geochim. Cosmochim. Acta, 54, 1343-1351.

McArthur J.M., Thomson J., Jarvis I., Fallick A.E. and Birch G. F. (1988) Eocene to Pleistocene Phosphogenesis off western South Africa. Mar. Geol., 85, 41-63.

McClellan G.H. (1980) Mineralogy of carbonate fluorapatites. Jour. Geol. Soc. London, 137, 675-681.

McClellan G.H. and Lehr J.R. (1969) Crystal chemical investigation of natural apatites. Am. Min., 54, 1374-1391.

McClellan G.H. and Saavedra F.N. (1986) Chemical and mineral characteristics of some Cambrian and Precambrian phosphorites. In: Phosphate Deposits of the World, 1, Proterozoic and Cambrian Phosphorites, Cook P.J. and Shergold J.H. (eds.), Cambridge Univ. Press, 244-267.

McConnell D. (1973) Apatite. Its Crystal Chemistry, Mineralogy, Utilization, and Geologic and Biologic Occurrences. Springer-Verlag, New York, 111 p.

McCrea J.N. (1950) On the isotopic chemistry of carbonate and paleotemperature scale. Jour. Chem. Phys., 18, 849-857.

McKelvey V.E. (1967) Phosphate Deposits. U.S. Geol. Surv. Bull., 1252-D, 1-21.

McKelvey V.E., Swanson R.W. and Sheldon R.P. (1953) The Permian phosphorite deposits of Western United States. In: Congr. Geol. Int. Compt. Rend., 19th, Algiers, 1952, XI(II), 45-64.

Nathan Y. (1984): The mineralogy and geochemistry of phosphorites. In: Phosphate Minerals, Nriagu J.O. and Moore P.B. (eds), Springer Verlag, Berlin, 275-291.

Nathan Y. and Nielsen H. (1980) Sulfur isotopes in phosphorites. In: Marine Phosphorites, Bentor Y.K. (ed.), Soc. Econ. Paleont. Miner., Spec. Publ., 29, 73-78.
Nelson B.K., Deniro M.J., Schoeninger M.J. and De Paolo D.J. (1986) Effects of diagenesis on strontium, carbon, nitrogen and oxygen concentration and isotopic composition of bone. Geochim. Cosmochim. Acta, 50, 1941-1949.
O'Brien G.W. and Veeh H.H. (1980) Holocene phosphorite on the east Australian continental margin. Nature, 288, 690-692.
O'Brien G.W., Veeh H.H, Cullen D.J. and Milnes A.R. (1986) Uranium-series isotopic studies of marine phosphorites and associated sediments from the East Australian continental margin. Earth Plan. Sci. Lett., 80, 19-35.
O'Brien .G.W, Veeh H.H., Milnes A.R. and Cullen D.J. (1987) Sea-floor weathering of phosphate nodules off East Australia: Its effect on uranium oxidation state and isotopic composition. Geochim. Cosmochim. Acta, 51, 2051-2064.
O'Neil J.R., Clayton R.N. and Mayeda T.K. (1969) Oxygen isotope fractionation in divalent metal carbonates. Jour. Chem. Phys., 51, 5547-5558.
Parrish J.T. (1982) Upwelling and petroleum source beds, with reference to Paleozoic. Amer. Ass. Petrol. Geol. Bull., 66, 750-774.
Parrish J.T., Ziegler A.M. and Scotese C.R. (1982) Rainfall patterns and the distribution of coals and evaporites in the Mesozoic and Cenozoic. Palaeogeogr, Palaeoclimatol., Palaeoecol., 40, 67-101.
Penrose R.A.F. Jr. (1888) Nature and origin of deposits of phosphate of lime, with an introduction by N.S. Shaler. U.S. Geol. Surv. Bull., 46, 143 p.
Perry Jr. E.C. (1990) Comment on "The implication of the oxygen isotope record in coexisting cherts and phosphates" by J. Karhu and S. Epstein. Geochim. Cosmochim. Acta, 54, 1175-1179.
Peterman Z.E., Hedge C.E. and Tourtelot H.A. (1970) Isotopic composition of strontium in seawater throughout Phanerozoic time. Geochim. Cosmochim. Acta, 34, 105-120.
Piepgras D.J., Wasserburg G.J. & Dasch E.J. (1979) The isotopic variations in seawater. Earth Plan. Sci. Lett., 45, 223-236.
Piper D.Z and Kolodny Y. (1987) The stable isotopic composition of a phosphate deposit: $\delta^{13}C$, $\delta^{34}S$ and $\delta^{18}O$. Deep Sea Res., 34, 897-911.
Price N.B. and Calvert S.E. (1978) The geochemistry of phosphorites from the Namibian shelf. Chem. Geol., 23, 151-170.
Riggs S.R. (1984) Paleoceanographic model of Neogene phosphorite deposition, U.S. Atlantic continental margin. Science, 223, 123-131.
Roe K.K., Burnett W.C., Kim K.H. and Beers M.J. (1982) Excess protactinium in phosphate nodules from a coastal upwelling zone. Earth Plan. Sci. Lett., 60, 39-46.
Romankevich Ye.A. and Baturin G.N. (1972) Composition of the organic matter in phosphorites from the continental shelf of southwest Africa. Geokhimia, 719-726 (Engl. Transl., 464-470).
Savin S.M. (1977) The history of the earth's surface temperature during the past 100 million years. Annu. Rev. Earth Plan. Sci., 5, 319-355.

Schmitz B., Åberg G., Werdelin L., Forey P. and Svend E.B.A. (1991) $^{87}Sr/^{86}Sr$, Na, F, Sr, and La in skeletal fish debris as a measure of the paleosalinity of fossil-fish habitats. Geol. Soc. Am. Bull., 103, 786-794.
Shaw H.F. and Wasserburg G.J. (1985) Sm-Nd in marine carbonates and phosphorites: Implications for Nd isotopes in seawater and crustal ages. Geochim. Cosmochim. Acta, 49, 503-518.
Sheldon R.P. (1980) Ancient Marine Phosphorites. Ann. Rev. Earth and Plan. Sci., 9, 251-284.
Sheldon R.P. (1964) Paleolatitudinal and Paleogeographic Distribution of Phosphorite. U.S. Geol. Surv. Prof. Pap., 501-C, 106-113.
Shemesh A. (1990) Crystallinity and diagenesis of sedimentary apatites. Geochim. Cosmochim. Acta, 54, 2433-2438.
Shemesh A., Kolodny Y. and Luz B. (1983) Oxygen isotope variations in phosphate of biogenic apatites. II. Phosphorite rocks. Earth Plan. Sci. Lett., 64, 405-416.
Shemesh A., Kolodny Y. and Luz B. (1988) Isotope geochemistry of oxygen and carbon in phosphate and carbonate of phosphorite francolite. Geochim. Cosmochim. Acta, 52, 2565-2572.
Shemesh A. and Kolodny Y. (1988) Oxygen isotope variations in phosphorites from the southeastern Tethys. Isr. Jour. Earth Sci., 37, 1-15.
Soudry D. (1987) Ultra-fine structures and genesis of the Campanian Negev high-grade phosphorites (southern Israel). Sedimentolology, 34, 641-660.
Soudry D. and Champetier Y. (1983) Microbial processes in the Negev phosphorites (southern Israel). Sedimentolology, 30, 411-423.
Staudigel H., Doyle P. and Zindler A. (1985/86) Sr and Nd isotope systematics in fish teeth. Earth Plan. Sci. Lett., 76, 45-56.
Stein M., Starinsky A. and Kolodny Y. (1982) Behaviour of uranium during phosphate ore calcination. Jour. Chem. Tech. Biotechnol., 32, 834-847.
Strakhov N.M. (1970) Principles of Lithogenesis, 3, Plenum, 577 p.
Strutt R.J. (1908): The accumulation of helium in geological time. Roy. Soc. London, Proc., A81, 272-277.
Tudge A.P. (1960) A method of analysis of oxygen isotopes of orthophosphates: its use in the measurement of paleotemperatures. Geochim. Cosmochim. Acta, 18, 81-93.
Tugarinov A.I., Zykov S.I. and Bibikova E.V. (1963) Determination of the age of sedimentary rocks by the lead-uranium method. Geochem., 3, 284-300.
Turner D.R. and Whitfield M. (1979) Control of seawater composition. Nature, 281, 468-469.
Urey H.C., Lowenstam H.A., Epstein S. and McKinney U. (1951) Measurements of paleotemperatures and temperatures of the Upper Cretaceous of England, Denmark and the south-eastern U.S. Geol. Soc. Am. Bull., 62, 399-416.
Vaasjoki M., Aikas O. and Rehtijarvi P. (1980) The age of mid-Proterozoic phosphatic metasediments in Finland as indicated by radiometric U-Pb dates. Lithos, 13, 257-262.
Veeh.H.H; (1982) Concordant ^{230}Th and ^{231}Pa ages of marine phosphorites. Earth Plan. Sci. Lett., 57: 278-284.
Veeh H.H., Burnett W.C. and Soutar A. (1973) Contemporary phosphorites on the continental margin of Peru. Science, 181, 844-845.

Veeh H.H, Calvert S.E. and Price N.B. (1974) Accumulation of uranium in sediments and phosphorites on the South West African shelf. Mar. Chem., 2, 189-202.

Veeh H.H. and Burnett W.C. (1978) Uranium-series dating of insular phosphorite from Ebon atoll, Micronesia. Nature, 274, 460-462.

Veeh H.H. and Burnett W.C. (1982) Carbonate and phosphate sediments. In: Uranium-Series Disequilibrium: Application To Environmental Problems, Ivanovich M. and Harmon R.S. (eds), Clarendon Press, Oxford, 459-480.

Veizer J., Fritz P. and Jones B. (1986) Geochemistry of brachiopods: Oxygen and carbon isotopic records of Paleozoic oceans. Geochim. Cosmochim. Acta, 50: 1679-1696.

Vengosh A., Kolodny Y. and Tepperberg M. (1987) Multi-phase oxygen isotopic analysis as a tracer of diagenesis: The example of the Mishash Formation, Cretaceous of Israel. Chem. Geol., 65, 235-253.

Waples D.W. (1982) Phosphate-rich sedimentary rocks: Significance for organic facies and petroleum exploration. Journ. Geochem. Explor., 16, 135-160.

Williams L.A. and Reimers C. (1983) Role of bacterial mats in oxygen-deficient marine basins and coastal upwelling regimes: Preliminary report. Geology, 11, 267-269.

Wright J., Seymour R.S. and Shaw H.F. (1984) REE and Nd isotopes in conodont apatite: variations with geological age and depositional environment. Geol. Soc. Am., Spec. Pap., 196, 325-340.

Yurtsever Y. and Gat J.R. (1981) Atmospheric waters. In: Stable Isotope Hydrology, Chap. 6, Gat J.R. and Gonfiantini R. (ed.), Tech. Report Series, Inter. Atomic Energy Agency, Vienna 210, 103-142.

ORIGIN AND DIAGENESIS OF CHERTS: AN ISOTOPIC PERSPECTIVE

L. Paul KNAUTH
Department of Geology, Arizona State University
TEMPE, Arizona 85287 (U.S.A.)

Cherts most commonly consist of microcrystalline quartz which has replaced precursor sedimentary materials. They occur in a variety of sedimentary associations of all ages and are thus attractive targets for isotopic studies. The origin of most chert types remains controversial, but isotopic investigations have allowed a greatly improved understanding of the timing and conditions under which silica diagenesis proceeds. Isotopic data on cherts also play a key role in the ongoing controversies regarding the isotopic history of ocean water and the climatic history of the earth.

At this stage, there have been a few broad surveys of the authigenic silica system, and several aspects of chert diagenesis have been examined in detail. However, the field is still largely unexplored due to the great diversity of chert types and the complexity of the silicification process. Considering the extraordinary range of isotopic variation that has been so far encountered, it seems likely that major discoveries and surprises are still ahead. This contribution presents an overview of the low temperature silica system and summarizes isotopic results for the major chert types.

GEOLOGY OF CHERTS AND OTHER AUTHIGENIC SILICA

Authigenic silica occurs in sediments as 3 types of opal and 5 types of quartz (Table 1). The room-temperature solubility values are approximately independent of pH below 9. Any diagenetic fluid carrying more than 6 mg/g dissolved silica is potentially a silici-fying fluid. At values above about 80 mg/g, it is likely that one of the more amorphous phases (opal-CT, opal-A) will precipitate instead of quartz.

Table 1

SILICA PHASE	CHARACTERISTICS	SOLUBILITY AT 25°C (Approx. ppm)
OPAL		
Opal-A	Isotropic, amorphous silica. Diffuse x-ray peak at 4.1 Angstroms. Up to 12 wt.% H_2O.	120
Opal-CT	V. low biref. Diffuse x-ray pattern containing weak peaks characteristic of α-cristobalite and α-tridymite. 0.5 to >1 Wt. % H_2O	80
Opal-C	X-ray pattern containing well defined peaks of alpha cristobalite. V. rare in sediments. H_2O content unknown.	?
QUARTZ		
Granular Micro-crystalline quartz	8-micron grain size, intergrown grains not ordered over the long range. Irreg. extinction. 0.2 - 1.2 wt.% H_2O.	>6?
Megaquartz	Pin point extinction, regular grain boundaries. Each grain is a crystal of α-quartz. H_2O as fluid inclusions	6
Fibrous quartz:	H_2O content up to 1.0 wt.%, or more.	>6?
Length-fast	Bundles of fibrous quartz length fast over entire length of fiber.	
Length-slow	Bundles of fibrous quartz length slow over entire length of fiber.	
Zebraic	Bundles of fibrous quartz which alternate length-fast and length-slow along length of fiber.	

Early observers of nodular cherts in limestones speculated that silica gels had precipitated directly from ocean water and transformed to quartz during burial. However, surface and shallow ocean waters today contain less than 1 mg/g silica due to efficient removal by silica-secreting diatoms. Therefore, direct precipi-tation of chert from modern ocean water does not occur.

Prior to the mid-Tertiary explosion of diatom populations, it is likely that ocean water was held comparably low in dissolved silica by the abundant silica secreting radiolarians and sponges. The best argument for this is the fact that chert does not show up as part of Phanerozoic evaporite sequences. It is therefore unlikely that any of the Phanerozoic cherts were direct precipitates from ocean water. Precambrian cherts are more problematical; direct precipita-tions of opal or quartz from the Precambrian oceans may have occurred and may account for varieties that do not occur in the Phanerozoic (such as some chert beds in iron formations).

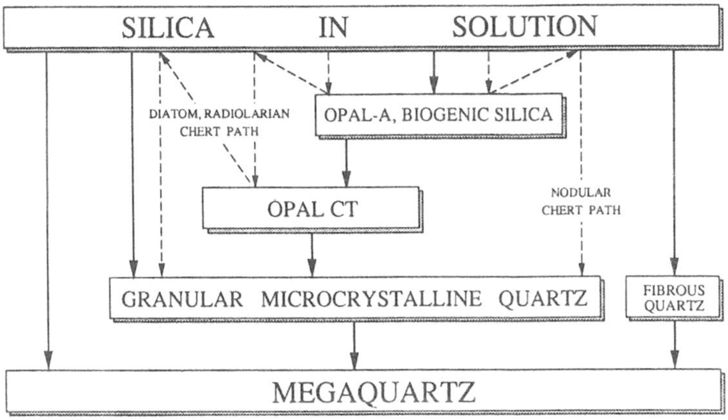

Figure 1: Diagenetic pathways for authigenic silica transformations. Granular microcrystalline quartz is the most common form of quartz in chert. In biogenic siliceous oozes,the initially precipitated opal-A dissolves and reprecipitates as opal-CT. Opal-CT subsequently dissolves and reprecipitates as microquartz. In platform carbonates, microquartz can directly replace limestone during early diagenesis. Other pathways are possible for other types of chert. Megaquartz can form as a direct precipitate from low temperature waters, or can be a metamorphic recrystallization of microquartz

Tertiary deep ocean sediments contain extensive layers of biogenic siliceous oozes. Eocene radiolarian oozes are commonly encountered during deep sea drilling, and the Pacific basin is rimmed with Miocene diatomaceous deposits. Prompted by the Deep Sea Drilling Project and the demonstrated oil potential of the Miocene oozes, the most extensive efforts to understand silica diagenesis have been focused on these deposits. Studies have repeatedly shown that siliceous oozes are initially laid down as biogenic opal-A and are then transformed during later burial to opal-CT and then quartz (Fig. 1). The bedded chert produced in this way is composed of quartz that formed during deep burial at temperatures up to 80°C, or higher. It can take more than 60 million years for the ooze to completely convert to quartz. Isotopic analyses of these types of chert yield information mainly about late diagenesis; determination of paleoclimatic temperatures is not possible unless it can be independently demonstrated that the transformation from opal to quartz occurred during shallow burial.

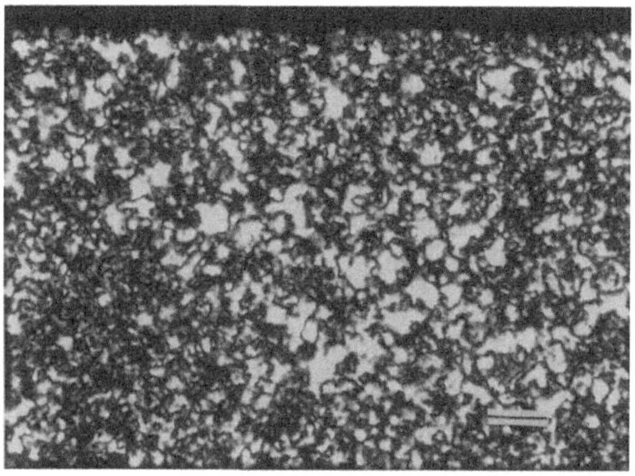

Figure 2: Photomicrograph of granular microcrystalline quartz in Cretaceous chert (Edwards Formation, Texas). Quartz grains are intergrown, have irregular extinction and irregular grain boundaries. Crossed nicols. Bar = 40 microns

Figure 1 gives alternative paths for the paragenesis of silica phases in sediments. The sequence deduced for the siliceous oozes does not necessarily apply to other types of cherts. Biogenic opal-A is coated with either metal oxides and/or organic matter to protect the opal from dissolving in sea water. This armoring greatly impedes the progress of silica transformations during burial. In addition, while pore fluids in siliceous oozes can achieve dissolved silica levels above the 80-120 mg/g values necessary for precipitation of opal-CT, this is rarely the case in other depositional settings. For example, ground waters moving through a carbonate unit may dissolve opaline tests or volcanic shards until the dissolved silica rises above 6 mg/g. At that point, quartz can nucleate and grow without ever going through any intermediate phases.

Table 2

	OPAL		Quartz				Marine	FW
OCCURRENCE	OPAL-A	OPAL-CT	OPAL-C	GMC*	Fibrous	Megaquartz		
Modern Organisms	x				?		x	x
Siliceous Oozes								
modern	x	?					x	x
ancient (Bedded Chert)		x		x	x	x	x	x
Nodular Chert in Carbonate								
Cenozoic	x	minor		x	x	x	x	x
Pre-Cenozoic				x	x	x	x	minor
Cherts Associated with Volcanic Rocks	x	?	x	x	x	x	x	x
Silicified wood	x	x		?	x	x		x
Iron Formations				x	x	x	?	?
Geodes					minor	x	?	x
Silica Cement								
Cenozoic	x	x		?	x	x	x	x
Pre-Cenozoic				?	x	x	x	x
Silicified Fossils								
Cenozoic	x	x		x	x	x	x	x
Pre-Cenozoic				x	x	x	x	x
Porcellanite (siliceous shales)				x			x	x
Inorganic Opal	x	x	x				?	x
"Magadii" cherts"	?	?		x	x	x		x
Silicified Erosion Surfaces	x		x	x	x	x		x
Fracture-Fills	x				x	x	x	x

*GMC = Granular Microcrystalline Quartz

Cherts are most commonly composed of interlocking, intergrown grains of α-quartz called granular microcrystalline quartz (microquartz). The grains average about 8 microns in size and display irregular extinction (Fig. 2). According to Micheelsen (1966), individual grains are made of plates about 600 angstroms thick and are stacked parallel to (0001). Each plate is the Brazil twin of the other and contains numerous lattice faults and dislocations studded with hydroxl groups. Molecular water and hydroxyl also line the micro-plate surfaces. With time, this structure is probably annealed out and probably tends to approach more ordered grains of α-quartz. However, many cherts as old as Archean have preserved their microquartz texture and are petrographically indistin-guishable from cherts as young as the Cretaceous. Once formed, this intergrowth of quartz is highly impermeable and inert. Metamorphism can induce wholesale recrystallization into larger, individual grains of α-quartz with well defined grain boundaries and regular extinction called megaquartz. Simple petrographic examination can therefore readily delineate altered, metamorphosed chert from that which may be the original quartz precipitate. A more detailed discussion of silica phases and silica diagenesis is given in the useful review by Hesse (1989).

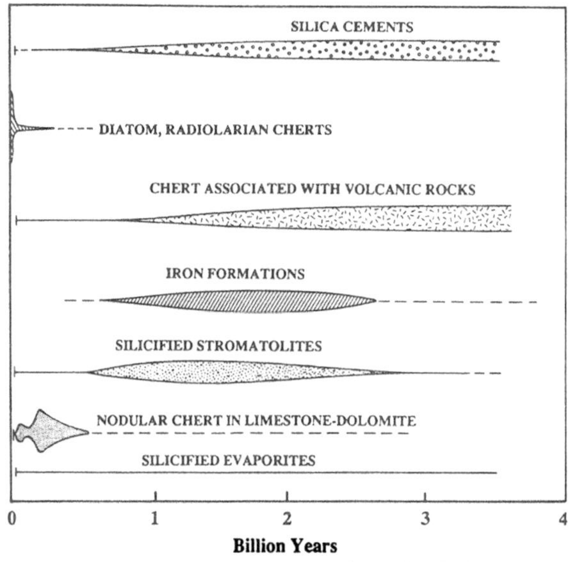

Figure 3: Evolution of authigenic silica associations with time. The width of the field for each chert type is a subjective impression of abundance during geologic time. The impression is based largely on the ease with which samples of each type can be obtained from the geologic column

Cherts are found in a number of different associations (Table 2). The paragenetic sequence of silica phases is not well known for most of these occurrences. Figure 3 gives an impression of the relative abundances of the major chert types over geologic time. Much of the variation is due to biological evolution. For example, the probable silica source of nodular cherts in limestones is siliceous sponges that co-existed with carbonate-secreting organisms. Such sponges were abundant in shallow platform sediments during the Paleozoic and Mesozoic, but are now abundant only in deep-sea settings. The near absence of nodular cherts in Neogene carbonates is probably due to the paucity of silica sponges in equivalent depositional environments that gave rise to such a profusion of Paleozoic nodular cherts. The abundance of bedded cherts derived from Cenozoic siliceous oozes is likewise attributable to the Neogene explosion of diatoms.

OXYGEN ISOTOPE SYSTEMATICS

The basic dissolution-precipitation reaction for the silica-water system is:

$$SiO_2 + 2H_2O = H_4SiO_4 \text{ (silicic acid)}$$

At equilibrium, O atoms are continually exchanged back and forth between water molecules and SiO_2 molecules via the mechanism shown. Isotopic fractionation should therefore occur and should be temperature dependent. It is important to note that isotope equilibration and exchange can occur only during events involving dissolution-precipitation of silica. If quartz simply recrystallizes without the addition or subtraction of $\delta^{18}O$, there will be no change in isotopic composition. For this reason, it is possible for cherts to be metamorphosed at high temperatures without undergoing any major change in $\delta^{18}O$, providing the water/rock ratio is low.

Quartz precipitates abundantly in sedimentary rocks, but the rate of precipitation is apparently too slow for meaningful laboratory synthesis at low temperatures. It has thus not been possible to calibrate the O isotope fractionation with temperature for quartz-water below 200°C. Various theoretical estimates and extrapolations from high temperature experiments have been made as summarized by Kyser (1987). Opal is readily synthesized, but the water content ranges up to 12 wt.% and cannot be easily removed without exchanging water O with the tetrahedrally bound silicate O. In addition, opal is amorphous and thus may not have Si-O bonding regular enough to yield systematic isotopic fractionation. The desired approach of grounding a method in theory, calibrating it in the lab, and then applying it to nature cannot yet be applied to the silica system. However, the data for natural samples display many systematics that can be usefully and quantitatively interpreted.

The fractionation with temperature expression is probably approximated by a standard curve of the form:

$$1000\ln\alpha = A\,(10^6)(T^{-2}) + B \qquad (T = {}^{\circ}K)$$

If A = 3.09 and B = -3.29 (Knauth and Epstein, 1976), then this expression can be used to obtain geologically reasonable temperatures for cherts of all types over a broad range of diagenetic temperatures.

Siliceous Oozes

The $\delta^{18}O$ of marine diatoms and radiolarians is about +34 to +39 per mil, some of the most ^{18}O-enriched material ever measured. Fairly pure layers of this material can range up to hundreds of meters thick in areas below the Carbonate Compensation Depth. Silica diagenesis may begin immediately upon burial, but substantial conversion into opal-CT may not occur until temperature rises above 45°C. In the typical case, the deep marine setting precludes entry of ^{18}O-depleted meteoric waters into the sediments during diagenesis. The pore fluids are ocean water which can be expected to become slowly enriched in ^{18}O by reequilibration with the ^{18}O-enriched silica. The amount of the enrichment depends upon the water/rock ratio (W/R), defined as the atomic amount of O in the water divided by the atomic amount of O in the silica that equilibrates with this water.

At the sediment-water interface W/R approaches infinity, but the value must drop once the pore fluid is effectively cut off from the ocean reservoir. In the limiting case where there is little pore fluid, the sediment pile contains essentially all the O of the system. A tiny amount of water that equilibrates continuously with the silica as temperature increases will follow the curve for W/R = 0 shown in figure 4. In this case, ^{18}O of the sediment pile will not change because essentially all the O is in the silica. In the opposite limiting case where the amount of water is so great that essentially all the O of the system is in the pore water, ^{18}O of the water will remain constant and ^{18}O of the silica will decrease along the curve shown in the figure for W/R = infinity. Obviously, the natural situation will fall between the limiting values shown in figure 4. As the sediment pile compacts and dewaters, the W/R ratio will decrease and $\delta^{18}O$ of water and silica will evolve along

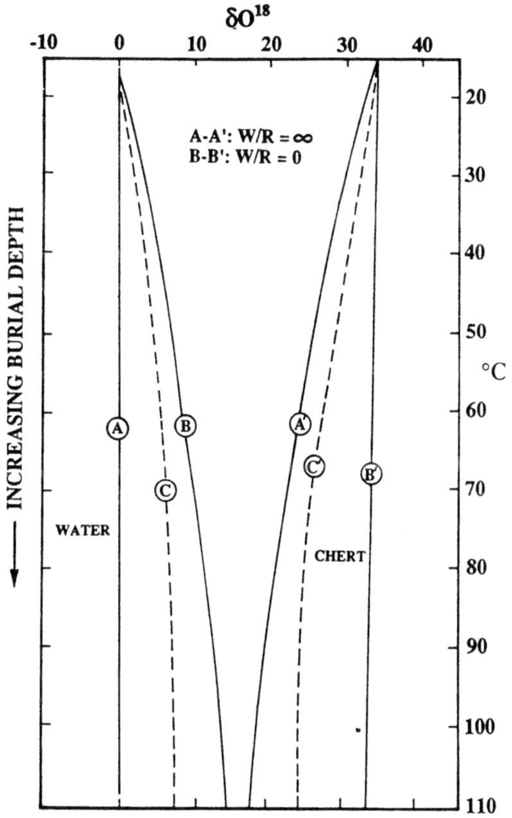

Figure 4: Closed system isotopic equilibration of chert with water during increased burial/temperature as a function of water/rock (W/R) ratio. When W/R is very low, $\delta^{18}O$ of the sediment pile will not change (curve B'), and pore waters will become enriched in ^{18}O along curve B. When W/R is very large, authigenic silica will precipitate along curve A'. Intermediate W/R ratios yield more realistic curves such as C-C'. A graph of this sort is most appropriate for diagenesis of deep sea siliceous oozes where intrusions of low-^{18}O meteoric waters are unlikely, and the system can evolve as shown

intermediate values shown by the dashed curves in figure 4. The $\delta^{18}O$ of opal-CT forming during burial can therefore be expected to be lower than $\delta^{18}O$ of the precursor biogenic silica. Upon deeper burial, the opal-CT transforms to quartz and its $\delta^{18}O$ can be expected to be lower still.

The general case of decreasing $\delta^{18}O$ of deep sea diagenetic silica with increasing diagenetic grade has been observed in all cases examined so far. The $\delta^{18}O$ values for opal-CT in core samples from buried deep sea oozes are always several per mil lower than co-existing remnants of the precursor biogenic silica. Quartz chert, which forms during deeper burial, is generally several per mil lower still (e.g. Kolodny and Epstein, 1976, and references cited therein). The same has been observed for the Miocene Monterey Formation, California, where diatomaceous oozes were buried to various depths and then rapidly uplifted along a tectonically active continental margin (e.g. Pisciotto, 1981).

Reliable temperature estimates based on $\delta^{18}O$ of the opal-CT and/or quartz chert are not available because W/R and $\delta^{18}O$ of the pore fluids are unknown. Indeed, much of the published work assumes $\delta^{18}O = 0$ for the water throughout the burial history and the W/R ratio is rarely considered. In addition, silica diagenesis may locally proceed at greatly enhanced rates in response to controls not yet understood. For example, many black, "glassy", quartz cherts in the Monterey Formation have $\delta^{18}O$ values above +28 per mil and cannot have formed at the 80°C temperature normally thought necessary to convert opal-CT to quartz. These are obviously an earlier generation of cherts that formed outside the normal burial diagenetic scheme of:

opal-A (biogenic silica) --> opal-CT --> Quartz.

This sequence is probably a limiting, general case upon which earlier transformations can be superimposed. Oxygen isotopes can be used to effectively determine which are the earlier (more ^{18}O enriched) phases. If $\delta^{18}O$ of the fluid can be measured or estimated, or if the W/R can be estimated, then actual temperatures can be estimated from a quantitatively constructed version of figure 4.

Bedded Cherts

Siliceous oozes in various stages of conversion to quartz chert can be found throughout Tertiary sections, especially in deep sea settings. Subduction has largely removed the older oceanic crust, so the only remnants of ancient siliceous oozes are certain bedded cherts which have been deformed and accreted along ancient continental margins. These are generally radiolarian cherts and are found in Mesozoic and Paleozoic strata. These rocks are too old to contain any of the precursor phases; all opaline material has had sufficient time and/or temperature to convert to quartz. The Devonian Arkansas-Caballos Novaculite of Arkansas, Oklahoma, and Texas is a famous example.

Jones and Knauth (1979) analyzed numerous samples of the novaculite in Arkansas. A wide range of $\delta^{18}O$ values was observed (Fig. 5). Most of the variation was attributed to chert crystallization at various burial depths. In places the novaculite has been strongly metamorphosed. During regional metamorphism temperatures could rise to more than 300°C. At this temperature, the fractionation between quartz and water is less than 6 per mil. The $\delta^{18}O$ of metachert will depend upon $\delta^{18}O$ of the metamorphic fluid, the W/R ratio, and the degree to which the fluid circulates through the region (thereby allowing it to exchange with rocks of different $\delta^{18}O$). The general effect is that $\delta^{18}O$ of the chert tends to decrease. However, unless a large amount of fluid is circulated throughout the chert beds, $\delta^{18}O$ of the chert will not be significantly lowered. For example, if T = 300°C and W/R is near zero, the water will simply equilibrate with the chert to a $\delta^{18}O$ value about 6 per mil lower than the chert. As the chert recrystallizes, the water will move through the rocks and continually reequilibrate with the chert. However, neither the water or the chert will be further altered isotopically; the reservoir of O is in the rock and there is insufficient water to add or substract ^{18}O from it. Jones and Knauth (1979) argued that regional metamorphism of the Arkansas bedded cherts had lowered $\delta^{18}O$ of the chert by 0 to 7 per mil. Many recrystallized cherts were isotopically similar to unmetamorphosed examples from the same terrane. Metamorphism under relatively dry conditions is implied. The most extreme lowering was for a contact metamorphosed sample which has been lowered to a $\delta^{18}O$ value of +18 per mil.

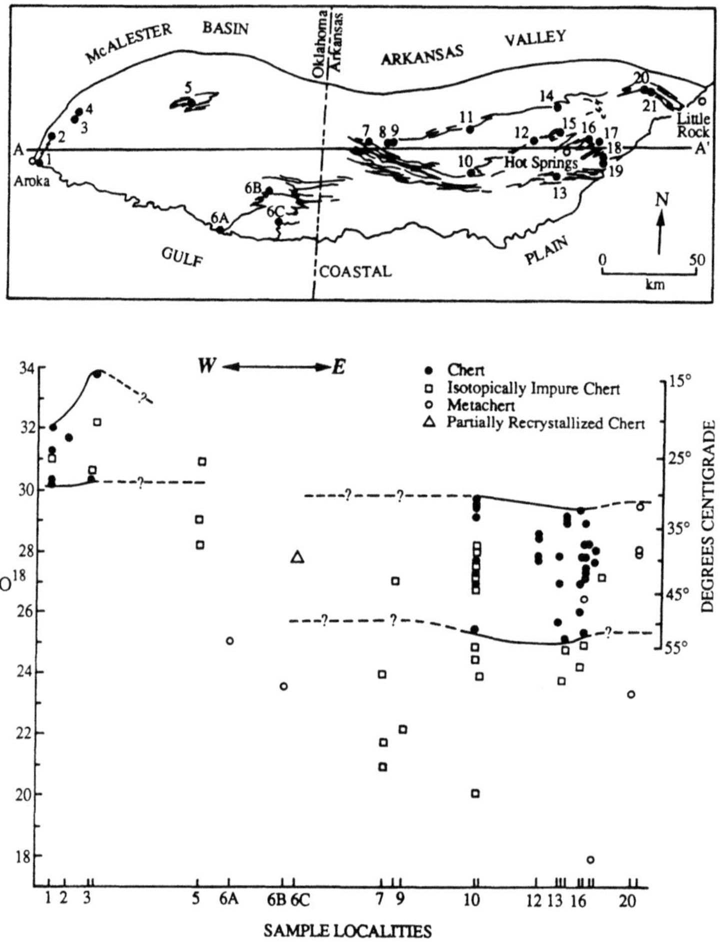

Figure 5: $\delta^{18}O$ of Devonian Novaculite, a bedded chert produced by diagenesis of a siliceous ooze. Outcrop and localities are in the top part. Sample localities are projected on line A-A' and plotted against $\delta^{18}O$ in the bottom half. Varied $\delta^{18}O$ values at each locality are interpreted as microquartz crystallization at various depths. Cherts on the western side are enriched in ^{18}O relative to the eastern side, indicating precipitation at shallower burial depth. Recrystallized cherts are not strongly depleted in ^{18}O indicating that metamorphic W/R ratios were low (From Jones and Knauth, 1979)

In general, bedded cherts found in Phanerozoic eugeosynclinal sequences probably originated as radiolarian oozes. Oxygen isotope variations for bedded chert samples composed of microquartz are probably caused by crystallization at various burial depths. Megaquartz samples have $\delta^{18}O$ values related to metamorphic reequilibration of precursor microquartz, but these values are not necessarily much lower than those of the original chert.

Magadi-Type Cherts

The world's youngest cherts are found in Lake Magadi, Kenya, and have been thoroughly described by Eugster (1969). Lake Magadi is located in the rift valley of East Africa and contains an alkaline lake perched within young volcanic rocks. Eugster proposed that Na is leached out of an authigenic hydrous Na-silicate phase by meteoric waters to produce hydrous Si which then becomes chert. However, O'Neil and Hay (1973) found that $\delta^{18}O$ of the chert ranged from +31 to +44.1 per mil. The +44 values are probably the heaviest terrestrial $\delta^{18}O$ values ever measured and are hard to reconcile with leaching by meteoric waters. If Eugster is correct, then the original Na silicate was an ^{18}O-enriched evaporite phase which was somehow subsequently leached free of Na by meteoric fluids without O isotopic exchange. An alternative is that many cherts in Lake Magadi simply precipitate during seasonal evaporation, and that these were the types of cherts analyzed by O'Neil and Hay (1973). In any case, Magadi cherts are unusual and probably will not be encountered often in the geologic record. They are most likely to have occurred in interior basin deposits.

Cherts in Iron Formations

Throughout much of the Precambrian, cherts are found interbedded with layers of Fe oxide and/or siderite. The origin of iron formations remains one of the most perplexing problems in geology, and many explanations have been offered for these remarkable rocks. Most have been metamorphosed, and the diagenetic sequence that gave rise to quartz chert cannot yet be specified.

Quartz may have been a primary precipitate from the oceans or it could be a transformation product of precursor opaline phases. This question is of profound importance in interpreting $\delta^{18}O$ data. Do the data tell us about the initial sedimentation, early diagenesis, deep burial diagenesis, or only metamorphism?

The most extensive study was done for the Brockman Iron Formation in Australia by Becker and Clayton (1976). They measured $\delta^{18}O$ of co-existing quartz and magnetite and concluded that the phases had equilibrated at temperatures approaching 350°C. The $\delta^{18}O$ of the quartz was about +21 to +22 per mil. These are much lower δ-values than observed for typical Phanerozoic cherts. One possible reason for this is that the quartz was preceded by more ^{18}O-enriched opaline precursors, and then lowered wholesale during metamorphism (Becker and Clayton, 1976). Alternatively, the relative amount of silica may have been so large that little ^{18}O was added or subtracted by metamorphic fluids and that the quartz today is little different from the primary values. The low $\delta^{18}O$ can then be alternatively explained in terms of higher Precambrian temperatures, lower $\delta^{18}O$ of the oceans, or input of low ^{18}O meteoric waters into the original sedimentary basin.

Unlike the Brockman Iron Formation, quartz in the Gunflint Iron Formation displays a large variation in $\delta^{18}O$. This quartz occurs interbedded with iron oxides, siderite, and replaces stromatolites and limestone, all in a bewildering variety of petrographic textures. Unpublished $\delta^{18}O$ values range from +17 to +26 per mil. Metamorphism is locally the cause of some of this variation, but interpretations of the data are still preliminary. The isotopic investigation of iron formations is still in its infancy. It is clear that many examples are likely to display large variations in $\delta^{18}O$, and this will undoubtedly be useful in understanding the origin(s) of these important rocks.

Nodular cherts

Nodules, thin beds, stringers, and irregular accumulations of quartz chert occur in carbonate rocks ranging from Archean to

Miocene time. In Phanerozoic rocks this chert occurs in limestones and dolostones. In Precambrian rocks it occurs in stromatolitic carbonates, but some is also found in siderite beds associated with iron formations. The outlines of former spicules are common in most Phanerozoic examples (best seen on wet fractured surfaces with a hand lens), and it is likely that sponges were the source of silica. A typical platform carbonate deposit probably consisted originally of aragonite and high-Mg calcite shell debris together with abundant sponge spicules of opal-A. During early diagenesis, the metastable carbonate phases were converted to low-Mg calcite and/or dolomite. At some point during this "stabilization" episode the biogenic silica was mobilized and reprecipitated into quartz chert. The timing of silicification was variable, but good arguments can be made that it was typically pre-compaction of the sediments, prior to most dolomitization, and before loss of significant permeability.

The chert typically replaces the metastable carbonate phases on a fine scale; fossils and early sedimentary textures are often remarkably well preserved inside cherts. Even the youngest Tertiary nodular cherts are composed of microquartz. While opal-CT may, in some cases, precede quartz precipitation, it appears that quartz was typically the primary precipitate. It is likely that the diagenetic fluids were saturated with respect to quartz, but rarely rose to the 80 mg/g values necessary for opal-CT precipitation. Vugs and dissolution cavities are often filled with fibrous quartz and drusy quartz, but many of these may relate to later subaerial exposure events that are being increasingly recognized in ancient, epeiric sea platform carbonates. Isotopic analyses of microquartz probably yield information about the earliest diagenetic events. Samples laced with fibrous quartz and megaquartz vug-fills yield δ-values that are harder to interpret because this material can fill void spaces at any time in the burial-uplift history of the unit.

The water content in nodular cherts ranges downward from about 1 wt.% and averages about 0.5 wt.%. In older cherts, any original hydroxyl bonded to Si has very likely changed into molecular H_2O via the mechanism shown in figure 6. The H_2O may escape, remain trapped as isolated molecules at dislocations, or combine with other H_2O molecules to form fluid inclusions. Fluid inclusions of the original water in which the chert precipitated may also be present. Water extracted from cherts may therefore have the combined δD values of the primary diagenetic fluid and fractionated water that was originally bonded as hydroxyl. The hydroxyl in cherts is probably depleted in D by 20-50 per mil relative to the water in which the chert forms because this is the magnitude of the OH-H_2O fractionation in clay minerals.

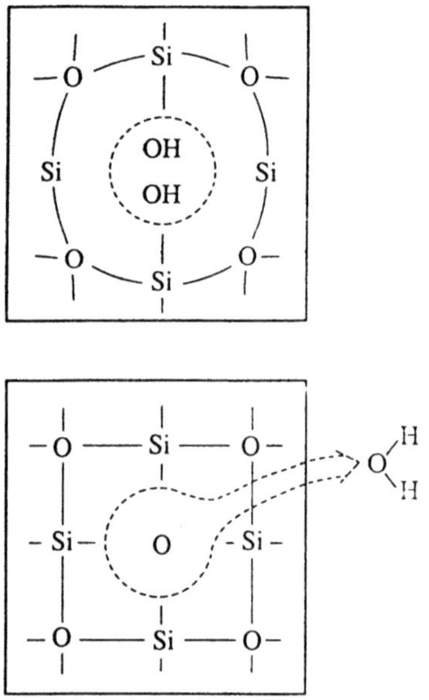

Figure 6: Dehydration of hydrous silica. Two hydroxyls can combine to produce H_2O. One hydroxyl O is left behind to Si, thereby annealing out the former defect. The H_2O may leave the system, or collect with other H_2O molecules into a tiny fluid inclusion. δD of this water is determined by the hydroxyl fractionation of the original precipitate

Figure 7 shows combined δD-$\delta^{18}O$ measurements for cherts of various ages. The $\delta^{18}O$ is for the silicate O and δD is for water extracted by dehydration at 1000°C under vacuum. Most of the samples are nodular cherts in carbonates. Cherts within specified age limits yield domains of δ-values which are stretched out roughly parallel to the meteoric water line. In Knauth and Epsteins' (1976) interpretation the cherts can be treated similar to the clay minerals; ^{18}O is preferentially concentrated during precipitation into the silica and D is preferentially excluded from the hydroxyl. On a δD-$\delta^{18}O$ diagram, chert data therefore plot to the lower right of the water with which they last equilibrated.

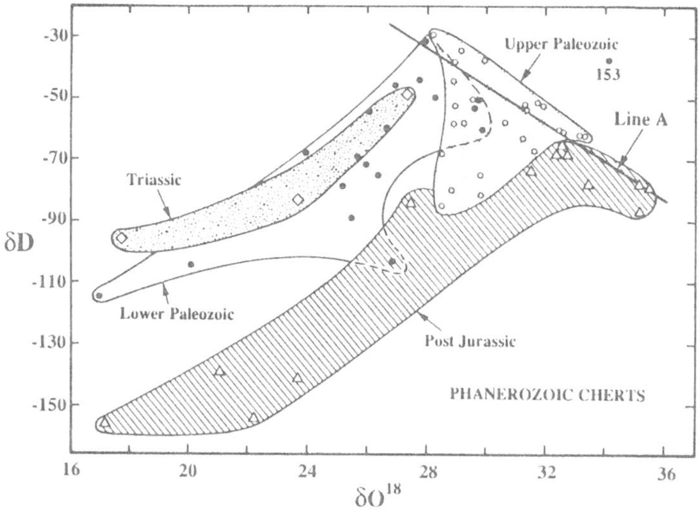

Figure 7: δD-$\delta^{18}O$ of Phanerozoic cherts in carbonates. The data for selected age groups are elongated parallel to the meteoric water line. The shift of the elongated domains toward and away from the meteoric water line with time can be interpreted in terms of large changes in climatic temperatures over geologic time. The "cut-off" in δ-values to the upper right is understood if line A is the approximate locus of δ-values for cherts in isotopic equilibrium with sea water. The lower δ-values reflect original precipitation of chert in mixed marine-meteoric diagenetic waters (From Knauth and Epstein, 1976)

The maximum δ-values for cherts of various ages appear to approach a limit, shown as line A in figure 7. Line A is interpreted as the trajectory on a δD-$\delta^{18}O$ diagram that chert will take as it equilibrates with ocean water (approximated by SMOW) at various temperatures. In coastal diagenetic environments δ-values for the groundwaters will decrease landward as lighter meteoric waters are mixed with marine waters. Cherts which form in mixed marine-meteoric fluids at a constant temperature will plot progressively farther away from line A in proportion to the magnitude of the light meteoric water component. An individual line parallel to the meteoric water line should result for cherts which form at the same temperature. For these reasons, Knauth and Epstein (1976) argued that cherts of the various ages shown had formed at various temperatures in mixed meteoric-marine diagenetic fluids. Those plotting nearest line A were those which contained the greatest marine component. The scatter within a domain probably arises because of temperature variations during the time shown and because some water is derived from submicroscopic fluid inclusions containing unfractionated remnants of the original diagenetic fluid. Additionally, pyrolosis of trace amounts of organic matter in the chert occurs during sample heating, and this yields water which contaminates the silica water.

Cherts of various ages plot in different positions on this diagram, and this can be interpreted in several ways. If ocean water was isotopically different in the past, then the position of the meteoric water line would also be different. Thus, if the oceans were several per mil depleted in ^{18}O in the Cambrian relative to the upper Paleozoic, then the domain of chert δ-values could be concomitantly displaced as shown (assuming that the range of precipitation temperatures was similar). However, large changes in $\delta^{18}O$ of ocean water are required between the Upper Paleozoic and the Triassic (-4 per mil) and between the Triassic and the Miocene (+6 per mil) to explain the data in terms of changes in the isotopic history of sea water. Such large changes are improbable.

The average ground temperature below a few meters depth is an excellent average of the yearly climatic temperature, so temperatures deduced for nodular cherts that formed during early diagenesis should approximate climatic temperatures. Temperatures can be assigned to line A by using one of the quartz-water O isotope fractionation curves and assuming that $\delta^{18}O$ of pre-Pleistocene ocean water was approximately -1 per mil (the value for an ice-free world). Parallels to the meteoric water line which pass through the data domains can be determined from their intersection point on line A. Temperatures assigned in this way are obviously approximate. Knauth and Epstein (1976) argued that climatic temperature variations accounted for the major part of the variations shown in figure 7, and suggested that climatic temperatures decreased from Cambrian values on the order of 34°C to Upper Paleozoic values of 20°C. The Triassic returned to temperatures similar to those of the Cambrian and then declined to

Miocene values of 17°C. A few Precambrian values (not shown in Fig. 7) are also variable, and range from values similar to those of the Upper Paleozoic to as high as 70°C in the Archean. The chert-water O isotope fractionation with temperature equation of Knauth and Epstein (1976) was estimated, in part, on trial-and-error fits to line A that would yield reasonable temperatures. The actual temperatures are therefore not rigorous, but temperature differences between different ages are similar no matter which quartz-water temperature-fractionation expression is used.

Hydrogen isotopic analysis of cherts has not been extensively used. Organic matter and clay minerals are ubiquitous contaminants and many workers are reluctant to accept that small amounts of water can remain isotopically unaltered in cherts over geologic time. The amount of water is so small that any external fluids which penetrate the sample could obliterate the record and any recrystallization during metamorphism or hydrothermal alteration could profoundly affect δD. Nevertheless, the δD data for cherts of a given age tend to vary sympathetically with $\delta^{18}O$. Careful analyses of samples screened for purity and textural preservation appear to be meaningful. A further refinement is possible in which the constituent petrographic phases are separated and analyzed; previous analyses have generally been for coarse separates or whole rocks. Even if the H data are accepted, it has not been resolved how the data in figure 7 are best interpreted. The climatic temperature variations are larger than many are willing to accept. However, the fossil record does not provide a strong constraint on actual temperatures. Many modern organisms thrive in the higher temperature range suggested for the Phanerozoic, and the blue-green algae (cyanobacteria) that make up the predominant fossils of the Precambrian are today found growing in abundance in hot springs at temperatures even warmer than the maximum suggested value of 70°C.

The strong indications that meteoric waters were involved in chert diagenesis in marine carbonate rocks prompted Knauth (1979) to propose a coastal mixing zone model for the origin of chert in limestone which explains many geochemical, petrographic, and field data for such cherts. Kolodny et al. (1980) have given independent arguments based on $\delta^{18}O$ and B geochemistry that meteoric waters were involved in the formation of nodular cherts in limestones. These arguments are independent of considerations involving H isotopes. Relative changes in δD and $\delta^{18}O$ will likely prove increasingly useful in unravelling silicification sequences within individual carbonate units.

An Application: $\delta^{18}O$ of Cherts versus Time

The first isotopic investigation of cherts was by Degens and Epstein (1962). They observed that $\delta^{18}O$ for cherts generally decreases for progressively older samples, just as it does for limestones. At the time, it was assumed that limestones slowly equilibrate with later, ^{18}O-depleted groundwaters. Therefore, the older the sample, the more time there was for exchange and the lighter was the calcite. Since the chert data paralleled the carbonate data, it was concluded that the cherts had also lost their isotopic memory.

Perry (1967) analyzed samples from thick, bedded chert sequences and observed that $\delta^{18}O$ for these trended back through geologic time along a different trajectory than that observed by Degens and Epstein (1962). He suggested that his cherts had preserved their $\delta^{18}O$ values from the time of deposition and that these could be used to infer important information about earth history. His Precambrian samples had $\delta^{18}O$ as low as +14.1 per mil, and Perry argued that this could not be a temperature effect because the temperatures would be unreasonably high. He suggested that $\delta^{18}O$ of the Archean ocean was 18 per mil lower than the modern ocean and that the oceans had become progressively enriched in ^{18}O as recorded by the cherts. This important and provocative paper called attention to the possibility of a rich isotopic record in ancient sedimentary rocks; a record that contained important information about earth history. Figure 8 shows a compilation of $\delta^{18}O$ measurements of cherts over geologic time. All of these data have been obtained since Perry's (1967) original paper. The data are generally heavier than Perry's original values, and this has reopened the possibility that climatic temperature variations are largely responsible for the lower $\delta^{18}O$ values in the past.

It now appears that cherts can be expected to have a wide range of initial $\delta^{18}O$ values that vary with the conditions of formation. In deep sea cherts and other varieties that have opaline precursors, the final quartz may form over a range of burial temperatures and thus have $\delta^{18}O$ values ranging from relatively heavy, low temperature values to relatively light, higher temperature values. In nodular cherts which form in platform carbonates, the chert may form before significant burial at a relatively constant temperature, but a large range in $\delta^{18}O$ may result because variable amounts of low-^{18}O meteoric waters are involved. For older cherts, metamorphism may further lower $\delta^{18}O$ by exchange at elevated temperatures. Thus, it is the maximum $\delta^{18}O$ for cherts of a given age that is of most importance. This is the δ-value that would most closely correspond to hypothetical chert that formed in sea water, and is the most useful value for monitoring either changes in $\delta^{18}O$ of the past oceans or past climatic temperatures. Of course, in evaporite environments $\delta^{18}O$ of the water can rise above marine values as the lighter isotopes are preferentially removed. However, evaporite cherts appear to be rare, and the odd case can usually be recognized in compilations of large amounts of data.

The temperature scale in figure 8 is for the case where $\delta^{18}O$ of the ocean is held at -1 per mil over geologic time and is based on the chert-water fractionation with temperature expression of Knauth and Epstein (1976). In this interpretation, the early earth may have been as hot as 70°C. By 1.2 Ga b.p. temperatures decreased to lower than 30°C, increased by about 10°C around Cambrian time, and generally decreased throughout the Paleozoic. Temperatures rose in the Triassic and declined to values shown for the mid-Tertiary. These are similar to temperatures deduced from the combined H and O isotope data, but this interpretation does not depend upon δD and the additional uncertainties associated with that measurement. Nearly all of the post-Archean data are from North America and are for cherts that undoubtedly formed at various latitudes. However, surface temperature variations with latitude are greatly reduced in an ice-free world, and may amount to less than 10°C between the equator and 60° latitude (cf. Barron and Washington, 1985; Fig. 1). It is therefore unlikely that the large variations shown are caused by different paleolatitudes of the samples. There are other concerns about interpreting the chert data purely in terms of temperature, but this remains the simplest interpretation for the major part of the variations shown.

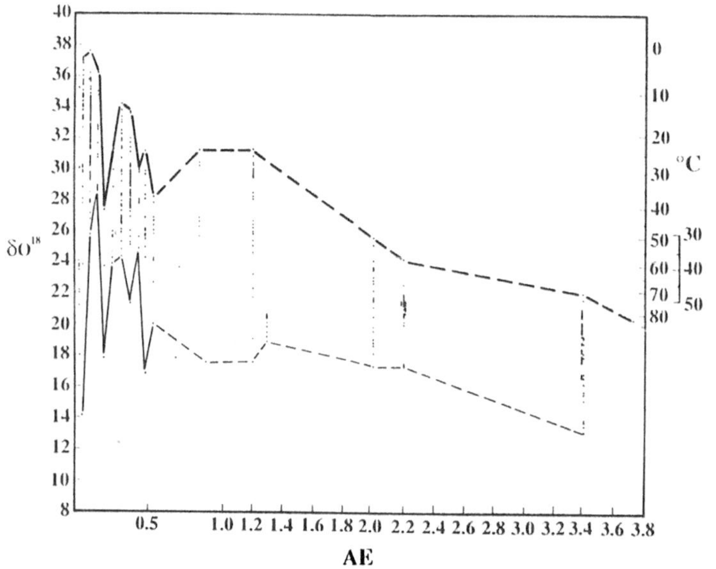

Figure 8: Changes in $\delta^{18}O$ of cherts with time. Phanerozoic data are mostly from nodular cherts in carbonates, but also include bedded cherts. Precambrian data are from silicified stromatolites, but also include iron formation cherts and cherts of uncertain origin. Cherts known to be altered by metamorphism have been omitted. At any given age the large range of $\delta^{18}O$ values is due to variable amounts of meteoric water involved in chert precipitation, crystallization at different burial depths and post-depositional lowering of $\delta^{18}O$ during metamorphism. The high $\delta^{18}O$ at any age best approximates $\delta^{18}O$ of chert in isotopic equilibrium with sea water. Assuming that $\delta^{18}O$ of sea water was approximately constant at -1 per mil, "surface" temperature variations over earth history are reflected by changes in the upper bound to the data and can be estimated by temperature scale (0-80°C) given on the right. The lower data may reflect the range of meteoric water $\delta^{18}O$ that exist at a given time, a function of climatic temperatures and $\delta^{18}O$ of the oceans. The strong ^{18}O depletions for the earliest cherts suggest that sea water at 3.5 AE was several per mil lower than its value over most of the time. The 30-50°C temperature range shown at the right is for $\delta^{18}O$ of sea water = -2. According to these interpretations, Archean surface temperatures may have been as high as 50-70°C

The lower curve brackets the minimum $\delta^{18}O$ values and these values may be of some significance. If the variation of $\delta^{18}O$ for unrecrystallized nodular cherts of a given age is largely due to the amount of meteoric water involved in silica diagenesis, then the minimum value for the cherts is also sensitive to changes in $\delta^{18}O$ of the past oceans. It is therefore noteworthy that the minimum $\delta^{18}O$ is roughly constant at about +18 per mil except for a number of Archean samples. This may be telling us that the Archean ocean was several per mil lower in $\delta^{18}O$ and the temperatures would then reduce by 15-20°C. These reduced values still indicate much warmer Archean climates. Of course, it is possible that the minimum δ-values are simply for the more deeply buried or metamorphosed cherts and put no constraints on the overall interpretation. More data for unrecrystallized low ^{18}O cherts are clearly needed.

An additional explanation for the low ^{18}O Archean cherts is that they precipitated from hot fluids circulating around mid-ocean spreading centers. Paris et al. (1985) have suggested such a setting for the South African cherts which make up the bulk of the Archean data shown in figure 8. The entire Onverwacht-Fig Tree sequence is silicified on a scale unknown in younger sedimentary rocks. This could support the arguments of Paris et al. (1985), or it could simply mean that surface temperatures were as hot as suggested above, and that the earth's surface was essentially an open hydrothermal system in the Archean.

OTHER ISOTOPE SYSTEMATICS

Silicon

The masses of the three stable Si isotopes (^{28}Si, ^{29}Si, ^{30}Si) are too heavy for large fractionation effects, but the occurrence of Si in terrestrial rocks is so widespread that several attempts have been made to measure natural variations in its isotopic composition. The most recent survey (Douthitt, 1982) confirms that

$^{30}Si/^{28}Si$ varies by about 1 per mil for igneous rocks, but varies by several per mil for low temperature silica. Opal from both sponge spicules and siliceous sinters (opal deposited around hot springs) is about 2 per mil depleted in ^{30}Si relative to the igneous rocks, while opal phytoliths are about 3 per mil enriched. Cherts display a 3.1 per mil variation centered around igneous rock values.

Douthitt (1982) suggested a kinetic isotope fractionation for the rapidly-precipitated siliceous sinter and for biogenic silica precipitation. Since the cherts do not display the Si isotope variations of their supposed biogenic precursors, it was suggested that the silica re-equilibrates with dissolved Si of normal isotopic composition during the conversion to quartz. Si isotope variations appear to be of marginal use in understanding chert diagenesis, but the previous work has been very limited. An investigation of cherts of various types and various ages using extremely accurate measurements might yet prove useful.

Lead, Neodymium and Strontium

Extremely pure domains of SiO_2 probably exist in most cherts and opal deposits, but examination of thin sections readily reveals impurities in every example. The impurities can include carbonate, clay, micas, feldspar, organic matter, fluid inclusions, and a variety of terrigenous debris. These impurities occur in trace amounts in the purest cherts, but can be the major phases in some samples. Chemical analyses of cherts typically relate to these impurities rather than to atomic substitutions within the quartz crystals or opaline domains themselves. The low temperature SiO_2 structure contains dislocations, is variably ordered on the long range, and may be quite inflated relative to the quartz of mineralogy books. Atomic substitutions not normally thought possible for quartz may be a reality in some forms of chert and opal. Detailed microprobe analyses of micron sized areas have not been reported, and the crystal chemistry of low temperature silica is largely unknown. Heavy isotope studies of cherts involve whole rock samples and thus relate primarily to the impurities rather than the SiO_2 itself. The extent to which such studies of impurities yield information about silica diagenesis must be evaluated for each individual case.

Barrett (1980) interpreted Pb isotopes in a sequence of Jurassic radiolarian cherts overlying ophiolites in terms of a mixture of continentally and mantle derived Pb. The cherts immediately overlying the basement are rich in volcanogenic Pb which was laterally dispersed from an active ridge. As the area was progressively blanketed with radiolarian ooze the amount of volcanogenic Pb decreased relative to the amount of radiogenic Pb provided by fine-grained terrigenous detritus. A stratigraphy of upwardly decreasing mantle Pb is thus preserved in the chert layers. The Pb isotope reservoir at the time of deposition was preserved when the siliceous oozed transformed to chert. Many ancient cherts are of problematic origin and hydrothermal activity associated with basaltic volcanism is often proposed as a silica source. The Pb isotopes appear to be a potentially useful approach for evaluating such situations, but the method is largely unexploited.

Sm-Nd systematics are similarly promising for tracing initial environmental conditions of cherts. Weis and Wasserburg (1987a) showed that oceanic cherts that form far from terrigenous source areas have impurities (carbonate?) with Nd isotopes characteristic of the ocean they form in. Others that form near continents have Sm/Nd ratios typical of continental sources. Sm-Nd systematics are incredibly sensitive to source terranes; even modern Atlantic ocean water is different from its Pacific counterpart. The possibilities are good that cherts encase a detailed Nd isotopic record that could be put to great use.

The Rb-Sr method holds promise for age dating some types of chert. Weis and Wasserburg (1987a) have shown that Rb is strongly enriched relative to Sr in most forms of biogenic silica and in selected chert samples. The enrichment of Rb over Sr is so large that measurable variations are likely in the $^{87}Sr/^{86}Sr$ ratios due to ^{87}Rb decay. The site of the Rb is unknown, but is probably associated with impurity phases rather than with the SiO_2 itself.

The Rb-Sr age dating method can be applied for situations where the chert encases impurities with the same $^{87}Sr/^{86}Sr$ ratios (such as unaltered marine carbonate), or where exchange between impurities and diagenetic fluids homogenizes the $^{87}Sr/^{86}Sr$ ratios prior to being sealed up by chert precipitation.

Siliceous oozes that accumulate far from land should contain only marine authigenic phases such as carbonate, zeolite, or smectite. The Sr in these phases should have the same $^{87}Sr/^{86}Sr$ ratio as sea water. This has been documented for a mid-Pacific chert sample dredged from an outcropping Eocene carbonate horizon on the flanks of a guyot (Weis and Wasserburg, 1987a). Siliceous oozes which accumulate nearer to land contain terrigeneous clays carrying radiogenic Sr from their continental source areas. These will yield data scatter or mixing lines on a Rb-Sr evolution diagram. There is some evidence that $^{87}Sr/^{86}Sr$ ratios of the various impurity phases can become homogenized by interaction with pore fluids during deep burial of siliceous oozes prior to the depth at which opal-CT forms (Brueckner and Snyder, 1985). For such cases, Rb-Sr isotope analyses may yield the age at which opal-A converted to opal-CT during deep burial. Similarly, Shibata and Mizutani (1982) obtained whole-rock Rb-Sr isotope ages for quartz cherts in the Triassic of Japan which are 15 million years younger than the biostratigraphic ages. They suggested that these cherts originated during deep burial diagenesis of a siliceous ooze, and that the radiometric age was the time at which the quartz became a chemically closed system.

Several attempts have been made to date the world's oldest cherts in the Onverwacht Group, S. Africa. Hurley et al. (1972) obtained a Rb-Sr isochron which gave an age of 3.5 Ga, consistent with ages for the more readily datable crystalline rocks associated with the Onverwacht sequence. Weis and Wasserburg (1987b) suggested that Hurley et al.'s linear array was, instead, a mixing curve caused by variations in the amount of carbonate and muscovite in the samples. They obtained a Rb-Sr age of 2.12 Ga which they argued was the age of a previously unrecognized metamorphic event that recrystallized the Onverwacht cherts (Fig. 9). This result raises interesting questions about just when the geochemical signatures, including $\delta^{18}O$, were locked into these ancient samples. Can Sr isotopes in the muscovite and carbonate impurities in these cherts be homogenized during metamorphism without wholesale exchange of the O isotopes? How much of the early diagenetic isotope record can be read through the later events that had 3.5 Ga to occur? In dealing with Archean sedimentary rocks, the questions of metamorphic overprinting are critical; satisfactory answers will require integrated field, petrographic, and isotopic studies on a scale not yet attempted.

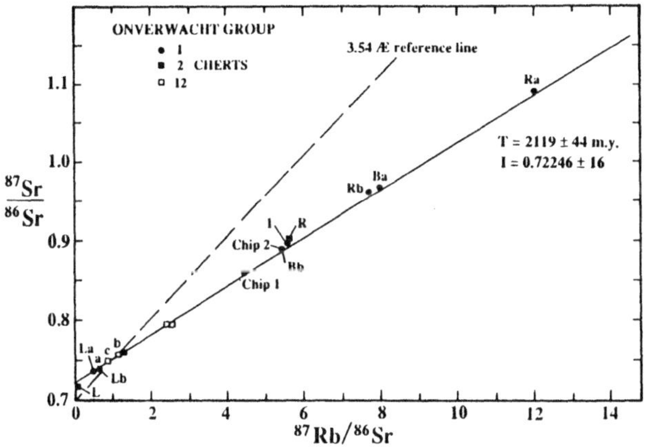

Figure 9: ^{87}Rb-^{87}Sr evolution diagram for cherts from the Onverwacht Group, South Africa. The letters B, L and R stand for bulk, leach, and residue. The letters a, b, and c correspond to different layers in sample 12. An additional sample plotting off the diagram has been used in the age and initial ratio calculation (After Weis and Wasserburg, 1987b)

CONCLUSION

Most previous isotopic work on cherts has involved broad surveys of limited numbers of chert types. Much emphasis has been given to the diagenesis of siliceous oozes, but these deposits are not generally representative of the cherts encountered in the pre-Mesozoic sedimentary record. Chert forms in siliceous oozes long after original deposition of the original, precursor biogenic opal, and isotopic studies of these cherts yields information primarily about deep burial diagenesis. The common nodular cherts in carbonates of all ages are typically early replacements of original carbonate and may not follow the paragenetic sequence known so well for siliceous oozes. Isotopic studies of these cherts may yield environmental information, including climatic temperatures and the meteoric/marine nature of early diagenetic fluids. Other chert types, including those associated with the abundant Precambrian iron formations, are poorly understood; results for siliceous oozes and nodular cherts may not apply to these rocks. Most of the older chert sequences have been metamorphosed to various degrees. Isotopic variations of these rocks may yield information only about conditions of metamorphism or they may remain unaltered in cases where the water/rock ratio was low. The large O isotopic variations in cherts of a given age are therefore due to quartz crystallization at various burial depths, variations in the amount of low-^{18}O meteoric waters in the diagenetic fluids, and lowering of $\delta^{18}O$ during metamorphism.

Water can be extracted from almost any chert, and its δD often varies sympathetically with $\delta^{18}O$ of the host silica. Combined δD-$\delta^{18}O$ analyses of nodular cherts in carbonates can be used to infer both temperature of formation and the relative amount of meteoric water that was present in the diagenetic fluid. Archean cherts are substantially depleted in ^{18}O relative to Phanerozoic examples. The variation of $\delta^{18}O$ of cherts over geologic time is not clearly understood, but much of it can probably be attributed to major changes in past climatic temperatures.

The Si isotopes vary little in terrestrial rocks, but the largest known effects have been observed in certain forms of opal. The isotopes of heavier elements are located in impurity phases in the silica and can be used as tracers and possibly as age indicators. Because of the large isotopic variations and the diversity of chert types, it is likely that isotopic analyses of cherts will be increasingly used, not only to better understand the chert problem, but to attack broader problems associated with the sedimentary sequences in which authigenic silica is found.

REFERENCES

Becker R.H. and Clayton R.N. (1976) Oxygen isotope study of Precambrian banded iron formation, Hamersley Range, Western Australia. Geochim. Cosmochim. Acta, 40, 1153-1165.

Barrett T.J. (1980) The Pb isotopic composition of Jurassic cherts overlying ophiolites in the North Apennines, Italy. Earth Plan. Sci. Lett., 49, 193-204.

Barron E.J. and Washington W.M. (1985) Warm Cretaceous climates: high atmospheric CO2 as a plausible mechanism. In: The carbon cycle and atmospheric CO2: Natural variations Archean to Present, Sundquist E.T. and Broecker W.S. (eds.), Geophys. Monograph, 32, Amer. Geophy. Union.

Brueckner H.K. and Snyder W.S. (1985) Chemical and Sr isotopic variations during diagenesis of Miocene siliceous sediments of the Monterey Formation, California. Jour. Sediment. Petrol., 55, 553-568.

Degens E.T. and Epstein S. (1962) Relationship between $^{18}O/^{16}O$ ratios in coexisting carbonates, cherts, and diatomites. Amer. Ass. Petrol. Geol. Bull., 46, 534-542.

Douthitt C.B. (1982) The geochemistry of the stable isotopes of silicon. Geochim. Cosmochim. Acta, 46, 1449-1458.

Eugster H.P. (1969) Inorganic bedded cherts from the Magadi area, Kenya. Contr. Mineral. Petrol., 22, 1-31.

Gregory R. and Taylor H.P. Jr. (1981) An oxygen isotope profile in a section of Cretaceous oceanic crust, Samail Ophiolite, Oman: Evidence for $\delta^{18}O$-buffering of the oceans by deep (>5 Km) seawater-hydrothermal circulation at mid-ocean ridges. Jour. Geophy. Res., 86, B4, 2737-2755.

Hesse R. (1989) Silica diagenesis: Origin of inorganic and replacement cherts. Earth Sci. Rev., 26, 253-284.

Hurley P.M., Pinson W.H., Nagy B. and Teska T.M. (1972) Ancient age of the Middle Marker horizon, Onverwacht group, Swaziland Sequence, South Africa. Earth Plan. Sci. Lett., 14, 360-366.

Jones D. and Knauth L.P. (1979) Oxygen isotopic and petrographic evidence relevant to the origin of the Arkansas novaculite. Jour. Sediment. Petrol., 49, 581-598.

Knauth L.P. (1979) A model for the origin of chert in limestone. Geology, 7, 274-277.

Knauth L.P and Epstein S. (1976) Hydrogen and oxygen isotope ratios in silica from the JOIDES Deep Sea Drilling Project. Earth Plan. Sci. Lett., 25, 1-10.

Knauth L.P. and Epstein S. (1976) Hydrogen and oxygen isotope ratios in nodular and bedded cherts. Geochim. Cosmochim. Acta, 40, 1095-1108.

Kolodny Y. and Epstein S. (1976) Stable isotope geochemistry of deep sea cherts. Geochim. Cosmochim. Acta, 40, 1195-1209.

Kolodny Y., Taraboulos A. and Frieslander U. (1980) Participation of fresh water in chert diagenesis: Evidence from oxygen isotopes and boron d-track mapping. Sedimentology, 27, 305-316.

Kyser T.K. (1987) Equilibrium fractionation factors for stable isotopes. In: Short Course Handbook, 13, Min. Ass. Canada, 1-76.

Micheelsen H. (1966) The structure of dark flint from Stevns, Denmark. Medd. Dansk Geol. Foren., 16, 285-368.

Muehlenbachs K. and Clayton R.N. (1976) Oxygen isotope composition of the oceanic crust and its bearing on sea water. Jour. Geophys. Res., 81, 4365-4469.

O'Neil J.R. and Hay R.L. (1973) $^{18}O/^{16}O$ ratios in cherts associated with the saline lake deposits of East Africa. Earth Plan. Sci. Lett., 19, 257-266.

Paris I., Stanistreet I.G. and Hughes J. (1985) Cherts of the Barberton Greenstone Belt interpreted as products of submarine exhalative activity. Jour. Geol., 93, 111-129.

Perry E.C. (1967) The oxygen isotopic chemistry of ancient cherts. Earth Plan. Sci. Lett., 3, 62-66.

Pisciotto K.A. (1981) Diagenetic trends in the siliceous facies of the Monterey Shale in the Santa Maria region, California. Sedimentology, 28, 547-571.

Shibata K. and Mizutani S. (1982) Isotopic ages of Jurassic siliceous shale and Triassic bedded chert in Unama, Central Japan. Geochem. Jour., 16, 213-223.

Weis D. and Wasserburg G.J. (1987a) Rb-Sr and Sm-Nd systematics of cherts and other siliceous deposits. Geochim. Cosmochim. Acta, 51, 959-972.

Weis D. and Wasserburg G.J. (1987b) Rb-Sr and Sm-Nd isotope geochemistry and chronology of cherts from the Onverwacht Group (3.5 AE), South Africa. Geochim. Cosmochim. Acta, 51, 973-984.

STABLE ISOTOPE GEOCHEMISTRY OF SULFATE AND CHLORIDE ROCKS

William T. HOLSER
Department of Geological Sciences
University of Oregon, Eugene, Oregon 97403, USA

Sulfate and chloride deposits are a prominent constituent of evaporite rocks, formed by the evaporative concentration of marine or non-marine waters. The geology and geochemistry of evaporite rocks has been well reviewed by Holser (1979a, b), Kendall (1984), Schreiber (1988) and Warren (1989), and will be summarized only briefly here.

The amount of concentration required to precipitate evaporite minerals ranges from about three times the concentration of normal modern sea water, when gypsum or anhydrite crystallize, to virtual dessication (60 to 80 times sea water) when the most soluble chlorides finally precipitate (Table 1). Rivers or other non-marine waters, starting from solutions that are much more dilute, require more intense evaporative regimes to reach corresponding saturations of sulfate and chloride minerals. It is clear that in either case the essential requirement for the formation of sulfate and chloride minerals is a high rate of net evaporation (evaporation minus rainfall) combined with a relatively restricted inflow of water. Evaporation, which depends on low atmospheric humidity aided by brisk wind speeds, is predominant in continental climates of middle to low latitudes. In such areas the restricted input necessary for evaporite deposition may be found in a variety of physical situations: (a) coastal intertidal and supratidal zones called sabkhas,

(b) small lagoons on coasts and atolls, (c) large deep-water marine basins, (d) sub-sea level basins with limited marine inflow, and (e) non-marine interior basins. One tectonic setting that has been particularly conducive to evaporite deposition is continental rifts. Beginning with dry interior rift valleys, sedimentation proceeds through marine evaporites controlled by restricted input of sea water, to deposition of normal marine deposits as inflow exceeds evaporation. Some "saline giants" accumulated in this manner in rifts of the North Atlantic and Gulf of Mexico during the Triassic-Jurassic, in the South Atlantic during the Cretaceous, and in the Red Sea in the Late Miocene. The tectonic control of extensive evaporite deposition on the craton, such as in the mid-Paleozoic on the North American craton and in the Permian on the European craton, is not so clear but may be incipient rifting or stretching. In contrast to these enormous deposits rich in salt rocks, more limited evaporites of predominantly Ca-sulfate mineralogy formed as sabkhas on shallow shelves such as those along present-day coasts of the Mediterranean, Red Sea and Persian Gulf.

These tectonic controls on restricted marine input are critical, shifty and transitory, short-lived geological accidents. Conditions may have been just right for an evaporitic episode for only a geological "instant" - at most a fraction of a million years. Though the interval of evaporite deposition was short, the rate of accumulation was extraordinarily rapid. Consider that once saturation had been reached, in a basin of whatever depth, a nominal rate of net evaporation, say two meters of water per year, would result in Ca-sulfate deposition at 1.2 mm/yr or salt at 25 mm/yr, compared with a typical rate of carbonate sedimentation of about 0.1 mm/yr. Consequently preformed evaporite basins tend to fill quickly to their level of input even during an episode of relatively short duration.

A facies classification of evaporite rocks is analogous to the facies classification of metamorphic rocks: it is based on the appearance of the indicated mineralogy, although this is due to the degree of evaporation rather than a maximum temperature or pressure. Table 1 outlines some of the characteristics of marine evaporite facies. A departure from this mineralogy or from its sequence indicates a differing chemistry of input waters, evidence of a non-marine component (Hardie, 1984), although non-marine evaporites are not commonly preserved in pre-Tertiary sediments (Holser, 1979c). Just as in metamorphic rocks, primary facies mineralogy may be accompanied by prograde or retrograde replacement of pre-existing minerals. Thus halite may crystallize directly, or replace earlier gypsum, and vice-versa; polyhalite may replace anhydrite which may in turn have replaced gypsum; primary calcitic limestone lamellae may be dolomitized. Along with the changes in mineralogy, the overlying or included brine also varies in density (Table 1) as well as chemistry. Because of its high density the brine will tend to become stratified, and because of stratification and low solubility of O, the deeper layers will tend to anoxic and their deposits laminated and rich in organic C and sulfides (Evans and Kirkland, 1988).

Table 1: Summary of Evaporite Facies (After Holser, 1979a)

Facies		Mineralogy	Measures of evaporation stage				
			Salinity wt %	Density g/ml	Percent H_2O evap.	Conc. Xs.w.	Activity H_2O
Potash-magnesia facies	Bittern subf.	Bischoffite, Tachyhydrite					
					99.2	120	
	Potash subf.	Carnallite, Sylvite, Kainite, Kieserite, Halite, Anhydrite or Polyhalite					
			380	1.31	98.7	78	
	$MgSO_4$[a] subf.	Epsomite, Bloedite, Halite, Polyhalite or Anhydrite					
			375	1.29	98.4	68	0.67
Halite facies		Halite, Anhydrite or Gypsum					
			300	1.20	91.0	12.2	0.76
$CaSO_4$ facies		Gypsum or Anhydrite Dolomite					
			150	1.10	72.0	3.6	0.93
Dolomite facies		Dolomite, Calcite					
			35	1.02	0	1.0	1.00
Marine facies		Calcite, Aragonite					

a usually altered to potash subfacies

In deep basins relatively isolated from the sea, brines will move quickly through the Ca-sulfate facies, the thickness of which will amount to only 0.6% of the basin depth before halite must start to crystallize, unless residual brines are also refluxed back to the sea (Holser, 1979a, p. 264). Burial under the succeeding weight of sediments leads to important changes in evaporite rocks. Primary deposition in the Ca-sulfate facies is commonly as gypsum, although at higher salinities and higher temperatures primary anhydrite may either precipitate directly or replace primary gypsum. However, as the temperature rises along the geothermal gradient during burial, any gypsum tends to dehydrate to anhydrite plus water; on being exhumed by erosion and exposed to cool fresh ground water the anhydrite cycles back to gypsum. Higher temperatures under burial also lead to plastic flow in salt rock, which under buoyant or tectonic forces moves upward to form salt pillows and eventually intrudes the overlying sedimentary section as salt domes or even extrusions (Lerche and O'Brien, 1987). The plastically deforming salt rock carries along admixed anhydrite and other accessory minerals. At shallow depths the top of the salt dome is attacked by ground water, which easily dissolves the halite and leaves a "cap rock" of residual anhydrite or its gypsum alteration. This gypsum is further altered by S-reducing bacteria to form H_2S, pyrite and other sulfides, native S, and calcite (as a byproduct of the bacterial activity).

STABLE ISOTOPES IN SULFATE AND CHLORIDE DEPOSITS

The deposition and subsequent erosion of evaporite rocks is a primary feature of the sedimentary cycle of chloride (Holser et al., 1980), S and (to a lesser extent) C (Holser et al., 1988). Isotopic studies have been helpful in untangling the complex cyclical history, particularly of sulfate rocks, and in attacking other questions of interest to geologists. Some of the questions for which isotopes are of potential value are:

(1) What was the time (age date) of primary deposition?

(2) Was the deposition from a marine, or from a non-marine source?

(3) Is the present mineralogy primary, or is it a diagenetic or later replacement of the original deposit?

The main features of stable isotopes in evaporite rocks have been reviewed by Holser (1979b), Arthur et al. (1983) and Pierre (1988, 1989). The most informative stable isotope system is that for S; the discussion will include the related events in the C isotope system. The $^{87}Sr/^{86}Sr$ ratio can be used as a tracer in the sedimentary cycle, directly on Recent sediments or on any Rb-poor mineral phase in any old geologic material. The discussion concludes with consideration of the isotope systems of H and O in water.

THE MARINE ISOTOPE SYSTEMS AND AGE CURVES

The exogenic geochemical cycles (sedimentary rocks, world ocean and atmosphere) of S and C are critically reviewed by Holser et al. (1988), and the corresponding cycle of Sr is discussed by Palmer and Elderfield (1985) and Wadleigh et al. (1985). The ocean is substantially mixed for each of these elements, so an isotope ratio for sea water reflects the balance of isotope transfers from rivers and ridges and out to the sediments. As indicated in figure 1, biological fractionations dominate the marine geochemistry of both S and C. The S of sea-water sulfate is fractionated by about -40 per mil by S-reducing bacteria that are ubiquitous in marine muds, and the sulfide product may be fixed as pyrite:

$$8(SO_4)^{2-} + 2Fe_2O_3 + H^+ + 15CH_2O \text{ -----> } 4FeS_2 + 15HCO_3^- + 8H_2O \quad (1)$$

This reaction favors light S (low $\delta^{34}S$), consequently an increase in the proportion of S reduced to sulfide raises the $\delta^{34}S$ of the

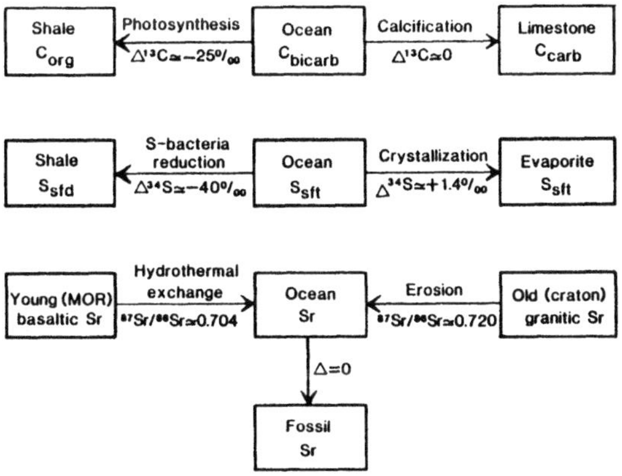

Figure 1: Schematic representation of the relations of the isotopes of C, S and Sr in the oceans (Holser et al., 1986)

sulfate remaining in the ocean; a decrease in S reduction reduces ^{34}S. The $\delta^{34}S$ can be sampled, anywhere in the world, by contemporary sulfate deposition (involving only a small fractionation), which thus maintains a record of secular shifts in sulfide-sulfate transfers. The record of such shifts during Phanerozoic time is embodied in the "sulfur isotope age curve" of figure 2a (Holser, 1984).

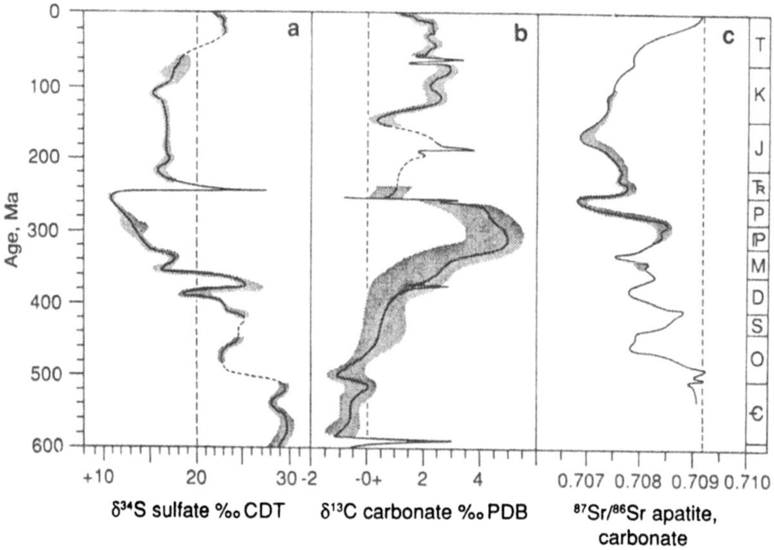

Figure 2: Age curves of (a) S isotopes in evaporite sulfate, (b) C isotopes in carbonates, and (c) Sr isotopes in carbonates and fossil apatite. Adapted from Holser (1984), taking into account new data for Carboniferous to Triassic for C (Popp et al., 1986; Baud et al., 1989) and Sr (Popp et al., 1986; Holser and Magaritz, 1987; Brookins, 1988), for Jurassic C (Jenkyns and Clayton, 1986), and for Tertiary C (Loutit et al., 1983; Vincent et al., 1985; Shackelton, 1986), S (Burdett et al., 1989) and Sr (Elderfield, 1986; McKenzie et al., 1988)

The photosynthesis of bicarbonate to organic C is accompanied by a fractionation that is similar (the organic product also prefers the light isotope) but smaller (Fig. 1). The fraction of C involved in organic synthesis is also smaller, and the consequent secular shifts in $\delta^{13}C$ of the residual bicarbonate (plus atmospheric CO_2 equilibrated with sea water) are correspondingly smaller (Fig. 2b). The marine bicarbonate is sampled (with negligible to minor fractionation) by limestones or dolostones, including those interlaminated with sulfate evaporites (e.g., Magaritz et al.,

1983).

In contrast to this record of high but variable $\delta^{34}S$ and $\delta^{13}C$ exhibited by the residual oxidized components of sea water, sulfate and carbonate of other origins will show very different and mostly much lighter values of $\delta^{34}S$ and $\delta^{13}C$. For example, oxidation of pyrite on an outcrop of shale will mimic without fractionation the very light S of the pyrite ($\delta^{34}S$ -30 to +5 per mil). The sulfate with low $\delta^{34}S$ may be fixed on the outcrop as gypsum, or move into the river system, perhaps ending as a lacustrine sulfate deposit. Similarly, oxidation of organic C retains a light isotope ratio ($\delta^{13}C$ -30 to -10 per mil) in the product CO_2. For example, the bicarbonate bypro-duct of sulfur reduction (Eq. 1) may reduce $\delta^{13}C$ in that milieu (e.g. Pierre and Rouchy, 1988).

Strontium is a different story: it does not fractionate, so its isotope ratio in sea water represents a balance of the input fluxes of Sr from old continental granites enriched in ^{87}Sr, and Sr exchanged from young basalts like those of the mid-ocean ridges depleted in ^{87}Sr (Palmer and Elderfield, 1985; Chaudhuri and Clauer, 1986; Veizer, 1989; Fig. 1). The sea water isotope ratio of Sr may be sampled by limestones, phosphate or sulfate rocks, the latter accept a few tenths percent Sr in place of Ca (Holser, 1979b). The compiled Sr-isotope age curve (Fig. 2c) presumably records the history of varying input sources with time.

The age curves for S, C and Sr are an empirical record of changes in sea water, verified where possible by analyses from two or more basins that would be expected to sample the world ocean independently of any local anomalies. Their validity has also been tested in varying degrees by computer modelling in which the empirical age data serve as input or comparative output (Holser et al., 1988), but these calculations depend on many assumptions and do not furnish robust constraints.

The $\delta^{18}O$ values in the sulfate ion of sea water were also evaluated as a balance of inputs and outputs, equilibration with ^{18}O of the water being so slow as to be ineffectual (Holser et al., 1979; Zak et al., 1980). On this basis, an age curve for $\delta^{18}O$ of sulfate in marine evaporites was proposed (Claypool et al., 1980). However, some studies of this system suggest that other factors such as local reduction-oxidation cycling, or diagenetic reaction with meteoric waters, may be dominant (Cortecci et al., 1981; Perry et al., 1982; Pierre, 1985; Richardson and Hansen, 1988). Until these additional complicating factors have been better evaluated, there is considerable question as to whether an age curve for $\delta^{18}O$ in marine sulfates can be defined and applied to geological problems.

APPLICATIONS OF THE AGE CURVES

An isotope age curve is simply our best estimate of the isotope ratio in (world-mixed) sea water, or in sediments directly deposited from that sea water. If a new evaporite of questionable provenance is sampled, its isotope ratio(s) may be compared with the pre-determined isotope age curve. If the stratigraphic age of the new samples is known, and the ratio fits the age curve reasonably well, the evaporite is probably of marine origin. If the age is known, and the ratio does <u>not</u> fit the age curve, the evaporite (and perhaps the sedimentary section as a whole), is probably non-marine. If the sulfate is not of marine origin, what is it? It could be from the erosion of an older sulfate evaporite within the drainage basin, in which case it would have retained the isotopic signature of the original marine sulfate rock. Or it could be sulfate fed to the drainage basin from the surface weathering of sulfides, in which case it should carry the light values of sulfide of igneous ($\delta^{34}S$ = 0 to +5 permil) or sedimentary ($\delta^{34}S$ = -30 to -5 per mil) origin; hydrothermal sulfides vary extremely widely (Hoefs, 1980).

Alternatively, the newly sampled evaporite might reasonably be supposed to be marine, as verified by its marine mineralogical sequence (Table 1) or by a normal bromide content of interbedded chloride deposits (Holser, 1979b). However, its age may be in question, perhaps between two or three choices, in such a case the decision may be made by a match of sample data with the appropriate age curve: "isotope chronostratigraphy" (Williams et al., 1988). The multi-valued aspect of the age curves (Fig. 2) means that isotope age curve dating is the more helpful as the range of choices is restricted, a characteristic that it shares with paleomagnetic stratigraphy. And of course its precision is best where the curve is most precisely defined by previous work, and where the isotope ratios are changing most rapidly (Elderfield, 1986). The precision is also improved if two isotope systems may be compared, as in the S-Sr plot of figure 3.

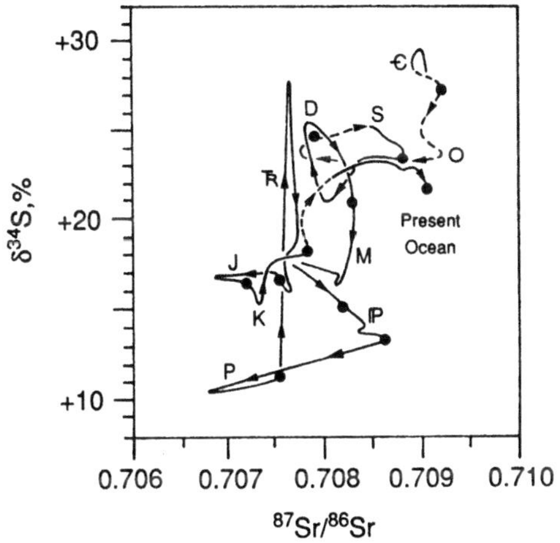

Figure 3: Cross-reference plot for marine isotope ratios of Sr vs. S (data from Fig. 2). Solid circles are boundaries of the geological periods, labelled as in Figure 2. Arrows point forward in time

Some examples may clarify the method. The Fort Dodge Gypsum covers an area of over 100 km^2 in Webster County, Iowa, in an inlier isolated by block faults. The gypsum overlies Pennsylvanian (Desmoinesian) shale and is covered by late Wisconsin glacial drift (Gwynne, 1957). No fossils had ever been found in the gypsum or in the rare interbedded shales, and for want of better information it had been lithologically correlated with the nearest evaporites, of Permian (Wolfcampian) age, occurring 300 km southwest in Kansas. Analyses of S isotopes in gypsum from a quarry face gave $\delta^{34}S$ = +16.1 $\pm$ 0.2 per mil. This value is quite definitely not Permian, and the deposit is probably of Mesozoic age (Middle Triassic to Early Cretaceous) (Fig. 2). In another case, deep drilling in the

Gulf of Paria, Trinidad, unexpectedly penetrated 700 m of anhydrite underlying Tertiary sediments. Sulfur isotope analyses determined the evaporite to be of pre-Late Cretaceous age (Claypool et al., 1980), and subsequently this was confirmed by micropaleontology of interbedded sediments.

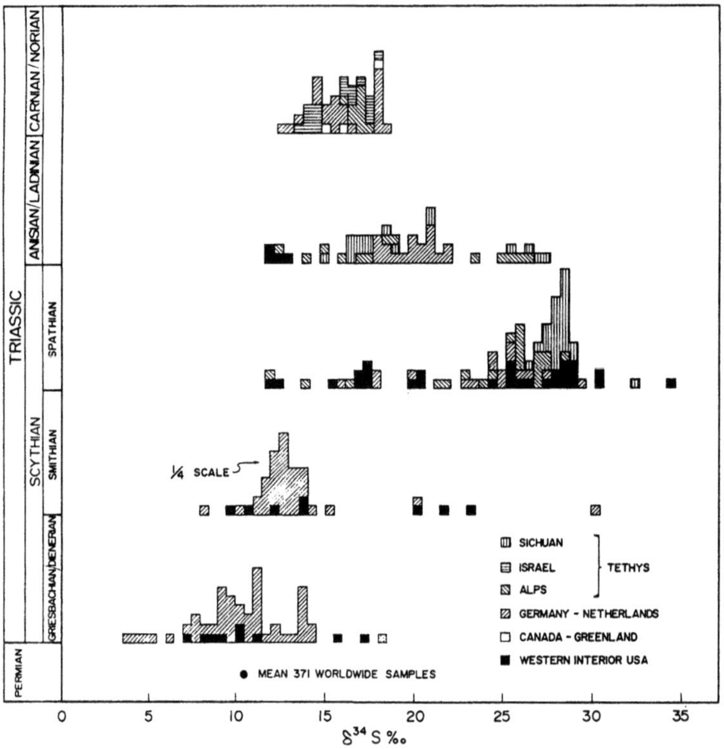

Figure 4: Detail of the sharp rise in the age curve for S in the Triassic. Each square is one analysis, except for the Smithian which is factored by 4; the time scale for the Scythian Stage (Lower Triassic) is greatly expanded. From Holser (1984)

In Europe the very widespread Zechstein evaporites of Late Permian age with $\delta^{34}S$ near +11 per mil are accompanied in some areas by evaporites of Early Triassic (Upper Buntsandstein) age with strongly contrasting $\delta^{34}S$ near +27 per mil (Fig. 4). In an overthrust zone Nielsen (1972) was able to verify by S isotopes that Zechstein gypsum actually <u>overlies</u> Buntsandstein (younger) evaporites. In a much more complex area of Alpine tectonics, complicated structures involving the two salts were untangled by their distinctive S isotope ratios (Pak and Schauberger, 1981; Pak, 1981;

Spotl 1988a, 1988b).

In the Tertiary (Oligocene-Miocene) Ebro Basin of northern Spain, a sequence of evaporites more than 3 km thick was known to be non-marine from the nature of the associated sediments. Within the Tertiary drainage Triassic evaporites had $\delta^{34}S = +14.7 \pm 0.7$ per mil, and the Tertiary gypsums varied from that level down to $\delta^{34}S = +9$ per mil, so it could be concluded that the source of the Tertiary evaporites was mainly from outcrops of Triassic gypsum, diluted in some cases by weathering of sulfide deposits (Birnbaum and Coleman, 1979). There is also isotopic evidence for recycling of Triassic evaporites into late Tertiary gypsum of the Granada Basin in southern Spain (Pierre, 1988).

The sulfate ion is a very stable unit, and consequently in the absence of local biological reduction, second-cycle brines or evaporites reflect their first-cycle derivation from the ocean. Thus, anhydrites filling porosity in the San Andres Limestone in Texas, and in the Lyons Sandstone in the Denver Basin were shown to be of Permian and not of later age (Holser and Kaplan, 1966). This same stability of the sulfate ion allows S isotopes to trace the origin of sulfate contents of sabkha brines (Robinson, 1985) formation waters (Muller et al., 1966), ground waters (Bassler, 1970), springs (Michel and Nielsen, 1977) and surface waters (Nielsen and Rambow, 1969; Hitchon and Krouse, 1972), and in particular to determine the relative importance of marine evaporites in their origin.

There is much current interest in the application of Sr isotope ratios to age determination and stratigraphy, especially in the Cenozoic where the age curve is both steep (Fig. 2c) and so well defined that a precision of ± 0.5 Ma has been claimed (Elderfield, 1986; McKenzie et al., 1988). Most attention has been given to the Sr isotope record in carbonate sediments, but a few results in sulfate rocks (F. Albarède, pers. comm., 1984; Zierenberg and Shanks, 1986; Brookins, 1988; M. A. Arthur, pers. comm., 1989) suggest a great potential for application in evaporites. Some limited data indicate that the Sr isotope ratio in restricted evaporite basins is especially susceptible to the influence of continental runoff (Clement and Holser, 1988; Muller et al., 1990).

The C-isotope age curve (Fig. 2b) has been applied to stratigraphy in the Late Permian, where paleontological correlations are inhibited by scarce and endemic fauna, especially in the evaporite basins that are characteristic of that time. Just below the base of the evaporite section in the Zechstein Basin, $\delta^{13}C$ in the interlaminated carbonates was found to rise abruptly to high positive values, and this high level of $\delta^{13}C$ continues through the overlying evaporites to the end of the Permian (Magaritz et al., 1981). Subsequently we looked for and found an analogous sharp rise just below the Castile evaporite in the Permian Basin of West Texas (Magaritz et al., 1983) and at the contact of the Gröden and Bellerophon Formations in the Tethyan section of the Southern Alps (Holser and Magaritz, 1985). In Texas the rise could be timed by varved sedimentation to have been accomplished in less than 5 ka (Magaritz et al., 1983), giving unprecedented precision to this world-wide correlation (Holser, Magaritz and Clark, 1986). In contrast to these primary carbonates of evaporitic series, diagenetic carbonates replacing the sulfate rocks show very low and variable $\delta^{13}C$ owing to their origin as organic carbon feeding the bacterial sulfate reduction (Eq. 1; Pierre and Rouchy, 1988).

There <u>are</u> problems. Residual variations remain both within and between contemporaneous basins, of 1 or 2 per mil for $\delta^{34}S$, and of 0.0002 for $^{87}Sr/^{86}Sr$ (Fig. 2). In some instances the residual variations of $\delta^{34}S$ have been attributed to intense sulfur reduction within the sampled basin (raising ^{34}S of sulfate locally), or to crystal fractionation in late-stage evaporites (decreasing $\delta^{34}S$) (Holser and Kaplan, 1966; Nielsen, 1972; Pierre, 1985). Residual variations in $^{87}Sr/^{86}Sr$ have been attributed to contamination by shaley radiogenic Sr or to diagenesis (Hess et al., 1986). But it is difficult to escape the conclusion that we do not yet understand these variations.

This skepticism is reinforced by a couple of dramatically anomalous results. Sulfur-isotope ratios in the lower Oligocene evaporites of the Rhinegraben vary widely from +10 to +23 per mil, but many from the potash facies are near +12 per mil (Nielsen,

1967), compared with values near +20 per mil in other marine Tertiary evaporites (Claypool et al., 1980). Nielsen (1972) suggests that during the Oligocene deep circulating brines dissolved Permian Zechstein evaporites 300 km distant, and eventually flowed into the Rhine Basin. During much of the evaporite deposition the sulfate content of the Permian brines was not sufficient to debase the incoming marine sulfate, but did have a strong influence during a sulfate-scarce late stage of evaporation. This proposal has yet to be confirmed by other evidence.

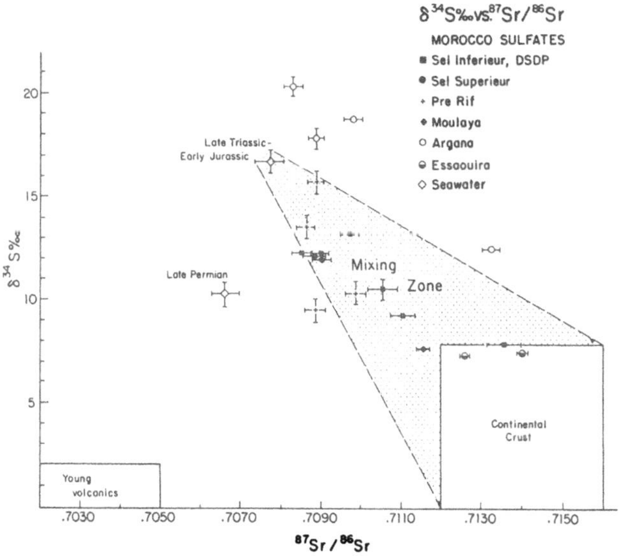

Figure 5: Isotope ratios of Sr vs. S in Moroccan evaporites. Analyses from various evaporite basins are shown relative to expected values for marine sources of Late Permian and Late Triassic-Early Jurassic ages, and for young volcanic and old continental sources. The shaded area represents possible values from mixing of Late Triassic-Early Jurassic marine waters with continental sources. From Clement and Holser (1988)

Another case proclaims the peril of conclusions with too little data. An initial geochemical study of an evaporite deep-sea core from a diapir off the coast of Morocco concluded that the evaporites were of Permian age, for a nice fit of its $\delta^{34}S = +10.1 \pm 0.9$ per mil, but in the face of much geological evidence that the salts were actually Upper Triassic to Lower Jurassic deposits (Holser et al., 1984). Fortunately geology prevailed in a later more detailed study of numerous evaporites from onshore Morocco, in which measurements of Sr isotopes were combined with those of S isotopes

(Clement and Holser, 1988). As shown in figure 5, some of the onshore evaporites had $\delta^{34}S$ down to less than +5 per mil, values not approached by marine waters of any geological age. These very low values of $\delta^{34}S$ were accompanied by levels of $^{87}Sr/^{86}Sr$ up to 0.714 - much higher than any known marine Sr (Fig. 2c). A plot of S against Sr isotopes shows a rough negative correlation (Fig. 5), which indicates a varying mixture in the evaporites of non-marine components and Upper Triassic-Lower Jurassic sea water. Apparently the offshore diapir mimicked the $\delta^{34}S$ of Permian sea water by a chance admixture of light non-marine S. Even so, it was difficult to construct a believable model that accounted for the very large volume of some of the salt deposits (and their Br content) as well as the S and Sr isotope data (Clement and Holser, 1988).

THE WATER SYSTEM

The isotope ratios of H and O co-vary in the H_2O system, and considered together they can supply interesting information about the evaporation, mixing, and crystallization of water. Figure 6 outlines some of the essential features of the water isotope system in evaporitic settings, in a plot of δD (= ^2H) against $\delta^{18}O$, on a scale relative to Standard Mean Ocean Water (SMOW). At the center of this system is the ocean, which is both the standard of measure (SMOW) and the principal reservoir. Present-day sea water is fairly well mixed and near SMOW in isotope ratio, although $\delta^{18}O$ may be lowered by a few per mil in coastal waters that have been freshened by, e.g., glacial melt (reviewed in Hoefs, 1980).

In the global water cycle, water evaporated from the sea is lighter in both δD and $\delta^{18}O$, and by successive evaporations both become lighter as the air masses sweep to higher elevations and higher latitudes. By rainout the resulting meteoric water lies somewhere along the line shown in figure 6, with polar latitudes in the lower left. As a consequence, liquid water remaining after evaporation increases, at first, in both δD and $\delta^{18}O$. The result complex, depending on the humidity and isotope ratios of D and O

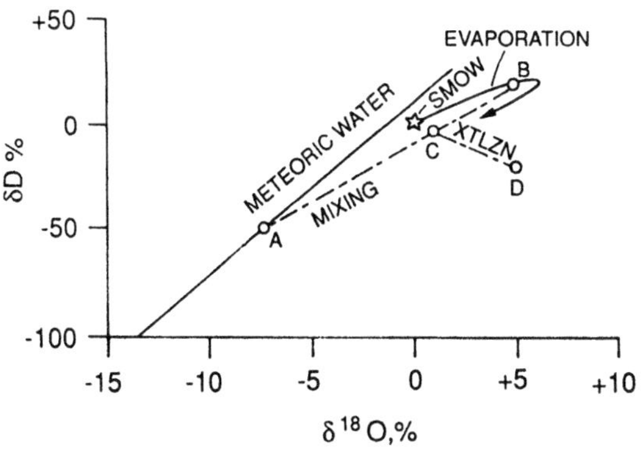

Figure 6: Schematic representation of the isotope systems for water in evaporative regimes. The evaporative loop for ocean water is as measured by Nadler and Magaritz (1980), but this is only one example of possible trajectories (Sofer and Gat, 1975). An example is given of a mixing line between an evaporated brine of composition "B" (near gypsum precipitation) and meteoric water of composition "A", in such proportions as to give a composition "C". When such a mixed water crystallizes in gypsum, δD will be lighter and $\delta^{18}O$ will be heavier at composition "D"

in the ambient air, as well as on the cationic composition of the solution, which reduces the activity of H_2O and its isotopic components (Sofer and Gat, 1975). A typical course of evaporation of sea water on the δD-$\delta^{18}O$ plot is shown in figure 6: at first it moves to more positive δD and $\delta^{18}O$, but along a slope shallower than the meteoric water line. Then it reaches equilibrium (in the $CaSO_4$ facies), first for δD and then for $\delta^{18}O$, and finally in concentrated brines (halite facies) it loops back to lower values of both δD and $\delta^{18}O$. The looped evaporation curve in figure 6 is only one observed example (Nadler and Magaritz, 1980) of a variety of possible courses of δD- $\delta^{18}O$ during evaporation under different conditions (Sofer and Gat, 1975)[1]. An evaporite brine with a δD and $\delta^{18}O$ such as at point "B" in figure 6, may be lightened in both isotope ratios by the effect of meteoric water, such as at point "A", to give a resulting isotopic composition at point "C". Incorporation of water molecules into the structure of a gypsum crystal fractio-

1 The evaporative loop illustrated by Holser (1979b, p. 329) and extensively cited in the interpretations by Knauth and Beeunas (1986) was purely diagramatic.

nated in $\delta^{18}O$ by +4 per mil and δD by -16 per cent, to reach point "D" in the typical example displayed in figure 6. Vice versa, a plot of δD and $\delta^{18}O$ in a gypsum sample may be translated by the same vector to the upper left, to deduce the δD-$\delta^{18}O$ composition of the "mother liquid" from which it had been crystallized.

Interstitial brines at the inner edge of a modern salt pan lie generally in the region of point "C", and are interpreted as resulting from mixing of evaporative brines with continental ground waters (Pierre and Ortlieb, 1980; Pierre et al., 1984; Knauth and Beeunas, 1986). Analyses of gypsums of Tertiary age from the Mediterranean region (Rouchy and Pierre, 1979; Pierre and Fontes, 1979) and from Poland (Halas and Krouse, 1982) plot their mother liquid on the meteoric water line, or slightly below it in the mixing area. Such results may be interpreted as either: primary crystallization from brines of mixed marine-meteoric heritage (as in the above-cited modern examples), or re-gypsification in the presence of later meteoric ground waters. Some of the gypsums from Poland scatter along a mixing trend that intercepts the meteoric water line at about δD = -110 per mil, $\delta^{18}O$ = -15 per mil, which is so very low that the corresponding meteoric water can only have been produced by a glacial climate. Hence the Ca sulfate that was precipitated in the Miocene was most recently gypsified during the Pleistocene (Halas and Krouse, 1982; see also Bath et al., 1987).

Alternatively the isotopic composition of old evaporite brines may be studied by analyzing fluid inclusions in salt rock. Microscopic inclusions of brine are trapped during primary crystallization of halite, especially along cubic growth planes, where they may be concentrated in a "chevron" texture. Subsequently these microscopic inclusions are reconstituted into mm-sized cubic cavities during re-crystallization of the halite, and under some conditions the brine is expelled to the intercrystalline grain boundaries (Roedder, 1984). In view of the variety of types of inclusions and other sites where water may be found in the salt rock, a clean unfractionated sample of inclusion fluid for isotopic analysis is difficult to recover and validate.

Despite these difficulties, interesting results were obtained

(J. Wright, pers. comm., 1989). Chemical compositions of fluid inclusions in halite from the Permian of Texas and New Mexico (Roedder, 1984; Knauth and Beeunas, 1986) generally lie in the same field as the brines of modern salt pans: under the hook of the evaporation curve and scattered down and to the left along possible mixing lines. Compositions of the primary chevron inclusions mostly lie closer to the evaporation curve. The isotope data have been interpreted as from primary Permian waters - a mixture of evaporated brines with contemporary meteoric waters (Knauth and Beeunas, 1986), but the variety of compositions found by analysis of fluid inclusions from the same formations cautions that the origin may be more complex and varied (Roedder, 1984; Holland et al., 1986; Roedder et al., 1987; Stein and Krumhansl, 1988). The resolution of these questions is critical for the evaluation of the integrity of salt rock as vaults for the storage of radioactive waste. Fluid inclusions are more rarely observed in anhydrite (McKibben et al., 1988), but isotope analyses are lacking.

SAMPLING PROTOCOLS

The application of isotope methods to geological problems of sulfate and chloride rocks imposes certain requirements on sampling. Sampling for comparison with the S isotope age curve is not particularly sensitive to the evaporative facies (Table 1) in which the sulfate is deposited. For routine analysis about 0.1 g of anhydrite is required, which means that a small hand specimen of anhydrite or gypsum rock suffices. A hand specimen is even sufficient for salt rock, as most such material contains at least 5 percent anhydrite. For this application sampling is generally insensitive to the form of Ca sulfate - primary gypsum or anhydrite, anhydrite replacing gypsum, or gypsum replacing anhydrite. The sampling is also relatively insensitive to the textural form of the gypsum or anhydrite: laminated, nodular, etc. In cores, anhydrite may be recognized more easily if the surface of the core is altered to a dead white adherent coat of gypsum - by exposure to water for a few hours or to a 4 M solution of K_2SO_4 for 15 or 20 minutes.

The most important precaution is to avoid areas of sulfate reduction, which are common in salt dome cap rock or in similar alteration near the outcrop of bedded sulfate rocks. The products of such reduction are sulfides and native S of low $\delta^{34}S$, leaving the remaining gypsum with high $\delta^{34}S$; contrarily, if the sulfide is back-oxidized in ground water to sulfate rather than only to native S, gypsum newly crystallized from such sulfate may be very light. Fortunately the effects of sulfate reduction are localized, so it is only necessary to move sampling to, e.g., the dome rock itself. For similar reasons it is well to avoid shaley or marly interbeds, which are likely to contain pyrite that may oxidize to sulfate on the outcrop.

The formation should be sampled at five or ten levels, and the computed mean value, discounting outlying values such as occasionally occur, should be used to calculate the age from the time-dependent isotope variation curve. For the most part, the precisions of preparation and mass spectrometry are better than the variations of up to 1 per mil or a little more that occur naturally within a formation, consequently it is usually economical to analyze more samples rather than making duplicate preparations from each sample.

Sampling for the Sr isotope age curve presents an additional problem. In order to avoid contamination by radiogenic Sr decayed from Rb in old detrital silicates (particularly clay or mica), it is recommended that samples have less than five percent insoluble residue of detritus. Alternatively the Sr isotope ratio in the sulfate phase may be derived by measuring both the isotope ratio and amount of insoluble residue in a series of samples, and extrapolating to zero insoluble residue (Lord et al., 1988). For sulfate rocks this is not as common a problem as it is with the carbonate rocks that are more usually sampled for Sr isotopes.

Sampling for analysis of fluid inclusions is concerned with salt rocks; gypsum and anhydrite have been little studied. A preliminary search through available material serves two purposes: (1) it allows the analytical procedure to concentrate on those samples richest in inclusions, and (2) it provides a qualitative judgement as to whether the inclusions are primary (microscopic with a chevron texture), recrystallized (cubes 0.1 mm or larger randomly distributed within a halite crystal), or grain-boundary inclusions. Inclusions may be encountered by chance under the microscope in thin (or thick) sections. However, any salt rock may be prepared easily for routine examination by the following procedure. Flatten the side of a length of core or hand specimen using large flat double-cut steel files - starting with a bastard cut and smoothing with a second cut. Then wipe the surface with glycerine, mineral oil or an index oil, and look "into" the halite crystals with hand lens or binocular microscope. Observation may be improved by temporarily affixing a large cover glass.

FINAL REMARKS

At the present stage the most important application of isotope geochemistry to geological problems of sulfate and chloride rocks seems to be age determination and stratigraphic correlation with the isotope age curves of S and Sr. Future progress in this application will be aided by two new emphases: (1) the refinement of the S isotope age curve by determination of $\delta^{34}S$ in stratigraphically well-controlled carbonates (Burdette et al., 1989), and (2) the cross-correlation of isotope analyses of both S and Sr (e.g., Clement and Holser, 1988; Figs. 3, 5). Indeed, the potential for unequivocal answers to present ambiguities should be greatly enhanced by parallel studies of both isotope and chemical characteristics of fluid inclusions and their evaporite systems. Isotope studies alone will not solve the complex geological problems that confront us.

REFERENCES

Arthur M.A. (ed.) (1983) Stable isotopes in sedimentary geology. Soc. Expl. Paleont. Miner. Short Course Notes, 10.

Bassler R. (1970) Hydrogeologische, chemische und Isotopen-Untersuchungen der Grubenwasser des Ibbenburener Steinkohlenreviers. Zeitschr. Deutsch. Geol. Gesellsch. Sonderheft Hydrogeol. Hydrogeochem., 209-286.

Bath A.H., Darling W.G., George, I. A. and Milodowski A.E. (1987) $^{18}O/^{16}O$ and $^{2}H/^{1}H$ changes during progressive hydration of a Zechstein anhydrite formation. Geochim. Cosmochim. Acta, 51, 3113-3118.

Baud A., Magaritz M. and Holser W.T. (1989) Permian-Triassic of Tethys: Carbon isotope studies. Geol. Rundsch., 78, 649-677.

Birnbaum S. J. and Coleman M. (1979) Source of sulphur in the Ebro Basin (Northern Spain) Tertiary nonmarine evaporite deposits as evidenced by sulphur isotopes. Chem. Geol., 25, 163-168.

Brookins D.G. (1988) Seawater $^{87}Sr/^{86}Sr$ for the Late Permian Delaware Basin evaporites (New Mexico, U.S.A.). Chem. Geol., 69, 209-214.

Burdett J.W., Arthur M.A. and Richardson M. (1989) A Neogene seawater sulfur isotope age curve from calcareous pelagic microfossils. Earth Plan. Sci. Lett., 94, 189-198.

Chaudhuri S. and Clauer N. (1986) Fluctuations of isotopic composition of strontium in seawater during the Phanerozoic eon. Chem. Geol. (Isot. Geosci. Sect.),. 59, 293-303.

Claypool G.E., Holser W.T., Kaplan I.R., Sakai H. and Zak I. (1980) The age curves of sulfur and oxygen isotopes in marine sulfate and their mutual interpretation. Chem. Geol., 28, 199-260.

Clement G.P. and Holser W.T. (1988) Geochemistry of Moroccan evaporites in the setting of the North Atlantic Rift. Jour. Afr. Earth Sci., 7, 375-383.

Cortecci G., Reyes E., Berti G. and Casati P. (1981) Sulfur and oxygen isotopes in Italian marine sulfates of Permian and Triassic ages. Chem. Geol., 34, 65-79.

Elderfield H. (1986) Strontium isotope stratigraphy. Palaeogeogr. Palaeoclimat. Palaeoecol., 57, 71-90.

Evans R. and Kirkland D.W. (1988) Evaporitic environments as a source of petroleum. In: Evaporites and Hydrocarbons, Schreiber B.S. (ed.), Columbia University Press, New York, 256-299.

Gwynne C.S. (1957) Gypsum at Fort Dodge, Iowa. Earth Sci., 10, 5, p. 18, 20, 22.

Halas S. and Krouse H.R. (1982) Isotopic abundances of water of crystallization of gypsum from the Miocene Evaporite Formation, Carpathian Foredeep, Poland. Geochim. Cosmochim. Acta, 46, 293-296.

Hardie L.A. (1984) Evaporites: Marine or non-marine? Am. Jour. Sci., 284, 193-240.

Hess J., Bender M.L. and Schilling J.G. (1986) Evolution of the ratio of strontium-87 to strontium-86 in seawater from Cretaceous to present. Science, 231, 979-984.

Hitchon B. and Krouse H.R. (1972) Hydrogeochemistry of the surface waters of the Mackenzie River drainage basin, Canada--III. Stable isotopes of oxygen, carbon and sulphur. Geochim. Cosmochim. Acta, 36, 1337-1357.

Hoefs J. (1980) Stable Isotope Geochemistry, 2nd ed., Springer-Verlag, Berlin, 208 p.

Holland H.D., Lazar B. and McCaffrey M. (1986) Evolution of the atmosphere and oceans. Science, 320, 27-33.

Holser W.T. (1979a) Mineralogy of evaporites. Miner. Soc. Am. Rev. Miner., 6, 211-294.

Holser W.T. (1979b) Trace elements and isotopes in evaporites. Miner. Soc. Am. Rev. Miner., 6, 295-346.

Holser W.T. (1979c) Rotliegend evaporites, Lower Permian of northwestern Europe, geochemical confirmation of the non-marine origin. Erdöl Kohle Erdgas, 32, 159-162.
Holser W.T. (1984) Gradual and abrupt shifts in ocean chemistry during Phanerozoic time. In: Patterns of Change in Earth Evolution, Holland H.D. and Trendall A.F. (eds.), Dahlem Konferenzen, Springer-Verlag, Berlin, 123-142.
Holser W.T., Hay W.W., Jory D.E. and O'Connell W.J. (1980) A census of evaporites and its implications for oceanic geochemistry. Geol. Soc. Am., Abstr. with Progr., 12, 449.
Holser W.T. and Kaplan I.R. (1966) Isotope geochemistry of sedimentary sulfates. Chem. Geol., 1, 93-135.
Holser W.T., Kaplan I.R., Sakai H. and Zak I. (1979) Isotope geochemistry of oxygen in the sedimentary sulfate cycle. Chem. Geol., 25, 1-17.
Holser W.T. and Magaritz M. (1985) The Late Permian carbon isotope anomaly in the Bellerophon Basin, Carnic and Dolomite Alps. Jahrb. Geol. Bundesanst., Wien, 128, 75-82.
Holser W.T. and Magaritz M. (1987) Events near the Permian Triassic boundary. Mod. Geol., 11, 155-180.
Holser W.T., Magaritz M. and Clark D.L. (1986) Carbon-isotope stratigraphic correlations in the Late Permian. Am. Jour. Sci., 286, 390-402.
Holser W.T., Magaritz M. and Wright J. (1986) Chemical and isotopic variations in the world ocean during Phanerozoic time. In: Global Bio-Events, Walliser O. (ed.), Springer-Verlag, Berlin, 63-74.
Holser W.T., Saltzman E.S. and Brookins D.G. (1984) Geochemistry and petrology of evaporites cored from a deep-sea diapir at Site 546 offshore Morocco. Init. Repts. Deep Sea Drill. Proj., 79, 509-540.
Holser W.T., Schidlowski M., Mackenzie F.T. and Maynard J.B. (1988) Geochemical cycles of carbon and sulfur. In: Chemical Cycles in the Evolution of the Earth, Gregor C.B., Garrels R.M., Mackenzie F.T. and Maynard J.B. (eds.), John Wiley and Sons, New York, 105-173.
Jenkyns H.C. and Clayton C.J. (1986) Black shales and carbon isotopes in pelagic sediments from the Tethyan Lower Jurassic. Sedimentology 33, 87-106.
Kendall A.C. (1984) Evaporites. In: Facies Models. R.G. Walker (ed.), 2nd ed., Geol. Assoc. of Canada, Toronto, 259-296.
Knauth L.P. and Beeunas M.A. (1986) Isotope geochemistry of fluid inclusions in Permian halite with implications for the isotopic history of ocean water and the origin of saline formation waters. Geochim. Cosmochim. Acta, 50, 419-433.
Lerche I. and O'Brien J.J. (eds.) (1987) Dynamical Geology of Salt and Related Structures, Academic Press, Orlando, 832 p.
Lord B.K., Jones L.M. and Faure G. (1988) Evidence for the existence of the Gondwana ice sheet in the ^{18}O depletion of carbonate rocks in the Permian formations of the Transantarctic Mountains. Chem. Geol. (Isot. Geosc. Sect.), 72, 163-171.
Loutit T.S., Pisias N.G. and Kennett J.P. (1983) Pacific Miocene carbon isotope stratigraphy using benthic foraminifera. Earth Plan. Sci. Lett., 66, 48-62.
Magaritz M., Anderson R.Y., Holser W.T., Saltzman E.S. and Garber J. (1983) Isotope shifts in the Late Permian of the Delaware Basin, Texas, precisely timed by varved sediments. Earth Plan.

Sci. Lett., 66, 111-124.
Magaritz M., Turner P. and Kading K.C. (1981) Carbon isotope change at the base of the upper Permian Zechstein sequence. Geol. Jour., 16, 243-254.
McKenzie J.A., Hodell D.A., Mueller P.A. and Mueller D.W. (1988) Application of strontium isotopes to late Miocene-early Pliocene stratigraphy. Geology, 16, 1022-1025.
McKibben M.A., Williams A.E. and Okubo S. (1988) Metamorphosed Plio-Pleistocene evaporites and the origins of hypersaline brines in the Salton Sea geothermal system, California: Fluid inclusion evidence. Geochim. Cosmochim Acta, 52, 1047-1056.
Michel G. and Nielsen H. (1977) Schwefel-Isotopenuntersuchungen an Sulfaten ostwestfalischer Mineralwasser. Fortschr. Geol. Rheinland Westfal., 26, 185-227.
Müller D.W., Mueller P.A. and McKenzie J.A. (1990) Strontium isotope ratios as fluid tracers in Messinian evaporites of the Tyrrhenian Sea (western Mediterranean Sea). Proc. ODP, Sci. Results 107, 603-614.
Müller G., Nielsen H. and Ricke W. (1966) Schwefel-Isotopen-Verhältnisse in Formationswassern und Evaporiten Nord- und Suddeutschlands. Chem. Geol., 1, 211-220.
Nadler A. and Magaritz M. (1980) Studies of marine solution basins--isotopic and compositional changes during evaporation. In: Hypersaline Brines and Evaporitic Environments, Nissenbaum A. (ed.), Elsevier, Amsterdam, 115-129.
Nielsen H. (1967) Sulphur isotopes in the Rhinegraben evaporite sulphates. Abhandl. Geol. Landesamt Baden-Wurttemberg, 6, 27-29.
Nielsen H. (1972) Sulphur isotopes and the formation of evaporite deposits. In: Geology of Saline Deposits, Richter-Bernburg G. (ed.), UNESCO, Paris, 91-102.
Nielsen H. and Rambow D. (1969) S-Isotopenuntersuchungen an Sulfaten hessischer Mineralwasser. Notizbl. Hess. Landes-Amt Bodenforsch., 97, 352-366.
Pak E. and Schauberger O. (1981) Die geologische Datierung der ostalpinen Salzlagerstatten mittels Schwefelisotopenuntersuchungen. Verhandl. Geol. Bundesanst., Wien, 1981, 185-192.
Palmer M.R. and Elderfield H. (1985) Sr isotope composition of sea water over the past 75 Myr. Nature, 314, 526-528.
Perry E.C.Jr., Grundl T. and Gilkeson R.H. (1982) H, O, S isotopic study of the ground water in the Cambrian-Ordovician aquifer system of northern Illinois. In: Isotope Studies of Hydrologic Processes, Perry E.D. and Montgomery C. (ed.), Northern Illinois University Press, De Kalb, 35-43.
Pierre C. (1985) Isotopic evidence for the dynamic redox cycle of dissolved sulphur compounds between free and interstitial solutions in marine salt pans. Chem. Geol., 53, 191-196.
Pierre C. (1988) Applications of stable isotope geochemistry to the study of evaporites. In: Evaporites and Hydrocarbons. Schreiber B.C. (ed.), Columbia University Press, New York, 300-344.
Pierre C. (1989) Sedimentation and diagenesis in restricted marine basins. In: Handbook of Environmental Isotope Geochemistry, 3, The marine Environment, Fritz P. and Fontes J.C. (eds.), Elsevier Sci. Pub., Amsterdam, 257-315.
Pierre C. and Fontes J.C. (1979) Oxygène 18, carbone 13, deutérium et soufre 34: marqueurs géochimiques de la diagenèse et du paléomilieu évaporitiques du Messinien de la Méditerranée.

Bull. Mus. Nat. Hist. Nat., Paris, (4), C, 1, 3-18.
Pierre C. and Ortlieb L. (1980) Sédimentation et diagenèse dans trois lagunes évaporitiques de Basse Californie (Mexique). Sci. Terre, Nancy, 14, 129-156.
Pierre C., Ortlieb L. and Person A. (1984) Supratidal evaporitic dolomite at Ojo de Liebre Lagoon: Mineralogical and isotopic arguments for primary crystallization. Jour. Sediment. Petrol., 54, 1049-1061.
Pierre C. and Rouchy J.M. (1988) Carbonate replacements after sulfate evaporites in the Middle Miocene of Egypt. Jour. Sediment. Petrol., 58, 446-456.
Popp B.N., Anderson T.F. and Sandberg P.A. (1986) Brachiopods as indicators of original isotopic composition in some Paleozoic limestones. Geol. Soc. Am. Bull., 97, 1262-1269.
Popp B.N., Podosek F.A., Brannon J.C., Anderson T.F. and Pier J. (1986) $^{87}Sr/^{86}Sr$ ratios in Permo-Carboniferous sea water from the analyses of well-preserved brachiopod shells. Geochim. Cosmochim. Acta, 50, 1321-1328.
Richardson S.M. and Hansen K.S. (1988) Stable isotopes in the sulfate evaporites from southeastern Iowa: A clue to the history of groundwater? Eos, 69, 1515.
Roedder E. (1984) The fluids in salt. Am. Miner., 69, 413-439.
Roedder E., d'Angelo W.M., Dorrzapf A.F.Jr. and Aruscavage P.J. (1987) Composition of fluid inclusions in Permian salt beds, Palo Duro Basin, Texas, U.S.A. Chem. Geol., 61, 79-90.
Rouchy J.M. and Pierre C. (1979) Données sédimentologiques et isotopiques sur les gypses des séries évaporitiques messiniennes d'Espagne méridionale et de Chypre. Rev. Géol. Dynam. Géog. Phys., 21, 267-280.
Schreiber B.C. (edit.) (1988) Evaporites and Hydrocarbons. Columbia University Press, New York, 475 p.
Shackelton N.J. (1986) Paleogene stable isotope events. Palaeogeogr. Paleoclimat. Palaeoecol., 57, 91-102.
Sofer Z. and Gat J.R. (1975) The isotope composition of evaporating brines: Effect of the isotopic activity ratio in saline solutions. Chem. Geol., 26, 179-186.
Spotl C. (1988a) Sedimentologisch-fazielle Analyse tektonisierter Evaporitserien - eine Fallstudie am Beispiel des alpinen Haselgebirges (Permoskyth, nordliche Kalkalpen). Geol. Paläont. Mitt., Innsbruck, 15, 59-69.
Spotl C. (1988b) Schwefelisotopendatierungen und fazielle Entwicklung permoskythischer Anhydrite in Salzbergbauen von Durrnberg/Hallein und Hallstatt (Oesterreich). Mitt. Ges. Bergbaustud. Oesterr., 34/35, 209-229.
Stein C.L. and Krumhansl J.L. (1988) A model for the evolution of brines in salt from the lower Salado Formation, southeastern New Mexico. Geochim. Cosmochim. Acta, 52, 1037-1046.
Veizer J. (1989) Strontium isotopes in seawater through time. Ann. Rev. Earth Plan. Sci., 17, 141-167.
Vincent E., Killingley J.S. and Berger W.H. (1985) Miocene oxygen and carbon isotope stratigraphy of the tropical Indian Ocean. Geol. Soc. Am. Mem., 163, 103-130.
Wadleigh M.A., Veizer J. and Brooks C. (1985) Strontium and its isotopes in Canadian rivers: Fluxes and global implications. Geochim. Cosmochim. Acta, 49, 1727-1736.
Warren J.K. (1989) Evaporite Sedimentology. Prentice-Hall, Englewood Cliffs, New Jersey, 285 p.
Williams D.F., Lerche I. and Full W.E. (1988) Isotope Chronostra-

tigraphy: Theory and Methods. Academic Press, San Diego, California, 345 p.
Zak I., Sakai H. and Kaplan I.R. (1980) Factors controlling the ^{18}O and ^{34}S isotope ratios of ocean sulfates, evaporites and interstitial sulfates from modern deep sea sediments. In: Isotope Marine Chemistry, Uchida Rokakuho Publ. Co. Ltd., Tokyo, 339-373.

HISTORY OF MARINE EVAPORITES: CONSTRAINTS FROM RADIOGENIC ISOTOPES

Sam CHAUDHURI(*)
Department of Geology, Kansas State University
Manhattan, Kansas 66506 (USA)

and

Norbert CLAUER
Centre de Géochimie de la Surface
1 rue Blessig, 67084 Strasbourg (France)

Ancient marine evaporite deposits represent important records of past marginal marine environments and climatic conditions, and therefore bear witness to chemical history of the past oceans. Borchert and Muir (1964) and Braitsch (1971) gave a list of marine evaporite minerals and summarized the tectonic and climatic conditions favorable for the formation of such deposits. Holland (1984) elaborated on the variations of the major chemical composition of sea water during the Phanerozoic Era and concluded from mineralogic and chemical data of marine evaporite deposits that the major composition of sea water remained nearly constant during the last 600 million years. This constancy contrasts with prominent shifts in the isotopic compositions of S and O in dissolved sulfate and in the isotopic compositions of marine Sr, Nd and C. The trend and significance of the isotopic excursions of these elements have been

(*) Presently at the Centre de Géochime de la Surface, Strasbourg (France)

described in companion contributions in this book and by Holser (1966), Holser et al. (1979), Claypool et al. (1980), Holland (1984), Shaw and Wasserburg (1985), Faure (1986), and Stille and Fischer (1990).

The present article gives a review of available information on Rb-Sr, K-Ca and K-Ar isotopic investigations of marine evaporite deposits to emphasize that radiogenic isotopes provide increased understanding of both the depositional and the post-depositional history of these marine sediments. Radiogenic isotope data may also help to determine the state of isolation of evaporite deposits from recent groundwater activities, as such information is critical to the assessment of the suitability of these deposits as repository of high-level radioactive wastes. Furthermore, analysis of radiogenic isotopes can lead to improved understanding of the relation between physical deformation and chemical modification of evaporite deposits and consequently can enhance knowledge of the development of salinity in continental subsurface waters. Finally, isotopic data can, in some instances, be useful in studies of stratigraphic correlations and tectonometamorphic evolutions of evaporite deposits.

PRIMARY STRONTIUM ISOTOPIC SIGNATURES OF EVAPORITES

The $^{87}Sr/^{86}Sr$ of modern sea waters has been found to be widely uniform at about 0.70901 (Burke et al., 1982), but the past ocean waters experienced excursions in the Sr isotopic composition through time in response to irregular changes in the balance between contribution of Sr with high $^{87}Sr/^{86}Sr$ from continents and input of Sr with low $^{87}Sr/^{86}Sr$ from submarine activities. Since the work of Peterman et al. (1970) on paleontologically well-dated carbonate fossils, many studies, especially that of Burke et al. (1982), have contributed to the knowledge of the Sr isotopic excursion during the Phanerozoic (see figure in Holser's companion contribution). This secular variation of the marine $^{87}Sr/^{86}Sr$ ratio during the

Phanerozoic can serve as a basis for the study of chemical evolution of ancient marine evaporite deposits.

Models for deposition of ancient marine evaporite deposits most frequently invoke the popular bar theory of Ochsenius (1877), or some modified version of it, which postulate a continuous supply of open ocean water across a barrier into an evaporating basin adjacent to the coast. As an alternative to the barred-basin model, Hsü (1972) advocated that ancient giant saline deposits could have formed from dessication of small ocean basins. Following the popular model of deposition in a barred basin, marine evaporite minerals should have the same Sr isotopic composition as open ocean water at the time of deposition, and any measurable deviation from this specific marine Sr isotopic value for an evaporite deposit should be related to post-depositional chemical modification. However, this generalization of the same Sr isotopic value for both water of the open ocean and water of an evaporating basin along a continental margin should be viewed with some caution. Indeed, if saline deposits formed from dessication of small ocean basins, as suggested by Hsü (1972), their primary Sr isotopic signatures could be different from that of contemporaneous open ocean water due to increased influence of input of continental Sr in the depositional basin. Moreover, the rate of precipitation can be faster than the rate of mixing of dissolved Sr in an evaporating basin dominated by restricted water circulation, in which case the evaporitic minerals can be spatially and temporally varied in their Sr isotopic compositions due to varied inputs of continental Sr in the basin.

STRONTIUM ISOTOPIC SIGNATURES OF ANCIENT BEDDED EVAPORITES

For lack of data from any representative modern bedded evaporite deposits, no conclusive argument can be made whether or not Sr isotopic difference could exist between the waters of evaporating basins and the open ocean. Nevertheless, the results of two studies suggest that primary Sr isotopic signatures of evaporites may not be the same as that of contemporaneous open ocean water. Clauer (1976) found that two sulfate deposits at different depths in the Messinian (Late Miocene) evaporite sequence of the Mediterranean Sea to be isotopically similar at an $^{87}Sr/^{86}Sr$ ratio of 0.70814, in clear contrast to 0.70890 for the contemporaneous open ocean water. Clauer proposed that the Sr isotopic deviation of the evaporite deposits could have been induced by the introduction of continental waters into the evaporating basin, which also appeared to be reflected in the O isotope composition of the sulfate deposits. Recently, we measured the Sr isotopic composition of three other Messinian sulfate deposits from different depths of the same Messinian sequence. The measurements were precise enough, with values ranging from 0.708730 $\pm$ 0.000008 to 0.708781 $\pm$ 0.000009, to suggest that the Sr isotopic compositions of the evaporite sulfates are slightly varied and that, as proposed by Clauer (1976), the Sr isotopic composition of the Messinian evaporitic deposits in the Mediterranean Sea differed from that of the contemporaneous open ocean water. The difference most likely relates to the paleohydrology of the evaporating basin and not to any post-depositional modification. Existence of a Sr isotopic difference between water of an evaporating basin and open sea can be defended on evidences presented by Aaberg and Wickman (1987) on Sr isotopic compositions of waters of the Bothnian Bay which has a restricted circulation relative to the open Baltic Sea. The authors found that the $^{87}Sr/^{86}Sr$ ratios of the water in the bay were between 0.70935 and 0.70951. The ^{87}Sr-enrichment of the Bothnian water relative to that of the open ocean is induced by the high $^{87}Sr/^{86}Sr$ ratios between 0.71772 and 0.73056 for the tributaries around the bay.

Evaporite deposits are known to be highly susceptible to post-depositional physical and chemical modifications. Hence, minerals in ancient evaporite deposits, especially in diapiric evaporite deposits, are suspect of varied isotopic modifications. The potential for modification of the depositional Sr isotopic signatures is also high for those mineral deposits with low Sr contents, so that halite or other highly soluble salt deposits known for their low Sr contents can be subject to greater degree of post-depositional Sr isotopic modifications than gypsum or anhydrite deposits which are known to be very high in Sr contents. Several recent studies on Sr isotopic compositions of sulfate and associated chloride, and mixed salts in ancient diapiric and bedded evaporite deposits, have been made to explore the use of Sr isotopes as tracers to constrain the evolutionary history of ancient evaporite deposits.

Chaudhuri et al. (1986) reported that the $^{87}Sr/^{86}Sr$ ratios of marine-bedded anhydrite and halite samples from Lower Permian evaporite deposits in central Kansas, to be varied between 0.70815 and 0.70885, and between 0.70848 and 0.70897, respectively. The Sr isotopic ratios for the anhydrite samples are higher than that of contemporaneous sea water, whose values ranged between 0.70740 and 0.70810, as suggested by the refined sea-water curve of Burke et al. (1982). The Sr contents of the anhydrite samples ranged from 1,638 to 3,842 $\mu g/g$, which is well within the values given by Ichikuni and Musha (1978) for primary marine sulfate deposits. Chaudhuri et al. claimed that the Sr isotopic signatures of the anhydrite samples from Kansas are primary and that their variations reflect the variations in $^{87}Sr/^{86}Sr$ ratios of waters of the evaporating basin, as a result of varied influence of the input of continental Sr to the basin.

The Br contents of the analyzed halite samples ranged from 46 to 100 $\mu g/g$. Although a linear relationship between the Br contents and the $^{87}Sr/^{86}Sr$ ratios was not too convincing, Chaudhuri et al. noted that the salt with highest Br content had the highest $^{87}Sr/^{86}Sr$ ratio and that those with low Br contents generally had low $^{87}Sr/^{86}Sr$ ratios. At least some of the halites, whose Br contents were less than that of primary halite, which is known to have a Br content of about 65 $\mu g/g$, appear to have been recrystallized. But, as evident from data, the effect of such recrystallization was to lower the $^{87}Sr/^{86}Sr$ ratios of the salts. Chaudhuri et al., therefore, claimed that, unless the salts were recrystallized also at other earlier time or times, the apparently high $^{87}Sr/^{86}Sr$ ratios of the salts relative to contemporaneous sea water reflect the waters in the evaporating basin. As the data suggest, the brines from which halite precipitated were slightly enriched in ^{87}Sr as compared to brines from which the sulfate minerals precipitated. They attributed this difference to increased influence of input of continental

Sr to the basin with Sr-poor brines when the precipitation of halite occurred as compared with Sr-rich brines when the precipitation of sulfate minerals occurred.

Brookins (1988) recognized Sr isotopic variations of waters of an evaporating basin from studies of marine bedded anhydrite and gypsum deposits in the Castille, Salado, and Rustler Formations, occurring in stratigraphically ascending order of the Late Permian Ochoan Group in the Delaware Basin of New Mexico (U.S.A.). The $^{87}Sr/^{86}Sr$ ratios of sulfate deposits in the Rustler Formation ranged from 0.70692 to 0.70707, whereas those in the Salado and Castille Formations were nearly uniform at about 0.70688. On the other hand, secondary selenite veins occurring in the overlying siltstone and sandstone of the Dewey Lake Red Beds of the Ochoan Group were high in the $^{87}Sr/^{86}Sr$ ratios with values between 0.70722 and 0.70861. The Sr isotopic values of the sulfates in the Rustler, Salido and Castille Formations were found to be well within the Sr isotopic values of contemporaneous sea water, which have rapidly increased from about 0.7068 at the beginning of the Late Permian time to about 0.7073 during the Permo-Triassic time (see figure 2 in Hoser's companion contribution). Like the temporal changes in the Sr isotopic composition of the sea waters, the bedded evaporite deposits of the Ochoan Group also increased in the $^{87}Sr/^{86}Sr$ ratios from low values during the Salado-Castille time to high values during the Rustler time. The internal variations in the Sr isotopic values (0.70692-0.70707) of the bedded anhydrite and gypsum deposits of the Rustler Formation is evident. Although he was unable to find a satisfactory explanation for the apparent variations in the Sr isotopic compositions, Brookins (1988) maintained that these variations were unlikely to have occurred from reaction with Sr released from a recrystallization process of clay minerals in clastic interbeds, as the amount of clays present is insufficient to reset the isotopic composition of anhydrites which contain a few thousands $\mu g/g$ of Sr. We tend to believe that such small variations can arise from variations in the isotopic composition of waters in an evaporating basin. The aspect of how primary Sr isotopic records could have vary during evaporite deposition is discussed later.

The study of Posey and Fisher (1989) on evaporite deposits of

Lower Permian Wolfcampian and Leonardian series in the Palo Duro Basin of Texas is an additional illustration of an apparent retention of syn-depositional Sr isotopic signatures by ancient bedded marine anhydrite deposits. The authors noted that the $^{87}Sr/^{86}Sr$ ratios of both bedded nodular mosaic anhydrites (0.70753-0.70824) and replacive nodular anhydrites (0.70796-0.70841) corresponded to that of Early Permian sea water (0.70740-0.70815), in contrast to those of vein-filling anhydrites (0.70828-0.70895) which obviously contained post-depositional Sr. The Sr isotopic compositions of the bedded anhydrite in the Palo Duro Basin of Texas, like similar anhydrite and gypsum deposits of other ages elsewhere (Clauer, 1976; Chaudhuri et al., 1986; Brookins, 1988), appeared to have varied randomly through time. Again, the variations in the Sr isotopic compositions of the bedded sulfate deposits in the Palo Duro Basin can be due to fluctuation in the Sr isotopic composition of the waters in the evaporating basin and not to any global excursion in the isotopic composition of sea water.

STRONTIUM ISOTOPIC SIGNATURES OF DIAPIRIC EVAPORITES

Diapiric evaporite deposits often record a long history of physical and chemical modifications involving complex episodes of water-rock interactions. An effect of these modifications is the formation of cap rocks over some of the salt domes. Perhaps the most thoroughly investigated diapiric evaporite deposits are the salt domes and associated cap rocks in the Gulf Coast of the U.S.A. Many of these salt diapirs contain cap rocks with a sequence of salt-dissolution related anhydrite accumulation directly overlying the halite diapir and of hydrocarbon-reaction related calcite, gypsum and sulfur above the anhydrite deposits. The salt diapirs in the Gulf Coast appear to have different ages, but the domes are believed to have originated from the extensive late Middle Jurassic Louann Formation (Salvador, 1987). Recent studies of Posey et al. (1987) and Land et al. (1988) have elucidated the Sr isotopic history of salt domes and associated cap rocks in the Gulf Coast area.

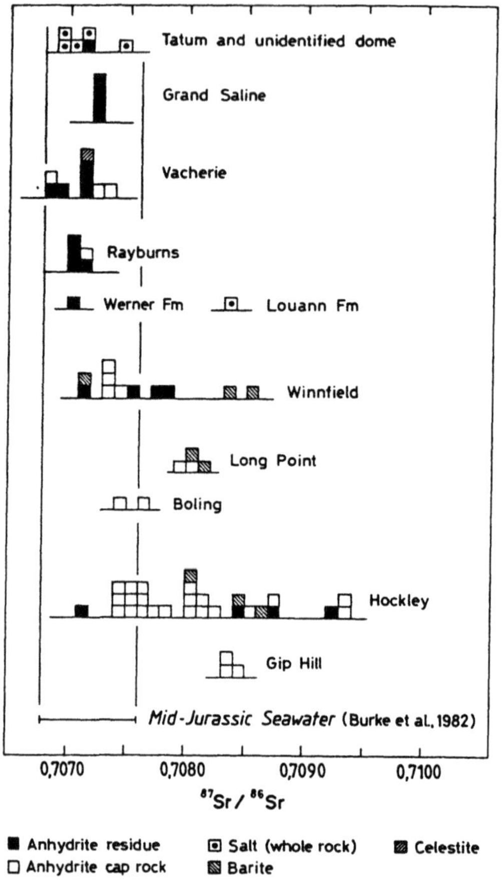

Figure 1: Sr isotopic compositions of salt domes and cap rocks from the Gulf-Coast region (Posey et al., 1987)

Posey et al. (1987) presented a summary of Sr isotopic ratios of a number of anhydrite cap rocks over salt domes from different locations in the Gulf-Coast region (Fig. 1). The $^{87}Sr/^{86}Sr$ ratios of several cap-rock anhydrites were found to be the same as that of the Middle Jurassic sea water (0.7068-0.7076) as defined by Burke et al. (1982), suggesting that the Sr in these cap-rock anhydrites derived from dissolution of Louann salt minerals. But the $^{87}Sr/^{86}Sr$ ratios of many other cap-rock anhydrites exceeded the Middle Jurassic marine Sr isotopic values. These high values could have resulted from invasion of ^{87}Sr-enriched formation waters into the salt domes, probably during diapirism. Posey et al. believe that either feldspar or clay within the clastic section could have been the source of the radiogenic ^{87}Sr in the cap-rocks. Regardless of the

Sr isotopic values, the Sr contents (550-1050 $\mu g/g$) of the cap-rock anhydrites are low in comparison to the Sr contents of many bedded evaporite deposits, which are frequently in the range of a few thousands of $\mu g/g$.

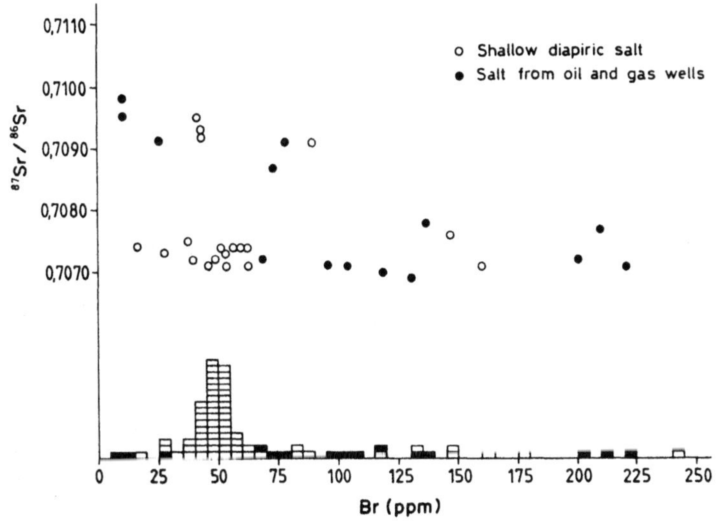

Figure 2: Sr isotopic compositions and Br contents of diapiric salts from the Gulf Coast area (Land et al., 1988)

In their study of Gulf Coast salt deposits, Land et al. (1988) observed that, although the $^{87}Sr/^{86}Sr$ ratios of some anhydrites and halites were nearly the same as that of the Middle Jurassic sea water, the isotopic values of these deposits were distinctly varied and that the variations occurred within a very short time as the samples analyzed came from a core less than 200 m thick. Conside-ring that wide isotopic shifts in short times are unlikely to be related to any global sea water isotopic effect, Land et al. concluded that the isotopic variations must be due to syn- or post-depositional introduction of continental Sr. They also noted that the $^{87}Sr/^{86}Sr$ ratios of many samples of diapiric halite were commonly between 0.7075 and 0.7095. The ^{87}Sr-enriched halites were found to be either depleted or similar in Br contents relative to primary marine salts, the latter presumably having more than 60 $\mu g/g$ of Br (Fig. 2). The ^{87}Sr-enriched salts with low Br contents were clearly the products of recrystallization of primary halites in presence of radiogenic ^{87}Sr-bearing fluids. The authors suggested that those salts with only slightly elevated $^{87}Sr/^{86}Sr$ ratios relative to sea water, but with Br contents similar to primary marine salts, either deposited from or recrystallized in ^{87}Sr-enriched fluids. The deeply buried ^{87}Sr-enriched salt minerals with very high Br contents (200 $\mu g/g$ or higher) were related to interactions with dense Br-rich brines generated at shallow depths as a result of salt recrystallization during diapirism. The study of Land et al. emphasized that many deformed or deeply buried salts probably evolved through complex rock-water interactions which can result in high Br contents for deep ground waters.

SOURCES OF CONTINENTAL STRONTIUM IN EVAPORATING BASINS

If Sr isotopic excursions of evaporating waters occur during the time span of the deposition of a few tens or hundreds of meters of evaporite deposits, as suggested by data from many bedded marine evaporite deposits described above, the isotopic variations are unlikely to be related to any global sea water Sr isotopic changes. The $^{87}Sr/^{86}Sr$ ratios of some bedded evaporite deposits were found to be different from that of the contemporaneous sea water. These isotopic perturbances can represent primary isotopic records of the evaporite deposits. To have such varied isotopic signatures, the waters in an evaporating basin would have to receive some continental Sr during the precipitation of the minerals. What is not clearly understood is the mode of transport of sufficient amounts of continental Sr to a marine water dominated evaporating basin causing the Sr isotopic compositions of the waters and precipitating minerals to be different from that of the contemporary sea water. The magnitude of the difference in the Sr isotopic composition between the water in the evaporating basin and the contemporaneous sea water would depend on the Sr isotopic composition and the relative contribution of Sr from the marine and continental sources. In a system with mixing of two-components, the resultant isotopic ratio R is expressed as:

$$R = (f_A \cdot i_A) + (f_B \cdot i_B)$$

where f represents fraction of Sr, i the Sr isotopic ratio, and subscripts A and B the two components. Considering this equation, the waters of an evaporating basin can contain $^{87}Sr/^{86}Sr$ ratios significantly different from that of the contemporaneous sea water, provided the $^{87}Sr/^{86}Sr$ ratios of the continental Sr are either much higher or much lower than that of sea water when the net contribution of Sr from the continental source is low, or only slightly above or below that of sea water when the relative contribution of continental Sr is high. Different sources of continental Sr to an evaporating basin need to be considered and their potentials of modifying the Sr isotopic compositions of the basin waters relative to that of the contemporaneous ocean may be assessed using the above isotopic mass balance equation.

River inflow represents one potential source of continental Sr entering an evaporating marginal marine basin, but the conditions necessary for evaporite deposition limit the river inflow to a small factor. Run-offs can have varied Sr contents and Sr isotopic compositions depending upon the age and the lithology of the rocks surrounding an evaporiting basin (see companion paper by Chaudhuri and Clauer, this volume). In most instances, run-offs have low Sr contents and low $^{87}Sr/^{86}Sr$ ratios; the average Sr isotopic value of run-off being slightly higher than that of sea water. The generally small difference in the Sr isotopic values between sea and continental water and the input of only limited amounts of run-off, are too small a factor to cause the evaporating waters to be measurably different in Sr isotopic composition from contemporaneous sea water.

Run-out or subsurface flow of waters is another potential source of continental Sr that may be added to the waters of an evaporating basin. Many moderately deep subsurface waters are commonly enriched in ^{87}Sr relative to sea water and have Sr contents which can be as high as 2,800 mg/l (see companion paper by Chaudhuri and Clauer, this volume). An isotopic mass balance calculation shows that 0.5% or less of subsurface water having a Sr content of about 2,000 mg/l with an $^{87}Sr/^{86}Sr$ ratio of about 0.7095 mixed with 99.5% of contemporaneous sea water in an evaporating basin can explain the Sr isotopic variations of the bedded

Permian evaporite deposits of Kansas and New Mexico. In fact, the importance of the addition of continental subsurface water to an evaporating basin as a cause for Sr isotopic variations of the evaporite deposits is difficult to assess because of the uncertainty in knowledge of the paleotopographic and paleohydrologic conditions of ancient evaporites.

Terrigenous clay-size and fine silt-size particles depositing into an evaporation basin could serve as an important source of continental Sr in the waters of the basin. A large amount of these particles might be transported by the wind, and the overall arid climate around evaporating basins can induce the conditions for high amounts of salts to be adsorbed on the surface of these fine particles which are, therefore, able to carry significant amounts of Sr simply adsorbed onto the surface of the minerals. We consider that the deposition of wind-transported, salt-encrusted fine particles in an evaporating basin might represent the major cause for the Sr isotopic excursions of many bedded evaporite deposits.

CHRONOLOGY OF EVAPORITE DEPOSITS

Geochronologic investigations of evaporite deposits are few for several reasons: (1) the abundance of evaporite deposits is small in comparison to that of other common sedimentary rocks and hence only in a few instances the stratigraphic positions of evaporite deposits are in question, (2) most common evaporite minerals are not amenable to dating by any conventional isotopic method due to their very unfavorable parent to daughter isotopic ratios and hence the dating of a deposit depends on the presence of some less common K-bearing salts, and (3) the origin of salt minerals amenable to isotopic dating is often not clearly understood and some of these minerals may be even always secondary, in which case the isotopic dates bear no relation to the depositional age of the evaporite. However, any potential failure of the isotopic data to yield a depositional age should not be construed as the isotopic data being of little value. On the contrary, the data can shed light on the post-depositional chemical history of the rocks and can give increased understanding of the origin of salinity of continental subsurface waters.

Because of its high K content, sylvite (KCl) became a subject of isotopic analyses as early as 1958, when Polevaya et al. (1958) reported K-Ar and K-Ca ages of one recrystallized and one primary Permian sylvite samples, and of one crystallized Cambrian sylvite. The two recrystallized sylvites gave K-Ar dates of about 165 and 162 Ma, but the primary lower Permian sylvite yielded a higher K-Ar date of 236 Ma, still lower than its stratigraphic age. The K/Ca dates were higher than the K/Ar dates for the recrystallized sylvite samples, but the dates still appeared anomalously low with respect to the stratigraphic ages.

Since the work of Polevaya et al.(1958), many others also found that K-Ar and Rb-Sr dates on sylvite, carnallite, and other K-bearing salt minerals are anomalously low as compared to the stratigraphic ages (Pilot and Blank, 1967; Pilot and Rösler, 1967; Dalrymple and Lanphere, 1969; Oesterle and Lippolt, 1975, 1976; Lippolt and Raczek, 1979; Brookins et al., 1980; Register and Brookins, 1980). Different explanations have been offered for the anomalously low dates for sylvite. Dalrymple and Lanphere (1969) and Huff and Wampler (1990) favored a diffusion-related loss of ^{40}Ar, whereas Macintyre (1978) recommended dissolution-recrystallization related loss of ^{40}Ar. In many instances, sylvites are secondary in origin, forming from transformation of carnallite in presence of fresh water (Holser, 1979b), and their isotopic dates may then represent the time of recrystallization and by extension the time of introduction of fresh water into the evaporite deposits.

Pilot and Blank (1967) and Pilot and Rösler (1967) reported results of K-Ar analyses on 20 langbeinite, 5 polyhalite, 12 carnallite, and 10 sylvite samples from Permian Zechstein. The dates ranged from about 280 to 50 Ma for the langbeinite, about 230 to 190 Ma for the polyhalite, about 160 to 10 Ma for the sylvite, and about 120 to 10 Ma for carnallite. Nearly two-thirds of the samples of langbeinite yielded K-Ar dates which corresponded to the deposition of the Zechstein, but all other minerals gave varying degrees of anomalously low ages. These low ages could have been, at least in part, related to Tertiary basaltic intrusion in the vicinity of the deposits. The observations of Pilot and his colleagues that evaporite minerals yield randomly anomalous ages have been confirmed by many subsequent studies.

Wardlaw (1968) analyzed K-Ar dates on 19 sylvite samples from potash zones in the Middle Devonian Prairie Evaporite Formation in Saskatchewan, Canada. The K-Ar dates of the minerals ranged from about 344 to about 56 Ma. As the author noted, this spectrum of young dates, in contrast to the depositional age of about 380 Ma for the evaporite rocks, is related to complex diagenetic history of the evaporite minerals, which at least included reaction of sylvite with Mg-Cl brine to produce carnallite, formation of sylvite from carnallite by leaching of Mg chloride, and recrystallization of early-formed sylvite.

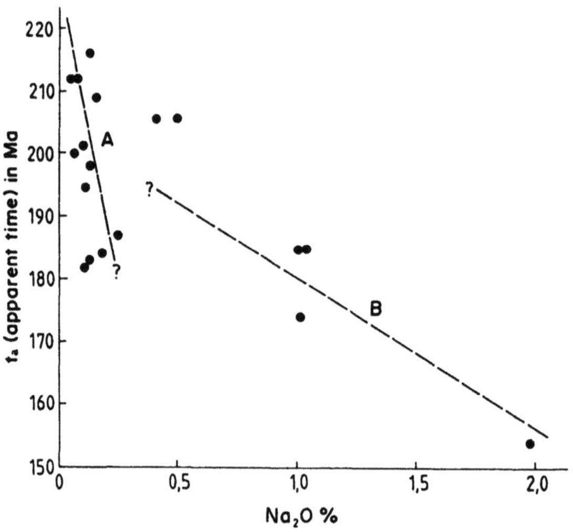

Figure 3: K-Ar apparent ages vs. Na contents of polyhalite samples from Late Permian evaporitic deposits in southeastern New Mexico (Brookins et al., 1980)

To understand the evolution of hydrous salts in relation to recent groundwater activities, Brookins et al. (1980) studied pure polyhalite, $K_2MgCa_2(SO_4)_4 \cdot 2H_2O$, from the Late Permian evaporitic deposits in southeastern New Mexico. The K-Ar dates of these pure polyhalite samples varied between 216 and 198 Ma, which are still less than the depositional age of the minerals. Brookins (1982) observed a correlation between K-Ar ages and Na contents of the poyhalites (Fig. 3). This Na content was thought to record the occurrence of halite or mix crystals of (Na,K)Cl salt in the polyhalite samples. The process responsible for the lowering of the age is still not clear, but when the correlation is extended to zero content of Na, the K-Ar date of the polyhalite becomes very similar to the stratigraphic age of the evaporite deposits.

Huff and Wampler (1990) measured the K-Ar ages of several samples of carnallite, $KMgCl_{3.63}.H_2O$, from Pennsylvanian Paradox Formation in Utah, U.S.A. The K-Ar dates ranged from about 27 to 9 Ma. The authors claimed that the older ages correspond to the igneous intrusions in the Colorado Plateau between 30 and 20 Ma ago, and the younger dates relate to the continuous loss of radiogenic ^{40}Ar as these carnallites remain non-encapsulated by other evaporite minerals.

Evaporite minerals yielding K-Ar isotopic ages in agreement with the depositional age are probably few. Langbeinite, $K_2Mg_2(SO_4)_3$, has been occasionally found to yield K-Ar ages which correspond to the time of deposition. Schilling (1973) reported a K-Ar age of 240 $\pm$ 10 Ma for a pure langbeinite from Late Permian deposits in southeastern New Mexico, which is in agreement with the stratigraphic age, but two associated sylvites gave values of about 74 and 18 Ma. Oesterle and Lippolt (1975, 1976) found concordant K-Ar and Rb-Sr dates for each of three different langbeinite samples. They reported dates of 230, 150 and 100 Ma, none of which corresponds to sedimentation times. Considering that the langbeinite minerals have high retentivity of both radiogenic ^{40}Ar and ^{87}Sr, the authors felt that the isotopic dates of langbeinite might be indicative of recrystallization processes related to different tectono-metamorphic activities in Western Europe.

Most of the common evaporite minerals have limited amount of radiogenic ^{87}Sr due to high amounts of Sr relative to Rb and are therefore unsuitable for dating by the Rb-Sr method. But sylvite or mix crystals of (Na,K) chloride salts and (K-Mg) sulfate salts, such as carnallite, langbeinite and rinneite, can have sufficient radiogenic ^{87}Sr relative to normal Sr due to very high Rb/Sr ratios. However, the low Sr content of 1 $\mu g/g$ or less and the complex evolution processes of these minerals often lead to ambiguous interpretations of the dates. Lippolt and Raczek (1979a) determined the Rb-Sr isotopic dates on two carnallitic salt rock samples, only a few meters apart from each other, of Permian evaporite deposits in northern Germany. The model Rb-Sr ages of these two samples were found to be different and anomalously low at 96 and 82 Ma. In another study, Lippolt and Raczek (1979b) observed Rb-Sr model ages ranging from about 84 to 20 Ma on four samples of rinneite which is known to occur as metasomatic product in Permian salt deposits in Germany. A sample of carnallite associated with the youngest rinneite (20 Ma) gave a much younger age of about 8.5 Ma. The meanings of these differing ages for evaporite minerals in the proximity of each other are not clear.

Register and Brookins (1980) reported a Rb-Sr isochron date of about 214 Ma from some scattered data on evaporite minerals of the late Permian Salado Formation in southeastern New Mexico. Model ages were calculated based on the assumption of the initial $^{87}Sr/^{86}Sr$ ratio being the same as that of the contemporaneous sea water. These model ages ranged from 270 to nearly 0 Ma, and clearly the Rb-Sr isochron age of 214 Ma has an ambiguous meaning, suggesting that the chemical evolution of minerals in the evaporite deposit was very complex. Some mineralization event must have occurred in the last few million years in the late Permian evaporite deposit of eastern New Mexico, as evident from data on a halite sample whose Rb/Sr ratio was about 1 and $^{87}Sr/^{86}Sr$ ratio was 0.7076. This Sr isotopic ratio is close to that of the anhydrite deposits, suggesting that the Sr in the halite could have derived primarily from evaporite deposit.

To understand diagenetic recrystallization and mineralization in an evaporite deposit, Baadsgaard (1987) examined the K-Ar, Rb-Sr, and K-Ca isotopic systematics of a large number of sylvite, carnallite, and halite minerals and also of whole-rock samples from potash zones in the Middle Devonian Prairie Evaporite Formation in western Canada. The ranges of $^{87}Sr/^{86}Sr$ ratios were between 0.7085 and 0.7406 for the halite samples, between 0.7157 and 2.2256 for the sylvite samples, and between 0.7093 and 2.7529 for the carnallite samples. The Rb-Sr model ages of nearly all sylvites, with the exception of a few, were between 20 to 60 Ma (Fig. 4). By comparison, the Rb-Sr model ages of carnallites, were low between 20 and 2 Ma.

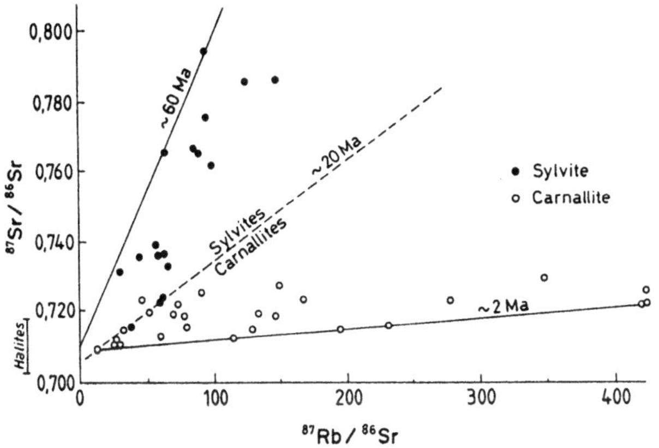

Figure 4: Rb-Sr isotopic compositions and the range of model ages for sylvites and carnallites of the Middle Devonian Prairie Evaporite in western Canada (Baadsgaard, 1987)

The $^{40}Ca/^{44}Ca$ ratios of halites were distinctly varied between 47.17 and 47.62, as compared to the value of 47.18 for the modern sea water. This suggests that the halites either locally recrystallized or precipitated in the presence of varied amounts of ^{40}Ca-bearing fluids. Because of the very low K/Ca ratios of the halites, their K-Ca model ages could not be calculated. But due to their favorable K/Ca ratios, sylvites and carnallites could be analyzed for the K-Ca isotopic dates. Like the Rb-Sr isotopic data, the K-Ca isotopic data were non-coherent failing to yield any distinct isochron (Fig. 5). As shown in the diagram, the calculated K-Ca model ages for the sylvites are lower than those for carnallites. The relative difference in the K-Ca model ages between the two minerals is opposite to that in the Rb-Sr model ages between the same two minerals. Baadsgaard noted that the disagreement between the two sets of results can be related to the different chemical behaviors of K, Rb, Ca, and Sr during the sylvite-carnallite diagenetic transformation or recrystallization. Baadsgaard also noted that whereas the Rb-Sr and the K-Ca dates of sylvites were less than 100 Ma, the K-Ar dates of sylvites were mostly between 344 and 207 Ma. He concluded that the older K-Ar ages in comparison to the other isotopic ages might be the reflection of local incorporation of varied amounts of radiogenic ^{40}Ar by the diagenetically formed sylvites.

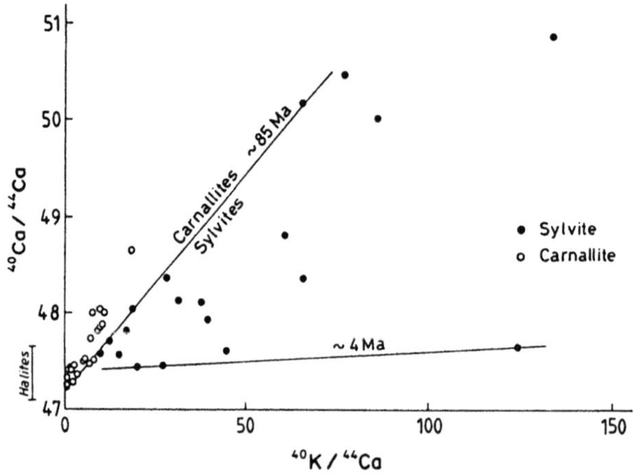

Figure 5: K-Ca isotopic data of sylvites and carnallites of the Middle Devonian Prairie Evaporite in western Canada (Baadsgaard, 1987)

The global marine Sr isotopic secular variation curve has also been used to determine the age of evaporite deposits from their Sr isotopic compositions (Posey and Fisher, 1989). But such a chronostratigraphic approach must be carefully considered. The Sr isotopic age-curve for the pre-Tertiary period is based on a limited number of scattered isotopic data at each geologic time. The accuracy for the isotope curve based age can therefore be severely limited even when the global marine Sr isotopic composition apparently shifted very rapidly. The utility of the isotope-age curve for chronostratigraphy of evaporite deposits is further limited by the Sr isotopic data of the deposit itself. We have already pointed out that even the primary $^{87}Sr/^{86}Sr$ records of bedded evaporite deposits may not bear the marine isotopic signature of the contemporary ocean waters.

SUMMARY

The $^{87}Sr/^{86}Sr$ ratios of minerals in marine, bedded evaporite deposits are often higher than that of contemporaneous sea water. The deviations are often explained in terms of secondary origin or recrystallization of these minerals in presence of radiogenic ^{87}Sr-bearing fluids, but such small isotopic variations recorded in ancient bedded evaporite deposits could also be due to the introduction of continental Sr into the evaporating basins during the time of deposition. A logical source for this continental Sr could be the release of continental salts from salt-encrusted fine clay- and silt-sized silicate particles, transported into the basins by wind.

The conventional isotopic dates of most evaporite minerals are often lower than their stratigraphic ages. If the loss of radiogenic isotopes occurs by a process of continuous diffusion, the anomalously low dates are geologically meaningless. On the other hand, if the loss relates to episodic events, then the data are useful to understand the halokinesis of an evaporite deposit. More studies are needed to define the criteria for recognition of the process by which isotopic disturbances might have occurred for any particular evaporite deposit. Undoubtedly, many evaporite minerals are secondary in origin and are related to several cycles of chemical disturbance. Groundwaters play an important role in the deformation of evaporite deposits, but connate fluids and crystallization waters become also important in bedded deposits to produce recrystallized or authigenic minerals. Knauth and Beeunas (1986) maintained from analyses of D and ^{18}O in fluid inclusions in salts that the fluids originate as highly evaporated sea water or synsedimentary brines mixed with meteoric water.

ACKNOWLEDGEMENTS

We thank P. Stille (Centre de Géochimie de la Surface, Strasbourg, France) for helpful comments on an earlier draft of this manuscript, and Mr. C. Hammel for drafting the figures.

REFERENCES

Aberg F. and Wickman F.E. (1987) Variations of $^{87}Sr/^{86}Sr$ in water from streams discharging into the Botham Bay, Baltic Sea. Nordic Hydrol., 18, 33-42.

Baadsgaard H. (1987) Rb-Sr and K-Ca isotope systematics in minerals from potassium horizons in the Prairie Evaporite Formation, Saskatchewan, Canada. Chem. Geol., 66, 1-15.

Borchert H. and Muir R.O. (1964) Salt Deposits: Their Origin, Metamorphism and Deformation of Evaporites. Van Nostrand D. Co., London, 338 p.

Braitsch O. (1971) Salt Deposits, Their Origin and Composition. Springer-Verlag, Heidelberg, 297 p.

Brookins D.G. (1982) Study of polyhalite from the WIPP site, New Mexico. In: The Scientific Basis for Nuclear Waste Management, Topp S.V. (ed.), Elsevier Sci. Publ., Amsterdam, 257-264.

Brookins D.G. (1988) Sea water $^{87}Sr/^{86}Sr$ for the Late Permian Delaware Basin evaporites (New Mexico, U.S.A.). Chem. Geol., 69, 209-214.

Brookins D.G., Register J.K.Jr. and Krueger H.K. (1980) Potassium-argon dating of polyhalite in southeastern New Mexico. Geochim. Cosmochim. Acta, 44, 635-637.

Burke W.H., Denison R.E., Hetherington E.A., Koepnick R.B., Nelson H.F. and Otto J.B. (1982) Variation of sea water $^{87}Sr/^{86}Sr$ throughout Phanerozoic time. Geology, 10, 516-519.

Chaudhuri S., Clauer N. and Clayton A. (1986) Rare-earth elements, sulfur isotopes, and strontium isotopes in a Lower Permian evaporite deposit in Kansas. Terra cognita, 6, 33.

Clauer N. (1976) $^{87}Sr/^{86}Sr$ composition of evaporitic carbonates and sulphates from Miocene sediment cores in the Mediterranean Sea (D.S.D.P., Leg 13). Sedimentology, 23, 133-140.

Claypool G.C., Holser W.T., Kaplan I.R., Sakai H. and Zak I. (1980) The age curves of sulfur and oxygen isotopes in marine sulfate and their mutual interpretation. Chem. Geol., 28, 199-260.

Dalrymple G.B. and Lanphere M.A. (1969) Potassium-Argon Dating. W.H. Freeman, San Francisco, California, 258 p.

Faure G. (1986) Principles of Isotope Geology. Wiley, New York, N. Y., 589 p.

Holland H.D. (1984) The Chemical Evolution of the Atmosphere and Oceans. Princeton University Press, Princeton, New Jersey, 441-459.

Holser W.T. (1966) Bromide geochemistry of salt rocks. In: Second Symposium on Salt, Rau J.L. (ed.), Northern Ohio Geol. Soc., Cleveland, Ohio, 1, 248-275.

Holser W.T. (1979a) Mineralogy of evaporites. In: Marine Minerals, Burns R.G. (ed.), Reviews in Mineralogy, Miner. Soc. Am., 6, 211-294.

Holser W.T. (1979b) Trace elements and isotopes in evaporites. In Marine minerals, Burns R.G. (ed.), Reviews in Miner., Miner. Soc. Am., 6, 295-346.

Holser W.T., Sakai H. and Zak I. (1979) Isotope geochemistry of oxygen in the sedimentary sulfate cycle. Chem. Geol., 25, 1-17.

Hsü K.J. (1972) Origin of saline giants: A critical review after the discovery of the Mediterranean evaporite. Earth Sci. Rev.,

8, 371-396.
Huff G.F. and Wampler J.M. (1990) K-Ar geochemistry of carnallite from salt-cycle six of the Paradox formation in Utah, U.S.A. Chem. Geol. (Isot. Geosc. Sect.), 80, 309-318.
Ichikuni M. and Musha S. (1978) Partitioning of Sr between gypsum and solution. Chem. Geol., 21, 359-363.
Knauth L.P. and Beeunas M.A. (1986) Isotope geochemistry of fluid inclusions in Permian halite with implications for the isotopic history of ocean water and the origin of saline formation waters. Geochim. Cosmochim. Acta, 50, 419-433.
Koepnick R.B., Burke W.H., Denison R.E., Hetherington E.A., Nelson H.F., Otto J.B. and Waite L.E. (1985) Construction of the sea water $^{87}Sr/^{86}Sr$ curve for the Cenozoic and Cretaceous: Supporting data. Chem. Geol. (Isot. Geosc. Sect.), 58, 55-81.
Land L.S., Kupecz J.A. and Mack L.E. (1988) Louann salt geochemistry (Gulf of Mexico sedimentary basin, U.S.A.): A preliminary synthesis. Chem. Geol., 74, 25-35.
Lippolt H.J. and Raczek I. (1979) Rinneite-dating of episodic events in potash salt deposits. Jour. Geophys., 46, 225-228.
Lippolt H.J. and Razec I. (1979) Cretaceous Rb-Sr total rock ages of Permian salt rocks. Naturwiss., 66, 422-423.
Macintyre R.M. (1978) Burying high-level wastes. Nature, 271, 605-606.
Ochsenius C. (1877) Die Bildung der Steinsltzlager und ihrer Mutterlaugensalze. Pfeffer, Halle, 172 p.
Oesterle F.P. and Lippolt H.J. (1975) Isotopische Datierung der Langbeinitbildung in der Kalisalzlagerstätte des Fuldabeckens. Kali und Steinsalz, 6, 391-398.
Oesterle F.P. and Lippolt H.J. (1976) Langbeinitedatierungen und Kalisalzmetamorphosen. Nachr. deutsc. geol. Ges., 15, 20-22.
Peterman Z.E., Hedge C.E. and Tourtelot H.A. (1970) Isotopic composition of strontium sea water throughout Phanerozoic time. Geochim. Cosmochim. Acta, 34, 105-120.
Pilot J. and Blank P. (1967) K-Ar-Bestimmungen von Salzgesteinen des Zechsteins. Z. angew. Geol., 13, 661-662.
Pilot J. and Rösler H.J. (1967) Altersbestimmung von Kalisalzmineralien. Naturwiss., 54, 490.
Polevaya N.I., Titov N.E., Belyaev V.S. and Sprintsson V.D. (1958) Application of the Ca method in the absolute age determination of sylvites. Geochem. Intern., 8, 897-906.
Posey H.H., Kyle J.R., Jackson T.J. and Hurst S.D. (1987) Multiple fluid components of salt diapirs and salt dome cap rocks, Gulf Coast, U.S.A. Applied Geochem., 2, 523-534.
Posey H.H. and Fisher R.S. (1989) A sulfur and strontium isotopic investigation of Lower Permian anhydrite, Palo Duro Basin, Texas, U.S.A. Applied Geochem., 4, 395-407.
Register J.K. and Brookins D.G. (1980) Rb-Sr isochron age of evaporite minerals from the Salado Formation (Late Permian), Southeastern New Mexico). Isochron/West, 29, 39-42.
Salvador A. (1987) Late Triassic-Jurassic paleogeography and origin of the Gulf of Mexico Basin. Amer. Ass. Petrol. Geol. Bull., 71, 419-451.
Schilling J.A. (1973) K-Ar dates on Permian potash minerals from southeastern New Mexico. Isochron/West, 6, 37-38.
Shaw H.F. and Wasserburg G.J. (1985) Sm-Nd in marine carbonates and phosphates: Implications for Nd isotopes in seawater and crustal ages. Geochim. Cosmochim. Acta, 49, 503-518.
Stille P. and Fischer H. (1990) Secular variation in the isotopic

composition of Nd in Tethys seawater. Geochim. Cosmochim. Acta, 54, 3139-3145.
Wardlaw N.C. (1968) Carnallite-sylvite relationships in the Middle Devonian Prairie Evaporite Formation, Saskatchewan. Geol. Soc. Am. Bull., 79, 1273-1294.

THE STABLE ISOTOPE COMPOSITION OF SEDIMENTARY IRON OXIDES WITH SPECIAL REFERENCE TO BANDED IRON FORMATIONS

Jochen HOEFS
Geochemisches Institut der Universität
Goldschmidtstrasse 1, 3400 Göttingen (BRD)

Iron-rich sedimentary rocks, which are deposited intermittently throughout earth history can be divided into four groups according to their different mineralogical facies: oxides, carbonates, silicates, and sulfides. The present chapter deals exclusively with the stable isotope composition of Fe oxide minerals with special emphasis on Banded Iron Formations (BIFs). This type of deposit is virtually the only one which is mined today for Fe.

As is well known BIFs occur on all continents in Precambrian shield areas. Some of the largest and best known deposits are found in the Labrador Trough in North America, the Hamersley Range in Western Australia, the Quadrilatero Ferrifero in Brazil, the Transvaal Basin of South Africa and the Krivoy Rog areas in the USSR. With respect to their geologic ages three main periods of formation can be differentiated: mid Archean (3.4-2.9 Ga), early Proterozoic (2.5-1.9 Ga), late Proterozoic (0.75-0.5 Ga). Early Proterozoic deposits far outweigh those of all other ages.

The origin of BIFs is still a major controversial subject as examplified, for instance, by a recent discussion of Garrels (1988) and Morris and Trendall (1988). Because H and especially O isotope data can provide insight into environments and temperatures of formation, this article summarizes the relevant literature to elucidate the depositional environment and the diagenetic-metamorphic history of BIFs. At first, however, some generalizations about the formation and the stability of the different iron oxide minerals are presented.

STABILITY AND OCCURRENCE OF IRON OXIDES

Under to day's weathering conditions Fe^{+2} is released from rocks and oxidized to Fe^{3+}. Because Fe^{3+} is extremely insoluble in water under naturally occurring conditions (only in very acid solutions does it reach higher concentrations) it will hydrolyze to form ferric oxides/hydroxides. The first Fe precipitate from solution is a highly hydrous oxide of indefinite water content. As is well known, such first precipitates are seldom stable and transform to stable phases by splitting out most of its water.

Stable phases are goethite (α-FeOOH), lepidocrocite (γ-FeOOH), hematite (α-Fe_2O_3) and maghemite (γ-Fe_2O_3). By far the most common in sediments are goethite and hematite. There has been considerable debate about the relative stabilities of goethite and hematite in rocks. According to Schwertmann (1969), environmental conditions play a decisive role in determining which of the two forms are produced. Fischer and Schwertmann (1975) presented experimental evidence that goethite and hematite are not interrelated genetically by a simple hydration-dehydration type of reaction, but are formed by separate pathways from $Fe(OH)_3$. In the case of some soils, there is evidence that much of the hematite forms directly from dehydration of the initial ferrihydrite precipitate (Fischer and Schwertmann, 1975). Nevertheless, it is a well established fact that in progressively older rocks hematite becomes more abundant than goethite. Therefore, it appears likely that much of the sedimentary hematite in older rocks was formed by dehydration of pre-existing goethite (James, 1966).

BRIEF CHARACTERIZATION OF BANDED IRON FORMATIONS

According to the nomenclature of James (1966), Fe-rich sedimentary rocks can be divided into two broad classes: ironstone, which occurs mainly as thin beds of Phanerozoic age, and BIFs, mainly of Precambrian age and representing by far the greatest concentration of Fe of sedimentary origin. Both have originated as chemically precipitated sediments. With respect to their isotopic composition, BIFs have only been studied so far, with the exception of the study by Hangari et al. (1980), who determined the isotope composition of siderite and magnetite from the Wadi Shatti District in Lybia, which are regarded to be equivalent to other minette-type Phanerozoic ironstones.

As defined by James (1954, p.239) a banded iron formation is a "chemical sediment, typically thin-bedded or laminated, containing 15% or more Fe of sedimentary origin, commonly but not necessarily containing layers of chert". James proposed a classification into 4 facies: oxide, carbonate, silicate and sulfide. In the following, we will restrict ourselves to the oxide facies: the cherty banded iron formation. Recent literature has been summarized for instance by Trendall and Morris (1983).

Fe oxides and SiO_2 together constitute over 90% of the rock. The main mineral components -magnetite, hematite and finely crystalline quartz (usually called chert)- are banded as to easily recognize Fe-rich and Fe-poor layers of varying thickness. Bands of varying scale and prominence represent the fundamental structures in all BIFs. Banding sometimes occurs on a scale of 1 mm or less (microbanding), but alternation on a larger scale also occurs (mesobanding and macrobanding; Trendall, 1983). The regularity in banding, with individual bands sometimes continuing over distances of tens or hundreds of kilometers, suggests deposition in large, quiescent water bodies. The favored model of BIFs involves a water column with an upper oxic layer overlying a much larger volume of anoxic water. Most investigators favor a marine origin, but there is a strong minority opting for a freshwater origin. Garrels

(1987), for instance, assumed that banded iron formations were deposited by evaporation of stream water in restricted basins. The microbanding is the result of initial deposition of iron minerals during evaporation followed by deposition of silica after saturation with that substance. If the concept of annual varves is accepted, the average varve thickness is about 0.15mm/y, which requires removal of 26 mg of Fe and 21 mg of SiO_2 for each cm^2 of the depositional basin by evaporation of about 500 cm of water (Garrels, 1987).

With the exception of small occurrences of bog ores found in small lakes of higher latitudes, there is no example in present-day environments of Fe deposition approaching in composition the Fe rich units of the geological past. Therefore, it is extremely difficult to obtain first-hand information on the primary phases that were originally precipitated. The principal problem seems to be to permit the deposition of Fe phases, but inhibit the precipitation of carbonates and prevent the introduction of clastic material. With respect to the Precambrian BIFs, there is a real dilemma: how to transport Fe to its sites of deposition if the hydrosphere and atmosphere were then oxidizing and how to precipitate it if they were reducing (Cloud, 1983)?

Based on thermodynamic stability calculations, one can, of course, attempt to deduce what the original depositional phases may have been. Therefore, on Eh-pH diagrams at 25°C and 1 atm total pressure, the stability fields of mineral assemblages can be calculated (Garrels and Christ, 1965). However, such equilibrium diagrams must be applied with caution to an evaluation of environmental conditions of the geological past, because it is common to find that the first formed phases are metastable species and it is only with the passage of time that these assemblages adjust through dehydration and crystallization to yield thermodynamically more stable phases.

A different and promising approach to obtain information on ancient sedimentary environments is to study the stable isotope composition of Fe oxide/hydroxide phases. As is well known, stable isotope ratios are especially useful for deducing possible sources of water and for calculating temperatures of mineral formation. Thus, O and H isotopes could be used to provide data on the isotopic composition of the waters present at the time of mineral formation and temperatures of formation could be calculated if fractionation factors between minerals and water at the respective temperatures are accurately known.

PRIMARY ISOTOPE COMPOSITION OF GOETHITE AND HEMATITE

The isotopic composition of most low temperature minerals (carbonates, silicates, sulfates, phosphorites) is generally well known. This cannot be said for the Fe oxide/hydroxide family. Only in recent years due to the efforts of Yapp and his coworkers, this deficiency is beginning to change. Yapp (1983; 1987a and b), Yapp and Pedley (1985) and Yapp and Poths (1986) analyzed the H, C, and C isotope composition of "museum quality" and synthesized goethites. The D values of 24 natural goethites varied between -202 and -94 per mil (Yapp, 1987a). Yapp and Pedley (1985) demonstrated that H isotope fractionation between goethite and water is approximately constant over the temperature range from 25 to 145°C, the fractionation factor being about 0.90. This suggests that goethite δD-values may be a direct temperature independent measure of the δD-values of the waters with which goethite last equilibrated. However, it might be impossible to decide whether goethites, which formed some million years ago, had preserved their original δD-values or had, instead, exchanges with modern waters. Yapp and Pedley (1985) suggested that substitution of Al^{3+} and/or Mn^{3+} for Fe^{3+} in the goethite structure might produce changes in the D/H fractionation factor.

The natural goethites analyzed so far have a $\delta^{18}O$ range of about 14 per mil. Scatter in the data could be due to different water compositions and/or to different temperatures of formation. Because Yapp (1987a) observed a temperature dependent $\delta^{18}O$ fractionation between 25°C, 44°C and 62°C and the corresponding water, he concluded that temperature is a dominant factor in determining the

O isotope composition of goethite. A plot of δD vs. $\delta^{18}O$-values of different hydroxyl bearing minerals reveals that the 25° goethite line is closest to the Meteoric Water Line while the 25° chert line is farthest for the Meteoric Water Line (Fig. 1). Figure 1 indicates that goethite can be used in combination with other low temperature minerals as a geothermometer.

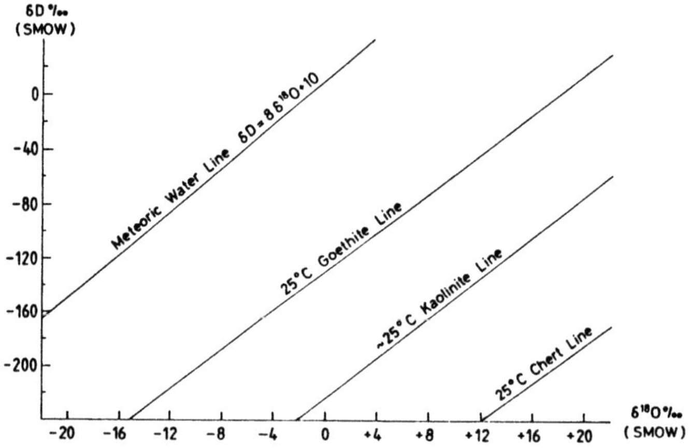

Figure 1: Variation of δD vs. $\delta^{18}O$ for different minerals relative to the meteoric water line (from Yapp, 1987)

One complication may arise from the fact that, as is well known, bacteria might be causal factors in the precipitation of Fe oxides/hydroxides. However, whether bacteria are necessary or directly involved in the precipitation of Fe is still unanswered (Nealson, 1982). Therefore, it is assumed that during Fe precipitation vital effects do not have any significant influence on the isotope composition of sedimentary oxides.

During their investigation of O and H isotope composition of natural goethites, Yapp and Poths (1986) noted that goethites contain variable amounts of C (2wt% CO_2). They concluded that C exists in two different forms: "trapped CO_2" and organic matter. Total ^{13}C-values range from -27.2 to -8.10 per mil. This large range could be explained by different ratios of trapped CO_2 to organic matter. According to Yapp (1987b), the amount of trapped CO_2 depends on the CO_2 partial pressures at the time of mineral formation, and thus represents a potential CO_2-geobarometer. It, therefore, seems possible that the analyses of "trapped CO_2" may give quantitative information on CO_2-partial pressures in the earth's geologic history.

Let us now discuss possible fractionation effects that might happen during the conversion of iron hydroxides to Fe oxides. Yapp (1983) demonstrated experimentally that partial dehydration of goethite to hematite is accompanied by a systematic H isotope fractionation, which results in a progressive decrease in the δD-values of the residual mineral water with increasing degree of hydration. Dehydration under open system conditions produces a smaller degree of isotope fractionation than under closed system conditions. There are no systematic experiments which studied the behaviour of O isotopes, but from comparison of data on natural goethites (Yapp, 1983) and on hematite from low temperature banded iron formations, there should be only a relatively small fractionation effect. This can also be deduced my mass balance considerations: even if dehydration goes to completion three quarters of O remain in the solid phase. This relationship does not speak for large fractionation effects.

ISOTOPE FRACTIONATION PROCESSES DURING BURIAL HISTORY OF BANDED IRON FORMATIONS

Interpretation of Temperatures

Cherty banded iron formations consist of two phases only: quartz and Fe oxides, which are stable over a wide range of temperatures and pressures. Therefore, petrologic criteria used generally to distinguish between diagenetic and metamorphic assemblages are inadequate in such rocks.

As is well known, stable isotope ratios can be used to calculate "temperatures of formation". However, there are a few problems when applying such geothermometers. The first condition which has to be fulfilled is isotope equilibrium. Secondly, it is generally assumed that the various Fe oxide minerals - magnetite, hematite, martite - do not show any difference in their O isotope fractionation behaviour. And indeed, by separating different concentrates of hematite, magnetite and martite from Fe ores in Minas Gerais, Hoefs et al. (1982) did not find any differences between these minerals. Thirdly, temperature calibrations must exist.

Based on spectroscopic and heat capacity measurements, Becker and Clayton (1972) gave a calculated magnetite-water curve. Such theoretical calculations are very difficult, because all vibrational frequencies of the crystalline lattice must be taken into account. Therefore, the most promising approach seems to be the experimental determination which, however, has not been performed below temperatures of 300°C, because exchange rates are so slow. Blattner et al. (1983) determined the O isotope fractionations between magnetite crystallized from hydrothermal steam at Wairakei (New Zealand), and the corresponding hydrothermal fluid at 175°C and 112°C. Their data support the earliest proposed magnetite-H_2O fractionation curve of O'Neil and Clayton (1964). Nevertheless, fractionation curves are still not very well defined at low temperatures. Therefore, the temperatures calculated from the quartz-Fe oxide fractionations will, to some extent, depend on the calibration curves that are used for the calculations. Recently, Muller et al. (1986) have compiled all available literature data on O isotope fractionations between quartz and Fe oxide in banded iron formations and have recalculated temperatures on the basis of the computer programme of Hoernes (1980). The thus obtained temperatures vary from 140 to 810°C., which encompass the whole range of temperatures which is found in the earth's crust. Temperatures for the least metamorphosed deposits vary from 200 to 300°C (Muller et al., 1986). Considering the uncertainties in temperature calibrations and in attainment of isotope equilibrium, these low temperatures should not be overinterpreted, but they indicate a low grade burial metamorphism.

Reconstruction of Depositional Environment

For an attempt to reconstruct the original sedimentary environment, the least metamorphosed banded iron formations represent the most suitable deposits. Based on O isotope fractionations, these are the Mesabi-Range, North America (Perry et al., 1973), the Hamersley Basin, western Australia (Becker and Clayton, 1976), the Kuruman deposit, South Africa (Perry and Ahmad, 1983), and the Urucum district in western Brazil (Hoefs et al., 1987). However, Becker and Clayton (1976) and Hoefs et al. (1987) concluded that isotope effects during diagenetic and metamorphic fluid/rock interactions are so pronounced that the initial isotopic composition of the sedimentary environment is no longer reflected in the SiO_2 and in the Fe-oxide phases. Gregory (1986) modelled these changes in terms of two major constraints on isotopic systems: conservation of mass and the kinetics of exchange equilibria.

It is convenient to treat fluid/rock interactions by simple endmember models: (a) closed systems, where interactions may occur between mineral phases and a coexisting fluid, all of which being present in constant relative proportions, and (b) open systems, where complex isotopic interactions occur between mineral phases and externally-derived fluids.

In cases, where open-system processes are suspected, it is useful to construct plots of $\delta^{18}O$ of one mineral vs. the $\delta^{18}O$ of a coexisting mineral (Gregory, 1986). Such graphs reflect bulk $\delta^{18}O$ variations, as well as $^{18}O/^{16}O$ fractionation data on coexisting minerals. Gregory (1986) has shown that for mineral assemblages with different bulk compositions equilibrated under closed system conditions, coexisting minerals should exhibit Δx-y values that define linear arrays of 45° slope on a δ-δ diagram which parallel the Δ = 0 fractionation line. In contrast, under open system conditions mineral pairs lie along steep (nearly vertical), positively sloped arrays indicating disequilibrium.

A summary of Fe oxide quartz data from Precambrian iron formations is shown in figure 2. There appear to be two major classes of data: mineral pairs from Minas Gerais (Brazil) and from Isua (Greenland) lie along slope 1 arrays, which is suggestive of equilibration under greenschist to granulite facies conditions, and mineral pairs from Hamersley and Weld Range (Australia), Kuruman (S. Africa) and Mesabi Range (North America) lie along

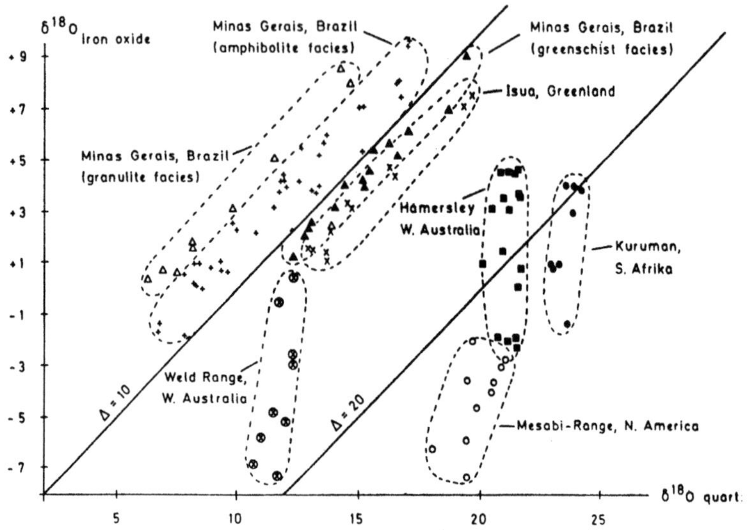

Figure 2: Closed and open system-behavior of quartz-Fe oxide pairs from various BIFs of different metamorphic grade. Closed system equilibrium arrays are indicated for the Iron Quadrangle deposits (Hoefs et al., 1982) and Isua, Greenland (Perry et al., 1978). The low grade deposits of Weld Range, West Australia (Perry and Ahmad, 1983), Hamersley, W. Australia (Becker and Clayton, 1976) and Mesabi Range, N. America (Perry et al., 1973) show open system disequilibrium arrays, that are characteristic of hydrothermally exchanged rocks (modified after Gregory, 1986)

sloped arrays indicative of disequilibrium exchange under open system conditions. These data indicate that under some high temperature conditions quartz readily exchanges ^{18}O with its surroundings. In low-grade, relatively undeformed areas, the data lie along steep disequilibrium arrays that are characteristic of hydrothermally exchanged rocks. These positions of the disequilibrium arrays indicate that different initial ^{18}O compositions are required for the precursors. The major conclusion that can be derived from the disequilibrium arrays shown in figure 2, is that the rocks have behaved as open systems and that fluid/rock interactions have erased the isotopic record of the surface history of the rocks. An open question is whether the samples which have been metamorphosed under higher temperatures and whose minerals appear to exhibit oxygen isotope equilibrium, initially passed through an open-system burial history analogous to the low temperature terranes. Nevertheless, in the following a scenario is presented which is by no means unique, but may represent plausible conditions.

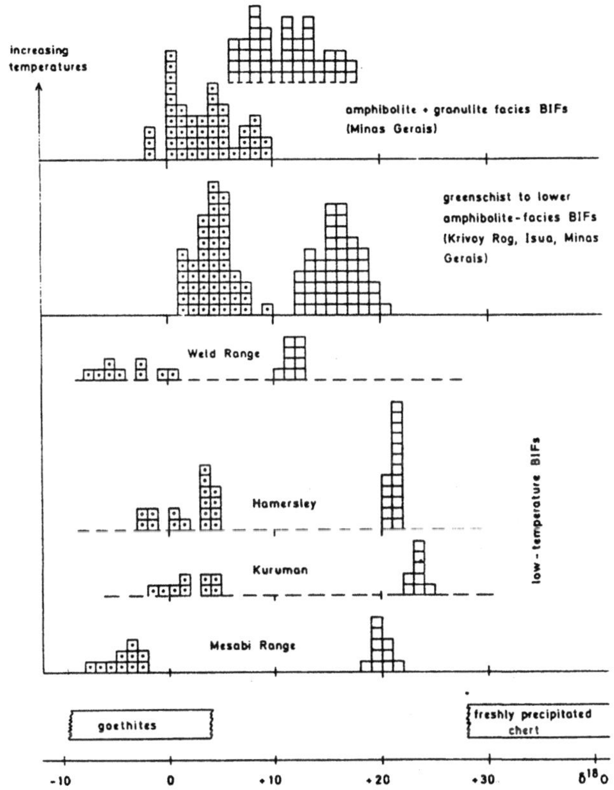

Figure 3: Variation of $\delta^{18}O$ values of cherts and Fe oxides from different BIFS

The initial ^{18}O composition of quartz and Fe oxide should be uniform when precipitated from ocean water or should have restricted compositional variations within single terranes when precipitated from evaporated fresh-water bodies. Assuming the original water composition was around 0 per mil and water temperatures were similar to those of today, then the amorphous precursor of the chert should have an isotopic composition around 35 per mil (Knauth and Epstein, 1976). A compilation of the O isotope variations of cherts and Fe oxides observed in BIFs of different metamorphic grade demonstrates that ^{18}O contents of cherts from the least metamorphosed deposits are at least 10 per mil lower (Fig. 3). This ^{18}O shift by 10 per mil or more, occurs obviously during phase transformations to an ordered SiO_2 polymorph via isotope exchange with a low ^{18}O fluid.

The ^{18}O contents of the original Fe phases are also unknown. Goethites from different geologic environments vary between +4.4 to -4.9 per mil with one low $\delta^{18}O$ value of -9.5 per mil (Yapp, 1987). This range is also found in the least metamorphosed banded iron formations (Fig. 3). Therefore, the Fe oxide system behaves much more inert than the SiO_2-system, which could be due to the fractionation behaviour of the Fe-oxide-water system. Although the exact fractionation curve between Fe oxide and water is still a matter of debate, in any case there should be a cross-over at rather low temperatures (Blattner et al. 1983) making the system rather insensitive towards temperature changes. Another reason why Fe oxides tend to maintain rather constant $\delta^{18}O$ values independent of metamorphic grade might be that the amount of O in the Fe oxides "buffers" the isotopic composition of SiO_2 which reaches in granulite facies banded iron formations from Minas Gerais unusual values as low as 6% per mil (Fig. 3).

Possible Paleoclimatoloical Significance

There has been much debate about the paleoclimatological significance of ^{18}O contents of banded iron formations. The observed ^{18}O decrease with geological time in cherts, together with the corresponding trends in carbonates and phosphorites, has lead to the conclusion that Precambrian oceans were depleted in ^{18}O (Perry et al 1973) relative to today's ocean or were warmer (Knauth and Epstein, 1976; Karhu and Epstein, 1986), the two alternatives not being exclusive. On the other hand, Muehlenbachs and Clayton (1976) and Gregory and Taylor (1981) have proposed that the $\delta^{18}O$ value of the ocean remains constant because of the buffering effect of ocean water-basalt interactions.

Fe oxide data discussed here do not add any decisive evidence to the pros and cons of this controversial subject. Nevertheless, the light $\delta^{18}O$ values from the Mesabi and Weld Range (Fig. 3) do not exclude the possibility of the precipitation in a fresh water environment, a conclusion which has been already made by Garrels

(1987) on totally different grounds. If this is so, an important implication is that the Precambrian was characterized by a hydrologic regime much like the present one.

CONCLUSIONS

The isotopic composition of goethites may provide insight into their environments of formation. After transformation into hematite, however, the primary record obviously erases; such a conclusion arises from analysis of the least metamorphosed banded iron formations.

Potentially Fe oxides besides quartz represent the most suitable minerals for geothermometry. Calculated temperatures from quartz-Fe oxide pairs of banded iron formations of all metamorphic grades vary from 140 to 810°C, which encompass the whole range of crustal temperatures. Because of ill-defined calibration curves at low temperatures calculated temperatures below 400°C should not be taken too seriously and may need refinement and revision.

REFERENCES

Becker R.H. and Clayton R.N. (1976) Oxygen isotope study of a Precambrian banded iron formation, Hamersley Range,Western Australia. Geochim. Cosmochim. Acta, 40, 1153-1165.

Bertenrath R., Friedrichsen H. and Hellner E. (1973) Die Fraktionierung der Sauerstoffisotope $^{18}O/^{16}O$ im System Eisenoxid-Wasser. Fortschr. Miner., 50, 32-22.

Blattner P., Braithwaite W.R. and Glower R.B. (1983) New evidence on magnetite oxygen isotope geothermometers at 175 and 112°C in Wairakei steam pipelines (New Zealand). Isot. Geosci., 1, 195-204.

Cloud P. (1983) Banded iron formation - a gradualists's dilemma, In: Iron formation: facts and problems, Trendall A.F. and Morris R.C., (eds.), Develop. in Prec. Geol., 6, Elsevier Publ. Amsterdam.

Fischer W.R. and Schwertmann U. (1975) The formation of hematite from amorphous iron. III-hydroxide. Clays and Clay Miner., 23, 33-37.

Garrels R.M. (1987) A model for the deposition of the microbanded Precambrian Iron Formations. Am. Jour. Sci., 287, 81-106.

Garrels R.M. (1988) Reply to a discussion of Morris and Trendall. Am. Jour. Sci., 288, 669-673.
Garrels R.M. and Christ C.L. (1965) Solutions, minerals and equilibria. Freeman, Cooper Co., San Francisco.
Gregory R.T. (1986) Oxygen isotope systematics of quartz-magnetite pairs from Precambrian Iron Formations: evidence for fluid-rock interaction during diagenesis and metamorphism. In: Fluid-rock interactions during metamorphism, Walther J.V. and Wood B.J. (eds.), Adv. in Phys. Geochem., 5, Springer Verlag Heidelberg, 132-153.
Hangari K., Ahmad S.D. and Perry E.C. (1980) Carbon and oxygen isotope ratios in diagenetic siderite and magnetite from Upper Devonian Ironstone, Wadi Shatti District, Libya. Econ. Geol., 75, 538-545.
Hoefs J., Muller G. and Schuster A.K. (1982) Polymetamorphic relations in iron ores from the Iron Quadrangle, Brazil: the correlation of oxygen isotope variations with deformation history. Contr. Mineral. Petrol., 79, 241-251.
Hoefs J., Muller G., Schuster A.K. and Walde D. (1987) The Fe-Mn ore deposits of Urucum, Brazil: an oxygen isotope study. Chem. Geol., 65, 311-319.
Hoernes S. (1980) A BASIC-program for the calculation of O-isotope temperatures. Contr. Mineral. Petrol., 22, 107-108.
James H.L. (1954) Sedimentary facies of iron-formations. Econ. Geol., 49, 325-293.
James H.L. (1966) Chemistry of the iron-rich sedimentary rocks. In: Date of Geochemistry, Chapter W., U. S. Geol. Surv. Prof. Paper, 440-W.
Knauth L.P. and Epstein S. (1976) Hydrogen and oxygen isotope ratios in nodular and bedded cherts. Geochim. Cosmochim. Acta, 40, 1095-1108.
Morris R.C. and Trendall A.F. (1988) A model for the deposition of the microbanded Precambrian iron formations. Am. Jour. Sci., 288, 664-669.
Muller G., Schuster A.K. and Hoefs J. (1986) The metamorphic grade of banded-iron formations: oxygen isotope and petrological constraints. Fortschr. Miner., 64, 163-185.
Nealson K.H. (1982) Microbiological oxidation and reduction of iron. In: Mineral deposits and the evolution of the biosphere, Holland H.D. and Schidlowski M. (eds.), Springer Verlag, Heidelberg,
O'Neil J.R. and Clayton R.N. (1964) Oxygen isotope geothermometry. In: Isotopic and cosmic chemistry, Craig H., Miller H. and Wasserburg G.J. (eds.), North Holland, Amsterdam.
Perry E.C. and Ahmad S.N. (1983) Oxygen isotope geochemistry of Proterozoic chemical sediments. Geol. Soc. Am. Mem., 161, 253-263.
Perry E.C., Ahmad S.N. and Swulius T.M. (1978) The oxygen isotope composition of 3,800 M.Y. old metamorphosed chert and iron formation from Isukasia, West Greenland. Jour. Geol., 86, 223-229.
Schwertmann U. (1969) Die Bildung von Eisenoxidmineralen. Fortschr. Miner., 64, 274-285.
Trendall A.F. (1983) The Hamersley Basin. In: Iron formations: facts and problems, Trendall A.F. and Morris R.C. (eds.), Develop. in Prec. Geol., 6, Elsevier Publ., Amsterdam.
Yapp C.J. (1983) Stable hydrogen isotopes in iron oxides-isotope effects associated with the dehydration of a natural goethite.

Geochim. Cosmochim. Acta, 47, 1277-1287.
Yapp C.J. (1987a) Oxygen and hydrogen isotope variations among goethites (α-FeOOH) and the determination of paleotemperatures. Geochim. Cosmochim. Acta, 51, 355-364.
Yapp C.J. (1987b) A possible goethite-iron. III- carbonate solid solution and the determination of CO_2 partial pressures in low temperature geologic systems. Chem. Geol., 64, 259-268.
Yapp C.J. and Pedley M.D. (1985) Stable hydrogen isotopes in iron oxides II. D/H variation among natural goethites. Geochim. Cosmochim. Acta, 49, 487-495.
Yapp C.J. and Poths H. (1986) Carbon in natural goethites. Geochim. Cosmochim. Acta, 50, 1213-1220.

ISOTOPIC RECORDS IN DETRITAL AND AUTHIGENIC FELDSPARS IN SEDIMENTARY ROCKS

Gunter FAURE
Department of Geological Sciences
The Ohio State University, Columbus, Ohio 43210 (USA)

When igneous and metamorphic rocks are exposed to chemical weathering at or near the surface of the earth, the minerals they contain decompose at different rates in accordance with Bowen's Reaction Series (Goldich, 1938). Consequently, quartz, muscovite and K-feldspar, which are more resistant to weathering than the ferromagnesian minerals and the plagioclase feldspars, may be incorporated into the solid weathering products. These contain, in addition, certain accessory minerals which also resist weathering: zircon, magnetite, ilmenite, spinel, garnet, tourmaline, sphene, etc. The grains of resistant minerals are concentrated by running water in deposits of sand and gravel which may become lithified into sandstones and conglomerates.

Geologists have traditionally used the presence of certain resistant minerals in sandstones to identify the sources from which the minerals originated. For example, purple zircons called hyacinth occur only in rocks of Archean age because the color results from radiation damage caused by the decay of U, Th and their unstable daughters (Poldervaart, 1955). Other resistant minerals that are derived from specific source rocks include diamonds, native Pt-group elements, glaucophane, certain kinds of garnets, tourmaline, spinel, and others.

Zircon is especially useful as an indicator of its provenance because it is not only highly resistant to chemical weathering and mechanical abrasion, but can also be dated by the isotopic U-Th-Pb method and by the fission-track method (Hurford et al., 1984). The technique for dating zircons by the isotopic U-Pb method has been refined such that small numbers of grains selected on the basis of color, shape and magnetic susceptibility can be dated (Gaudette et al., 1981). Even fragments of single zircon grains have been dated successfully by conventional wet chemical methods by Schärer and Allègre (1982), Gariepy et al. (1984) and by others, whereas Froude et al. (1983) used an ion microprobe called SHRIMP to identify 4.1 to 4.2 Ga detrital zircons from the Mt. Narryer quartzite in the Archean Yilgarn Block of Western Australia.

In this chapter we focus attention on the alkali feldspars which occur in sedimentary rocks both as detrital grains and as authigenic minerals or as authigenic overgrowths on detrital grains. The alkali feldspars are not only common constituents of sedimentary rocks, but also contain different kinds of isotopic records from which useful information about their provenance and about the geologic history of their host rocks can be derived.

THE OCCURRENCE OF FELDSPAR

Detrital Feldspars in Sedimentary Rocks

The feldspars are the most abundant group of minerals in the crust of the earth because they occur in most igneous and high-grade metamorphic rocks. They are aluminosilicates of Na, K, Ca and Ba, but also contain many other cations including Rb, Sr, Pb and even NH_4^+ (Gulbrandsen, 1974). The mineralogy of the feldspars is presented in textbooks and in specialized monographs by Barth (1969), Smith (1974), and Ribbe (1983).

Feldspar is a common constituent of many sand and gravel deposits and occurs as a detrital mineral in sandstones and conglomerates that formed by lithification of such deposits (Pettijohn, 1963). In addition, detrital feldspar occurs in unsorted weathering products, including till and soil, and has also been reported in deep-sea sediment from the South Pacific Ocean by Peterson and Goldberg (1962) who concluded that the feldspar in deep sea sediment was derived from volcanic centers on the ocean floor.

Detrital feldspar in terrigenous sediment tends to be enriched in K-feldspar compared to plagioclase because it is more resistant to chemical weathering and because plagioclase grains break along twinning planes more readily than do grains of K-feldspar (Pittman, 1969; Haldorsen, 1978). For example, Taylor and Faure (1981) reported that the ratio of K-feldspar to plagioclase in Wisconsinan till of Ohio and Indiana (U.S.A.) increases with increasing grain size and that the Rb/Sr ratios of these size fractions increase similarly. The same authors later observed the same relationships in Neogene till of the Transantarctic Mountains (Faure and Taylor, 1981; Faure et al., 1983).

The difference in the relative susceptibilities to chemical weathering of K-feldspar and plagioclase has led to suggestions that their abundances in detrital sedimentary rocks indicate paleoclimatic conditions (Todd, 1968) or reflect the advent of land plants during the Silurian Period (Basu, 1981). However, a study by James et al. (1981) of feldspar weathering in Holocene stream deposits did not support Todd's proposal because they found that plagioclase (Ab 80-95) is more altered than K-feldspar regardless of climatic conditions. Moreover, Basu's hypothesis was challenged by Graustein and Velbel (1981) because K-uptake by plant roots does not necessarily accelerate weathering of K-feldspar in soil and because of the absence of evidence that the K-feldspar/plagioclase ratios of all kinds of sandstones actually decreased during the Silurian Period when plants began to grow on land.

When detrital feldspar grains in sediment are buried, they may decompose as a result of diagenetic reactions with connate water. This process was called "burial metamorphism" by Aronson and Hower (1976) who demonstrated that the increase in the illite/smectite ratio with depth in Tertiary sediment deposited at the mouth of the

Mississippi River coincides with a decrease in the abundance of K-feldspar. They concluded from this evidence that K-feldspar was dissolving and that the K ions entered interlayer sites of smectite causing it to be converted to illite. Similarly, Land and Milliken (1981) demonstrated that plagioclase is converted to albite and that the abundance of K-feldspar decreases with depth in the sandstones of the Frio Formation (Oligocene) in Brazoria County (Texas). Gold (1987) reported similar observations from a study of Miocene sandstones in Louisiana.

Authigenic Feldspar in Sedimentary Rocks

It has been known for a surprisingly long time that many sedimentary rocks contain small amounts of authigenic feldspar (Drian, 1861). Both albite and the K-feldspar adularia occur in many sandstones, siltstones, shales, limestones, dolomites, and evaporites as authigenic constituents that formed at low temperature by reactions involving aqueous solutions (Füchtbauer, 1956; Kastner and Siever, 1979). The existence of authigenic feldspar in unmetamorphozed sedimentary rocks was somewhat controversial because of difficulties in distinguishing it from detrital feldspar by optical methods or by X-ray diffraction (Wright, 1968). In addition, it was by no means obvious how feldspars, which crystallize at high temperatures in igneous and metamorphic rocks, can also form by chemical reactions involving aqueous solutions at low temperatures.

The stability of the feldspars in contact with aqueous solutions is now explained by means of activity diagrams based on thermodynamics and the Law of Mass Action (Helgeson, 1972). Such diagrams were used by Kastner and Siever (1979) to demonstrate that bottom seawater and interstitial water in marine sediment may be supersaturated with respect to K-feldspar and may even reach saturation with respect to amorphous Si. In addition, albite and amorphous Si may also form from interstitial water, which explains why adularia and albite may coexist in equilibrium with secondary quartz in sedimentary rocks of marine origin.

However, the Ca-feldspar anorthite cannot form in this way because calcite precipitates before the activity of Ca increases sufficiently to be in equilibrium with anorthite (Bowers et al., 1984). The instability of anorthite in contact with natural solutions explains the observation by Land and Milliken (1981) that detrital plagioclase in the Frio Formation of Texas was progressively converted to albite with increasing depth. The Ca released by the decomposition of plagioclase subsequently precipitated as a carbonate cement.

The activity diagrams also indicate that detrital feldspar grains may be stabilized when they are emersed in solutions having the required activities of K, Na, H and H_4SiO_4. In such environments, secondary overgrowths of authigenic albite or adularia may form on detrital feldspar grains. Such overgrowths are common in the Permo-Triassic sandstones of England and elsewhere (Waugh, 1978). The presence of both detrital and authigenic feldspars in the same rock greatly enriches the isotopic record that is available for study, provided that the two varieties can be separated from each other. A method for separating overgrowths of secondary quartz, developed by Lee and Savin (1985), was used successfully by Girard et al. (1988) to separate diagenetic feldspar from its detrital substrate. In addition, Hearn (1987) proposed a quantitative technique to determine the abundances of authigenic and detrital feldspar in mineral concentrates using cathode luminescence and a scanning electron microscope.

Feldspar in Hydrothermal Deposits and in Volcanic Ash

Adularia is a common constituent of hydrothermal veins associated with granitic intrusives. It occurs in the quartz-fluorite veins of Europe (Pivec, 1973; Lippolt et al., 1985), is associated with calcite and zeolite in veins on the Lizard Peninsula of Cornwall, England (Halliday and Mitchell, 1976), and has been reported from veins in the geothermal areas of New Zealand (Steiner, 1970; Blattner, 1975). The occurrence of adularia in active geothermal systems is especially important because the temperature and chemical composition of the solution from which it is forming can be studied directly. Adularia has also been reported from spilites on the Iberian Peninsula of Europe (Munhà et al., 1980) and from the Precambrian basement rocks of Ohio (Bass, 1960; Faure and Barbis, 1983; Mensing and Faure, 1983). Additional occurrences of adularia in hydrothermal vein deposits were discussed by Bass and Ferrara (1969) and by Cerny and Chapman (1986).

Feldspar crystals of igneous origin occur also in deposits of volcanic ash and in bentonites derived from them. In addition, such deposits may contain cogenetic zircon, biotite, and plagioclase all which are datable by precise isotopic methods. Such dates reflect the time of crystallization and rapid cooling of the minerals and commonly indicate the time of deposition of the ash layers. Dates derived from phenocrystic minerals of bentonites are good estimates of the depositional ages of the sedimentary rocks with which they are interbedded and therefore help to define the geologic time scale (Odin, 1982; Baadsgaard and Lerbekmo, 1983; Baadsgaard et al., 1988; Yanagi et al., 1988). However, bentonites may be altered by hot brines which may cause conversion of smectite to illite and resetting of the isotopic dates derivable from the clay minerals (Elliott and Aronson, 1987). Evidence for the presence of brines derived from the isotope record of authigenic K-feldspar in Cambrian carbonate rocks was reported by Hearn and Sutter (1985) and by Hearn et al. (1987).

Finally, we mention the unexpected discovery of spherules of almost pure K-feldspar in the basal laminae of sediment deposited during the transition from the Cretaceous to the Tertiary periods at Caravaca, Spain (Smit and Klaver, 1981). These spherules may have formed from a melt that resulted from the impact of an asteroid on the earth. The spherules are very K-rich (K/K+Na = 0.99), but contain opaque inclusions with high concentrations of Cr, Ni, Co, As and Sb. The feldspar occurs in the form of crystal-lites radiating from the opaque inclusions. Sanidine spherules have also been found in K-T boundary layers in the Pacific Ocean (Kyte and Rasmussen, 1982), as well as in Colorado and New Mexico (Izett, 1987). However, spherules extracted from this layer in Umbria (Italy) are rich in siderophile elements (Smit and Kyte, 1984) and the origin of the different kinds of spherules in the boundary layers remains controversial (Cisowski, 1988; Montanari, 1986).

PROVENANCE DATING OF DETRITAL FELDSPAR

Detrital feldspar grains in sedimentary rocks crystallized at the time of formation of the igneous or high-grade metamorphic rocks from which they originated. Their ages therefore reflect events in the geologic history of their source rocks and may be used to identify their provenance. Feldspars are potentially datable by several isotopic methods including the Rb-Sr, K-Ar, K-Ca, $^{40}Ar/^{39}Ar$, and the common Pb method (Faure, 1986). Each of these relies on the validity of assumptions pertaining to the relevant parent-daughter decay schemes. However, when these methods are used to date detrital feldspar grains from a sedimentary rock, additional assumptions about the geology of the source area and about the state of preservation of the feldspar come into play. These assumptions include:

1. All feldspar grains recovered from a specific stratigraphic interval originated from the same source rocks which had a specific crystallization age.
2. The parent-daughter decay schemes in the feldspar grains were not disturbed as a result of chemical weathering of the source rocks or during transport of the feldspar grains to the site of deposition.
3. The feldspar grains were not altered during diagenesis (burial metamorphism) or by later interaction with circulating brines.
4. Overgrowths of authigenic feldspar are either not present or were removed prior to analysis for isotopic dating.

These assumptions are rarely, if ever, completely satisfied. In many cases, detrital feldspar grains in sedimentary rocks are mixtures derived from source rocks having different ages. The grains are commonly altered to kaolinite or illite as a result of weathering or during diagenesis after deposition, and they may contain overgrowths of authigenic feldspar or may be completely reconstituted to adularia or secondary albite by reactions with brines after diagenesis. These difficulties significantly restrict the use of detrital feldspar in sedimentary rocks for studies of their provenance.

For this reason, isotopic dating of detrital muscovite is an attractive alternative because this mineral is more resistant to alteration than K-feldspar and has higher concentrations of K and Rb than most feldspars. In addition, muscovite is somewhat easier to recover in pure form from sedimentary rocks because of its distinctive grain shape and larger grain size than detrital feldspar (Heller et al., 1985).

The intrinsic limitations on dating feldspar by isotopic methods are:

1. Most K-feldspars have low Rb/Sr ratios (< 1.0) because Sr is captured in place of K whereas Rb is only admitted by growing crystals of microcline or orthoclase. As a result, the amount of radiogenic ^{87}Sr in K-feldspars is low compared to muscovite, which increases the uncertainty of dates derived from them.
2. When mixtures of detrital feldspar grains, derived from sources having different ages, are dated by the Rb-Sr method they do not form proper Rb-Sr isochrons and dates derived from the slopes of linear data arrays have no real time significance.
3. K-feldspar does not retain radiogenic ^{40}Ar quantitatively, except volcanic sanidine in tuff beds or bentonites. Therefore, K-Ar dates of detrital feldspar underestimate the crystallization age.
4. Mixtures of detrital feldspars derived from sources having different ages yield intermediate K-Ar dates which reflect the proportions of mixing and have no real time significance.

In spite of the limitations on isotopic dating of detrital feldspar, useful information was obtained by Heller et al. (1985) from K-feldspar in sandstones of the Eocene Tyee Formation in Oregon (U.S.A.) and by Szabo and Faure (1987) who dated feldspar from the Berea Sandstone (Lower Carboniferous) of Ohio. In addition, Patterson and Tatsumoto (1964) interpreted the isotope compositions of Pb in detrital feldspars extracted from beach and river sands in North America.

Detrital feldspar in <u>till</u> is especially well-suited for dating by the Rb-Sr method because the feldspar grains are commonly less altered than they are in sandstones. Moreover, the preferential grinding of plagioclase relative to K-feldspar causes differences in the Rb/Sr ratios of the feldspar in the sand-size fractions of till (Faure and Taylor, 1981; Faure et al., 1983).

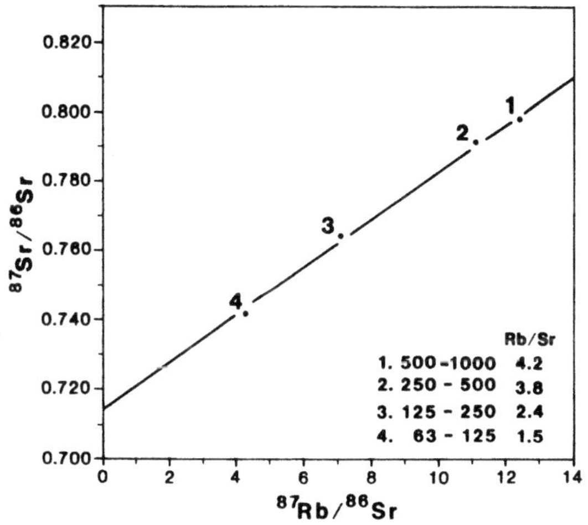

Figure 1: Size fractions of feldspar-quartz concentrates of till from the summit plateau of the Wisconsin Range in the Transantarctic Mountains. The slope of the line yields a date of 480 ± 21 Ma which is characteristic of the granitic basement rocks of the region. The Rb/Sr ratios decrease with decreasing grain diameters because of preferential grinding of plagioclase by the ice sheet (Faure et al., 1983)

An example of feldspar dating by the Rb-Sr method is the study by Faure et al. (1983) of till from the summit plateau of the Wisconsin Range in the Transantarctic Mountains (Mercer, 1968). A sample of this till was sieved into four size fractions between 63 μm and 1000 μm. These fractions were cleaned ultrasonically in dilute HCl and demineralized water for about 8 hours, followed by removal of ferromagnesian minerals on a magnetic separator. The resulting feldspar-quartz concentrates were analyzed for dating by the Rb-Sr method. The results form four colinear points on the Rb-Sr isochron diagram in figure 1. The slope of the line yields a date of 480 ± 21 Ma ($\lambda\ ^{87}Rb = 1.42 \times 10^{-11}\ y^{-1}$) and the initial $^{87}Sr/^{86}Sr$ ratio is 0.7144 ± 0.0030. Faure et al. (1983) concluded from these results that the feldspar in this till sample had originated from the local granitic basement rocks which are known to have crystallized during the Ross Orogeny in Cambro-Ordovician time. Therefore, this till may have been deposited by an ice cap that formed on the Wisconsin Range plateau, as originally proposed by Mercer (1968). The apparent absence of a component of Precambrian feldspar in this till means that it was probably not formed by ice which originated on the East Antarctic craton which contains granulite gneisses of Archean age (Grew and Manton, 1979).

A component of Precambrian feldspar is present in till studied by Faure and Taylor (1981) from Mt. Tuatara along the Byrd Glacier in the Transantarctic Mountains. Feldspar fractions at this locality yielded a date of 1100 $\pm$ 69 Ma which is the result of mixing of grains derived from the local granitic basement (about 500 Ma) and from the East Antarctic craton (about 2500 Ma). The presence of Precambrian feldspar in till deposited by the Byrd Glacier is consistent with the fact that it drains a large area of the ice-covered interior of East Antarctica.

THERMOCHRONOMETRY OF DETRITAL MICROCLINE

When detrital feldspar grains derived from granitic rocks of uniform age are buried after deposition, they may be reheated sufficiently to lose some of the radiogenic ^{40}Ar that had formed by decay of ^{40}K since they cooled through their closure temperature of about 130°C. Therefore, the K-Ar dates of detrital feldspar in sedimentary rocks may decrease with increasing depth in a core or with increasing age of their sedimentary host rocks. This tendency may be counteracted by the fact that feldspar derived from the top portion of a granite batholith may have older K-Ar ages than feldspar which originated from deeper levels in the batholith because the top of the batholith cooled through the closing temperature of microcline before the interior did. Consequently, the K-Ar dates of detrital feldspar released during the erosion of such a batholith may increase with the stratigraphic age of their sedimentary host.

The loss of radiogenic ^{40}Ar from K-feldspar grains caused by diffusion in response to heating can be revealed by the $^{40}Ar/^{39}Ar$ age spectrum obtained by stepwise release of Ar at increasing temperatures (Faure, 1986). An example of such a study is the work of Harrison and Bé (1983) who obtained $^{40}Ar/^{39}Ar$ age spectra for detrital microcline concentrates from drill cores of Tertiary sandstones in the San Joaquin Valley of central California (U.S.A.). The spectra derived from eleven samples yielded plateau dates that averaged about 78 Ma, whereas the fission-track dates of detrital zircons from the same cores yielded 83 Ma (Briggs et al., 1981). These dates refere to the age of the igneous rocks in the adjacent Tehachapi and Sierra Nevada Ranges from which the sediment originated. The dates represent the time elapsed since microcline and zircon cooled through their respective blocking temperatures when radiogenic ^{40}Ar and fission tracks began to accumulate.

The $^{40}Ar/^{39}Ar$ ratios of the very first gas released by the microclines yielded fictiteous dates of about 65 Ma which are attributable to loss of ^{40}Ar during slow cooling of the granitic pluton. However, the microcline recovered from the greatest depth of 6200 m lost additional radiogenic ^{40}Ar amounting to 18%, presumably caused by geothermal heating. Harrison and Bé (1983) used Fick's second law of diffusion to show that the observed loss of ^{40}Ar could have been caused by heating the perthitic microcline to 160°C starting at 200 ka in the past. However, the ^{40}Ar loss can also be attributed to heating for a longer period of time as a result of increasing depth of burial. If the geothermal gradient was about 30°C/km, the temperature to which the microcline was heated would have increased steadily to its present observed value of 157°C. The loss of ^{40}Ar caused by diffusion under these conditions would also amount to about 18%. Therefore, the $^{40}Ar/^{39}Ar$ method of dating detrital microcline provides not only a provenance date but can, under favorable circumstances, also reveal evidence regarding its thermal history.

OXYGEN ISOTOPE COMPOSITIONS OF DETRITAL AND AUTHIGENIC FELDSPARS

Authigenic feldspars which formed in sedimentary rocks from aqueous solutions at low temperatures may be enriched in ^{18}O compared to detrital feldspar of igneous origin (Savin and Epstein, 1970; Lawrence and Kastner, 1975). Therefore, the isotope composition of O can be used to distinguish between authigenic and detrital feldspars that may coexist in a sedimentary rock. In addition, the $\delta^{18}O$ values of cogenetic secondary quartz and authigenic feldspar can be used to determine both the $\delta^{18}O$ value of the water and the temperature of isotope equilibration using experimentally determined fractionation equations (Friedman and O'Neil, 1977).

The temperature-dependence of the isotope fractionation factors is expressed by equations of the form (Faure, 1986):

$$1000 \ln \alpha^{1}_{2} = \frac{A \times 10^6}{T^2} \quad B \tag{1}$$

where α^{1}_{2} is the fractionation factor of O in phases 1 and 2, A and B are experimentally determined constants, and T is the temperature in Kelvins. For albite (or K-feldspar), equation 1 takes the form (Matsuhisa et al., 1979):

$$1000 \ln \alpha^{f}_{w} = \frac{1.59 \times 10^6}{T^2} \quad 1.16 \tag{2}$$

and for quartz:

$$1000 \ln \alpha^{q}_{w} = \frac{3.34 \times 10^6}{T^2} \quad 3.31 \tag{3}$$

Equations 2 and 3 can be combined by making use of the approximation that:

$$1000\ln \alpha^{1}_{2} = \delta_1 - \delta_2 \qquad (4)$$

Therefore, the fractionation of O isotopes between feldspar and quartz in equilibrium with water of the same isotope composition at the same temperature is:

$$\delta^{18}O_q - \delta^{18}O_f = \frac{1.75 \times 10^6}{T^2} \quad 2.15 \qquad (5)$$

However, this possibility has not yet been realized partly because it is difficult to separate overgrowths of quartz and feldspar from their substrates.

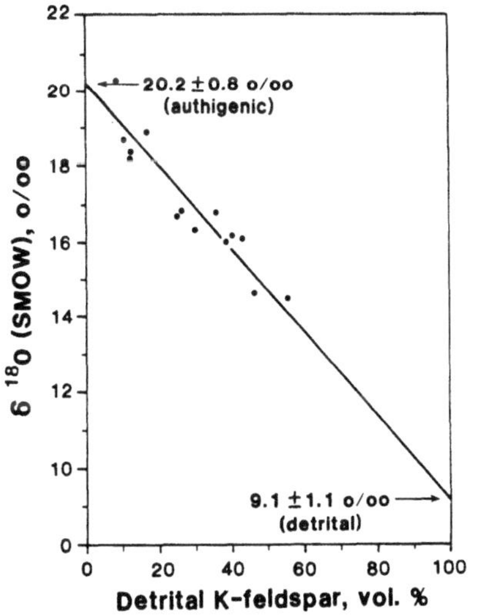

Figure 2: Relationship between $\delta^{18}O$ values of concentrates of authigenic K-feldspar overgrowths and the abundance of residual detrital K-feldspar in three arkoses from Angola. The concentrations of detrital feldspar were estimated by point counting based on cathodoluminescence. The extrapolated $\delta^{18}O$ values apply to the pure authigenic and detrital feldspar components (Girard et al., 1988)

Authigenic overgrowths of adularia were successfully concentrated by Girard et al. (1988) from Lower Cretaceous arkose in Angola, southwest Africa, and the $\delta^{18}O$ values and K-Ar dates of the concentrates were measured. The results indicated that both parameters varied linearly with the abundance of detrital K-feldspar remaining in the concentrates. The O data in figure 2 were extrapolated to the pure end-members whose $\delta^{18}O$ are +20.2 $\pm$ 0.8 permil for the authigenic feldspar and +9.1 ± 1.1 permil for the detrital component. The K-Ar dates of the two feldspar components were similarly resolved into 98 ± 16 Ma for the authigenic feldspar and 516 ± 26 Ma for the detrital component. These dates indicate that the authigenic overgrowths formed during the Cretaceous Period about 50 Ma after deposition. The detrital feldspar probably originated from granitic basement rocks which participated in the Pan-African event in Cambro-Ordovician time.

The average $\delta^{18}O$ value of the authigenic feldspar overgrowths can be used to calculate the $\delta^{18}O$ of the water from which they precipitated, assuming a temperature of 145 ± 4°C, which is the homogenization temperature of fluid inclusions in these overgrowths. The result obtained by Girard et al. (1988) is that the $\delta^{18}O$ value of the water was between +5.7 and +8.9 permil (SMOW). Evidently, the water was strongly enriched in ^{18}O compared to meteoric water or seawater, either as a result of prior evaporation at the surface or because of isotope exchange with oxygen-bearing minerals at elevated temperatures. In either case, the brines did not evolve in situ but invaded the rocks after deposition and caused the precipitation of feldspar overgrowths.

ISOTOPIC RECORDS OF AUTHIGENIC FELDSPAR

Authigenic K-feldspar and albite are common but minor constituents of many sedimentary rocks. These minerals can form on the bottom of the oceans, in which case they record both the time of sedimentation and the $\delta^{18}O$ or temperature of the water from which they formed. However, authigenic feldspar may also form after sediment deposition during diagenesis or by reactions caused by circulating brines. In such cases, the ages of the feldspars post-date the time of deposition, and their O isotope compositions depend on the $\delta^{18}O$ of the connate water or circulating brines and on the temperature at the time the reactions took place. Migrating brines may also cause the formation of adularia in igneous or metamorphic rocks at unconformities along which the primary feldspars may have been altered to clay minerals when the rocks were exposed to weathering. The dates derived from such authigenic feldspars also postdate the age of the basement rocks and may approach the time of brine alteration.

Authigenic sanidine in nearly monomineralic layers in the St. Peter Sandstone of Wisconsin was first dated by the K-Ar method by Krueger and Woodard (1972) and more recently by Marshall et al. (1986). The latter reported discordant K-Ca and K-Ar dates for two samples of 443 $\pm$ 12 and 403 $\pm$ 16 Ma (K-Ca) and 402 $\pm$ 13 and 377 $\pm$ 13 Ma (K-Ar), respectively. Information reviewed by Marshall et al. (1986) indicates that the St. Peter Sandstone (Middle Ordovician) is older than about 454 Ma. Consequently, even the K-Ca dates are younger than the depositional age of the rocks, and this indicates that the sanidine is not syngenetic but is either late diagenetic or epigenetic in origin. This is a very significant result because dates derived from such authigenic feldspar <u>cannot</u> be used to constrain the geologic timescale. In addition, Marshall et al. (1986) used the apparent loss of radiogenic ^{40}Ar from the two sanidines (about 8.4%) to show that Ar diffuses significantly faster at low temperatures than expected from an extrapolation of experimental data obtained at high temperatures.

The presence of authigenic feldspar in the crystalline basement rocks of Ohio (U.S.A.) was first indicated by Faure and Barbis (1983). They reported that whole-rock samples of granitic gneiss taken up to 6.5 meters below the unconformity yielded an anomalously low Rb-Sr isochron date of 699 $\pm$ 16 Ma, whereas samples from greater depth confirmed the Grenville age of these rocks with a date of 1173 $\pm$ 41 Ma. Subsequently, Mensing and Faure (1983) demonstrated by cathode luminescence that the feldspars below the unconformity contained both secondary low-temperature feldspar (adularia) and residual primary feldspar, whereas the feldspar more than 6.5 meters below the former erosion surface was composed only of microcline. The mixed adularia-microcline feldspars yielded a date of only 599 $\pm$ 69 Ma compared to 1162 $\pm$ 11 Ma for the unaltered microclines, in good agreement with the whole-rock date obtained by Faure and Barbis (1983). Mensing and Faure (1983) regarded the line formed by the microclines in figure 3 as an isochron but interpreted the adularia-microcline array as a mixing-line. Therefore, they concluded that the adularia is younger than the date indicated by the slope of this line.

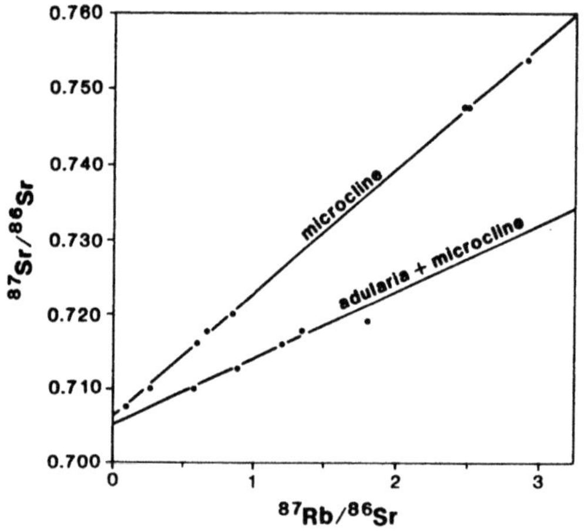

Figure 3: Rb-Sr dating of feldspars in granitic basement rocks of Ohio, U.S.A.. The microclines form an isochron yielding an age of 1168 ± 11 Ma consistent with the Grenville age of these rocks. However, the adularia + microcline concentrates, taken within 6.4 meters of the unconformity which separates the basement from the overlying Paleozoic sedimentary rocks, is a mixing line. Therefore, the age of the adularia is <u>less</u> than the 599 ± 69 Ma date indicated by its slope (Mensing and Faure, 1983)

The $\delta^{18}O$ of the feldspars from the Ohio basement rocks were increased as a result of the formation of adularia in such a way that $\delta^{18}O$ correlates positively with K and negatively with Na concentrations (Shieh and Mensing, 1986). Apparently, the exchange of O isotopes was accompanied by the replacement of Na by K ions in the feldspar. In addition, MacIntyre (1986) reported that K-Ar dates of the same feldspar concentrates increase with depth in the upper portion of the core from about 370 to 420 Ma. However, the unaltered microcline 6.4 meters below the unconformity yielded K-Ar dates between 750 to 850 Ma which are consistent with K-Ar dates of biotites from the Precambrian basement of Ohio, if allowance is made for the loss of radiogenic ^{40}Ar from the microcline by diffusion at low temperature (Bass, 1960; Lidiak et al., 1966).

EVIDENCE FOR HYDROTHERMAL BRINES IN THE APPALACHIANS

The isotopic records preserved by the adularia in the basement rocks of Ohio point to the invasion of hot saline solutions in Late Paleozoic time which affected not only the basement rocks but interacted also with the sedimentary rocks of Early Paleozoic age that overlie them. For example, Grant et al. (1984) reported that glauconite from the Lower Silurian Brassfield Formation of Ohio and Indiana yielded Rb-Sr and K-Ar dates of 370 ± 11 Ma and 362 ± 6 Ma, respectively. These dates are younger than the stratigraphic age of the Brassfield Formation, but they coincide with the K-Ar dates of the adularia in the underlying basement (MacIntyre, 1986). The dates also agree with Rb-Sr and K-Ar age determinations by Grant et al. (1984) of glauconites from the Davis Formation (Upper Cambrian) in the "Viburnum Trend" of Missouri which is famous for its Mississippi-Valley type Pb-Zn deposits. Evidently, the ages of the glauconites in Ohio, Indiana, and Missouri may have been reset by brines which participated in the deposition of galena and sphalerite of the ore deposits in Missouri.

Evidence for the circulation of hot brines in the Paleozoic rocks of the southern Appalachian Mountains was reported by Elliott and Aronson (1987). They obtained K-Ar dates for authigenic clay (1-2μm and <0.125μm) from Middle Ordovician bentonites ranging from 256 to 313 Ma, which they attributed to the illitization of smectite by brines driven out of the Appalachian Basin during the Alleghanian Orogeny in Late Carboniferous to Permian time. Although their dates are younger than those of Grant et al. (1984), Elliott and Aronson (1987), nevertheless, associated the illitization of bentonites with the formation of Mississippi-Valley type Pb-Zn deposits.

Another line of evidence for the presence of hot brines arises from $^{40}Ar/^{39}Ar$ dating of K-feldspar from marine carbonate rocks of Early Paleozoic age in the southern Appalachian Mountains (U.S.A.). Hearn and Sutter (1985) and Hearn et al. (1987) obtained $^{40}Ar/^{39}Ar$ partial-release spectra of detrital feldspars with authigenic overgrowths from several Cambrian formations in Pennsylvania, Maryland, Virginia, and Tennessee. They dated the detrital feldspars with their overgrowths and also the detrital feldspars alone, after removing the overgrowths by leaching with 5% HF. The age of the authigenic component (A_0) was then calculated for each incremental step by the equation:

$$A_t = A_0 M_0 + A_c (1-M_0) \qquad (6)$$

where A_t is the apparent age of the total sample, A_c is the apparent age of the core, and M_0 is the mass fraction of the authigenic overgrowths. Values of A_t and A_c were calculated by standard methods from the $^{40}Ar/^{39}Ar$ ratios of gas released at increasing temperatures and M_0 was determined by a modal analysis in which the total area of feldspar grains was compared to the area of the grains displaying cathode luminescence. Authigenic feldspars do not luminesce because they lack the trace elements that cause cathode luminescence.

The results reported by Hearn et al. (1987) indicate plateau dates ranging from 278 to 315 Ma for the authigenic feldspars. These dates are within the late Carboniferous-Early Permian Periods and agree with the K-Ar dates of illitization of Ordovician bentonites in the same area reported by Elliott and Aronson (1987). Hearn et al. (1987) demonstrated that the formation of authigenic K-feldspar required the introduction of K-bearing brines. In addition, a study of fluid inclusions within the overgrowths indicated homogenization temperatures between 100 to 200°C and salinities of 18 to 21 weight percent NaCl equivalent. These results are similar to conclusions derived from fluid inclusions of Pb-Zn deposits in the southern Appalachian States. Therefore, Hearn et al. (1987) suggested that the authigenic feldspar and the Pb-Zn deposits were probably formed by the same brines in late Paleozoic time. Moreover, the brine migration was apparently a regional phenomenon associated with the tectonic activity that accompanied the Alleghanian orogeny.

The concordant results reported by Elliott and Aronson (1987) and Hearn et al. (1987) apply to the southern and central Appalachian Mountains, whereas Grant et al. (1984) studied material from the midwestern region of the U.S.A., including Ohio, Indiana, Kentucky and Missouri. Therefore, the reequilibration of the Rb-Sr decay schemes in glauconite by brines in the midwest may have preceded the alteration of bentonites and the formation of authigenic feldspar in the southern Appalachian Mountains. Consequently, the so-called Mississippi-Valley type Pb-Zn deposits may have formed during a lengthy interval of time that lasted from the Early Devonian to the Early Permian Periods.

SUMMARY

Detrital as well as authigenic alkali feldspars are common constituents of sedimentary rocks. They can be dated by different isotopic decay schemes including the K-Ar, $^{40}Ar/^{39}Ar$, K-Ca, Rb-Sr and the Pb-Pb methods. In addition, the O isotope composition of authigenic feldspars can be used to determine the temperature of isotope equilibration or the isotope composition of the water with which they equilibrated.

The examples considered in this chapter include the provenance determination of detrital feldspar in Neogene till of the Transantarctic Mountains, the thermal history of detrital microclines in Tertiary sandstones in the San Joaquin Valley of California, as well as the origin of detrital feldspar and the subsequent formation of overgrowths of adularia in an arkose in Angola in southwest Africa. The late diagenetic or post diagenetic origin of authigenic feldspar in Angola applies also to authigenic sanidine layers in the Middle Ordovician St. Peter Sandstone of Wisconsin. Consequently, dates derived from authigenic feldspar cannot be used to construct the geologic timescale, unless it can be shown that the feldspar is syngenetic or early diagenetic in origin.

Authigenic feldspar in the crystalline basement rocks of Ohio shows evidence of brine migration during the Devonian Period. The same brines also appear to have altered glauconite in the early

Paleozoic rocks of Ohio and Indiana and in the Pb-Zn deposits of Missouri. However, studies of authigenic feldspar and bentonites in the southern Appalachians indicate that the Pb-Zn deposits in the region formed later in Carboniferous to early Permian time.

The studies cited in this chapter illustrate the broad range of important geologic problems that can be addressed by isotopic studies of feldspar in sedimentary rocks. Now that the potential of this line of research has been demonstrated, the field is wide open for much additional research.

ACKNOWLEDGEMENTS

I thank Norbert Clauer and Sam Chaudhuri for the opportunity to present the evidence pertaining to isotopic and geochemical studies of feldspar in sedimentary rocks. Mary Hill typed the manuscript and Susan Smith drafted the diagrams. This paper is contribution 87 of the Laboratory for Isotope Geology and Geochemistry (Isotopia).

REFERENCES

Aronson L.J. and Hower J. (1976) Mechanism of burial metamorphism of argillaceous sediment: 2. Radiogenic argon evidence. Geol. Soc. Am. Bull., 87, 738-744.

Baadsgaard H. and Lerbekmo J.F. (1983) Rb-Sr and U-Pb dating of bentonites. Can. Jour. Earth Sci., 20, 1281-1290.

Baadsgaard H., Lerbekmo J.F. and McDougall I. (1988) A radiometric age for the Cretaceous-Tertiary boundary based upon K-Ar, Rb-Sr, and U-Pb ages of bentonites from Alberta, Saskatchewan and Montana. Can. Jour. Earth Sci., 25, 1088-1097.

Barth T.W.F. (1969) Feldspars. Wiley-Interscience, New-York, N.Y., 261 p.

Baskin Y. (1956) A study of authigenic feldspars. Jour. Geol., 6, 132-155.

Bass M.N. (1960) Grenville boundary in Ohio. Jour. Geol., 68, 673-677.

Bass M.N. and Ferrara G. (1969) Age of adularia and metamorphism, Ouachita Mountains, Arkansas. Am. Jour. Sci., 267, 491-498.

Basu A. (1981) Weathering before the advent of land plants: Evidence from detrital K-feldspar in Cambrian-Ordovician arenites. Geology, 9, 132-133.

Blattner P. (1075) Oxygen isotopic composition of fissure-grown quartz, adularia, and calcite from Broadlands geothermal field, New Zealand. Am. Jour. Sci., 275, 785-800.

Bowers T.S., Jackson K.J. and Helgeson H.C. (1984) Equilibrium

activity diagrams. Springer-Verlag, Berlin, 397 p.
Briggs N.D., Naeser C.W. and McCulloh T.H. (1981) Thermal history of sedimentary basins by fission track dating. Nucl. Tracks, 5, 235-237.
Buyce M.R. and Friedman G.M. (1975) Significance of authigenic K-feldspar in Cambrian-Ordovician carbonate rocks of the Proto-Atlantic shelf of North America. Jour. Sedimen. Petrol., 45, 808-821.
Cerny P. and Chapman R. (1986) Adularia from hydrothermal vein deposits: Extremes in structural state. Can. Min., 24, 717-728.
Cisowski S.M. (1988) Magnetic properties of K/T and E/O microspherules: Origin by combustion? Earth Plan. Sci. Lett., 88, 193-208.
Drian M. (1861) Notice sur les cristaux d'albite renfermés dans les calcaires magnésiens des environs de Modane. Bull. Soc. géol. France, sér. 2, 18, 804-806.
Elliott W.C. and Aronson J.L. (1987) Alleghanian episode of K-bentonite illitization in the southern Appalachian basin. Geology, 15, 735-739.
Faure G. and Taylor K.S. (1981) Provenance of some glacial deposits in the Transantarctic Mountains based on Rb-Sr dating of feldspars. Chem. Geol., 32, 271-290.
Faure G., Taylor K.S. and Mercer J.H. (1983) Rb-Sr provenance dates of feldspar in glacial deposits of the Wisconsin Range, Transantarctic Mountains. Geol. Soc. Am. Bull., 94, 1275-1280.
Faure G. and Barbis F. (1983) Detection of neoformed adularia by Rb-Sr age determination of granitic rocks in Ohio. In: Leaching and diffusion in rocks and their weathering products, Augustithis S.S. (ed.), Theophrastus Publ., Athens, Greece, 307-320.
Faure G. (1986) Principles of Isotope Geology. Second edition. Wiley, New York, N.Y., 589 p.
Friedman I. and O'Neil J.R. (1977) Compilation of stable isotope fractionation factors of geochemical interest. In: Data of Geochemistry, sixth edition, Fleischer M. (ed.), U.S. Geol. Surv. Prof. Paper, 440.
Froude D.O., Ireland T.R., Kinny P.D., Williams I.S., Compston W., Williams I.R. and Myers J.S. (1983) Ion microprobe identification of 4,100-4,200 Myr-old terrestrial zircons. Nature, 304, 616-618.
Füchtbauer H. (1956) Zur Entstehung und Optik authigener Feldspäte. Neues Jahrb. Miner., Monatsh., 1, 9-23.
Gariepy C, Allègre C.J. and Lajoie J. (1984) U-Pb systematics in single zircons from the Pontiac sediments, Abitibi greenstone belt. Can. Jour. Sci., 21, 1296-1304.
Gaudette H.E., Vitrac-Michard A. and Allègre C.J. (1981) North American Precambrian history recorded in a single sample: High-resolution U-Pb systematics of the Potsdam sandstone detrital zircons, New York State. Earth Plan. Sci. Lett., 54, 248-260.
Girard J.P., Aronson J.L. and Savin S.M. (1988) Separation, K/Ar dating and $^{18}O/^{16}O$ ratio measurements of diagenetic K-feldspar overgrowths: An example from the Lower Cretaceous arkoses of the Angola margin. Geochim. Cosmochim. Acta, 51, 2207-2214.
Gold P.G. (1987) Textures and geochemistry of authigenic albite from Miocene sandstones, Louisiana Gulf Coast. Jour. Sedimen. Petrol., 57, 353-362.

Goldich S.S. (1938) A study of rock weathering. Jour. Geol., 46, 17-58.
Grant N.K., Laskowski T.E. and Foland K.A. (1984) Rb-Sr and K-Ar ages of Paleozoic glauconites from Ohio-Indiana and Missouri, U.S.A. Isot. Geosc., 2, 217-239.
Graustein W.C. and Velbel M.A. (1981) Comment and Reply on "Weathering before the advent of land plants: Evidence from detrital K-feldspars in Cambrian-Ordovician arenites." Geology, 11, 505-506.
Grew E.S. and Manton W.I. (1979) Archean rocks in Antarctica: 2.5 billion year uranium-lead ages of pegmatites in Enderby Land. Science, 206, 443-445.
Gulbrandsen R.A. (1974) Buddingtonite, ammonium feldspar, in the Phosphoria Formation, southeastern Idaho. U.S. Geol. Surv. Jour. Res., 2, 693-697.
Haldorsen S. (1978) Glacial comminution of mineral grains. Norsk Geol. Tidssk., 58, 241-243.
Halliday A.N. and Mitchell J.G. (1976) Structural, K-Ar and ^{40}Ar-^{39}Ar age studies of adularia K-feldspars from the Lizard Complex, England. Earth Plan. Sci. Lett., 29, 227-237.
Harrison T.M. and Bé K. (1983) ^{40}Ar/^{39}Ar age spectrum analysis of detrital microclines from the southern San Joaquin Basin, California: An approach to determining the thermal evolution of sedimentary basins. Earth Plan. Sci. Lett., 64, 244-256.
Hearn P.P. and Sutter J.F. (1985) Authigenic potassium feldspar in Cambrian carbonates: Evidence of Alleghenian brine migration. Science, 228, 1529-1531.
Hearn P.P. Jr. (1987) A quantitative technique for determining the mass-fractions of authigenic and detrital K-feldspar in mineral separates. Scan. Microsc., 1, 1039-1043.
Hearn P.P., Sutter J.F. and Belkin H.E. (1987) Evidence for Late Paleozoic brine migration in Cambrian carbonate rocks of the central and southern Appalachians: Implications for Mississipi Valley-type sulfide mineralization. Geochim. Cosmochim. Acta, 51, 1323-1334.
Helgeson H.C. (1972) Chemical interaction of feldspars and aqueous solutions. In: The Feldspars, McKenzie W.S. and Zussman J. (ed.), Proceedings NATO Advanced Study Inst., Manchester Univ. Press, England, 184-217.
Heller P.L., Peterman Z.E., O'Neil J.R. and Shafiqullah M. (1985) Isotopic provenance of sandstones from the Eocene Tyee Formation, Oregon Cost Range. Geol. Soc. Am. Bull., 96, 770-780.
Hurford A.J., Fitch F.J. and Clarke A. (1984) Resolution of the age structure of the detrital zircon populations of two Lower Cretaceous sandstones from the Weald of England by fission track dating. Geol. Mag., 121, 269.
Izett G.A. (1987) Authigenic "spherules" in K-T boundary sediments at Caravaca, Spain, and Raton Basin, Colorado and New Mexico, may not be impact derived. Geol. Soc. Am. Bull., 99, 78-86.
James W.C., Mack G.H. and Suttner L.J. (1981) Relative alteration of microcline and sodic plagioclase in semi-arid and humid climates. Jour. Sedimen. Petrol., 51, 151-163.
Kastner M. (1971) Authigenic feldspars in carbonate rocks. Am. Min., 56, 1403-1442.
Kastner M. and Siever R. (1979) Low temperature feldspars in sedimentary rocks. Am. Jour. Sci., 279, 435-479.
Krueger H.W. and Woodard H.H. (1972) Potassium-argon dating of sanidine-rich beds in the St. Peter Sandstone, Wisconsin.

Geol. Soc. Amer., Abstr. with Progr., 4, 568-569.
Kyte F.T. and Rasmussen K.L. (1982) Sanidine spherules in a Pacific K-T boundary. Meteoritics, 17, 239-240.
Land L.S. and Milliken K.L. (1981) Feldspar diagenesis in the Frio Formation, Brazoria County, Texas Gulf Coast. Geol., 9, 314-318.
Lawrence J.R. and Kastner M. (1975) $^{18}O/^{16}O$ of feldspars in carbonate rocks. Geochim. Cosmochim. Acta, 39, 97-102.
Lee M. and Savin S.M. (1985) Isolation of diagenetic overgrowths on quartz sand grains for oxygen isotopic analysis. Geochim. Cosmochim. Acta, 49, 497-501.
Lidiak E.G., Marvin R.F., Thomas H.H. and Bass M.N. (1966) Geochronology of the midcontinent region, United States. Jour. Geophys. Res., 71, 5427-5438.
Lippolt H.J., Mertz D.F. and Ziehr H. (1985) The Late Permian Rb-Sr age of a K-feldspar from the Wölsendorf mineralization (Oberpfalz, FR Germany). Neues Jahrb. Mineral., Monatsh., 2, 49-57.
MacIntyre R.M. (1986) K-Ar ages and MVT deposits. Terra cognita, 6, 228.
Marshall B.D., Woodard H.H. and DePaolo D.J. (1986) K-Ca-Ar systematics of authigenic sanidine from Waukau, Wisconsin, and the diffusivity of argon. Geology, 14, 936-938.
Matsuhisa Y., Goldsmith J.R. and Clayton R.N. (1979) Oxygen isotopic fractionation in the system quartz-albite-anorthite-water. Geochim. Cosmochim. Acta, 43, 1131-1140.
Mensing T.M. and Faure G. (1983) Identification and age of neoformed Paleozoic feldspar (adularia) in a Precambrian basement core from Scioto County, Ohio, U.S.A. Contrib. Mineral. Petrol., 82, 327-333.
Mercer J.H. (1968) Glacial geology of the Reedy Glacier area, Antarctica. Geol. Soc. Am. Bull., 79, 471-486.
Montanari A. (1986) Spherules from the Cretaceous-Tertiary boundary clay at Gubbio, Italy: The problem of outcrop contamination. Geology, 14, 1024-1026.
Munhà J., Fyfe W.S. and Kerrich R. (1980) Adularia, the characteristic mineral of felsic spilites. Contrib. Mineral. Petrol., 75, 15-19.
Odin G.S. (ed.) (1982) Numerical Dating in Stratigraphy. Wiley, Chichester, U.K., 1040 p.
Patterson C. and Tatsumoto M. (1964) The significance of lead isotopes in detrital feldspar with respect to chemical differentiation within the earth's mantle. Geochim. Cosmochim. Acta, 28, 1-22.
Peterson M.N.A. and Goldberg E.D. (1962) Feldspar distributions in South Pacific pelagic sediment. Jour. Geophys. Res., 67, 3477-3492.
Pettijohn F.J. (1963) Chemical composition of sandstones-excluding carbonate and volcanic sands. In: Data of Geochemistry, sixth edition, Fleischer M. (ed.), U.S. Geol. Surv. Prof. Paper, 440-S.
Pittman E.D. (1969) Destruction of plagioclase twins by stream transport. Jour. Sedim. Petrol., 39, 1432-1437.
Pivec E. (1973) Hydrothermal K-feldspar (adularia) from fluorite-bearing veins. Acta Universitatis Carolinae, Geologica, 3, 171-177.
Poldervaart A. (1955) Zircons in rocks. 1. Sedimentary Rocks. Am. Jour. Sci., 253, 433-461.

Ribbe P.H. (edit.) (1983) Feldspar mineralogy. Rev. in Min., Min. Soc. Am., 2, 350p.
Savin S.M. and Epstein S. (1970) The oxygen isotopic compositions of coarse grained sedimentary rocks and minerals. Geochim. Cosmochim. Acta, 34, 323-329.
Shieh Y.N. and Mensing T.M. (1986) Isotopic and chemical evidence for hydrothermal alteration in buried Precambrian basement rocks from the midcontinent, U.S.A. Terra cognita, 6, 213.
Smit J and Klaver G. (1981) Sanidine spherules at the Cretaceous-Tertiary boundary indicate a large impact event. Nature, 292, 47-49.
Smit J. and Kyte F.T. (1984) Siderophile-rich magnetic spheroids from the Cretaceous-Tertiary boundary in Umbria, Italy. Nature, 310, 403-404.
Smith J.V. (1974) Feldspar minerals. Springer-Verlag, Berlin.
Steiner A. (1970) Genesis of hydrothermal K-feldspar (adularia) in an active geothermal environment at Wairakei, New Zealand. Min. Mag., 7, 916-922.
Szabo Z. and Faure G. (1987) Isotopic studies of carbonate cements and provenance dating of feldspar in basin analysis: The Berea sandstones of Ohio. In: Geochemistry and Mineral Formation in the Earth Surface, Rodriguez-Clemente R. and Tardy Y. (eds.), C.S.I.C. and C.N.R.S., Madrid, Spain, 51-66.
Taylor K.S. and Faure G. (1981) Rb-Sr dating of detrital feldspar: A new method to study till. Jour. Geol., 89, 97-107.
Todd T.W. (1968) Paleoclimatology and the relative stability of feldspar minerals under atmospheric conditions. Jour. Sedimen. Petrol., 38, 832-844.
Waugh B. (1978) Authigenic K-feldspar in British Permo-Triassic sandstones. Jour. Geol. Soc. London, 135, 51-56.
Wright T.L. (1968) X-ray and optical study of alkali feldspars: II. An X-ray method for determining the composition and structural state from measurement of 2θ values for these reflections. Am. Min., 53, 88-104.
Yanagi T., Baadsgaard H., Stelck C.R. and McDougall I. (1988) Radiometric dating of a tuff bed in the middle Albian Hulcross Formation at Hudson's Hope, British Columbia. Can. Jour. Earth Sci., 25, 1123-1127.

ISOTOPIC COMPOSITIONS OF CLAY MINERALS AS INDICATORS OF THE TIMING AND CONDITIONS OF SEDIMENTATION AND BURIAL DIAGENESIS

Norbert CLAUER
Centre de Géochimie de la Surface
1 rue Blessig, 67084 Strasbourg (France)

Samuel M. SAVIN
Department of Geological Sciences
Case Western Reserve University, Cleveland, Ohio 44106 (USA)

and

Sam CHAUDHURI[*]
Department of Geology, Kansas State University
Manhattan, Kansas 66506 (USA)

The abundances of radioactive and radiogenic isotopes and of light stable isotopes can provide useful information about the origins of minerals of siliciclastic rocks and the nature of the diagenetic processes that have affected them. The K-Ar and Rb-Sr isotopic systems of sedimentary rocks and separated minerals have been used in many studies to determine the time of their deposition, although this approach remains controversial. While there are ambiguities in their application in stratigraphic studies, the K-Ar and the Rb-Sr systems have proven especially useful for determi-

---

(*) Presently at the Centre de Géochimie de la Surface, Strasbourg (France) ning the time and in some cases, the duration of diagenetic processes. In addition, $^{87}Sr/^{86}Sr$ ratios are useful for tracing the origin of Sr incorporated in diagenetic minerals. The isotopes of O, H and C can provide information about sedimentary provenance, origins of diagenetic fluids, the degree of openness of water-rock systems to the movement of fluids, and the temperatures at which diagenetic reactions have occurred.

Knowledge of the environments and conditions under which the isotopic systems of minerals become set or reset is critical to their application to sedimentary rocks. If an isotopic composition of a detrital mineral is to be used to determine its provenance, it is necessary that the mineral retains its original isotopic signature during transportation and deposition as well as during the post-depositional history of the sediments in which the mineral occurs. If an isotopic composition is used to tell us about the timing or conditions of sedimentation, it is necessary to analyze either detrital phases that have entirely equilibrated isotopically with the depositional environment or authigenic phases that have formed in equilibrium with that environment. In either case, the phases must subsequently retain the isotopic signatures acquired in the deposition environment if the analytical data are to reflect conditions or timing of sedimentation. Finally, if the isotopic compositions of minerals in sediments are to provide information about the diagenetic history of a rock, it is necessary that the phases analyzed equilibrated or re-equilibrated with the diagenetic environment.

This review summarizes the results of many isotopic investigations of siliciclastic rocks, and especially their clay constituents, the study of which has dominated research on diagenesis of shales and sandstones during the last two decades. Special problems arise in the isotopic studies of siliciclastic rocks, which typically are mixtures of detrital and secondary or diagenetic particles, so that the same mineral is often represented in both the detrital and the secondary components of the rock. Furthermore, in many sedimentary rocks the diagenetic and detrital constituents are intimately intermixed or closely associated with one another. Quartz and feldspar overgrowths on detrital quartz and feldspar cores, clay cements in clay-rich sandstones, and calcite cement in rocks containing detrital calcite are but a few examples of this. Further complications may arise when a diagenetic mineral formed during more than one episode; multiple generations of quartz, chlorite and carbonate minerals, for instance, are not uncommon. Consequently, if ambiguous or misleading interpretations of results are to be avoided in a geochemical or isotopic investigation of minerals in a sedimentary rock, it is necessary to precede the chemical or isotopic analyses by a detailed petrographic and mineralogic study. It is then necessary to separate the mineral constituents and to characterize thoroughly the separates to assess the relative proportions of diagenetic and detrital components.

In the present review, we will use the term epigenesis to describe any post-depositional formation of new minerals close to the sediment-water interface. Diagenesis is ment to include all of the post-depositional physical and chemical changes that occur in a mass of sediments during burial. These changes include dissolution or alteration of primary minerals, and formation of new minerals.

ISOTOPIC BEHAVIOR OF CLAY MINERALS DURING WEATHERING, TRANSPORTATION AND SEDIMENTATION

Isotopic analyses of detrital components in sedimentary rocks are generally done to obtain information about provenance. Epigenetic and diagenetic components in sedimentary rocks are thought to be analyzed for the purpose of determining either the time of deposition or of further specific diagenetic events, and the physical and chemical conditions that could prevail during the formation of the components analyzed.

Isotope Ratios of Terrigenous Clay Minerals: Evidence for Resistance to Alteration in the Marine Environment

Because clay minerals are extremely fine-grained, it is especially important to determine the extent to which, once formed, they remain as closed isotopic systems. In early Ar, Sr and O isotopic studies, it became apparent that these isotopic systems of clay minerals are usually extremely resistant to alteration during transportation on the continents as well as in the depositional environment, even when particle sizes are well below 1μm. Isotopic compositions may become markedly altered, however, when clay minerals have undergone mineralogical modification.

The K-Ar system

K-Ar ages of illite are resistant to alteration during weathering and riverine transport. Hurley et al. (1961) examined the K-Ar ages of illite separated from a soil developed on illite-rich Devonian shale, and in spite of the fact that the soil clay had a lower K content than the illite in the underlying shale, both yielded nearly the same K-Ar ages. In the same study, Hurley et al. reported that the clay-sized fractions of recent sediments from the Mississippi River Delta had K-Ar ages consistent with their presumed detrital origin as mixtures of illite derived from weathering of Paleozoic shales exposed in the North American mid-continent and mixed-layer illite/smectite derived from bentonites, especially within the Missouri River drainage. In a later study, Hurley et al. (1963) measured K-Ar ages of 200 to 400 Ma for illitic clays from a number of Cenozoic pelagic sediment cores from the Atlantic Ocean. They concluded that the sediments consisted primarily of detrital clay derived from the weathering of Paleozoic shales on the continent. Exposure of the clay to sea water for long periods of time seems not to have produced gross alteration of its K-Ar systematics. However, because the K-Ar ages of the detrital constituents prior to deposition in the ocean basin were not known, the data could not be used to rule out modest effects of exposure to sea water.

The Rb-Sr system

During weathering of a variety of parent rocks, the $^{87}Sr/^{86}Sr$ ratio of the weathered material closely resembles that of the parent material (Dasch, 1969; Clauer, 1976). Brass (1975) noted significant differences in the Rb/Sr and in the $^{87}Sr/^{86}Sr$ ratios not only among different clay size fractions in a soil profile, but also among different soil profiles developed on similar parent rocks. He proposed that these differences were due to differential loss of common and radiogenic Sr from the soil profiles during the weathering process. Clauer (1979) reported that the same type of clay minerals formed within weathering profiles developed on different parent-rock types, might have widely varying $^{87}Sr/^{86}Sr$ ratios.

The varied $^{87}Sr/^{86}Sr$ ratios of continental detritus are reflected in modern ocean sediments. As he found no evidence for exchange of Sr isotopes between sea water and detrital clay minerals in ocean sediments, Dasch (1969) related the $^{87}Sr/^{86}Sr$ ratios of a suite of Recent, widely distributed, detrital Atlantic Ocean sediments to their presumed continental sources. In a complementary study of Recent sediments from the Argentine Basin, Biscaye and Dasch (1971) also used both Rb/Sr and $^{87}Sr/^{86}Sr$ ratios as indicators of provenance. Thus, Sr isotopes can, in principle, be useful for distinguishing authigenic from detrital components in ocean sediments. However, this is difficult in predominantly terrigenous sediments containing small proportions of authigenic components.

The oxygen and hydrogen isotopes

Savin and Epstein (1970a) and Lawrence and Taylor (1971, 1972) showed that the aluminosilicate framework of pedogenic clay minerals forms in O and H isotopic equilibrium with the environmental water. Furthermore, pedogenic minerals seem to retain their original isotopic signatures through the processes of erosion, transportation and deposition. In contrast to the framework, O and H of the interlayer and adsorbed water of clay minerals are extremely labile.

Savin and Epstein (1970b) reported that no large-scale O isotope exchange occurred between the aluminosilicate framework of terrigenous clay minerals and seawater, and found inconclusive evidence for large-scale H isotope exchange. Yeh and Savin (1976) estimated that approximately 5 to 9% of the O of three ocean sediment samples composed primarily of detrital illite and smectite had apparently exchanged with seawater over periods from 14,000 yr to 2.7 Ma. However, they noted the possibility that what they termed <u>exchange</u> could, in fact, have been the growth of new diagenetic clay upon detrital grains. The amount of apparent exchange they observed ranged from 13 to 18% in the 0.1 to 0.5μm fractions of their samples and 16 to 43% in the finer than 0.1μm fractions. Yeh and Epstein (1978) estimated that between 8 and 28% of the hydroxyl H in one of the same (smectite-rich) sediments apparently exchanged in 2.7 Ma. Eslinger and Yeh (1981) found no evidence for any O or H isotopic exchange in Pleistocene sediments rich in illite, chlorite and vermiculite and some mixed-layer clays from the Aleutian Trench.

In apparent conflict with the results discussed above, James and Baker (1976) reported that significant amounts of O isotopic exchange occurred between poorly crystalline illite and water in the laboratory at 22°C in periods of a few months to a few years. O'Neil and Karaka (1976) also recorded very small amounts of exchange between illite and water at 100°C after two years, and observed that smectite exchanged more readily, probably because of its smaller particle size and the easier access of the water to the aluminosilicate structure by way of the expandable layers. More recent work has suggested that at temperatures which are not greatly above those of the earth's surface, H isotopic exchange may occur between clay minerals and water. Bird and Chivas (1988) presented convincing evidence that Australian Permian kaolinite had undergone post-formational H (but not O) isotopic exchange at temperatures below 80°C. Souchez et al. (1990) have also presented intriguing evidence suggesting that illite may exchange both O and H isotopes with water in basal ice of the Greenland Ice Sheet, presumably as the result of subglacial abrasion. Additional testing of this idea is needed.

The $\delta^{18}O$ and δD values of clays formed on the continents in weathering or diagenetic environments are generally different from those of authigenic marine clays. Therefore, the $^{18}O/^{16}O$ and D/H ratios of clays in sediments can be used both to distinguish between detrital and authigenic marine components and to trace the provenance of the detrital components. Savin and Epstein (1970b) used this approach to infer that the illite, smectite and chlorite in a globally distributed suite of ocean sediments they analyzed were primarily of detrital origin, being derived from the continents. Salomons et al. (1975) found a significant variation from river to river (15.8 to 21.7 per mil) in the $\delta^{18}O$ values of the <2μm sediment fractions of major rivers flowing into the Atlantic from western Europe. They were able to use the different $\delta^{18}O$ values of these sediment sources to trace the origin and movement of fine-grained sediments along the western European coast. Lawrence (1979) measured $\delta^{18}O$ values of a large collection of Recent sediments sampled west of the Mid Atlantic Ridge in the South Atlantic. His results confirmed those of the Sr isotope study of Biscaye and Dasch (1971) about the different sources of sediment deposited in the Brazil Basin. Tsirambides (1986) used $\delta^{18}O$ values of separated size fractions of a suite of sediments from the western Indian Ocean to distinguish detrital from authigenic constituents.

Sediments are often complex mixtures derived from multiple sources. The isotopic approach to determining sedimentary provenance will be most definitive when multiple isotopic (O, H, Sr) and mineralogic signatures are used.

Isotope Ratios of Clays in Ocean Sediments: Evidence for Isotopic Equilibration and Neoformation in the Marine Environment

While detrital clays in ocean sediments have isotopic signatures which reflect, at least predominantly, their formation on the continents, there is also much evidence for both neoformation of clays and for partial resetting of the isotopic systems of detrital clay components in the marine environment. Quantitative estimation of the overall extent of these processes is especially important to the question of the feasibility of isotopic determination of the time of sedimentation.

Glauconite

Glaucony (often referred to as glauconite pellets) is a heterogeneous mixture of mixed-layer illite/smectite and other mineral constituents (Burst, 1958a). Glauconite is an Fe- and K-rich highly illitic mixed-layer illite/smectite, which is commonly the predominant constituent of glauconitic pellets. Glaucony often forms penecontemporaneously with the deposition of associated sediments. The glauconite phase contains both K and Rb, and consequently was analyzed in some of the earliest attempts to determine sedimentation times by isotopic methods (Cormier, 1956; Wasserburg et al., 1956). It quickly became apparent, however, that interpretation of glauconite ages could be problematic. The isotopic ages of glauconites, especially those from Paleozoic and older rocks, are commonly 10 to 20% lower than the stratigraphic ages of the sediments in which they occur (Amirkhanov et al., 1958; Lipson, 1958; Goldich et al., 1959; Hurley et al., 1960). Comparisons between stratigraphic and measured isotopic ages of glauconitic materials have been discussed in several circumstances (e.g. Keppens et al., 1978a, b) and a sketch is given in figure 1. Harris and Fullagar (1989) summarized a number of possible causes for the low isotopic ages. Thompson and Hower (1973) studied the dissolution kinetics of glauconite and suggested that Ar was most readily lost from "open" K sites in the expandable interlayers of mixed-layer clays.

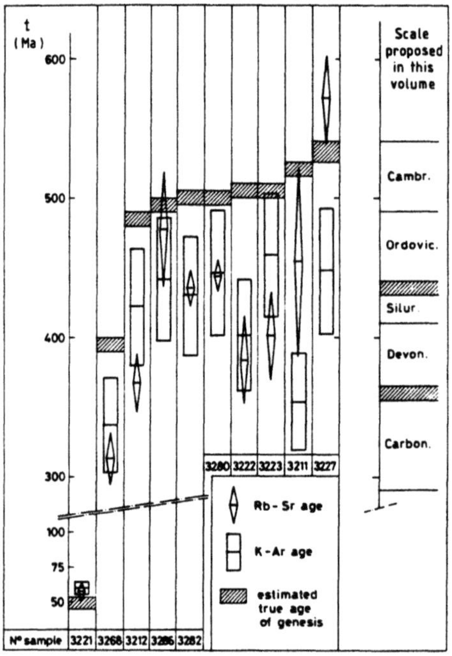

Figure 1: Comparison between stratigraphic and measured Rb-Sr and K-Ar isotopic ages for glauconites (Keppens and Pasteels, 1982)

Older glaucony pellets commonly contain fewer non-clay constituents than younger pellets, and the illite content of the mixed-layer illite/smectite is commonly higher in older pellets. These observations suggest that the pellets become diagenetically altered, with the gradual elimination of detrital components, after their initial formation (Odin and Matter, 1981). From a study of the mineralogy and K-Ar isotope systematics of glauconies in Recent sediments from the Atlantic Ocean, near the mouth of the Congo River, Odin and Matter (1981) noted that the detrital components of the pellets seem to be progressively removed, probably by dissolution, until the K_2O content reaches about 6 or 6.5%. They interpreted this process of glauconitization as de novo crystallization of authigenic smectite which then evolves progressively to form the glauconite (i.e., K-rich illitic) end member. A more detailed understanding of the glauconitization process comes from recent Sr, Nd and O isotope studies of the same west African off-shore glauconitic material and detrital clay from the associated mud previously studied (Keppens and O'Neil, 1984; Clauer et al., 1992a; Stille et al., 1992). The results of these studies indicate that the detrital clay from mud, which is initially present in the glauconitic pellets, gradually undergoes alteration until the K_2O content of the pellets reaches about 4.5%. On the basis of Sr and Nd isotopic data, Clauer et al. were able to determine that this early stage of glauconitization occurs by a process of dissolution and reprecipitation of the detrital clay material present in the pellets, in an environment in which the chemical and isotopic compositions were determined by the clay precursor, with no significant influence of seawater chemistry. In a subsequent step, the K_2O content continues to increase to as much as 7%, but in an environment which is influenced by sea water. This two-stage model of glauconitization is consistent with the observation that glauconitic pellets with K_2O contents less than 4.5% often yield isotopic ages higher than the stratigraphic ages. These high ages do not result from occurrence in the pellets of detritus that is not involved in the glauconitization process, but they do imply a genetic link to a detrital precursor clay that is progressively dissolved and modified into a Fe-smectite in a chemical system closed to sea water. This produces biases in the initial $^{87}Sr/^{86}Sr$ and $^{40}Ar/^{36}Ar$ ratios of the glauconitic minerals.

Smectites and mixed-layers illite/smectite

While $^{87}Sr/^{86}Sr$ ratios of bulk clays in ocean sediments may be used to distinguish detrital from authigenic clays, this method may not be sufficiently sensitive to detect minute small proportions of authigenic components in predominantly terrigenous sediments. However, in comparing the $^{87}Sr/^{86}Sr$ ratios of the predominantly detrital Atlantic Ocean sediments analyzed by Dasch (1969) with the clay mineralogy of the samples, Clauer (1976) observed that the samples with high smectite contents yielded $^{87}Sr/^{86}Sr$ ratios that were much lower than those of radiogenic Sr derived from the continents (Fig. 2). Some were as low as 0.704 (i.e., lower than that of sea water) indicating that the analyzed smectites could have formed by alteration of volcanic parent materials, either on the continents or in the ocean. The latter possibility is consistent with the conclusion of Clauer et al. (1982a, 1984), discussed below, that leaching experiments can be useful for detecting small amounts of authigenic minerals in predominantly terrigenous marine sediments.

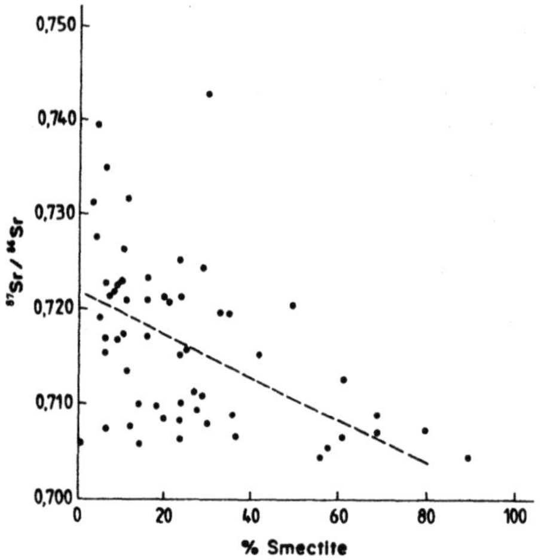

Figure 2: Variations of $^{87}Sr/^{86}Sr$ ratios with smectite contents of recent ocean sediments (Clauer, 1976)

Clauer et al. (1982a) studied the Rb-Sr system in a suite of deep-sea red clays cored in the southeastern Pacific Ocean where there is little or no detrital input. The clay mineral suite con-

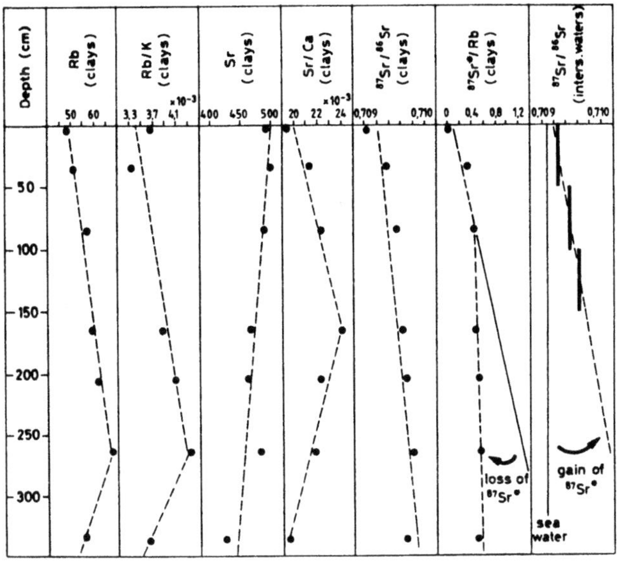

Figure 3: Variations of $^{87}Sr/^{86}Sr$, Rb/K and Sr/Ca ratios, and Rb and Sr contents of red clays from deep ocean in southeastern part of the Pacific Ocean (Clauer et al., 1982)

consists predominantly of nontronite (Fe-rich smectite) that very probably formed in the marine environment. The $^{87}Sr/^{86}Sr$ ratios of the nontronites are slightly, but significantly, lower than that of contemporaneous sea water. After removal of the marine Sr adsorbed on the nontronite particles by gentle leaching with 1N HCl, the $^{87}Sr/^{86}Sr$ ratios of the residual Sr in the clay lattices increases slightly with depth, from 0.7086 near the sediment-water interface to about 0.7115 at a depth of 265 cm below the interface (Fig. 3). As the nontronites collected at the water-sediment interface contain Sr with an $^{87}Sr/^{86}Sr$ ratio below that of sea water, Clauer et al. concluded that these clays originated by hydrothermal alteration of basalt, either in association with intraplate volcanic activity or along mid-ocean ridges, and that they were subsequently transported by bottom currents to the abyssal plains where they were deposited. The increase in the $^{87}Sr/^{86}Sr$ ratio of the nontronite vs. depth was interpreted to reflect the decay of ^{87}Rb into ^{87}Sr with time. Hoffert et al. (1978) and Clauer et al. (1984) drew similar conclusions in complementary studies. Buatier (1989) recently showed that the $^{87}Sr/^{86}Sr$ ratios of nontronites from the Galapagos Ridge are about 0.70814, while those of similar clays collected off the ridge in nearby hydrothermal mounds are slightly higher (0.70990), suggesting that the clay minerals of these mounds are primarily authigenic but that they contain also small amounts of detrital phases that were altered and probably partly or completely rehomogenized by hydrothermal fluids similar to those collected within the rift valley by diving devices.

Alteration of detrital clay in the marine environment may be apparent from careful petrographic studies as well as from isotopic data. Clauer et al. (1990) used XRD, transmission electron microscopy (TEM), chemical analyses including rare earth element analysis (REE), microchemical determinations with a scanning transmission electron microscope (STEM) and isotopic (Sr, O, H) analyses to investigate detrital smectites from Albian-Aptian and Tertiary black-shales collected in the Atlantic Ocean. They concluded that the detritus initially consisted of detrital smectite with flaky morphology and that most of the flakes had become altered into lath-like smectite particles with similar chemical compositions. Smectites of both morphologies have very similar major element compositions, flat REE patterns without Ce anomalies, and only small differences in their initial $^{87}Sr/^{86}Sr$ ratios and $\delta^{18}O$ and δD values. This chemical data led Clauer et al. to conclude that recrystallization had taken place in a restricted, almost closed environment, the chemistry of which was dominated by detrital clay. The time at which recrystallization occurred could not be determined directly from the Rb-Sr isotopic data, as all size fractions separated from each sample were mixtures of flaky detrital particles and lath-like authigenic particles. However, in correlating the apparent Rb-Sr ages of the different size fractions with the abundance of laths present, and by extrapolating to the pure lath-like authigenic end member, the time of formation (104-113 Ma for the Albian-Aptian and 58 Ma for the Tertiary samples) of the laths could be determined: for all samples studied, the ages calculated in this way agreed well with the stratigraphic ages of the sedimentary sequences, suggesting that the lath-like particles formed shortly after deposition.

ISOTOPIC DETERMINATION OF THE TIME OF SEDIMENTATION

Studies of the Sr isotope systematics of clays in ocean sediments indicate, as shown earlier, that equilibration of the isotopic systems can occur, to some degree, between predominantly detrital clay material and the marine environment. If the equilibrated Rb-Sr system or reset of the K-Ar system can be analyzed directly or indirectly, it ought to be possible to estimate more or less accurately the time of deposition of any sedimentary sequence containing such clay minerals.

Early Geochronologic Studies of Sediments

Bailey et al. (1962) examined the discrepancies between the stratigraphic ages of clay-rich sediments and the K-Ar ages of purified concentrates of $2M_1$ and $1M_d$ illite from underclays and shales of Pennsylvanian age. The K-Ar ages of the $2M_1$ polytypes were as much as 75 Ma greater than the depositional ages of the rocks, while the ages of the $1M_d$ polytypes were about half the stratigraphic ages of the rocks. The 2M polytypes of illite are known to form at high temperatures and, in shales that have not been markedly heated, they must be detrital. It is therefore not surprising that they yield K/Ar ages greater than stratigraphic ages. On the other hand, the 1M polytypes are of low temperature origin and could have formed during or after deposition. Bailey et al. suggested that after deposition the $1M_d$ polytype either lost ^{40}Ar by diffusion as a result of its very small particle size, or it underwent continued reorganization during which it fixed K and was converted to the $2M_1$ polytype. Neither of these suggestions is convincing in light of the results of subsequent studies. Indeed, Ar diffuses from clays only extremely slowly, as shown both by the study of natural samples in a contact metamorphic zone (Lee and Aronson, 1986) and by laboratory experiments (Odin et al., 1977). Clay minerals may often grow in the diagenetic environment long after deposition, at burial temperatures of 150°C or higher, and this by itself can result in K-Ar ages younger than stratigraphic ages. In addition, if reorganization of 1M illite into 2M illite had occurred in their samples as suggested by Bailey et al., the isotopic ages of the 2M clays would have become younger than the 1Md clays.

Hurley et al. (1963) measured the K-Ar ages of different size fractions, ranging from 4μm to <0.08μm, in an Ordovician illitic marine shale. The ages were found to decrease with decreasing particle size. The age of the coarsest fraction was greater than the stratigraphic age and that of the finest fraction was less. It follows from the results of both Hurley et al. and Bailey et al. (1962) that the ages of either bulk shales or separated size fractions of shales reflect the weighted average ages of all of the detrital and authigenic components in the samples analyzed. Because authigenic particles are typically concentrated in the finest fractions of clay-rich rocks, the ages of these fractions normally most closely approximate the timing of authigenesis or diagenesis (Clauer, 1976; Huon, 1985; Reuter, 1987; Liewig et al., 1987a; Mossmann et al., 1992).

Hurley et al. (1963) measured K-Ar ages of <2μm clay fractions of several Paleozoic shales that had been paleontologically well-dated. The ages of nearly all the samples were about 10 to 20% lower than the stratigraphic ages of the rocks from which they were separated. This discrepancy is similar to that which Hurley et al. (1960) observed for Paleozoic glauconites. But Hurley et al. (1963) also found that: (1) the K-Ar ages of many clays from Mesozoic and Cenozoic rocks are in fairly good agreement with stratigraphic ages, and (2) the K-Ar ages of clay fractions from very young rocks were as much as 4 times greater than the stratigraphic ages. Hurley et al. interpreted their results to reflect mixtures in varying proportions, of detrital and authigenic clay constituents. They attritubed the low K-Ar ages of clays from many Paleozoic rocks to the high proportions of diagenetic clay minerals they contained, whereas the high K-Ar ages of the younger rocks were attributed to the dominant influence of detrital constituents. In this interpretation, agreement between isotopic and stratigraphic ages could be the coincidental result of analysis of a mixture, in adequate proportions, of detrital and diagenetic components.

Low K-Ar Ages of Fine-Grained Clay Minerals

Two causes of low K-Ar ages of finer clay particles were proposed in early studies: (1) increasing diffusional loss of radiogenic ^{40}Ar with decreasing particle size, a suggestion which is no longer readily accepted, and/or (2) diagenetic addition of K to smectite and altered micas over an extended period of time. TEM observations (e.g. Hunziker et al., 1987) show that detrital clay particles often have irregular edges, whereas authigenic clays have sharp well defined edges. This morphological distinction has proven useful for differentiating between the two types of clay.

Initially, evidence in support of the hypothesis of diffusional loss of Ar from clay particles was very strong, as several experimental studies suggested the potential for loss of Ar by glauconite and related clay minerals in sedimentary environments (Amirkhanov et al., 1958; Evernden et al., 1961). However, as already noted, there was also substantial evidence in the early 1960's that fine-grained clays are retentive of radiogenic ^{40}Ar in sedimentary and diagenetic environments. For instance, Hurley et al. (1960) showed that the K-Ar and the Rb-Sr ages of glauconites can be similarly discordant with respect to stratigraphic age, even though the rates of diffusion of Ar and Sr differ greatly in clay minerals. Obradovich (1965) observed no differences in the K-Ar ages of different size fractions of a mineralogically homogeneous glauconite sample, and Clauer et al. (1984) did not find any difference in the Rb-Sr and K-Ar apparent ages of oceanic clay minerals from the southern Pacific Ocean, even at particle size as small as 0.2μm.

Hunziker et al. (1986) proposed that Ar diffusion might occur in detrital clay of any size fractions because the edges of detrital particles tend to be diffuse and ill-formed. As neoformed crystals, which predominate in the finer clay fractions, typically have clean, sharp edges, Hunziker et al. suggested that diffusion is minimized in such neoformed fine particles. If this idea is correct, the proportions of large, diffuse-edged, leaky crystals and small, tight crystals probably vary from sample to sample, reflecting the details of lithology, the degree to which recrystallization has occurred, and the sizes and locations of the clay particles within the rocks. Huon et al. (in prep.) question, on the basis of microstructural, mineralogical and morphological evidence, whether Ar diffuses in illite under diagenetic conditions, as proposed by Hunziker et al (1986).

Isotopic Age Determination of the Time of Sedimentation

By analysis of whole rocks

During the early 1960's, the Rb-Sr method was often applied to bulk sedimentary rocks (Compston and Pidgeon, 1962; Whitney and Hurley, 1964; Chaudhuri and Faure, 1967; Bonhomme and Prévôt, 1968). This approach, based on the assumption that the Sr isotopes of all of the rocks in any suite of samples had equilibrated with one another (and with the ambient fluids) during or shortly after deposition, was reexamined by Cordani et al. (1978; 1985). They showed that frequently the isotopic data for a suite of bulk samples did not plot along any line on an $^{87}Sr/^{86}Sr$ vs. $^{87}Rb/^{86}Sr$ diagram that could be interpreted as an isochron.

Acid-leaching of shales was done in a number of studies to remove and analyze Rb and Sr which were either contained in carbonates or were less tightly bound within silicate mineral structures. The rationale for this was the idea that acid-leachable Rb and Sr would be more likely to constitute a system that had equilibrated with ambient fluids at the time of sedimentation (Bofinger et al., 1968; Chaudhuri and Brookins, 1969; Clauer, 1976; Chaudhuri and Brookins, 1979). In some instances, the isotopic compositions of the leachates and the residues plotted along straight lines on Sr-evolution diagrams. While in some cases those lines could be interpreted as isochrons indicating the time of sedimentation, most investigators concluded that Rb-Sr data from argillaceous sediments could not be used to estimate the time of sedimentation. Clauer (1982) examined the potential of whole-rock Rb-Sr dating in detail, and he concluded that the method is not reliable, because the basic assumption of complete isotopic homogenization of all silicate and acid-leachable components of whole rocks, is never achieved during deposition.

By analysis of separated clay fractions

When isotopic compositions of separated size fractions of clays from a suite of related rocks are plotted on a Sr-evolution diagram, it is common for the data of the same size fractions to plot close to a straight line that can be interpreted as an isochron. However, as was found in early K-Ar studies, the Rb-Sr apparent ages of different size fractions separated from the same whole rock commonly decrease with decreasing particle size (Clauer, 1974; Dasch, 1969; Hofmann et al., 1974). This led to the conclusion that the Rb and Sr isotope systematics of clay minerals from shales primarily reflect the ages of the detrital components, that diagenetic alteration during burial chiefly affects the fine-grained clay components, and that stratigraphic dating of whole rocks using the Rb-Sr method is not feasible. Bonhomme et al. (1966) first emphasized the need for integrating the results of clay mineralogic analyses into geochronologic studies. Indeed, if it can be assumed that <u>in situ</u> formation or modification of clay is accompanied by Sr isotope equilibration with the environment, this certainly might be the most visible in the smaller size fractions. Agreements between stratigraphic and isotopic ages were found in a number of studies and were taken as evidence that Sr isotope equilibration had occurred (Bonhomme et al., 1970; Clauer, 1973; Harris and Bottino, 1974a, b; Harris, 1976; Harris and Baum, 1977; Keppens et al., 1978a, b).

Studies of clays in Recent sediments provide clear evidence that some of the detrital clays approach Sr isotopic equilibrium with the environments in which they have been deposited. Evidence also exists for Sr isotopic equilibration between some detrital clay minerals and interstitial fluids shortly after deposition. There are a number of examples in which glauconies formed by the alteration of detrital clays exhibit apparent age spectra which decrease, approching stratigraphic age, in a manner similar to that seen by Odin et al. (1979) for Recent glauconies. These include the Eocene Weches Formation in Texas (Ghosh, 1972), Miocene glauconies from Morocco (Tisserant and Odin, 1979), and illite, smectite and attapulgite assemblages from Tertiary formations of the Central Massif in France (Clauer, 1976). These examples indicate that the model of early diagenesis of clay minerals that is inferred from the study of Recent sediments, is applicable to the study of other authigenic clay components separated properly from their sedimentary rocks. The K-Ar and Rb-Sr isotopic systems and consequently the apparent depositional ages of clay fractions may also be affected by weathering and/or burial diagenesis.

It should also be mentioned here that determination of the time of deposition of Proterozoic sedimentary sequences by isotopic analysis of their clay minerals seems to need special care. Clauer et al. (1982b), for instance, found that the K-Ar ages of illites of Proterozoic age are always younger than the Rb-Sr ages obtained on the same aliquots. The reason for this is not clear, but a possibility exists that early diagenesis may persist for long intervals in some Proterozoic series that were never buried to great depths, as the result of very low sedimentation rates. The fact that the Rb-Sr and/or K-Ar methods often cannot be properly applied to Archean and Proterozoic sedimentary sequences for stratigraphic purposes, has led to attempts to date the time of deposition using the Sm-Nd method on authigenic clays (Stille and Clauer, 1986; Bros et al., 1992), and the Pb-Pb method on carbonates (e.g. Moorbath et al., 1987). The full potential of these relatively new isotopic methods of dating sediments has yet to be determined.

DETERMINATION OF THE TIMING AND MECHANISMS OF DIAGENESIS

Mechanism and Timing of Diagenetic Processes in Shales

When shales containing mixed-layer illite/smectite are buried deeply in a sedimentary basin, the proportion of illite layers in the clay increases with increasing depth of burial (Burst, 1958b; Powers, 1959, 1967; Weaver, 1959, 1960; Dunoyer de Segonzac, 1969; Perry and Hower, 1970; Hower et al., 1976; Boles and Franks, 1979; and others). This process has been examined in many sedimentary basins but has been studied most intensively in the sediments of the Gulf Coast. Combined K-Ar, Rb-Sr and light stable isotope studies have helped elucidate the processes that occur in deeply buried shales and have clarified the significance of ages obtained in geochronologic studies of shales.

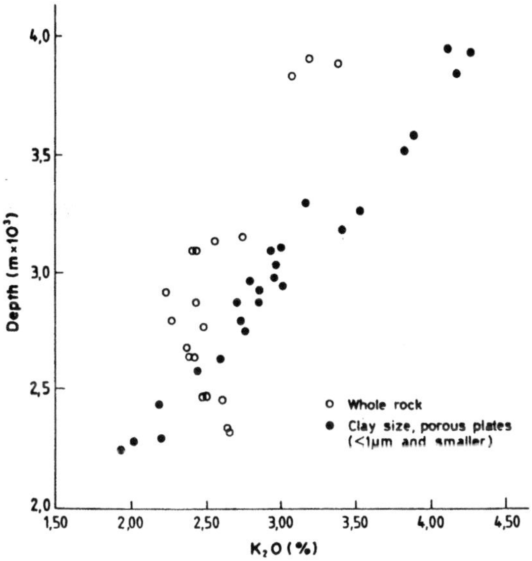

Figure 4: K content versus depth for clay fractions (<1μm) and bulk samples from Gulf Coast sediments (Perry and Hower, 1970)

Perry and Hower (1970) published a detailed examination of the mineralogical and chemical variations of deeply buried shales of Pleistocene to Eocene age from the Gulf Coast. In spite of progressive K-fixation with depth by the mixed-layer clays, accompanying the conversion of smectite to illite, the K contents of the whole rocks do not vary systematically with depth (Fig. 4). This suggested to the authors that each rock volume was closed to the movement of K, and that K was redistributed within each rock volume during burial diagenesis. Because the K-feldspar content of each sequence decreases with depth, Perry and Hower proposed that K-feldspar was the most likely source of the K fixed by the mixed-layer clay. In this interpretation, breakdown of detrital illite and feldspar supplied K and Al which were taken up in the conversion of smectite layers to illite layers in the mixed-layer illite/smectite. This was supported by the conclusions of a more detailed study by Hower et al. (1976) on samples from another well in the Texas Gulf Coast.

K-Ar isotopic evidence

Diagenetic reactions, as suggested by Perry and Hower (1970) and Hower et al. (1976), have a major effect on the K-Ar ages of whole rocks and clay fractions. This was confirmed in Perry's (1974) examination of the K-Ar systematics of separated clay size fractions from a sequence of Miocene shales buried to depths from 1,580 m to 5,525 m (estimated temperature 165°C) in the Louisiana Gulf Coast. In the <0.5μm fractions, mixed-layer illite/smectite is the predominant mineral with smaller amounts of kaolinite, detrital illite, and sometimes discrete chlorite. The proportion of illite layers in the <0.5μm mixed-layer illite/smectite increased from 20% at shallow depths to 80% in deeply buried sediments. This increase in the illite content is accompanied by a decrease in the K-Ar ages from 164 ± 3.3 Ma at 1,580 m to 100 ± 2.0 Ma at 5,525 m, but even the youngest of these ages is considerably greater than the depositional age of the sediments. The 0.5-2μm fractions contain less mixed-layer illite/smectite and more detrital illite, feldspar, and quartz than the finer fractions, and have K-Ar ages between 312 ± 6.2 Ma at a depth of 1,713 m and 129 ± 2.6 Ma at a depth of 5,980 m. The dispersion of ages among the size fractions of a sample decreases with increasing depth (Fig. 5), but at no depth are the K-Ar ages of the different size fractions concordant.

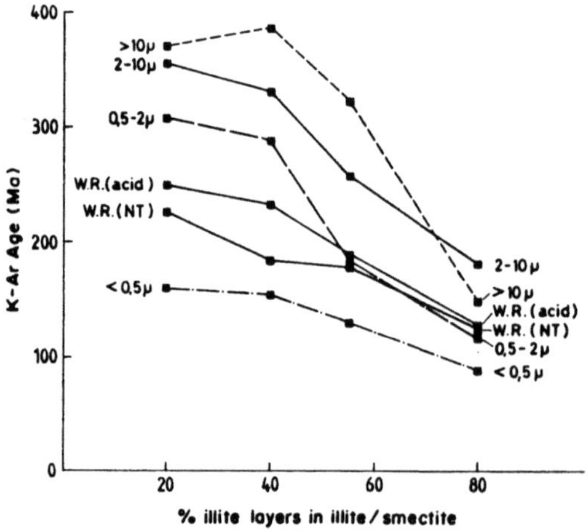

Figure 5: Variations of K-Ar ages with changing illite contents in smectite-illite minerals for different size fractions of sediments from the Texas Gulf Coast (Perry, 1974)

Weaver and Wampler (1970) and Aronson and Hower (1976) also measured K-Ar ages of clay fractions separated from deeply buried sediments in the Gulf Coast region. Weaver and Wampler, in their study of <0.2μm clay fraction, noted a decrease in the age from 165 ± 60 Ma at a depth of about 1,320 m to about 125 ± 20 Ma at a depth of 5,140 m. These measurements required large corrections for atmospheric Ar, causing the measured ages at all depths to overlap within the experimental uncertainty, but the results can be assumed to agree well with those of Perry (1974). Aronson and Hower (1976) measured K-Ar ages of Oligocene illitic clay in the Texas Gulf Coast well studied by Hower et al. (1976). They examined the <0.1μm clay fractions, i.e. finer than those examined by either Perry (1974) or Wampler and Weaver (1970), and reported lower ages ranging from 53.9 ± 2 Ma at a depth of about 1,185 m to 32.6 ± 2 Ma at a depth of about 5,250 m. The stratigraphic age of the sedimentary sequence ranges from approximately 28 to 35 Ma, so a complete resetting of the isotopic ages in the Gulf Coast region seems not to have occurred even in <0.1μm clay particles buried to depths as great as 5,250 m.

In principle, isotopic ages of diagenetic components of shales should indicate the time of diagenesis, but in practice the time of diagenesis of shale may be determined only with a good deal of uncertainty. This uncertainty is primarily the result of the difficulty of separating detrital from diagenetic clay components (Clauer, 1982). In spite of this, however, Aronson and Hower (1976) were able to use K-Ar data to estimate the mean time of K-fixation by the mixed-layer phase. They did this by estimating the incremental amount of K added to the fine clay fraction during diagenetic illitization and the incremental amount of ^{40}Ar relative to the ^{40}Ar content of the more shallowly buried fine clay fractions. The age they computed from these incremental values was 18 Ma, which is consistent with the time estimated from burial history curves for heating of the sediments to temperatures in the range of 80 to 160°C.

Rb-Sr isotopic evidence

When it is deposited, each Sr-bearing detrital component of a sediment has an ^{87}Sr/^{86}Sr ratio which reflects its origin. Each diagenetic component acquires an $^{87}Sr/^{86}Sr$ ratio which is determined by the Sr isotopic composition of the environment at the time it is formed or recrystallized. In Rb-containing phases the $^{87}Sr/^{86}Sr$ ratios, of course, evolve with time. In the early 1970's it became apparent that at least partial homogenization of Sr isotopes among minerals in a rock could accompany diagenesis. Perry and Turekian (1974) measured the $^{87}Sr/^{86}Sr$ ratios of different size fractions of clay from deeply buried Miocene shales in the Louisiana Gulf Coast (apparently the same well studied by Perry, 1974). They observed a trend toward Sr isotopic homogenization in the most deeply buried sediments, but $^{87}Sr/^{86}Sr$ ratios did not become completely homogenized even at a burial depth of 5,523 m (Fig. 6).

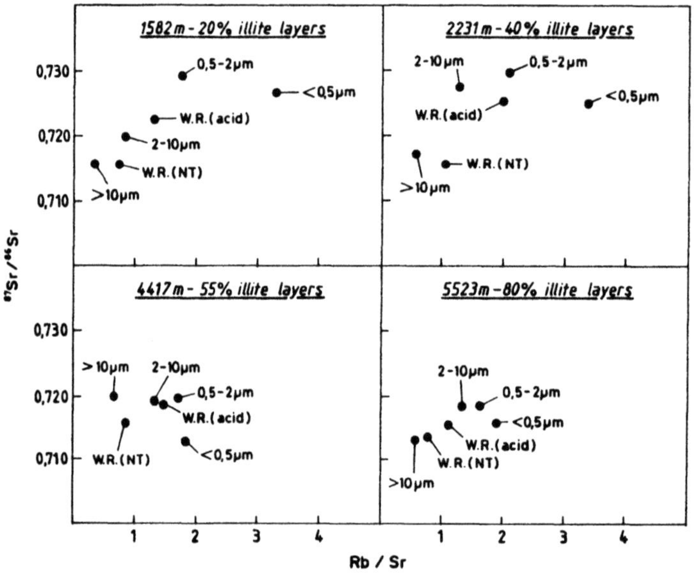

Figure 6: Variations of Rb-Sr isotopic compositions for different size fractions of sediments from different depths in the Gulf Coast area (Perry and Turekian, 1974)

In sharp disagreement with the inference by Perry and Turekian (1974) that homogenization of Sr isotopes among the different constituents within a shale sample is very slow, Morton (1985) argued that Sr isotopic homogenization among very fine (< 0.05μm) clay particles had occurred over a stratigraphic interval between 3,200 m and 4,900 m in a well in the Texas Gulf Coast during a period of 5 Ma or less. The data points for the samples Morton analyzed form a linear array on a Rb-Sr isochron diagram. Morton interpreted the data as indicating that isotopic homogenization of the finest fractions of all of the samples occurred at 23.6 ± 0.8 Ma, compared to a depositional age of 25 to 29 Ma, and that the initial $^{87}Sr/^{86}Sr$ ratio at the time of homogenization was 0.7088 ± 0.0004, similar to that of late Tertiary sea water (Fig. 7). Thus, in Morton's interpretation, homogenization of Sr isotopes of the finest clay fractions of Gulf Coast sediments would have occurred over a large stratigraphic interval shortly after deposition, probably in response to a change in the hydrologic regime which, in turn, affected the pore-water chemistry. That would have promoted the illitization of the mixed-layer illite/smectite to varying degrees throughout the stratigraphic interval. Morton also suggested that no further transformation of smectite layers to illite occurred subsequent to that early diagenetic episode. Although Morton's data are intriguing, his interpretation of the homogenization of Sr isotopes of fine-grained clay over an interval of thousands of meters during a brief episode has not met widespread acceptance. In an alternative view of the data, the apparent Rb/Sr isochron might suggest that a thick sequence of sediment, which was deposited and buried within a few million years, was rapidly heated as the result of burial and underwent diagenetic illitization during a relatively short interval. However, a detailed critique of Morton's conclusions has not yet been published.

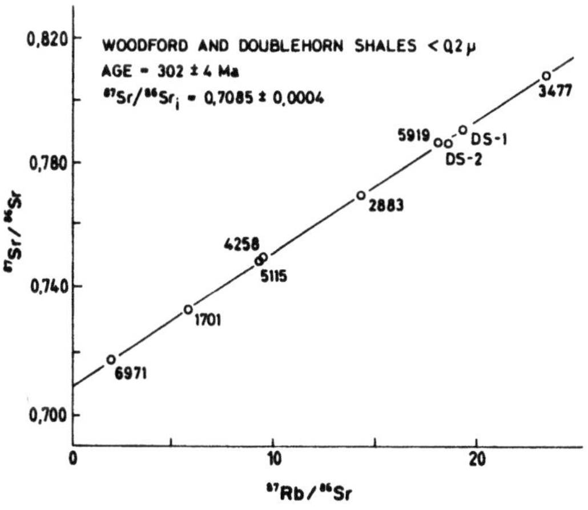

Figure 7: Rb-Sr isochron of very fine clay fractions (<0.2μm) from shales in the Texas Gulf Coast area (Morton, 1985)

Oxygen and Hydrogen isotope evidence

The O and H isotope geochemistry of minerals in shale sequences provide additional evidence about the nature of the burial diagenetic conversion of highly expandable illite/smectite to clay of lower expendability. Yeh and Savin (1977) examined the $\delta^{18}O$ values of separated size fractions of quartz and clay from three Gulf Coast wells, including the CWRU-6 well studied by Hower et al. (1976) and Aronson and Hower (1976). Yeh (1980) measured the δD values of separated clay size fractions from the same wells. In the shallowest sample of CWRU-6 studied (1,350 m, 44°C), each of the different clay size fractions had different $\delta^{18}O$ and δD values, indicating that there has been little or no isotopic exchange either among clay fractions within the sample or between pore fluids and clay fractions except, conceivably, the finest fraction. With progressive burial, the spread of $\delta^{18}O$ and δD values in a sample diminished (Fig. 8). The greatest decrease in the range of δ values occurred between 2,500 and 3,500 m, the depth range in which the greatest amount of illitization of illite/smectite occurred.

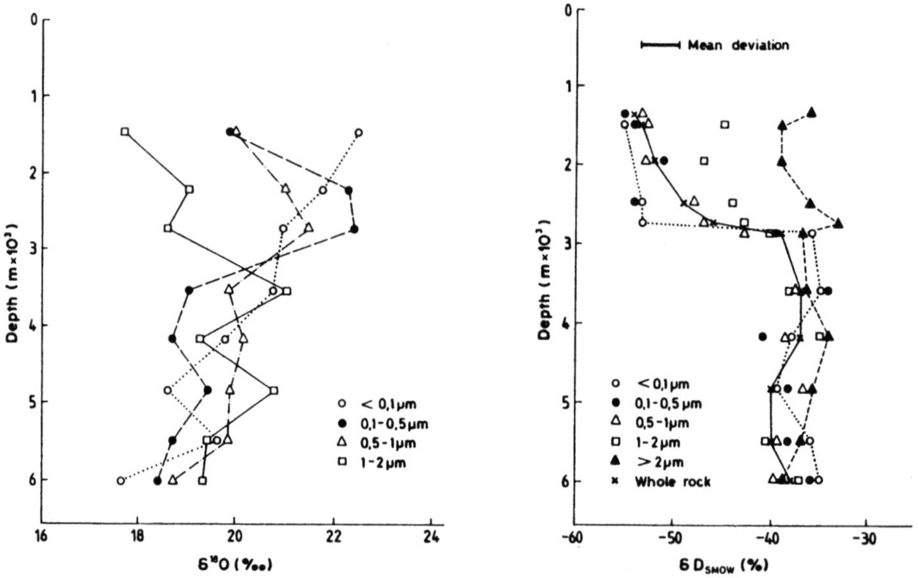

Figure 8: Oxygen- and hydrogen isotope variations with depth from various clay fractions in sediments from CWRU Gulf Coast Well N°6 (Yeh, 1980)

At every depth, there is a significant dispersion of $\delta^{18}O$

values of different size fractions of quartz. Below 2,500 m, the 2 to 1μm quartz grains are isotopically zoned. The inner portions of the grains have $\delta^{18}O$ values typical of silt-size detrital quartz in the Mississippi River drainage system and the outer portions of the grains are substantially enriched in ^{18}O. Yeh and Savin (1977) interpreted the data as confirming the diagenetic mechanism for illitization proposed by Hower et al. (1976). That is, the break down of detrital K-feldspar and mica provides K and Al necessary for conversion of smectite layers to illite layers. The Si released during this breakdown and/or the Si released from the tetrahedral sites of the smectite when it is converted to illite becomes available to form quartz in the shale. Yeh and Savin proposed that at least a portion of that Si formed diagenetic quartz overgrowths on detrital silt- and clay-size quartz grains.

Diagenetic Processes in Sandstones

In many sandstone reservoirs, the finest particle sizes of the clay fraction are dominated by diagenetic clay. Separation of diagenetic clays from sandstone is less difficult than that from shale, and this has facilitated a number of studies of diagenetic processes utilizing radioactive and stable isotope measurements. The significance and utility of many of these studies derive to a large extent from the fact that they were based on integrated evidence of basinal history, careful petrography and mineralogy, geochronology and stable isotope studies.

From the $\delta^{18}O$ values of a pair of cogenetic minerals which formed in isotopic equilibrium with one another and with the ambient fluid, one can, in principle, determine both the temperature of mineral formation and the $\delta^{18}O$ value of the ambient fluid. However, cogenetic mineral pairs are extremely rare in sandstones. From the $\delta^{18}O$ value of a single mineral, the conditions of its formation can be constrained to lie upon a trajectory on a diagram of temperature of formation vs. $\delta^{18}O_{fluid}$. When either the temperature of formation or $\delta^{18}O_{fluid}$ can be constrained using non-isotopic data, it is possible to use the $\delta^{18}O$ value of the mineral to constrain the other. This is the approach that has been taken in interpretating the O isotopic data obtained in several studies of sandstone diagenesis that are briefly summarized below.

Studies in the Rotliegendes of the North Sea

The Permian Rotliegendes Sandstone in the southern part of the North Sea basin has been the subject of recent geochemical studies focusing in part on the time of emplacement of the hydrocarbons in gas reservoirs (Lee et al., 1985, 1989). Lee et al. (1985) examined the variation with depth of the nature and extent of diagenetic changes in two wells from the Groningen gas field and the Broad Fourteens oil field. A basic assumption of their approach was that the smallest size fractions of diagenetic illite growing in sandstone pores as an alteration product of K-feldspar are most likely to have been derived from the edges and tips of the illite flakes. The smallest fractions should be those to be the last to have grown. Consequently, in the absence of any detectable contamination by detrital illite, the range of ages obtained from different clay size fractions ought to be indicative of the duration of the illite-forming process.

Lee et al. (1985) also postulated that the emplacement of gas in a sandstone would inhibit the progression of diagenetic alteration by displacing the formation water. In each field studied the intensity of both the alteration of detrital feldspar and the growth of diagenetic illite increases with depth over a relatively small interval (174 m in the Groningen well and 15 m in the Broad Fourteens well). The authors, therefore, suggested that the downward increase in the intensity of diagenetic alteration reflected the gradual emplacement of hydrocarbons in the reservoir and the downward migration of the gas-water contact. In this scenario, rocks which had been saturated with water for longer periods of time would have become more intensively altered.

The data for the Groningen gas field are consistent with this hypothesis. In the uppermost sample studied the K-feldspar was fresh and there was no diagenetic illite to date. Lower in the gas zone the diagenetic illite yielded K-Ar apparent ages of 143 Ma (0.2-0.5μm) and 151 Ma (<0.1μm); the depositional age of the Rotliegendes being about 270 Ma. The 0.2-0.5μm was contaminated with highly altered detrital K-feldspar, and Lee et al. suggested that the relatively low age of that fraction reflected Ar loss from the feldspar. They interpreted the age of the <0.1μm fraction as indicating the latest possible time of illite growth in the deeper gas zone sample, and by inference, the time of migration of gas into that level of the reservoir. Still lower in the section in the water zone, diagenetic illite gives K-Ar apparent ages of 133 Ma (0.2-0.5μm) and 120 Ma (0.2-0.5μm). The abundance of K-feldspar in both fractions of that sample being below the XRD detection limit (a few per cent), Lee et al. concluded that the formation of illite terminated much later in the water zone than in the gas zone.

Studies in the Brent of the North Sea

The uncertainty in the significance of K-Ar ages of different size fractions of illite from a reservoir sandstone is best illustrated by the disparate results on diagenetic illite from the Brent Formation in the North Sea. Hamilton et al. (1987) reported K-Ar ages between 46 and 59 Ma for the diagenetic illite of that oil-bearing sedimentary level. Hogg et al. (1987) also found scattered K-Ar ages between 45 and 60 Ma, whereas Jourdan et al. (1987) claimed, on the basis of K-Ar data, that the Brent Formation had a long and complex diagenetic history extending from Lower Cretaceous time (100 Ma) to the Eocene (40-50 Ma). In sharp contrast, Liewig et al. (1987a) observed that the average K-Ar age of the diagenetic illite from the Brent was measured to be about 41 Ma when the illite was separated following a gentle freeze-thaw disaggregation, but the K-Ar ages of similar size fractions of illite from similar nearby rocks collected varied between 130 and 50 Ma when the clays were separated following comminution by jaw crusher and ball mill. Glasmann et al. (1989) used gentle crushing and ultrasonic treatment to disaggregate Brent sandstones and found close K-Ar ages of about 38 Ma for the diagenetic illite fractions. Liewig et al. (1987a) concluded that sample preparation could introduce scatter in the dates due to the introduction of trace amounts of contaminant detrital components in the fractions analyzed. Clauer et al. (1992b) also examined the manner in which laboratory procedures might affect isotopic results of diagenetic clay minerals. As an alternative to diagenetic events of long duration inferred by Jourdan et al. (1987) for the Brent, Cocker et al. (1988) found that the difference between the K–Ar age of illite formed in the water zone and illite formed in the oil zone of a Brent reservoir was only 1 to 2 Ma for the 40 Ma event. These illites were separated following disaggregation by a gentle technique. Lee et al. (1985) used a different gentle disaggregation technique and concluded from their study of the hairy illites in the Rotliegendes that the duration of diagenetic illite formation in the water zone was long, but the evolutionary history of the Rotliegendes in the southern North Sea may have been different than that of the Brent sandstones located further North.

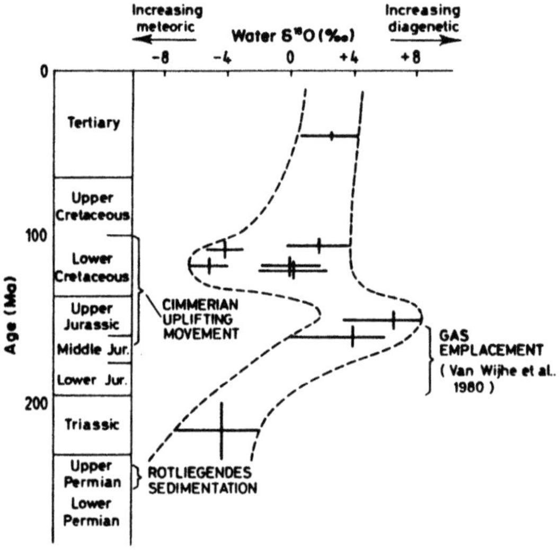

Figure 9: Oxygen isotopic compositions of fluids related to illite diagenesis in Rotliegendes sandstones, northeastern Netherlands (Lee et al., 1989)

In a survey of the K–Ar and O isotope systematics of a number of wells from several fields in the Rotliegendes sandstones, Lee et al. (1989) used the K-Ar age of diagenetic illite and the burial history curve for each sample to infer the depth at which any sample was buried when the illite formed. The temperature of illite formation was then calculated from the depth of burial and the estimated paleogeothermal gradient (taken to be similar to the modern one at each locality). The $\delta^{18}O_{fluid}$ was further estimated from calculated temperature and measured $\delta^{18}O$ value of the mineral. The evolution of the $\delta^{18}O$ values of illite-forming fluids in north-eastern Netherlands is shown in figure 9. The relatively high values (0 to +4 per mil) obtained for the $\delta^{18}O$ of the fluids from which most of the illites formed, indicate that the fluids had been significantly affected by interaction with rocks. However, $\delta^{18}O_{fluid}$ values were relatively low (-4 to -5 per mil) during the time of the Kimmerian uplift early in the Cretaceous period, indicating a significant meteoric component at that time. Lee et al. were also able to show a geographic pattern in the inferred $\delta^{18}O_{fluid}$ values at that time, low values occurring closest to an early Cretaceous outcrop of the Rotliegendes and high values occurring furthest away. Thus, the O isotopic data were useful in the determination of the paleohydrological pattern in response to tectonic evolution.

Studies in the Alberta Basin

The diagenesis of the Cretaceous and Tertiary sandstones of the Alberta Basin has been extensively studied by F. Longstaffe and his colleagues. Of particular significance to this review are the studies devoted to the evolution of the fluids responsible for the formation of diagenetic cements (Longstaffe, 1984, 1986; Longstaffe and Ayalon, 1987, 1990; Ayalon and Longstaffe, 1988). The Alberta basin is an especially attractive region for studies of this sort because the $\delta^{18}O$ values of meteoric water in the region are much lower than values expected for water which has been modified isotopically to a significant degree by diagenetic reaction with rocks.

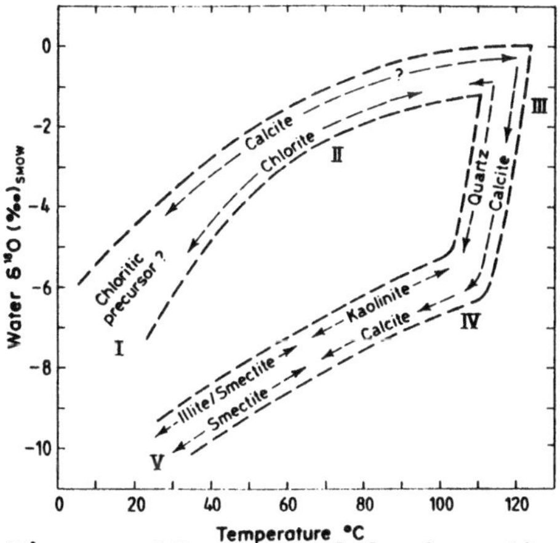

Figure 10: A model for the evolution of oxygen-isotope composition of pore fluids and diagenetic minerals in the basal Belly River sandstone, Alberta (Ayalon and Longstaffe, 1987)

The results of the above mentioned studies point out very clearly that different episodes of diagenesis may occur in a sandstone in the presence of fluids of very different nature. These different fluids enter the sandstone in response to the changing depositional and tectonic regimes in the sedimentary basin. Isotopic analyses of different diagenetic phases carefully documented by petrologic and mineralogic observations can be of prime importance in the reconstruction of the varied diagenetic episodes occurring during the evolution of a sedimentary basin. As an example, the upper Cretaceous basal Belly River Sandstone in Canada contains a complex suite of diagenetic minerals which includes, in order of formation, iron oxide, chlorite, quartz overgrowths, feldspar overgrowths, kaolinite, illite, illite/smectite and pyrite (Ayalon and Longstaffe, 1988). Using $\delta^{18}O$ values of several of the diagenetic phases, the conditions of formation of each could be constrained to lie upon a trajectory on a diagram of temperature vs. $\delta^{18}O_{fluid}$ (Fig. 10). Ayalon and Longstaffe (1988) concluded that the early diagenetic fluids were cool and were meteoric in origin. Later, phases formed after burial heating, in the presence of diagenetically modified ^{18}O-rich water. Still later, the formation cooled as the result of uplift and erosion, and the fluids from which minerals formed were again largely meteoric in origin.

Studies in the Powder River Basin

The Cretaceous Muddy Sandstone of the Powder River Basin (Law, 1983; Burrows, 1985; Savin et al., in prep.) is a fluvio-deltaic sequence with a depositional age of approximately 100 Ma containing abundant detrital illite/smectite mixed-layers. Burial history was relatively simple and present burial depths (approx. 1,000 to 2,500 m) that are probably close to maximum burial depths. Diagene-tic phases include quartz overgrowths on detrital quartz sand-grains as well as intergranular kaolinite and illite/smectite. The mineralogic and mineral preparation procedures used were fairly primitive when compared to those of some more recent studies discussed above.

The 0.05-0.1μm and <0.05μm fractions were dated using the K-Ar method to distinguish the diagenetic from detrital illite/smectite mixed-layers. Figure 11 is a plot of K-Ar age vs. percent illite layers in the illite/smectite for all except the most shallowly buried samples. Most of the coarser clay fractions, and a few of the finer fractions gave ages which were older than the depositional age, clearly indicating that they contain an abundant detrital component. The proportion of illite layers in the detrital illite/smectites varies widely from less than 5 to more than 70%. The K-Ar ages of most of the finer clay fractions and a few of the coarser ones were significantly younger than the age of deposition, suggesting that they were dominated by diagenetic material. The mineralogical and K-Ar data suggest that the conditions of formation of the diagenetic illite/smectite were fairly uniform over the area studied, at least with respect to the parameters that determine the percentage of illite layers in that mineral.

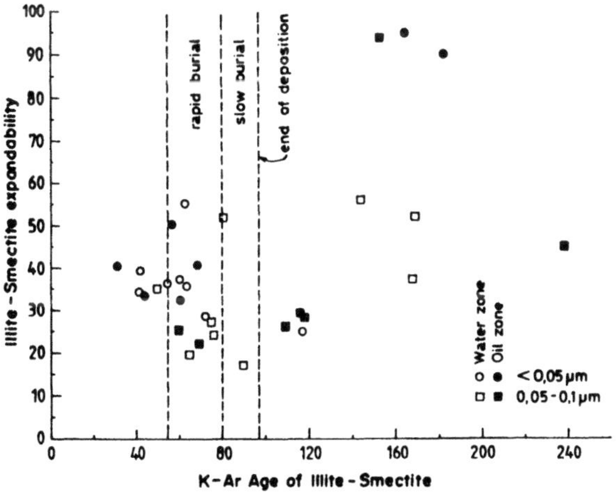

Figure 11: K-Ar age vs. percent illite in mixed-layer illite-smectite clays in Cretaceous Muddy sandstone in the Powder River basin (Savin et al., in preparation)

The ^{18}O values of the size fractions were also measured. Most of these fractions were mixtures of two or three of the three end-member components: detrital illite/smectite, diagenetic illite/smectite and kaolinite. Burrows (1985) and Savin et al. (in prep.) used plots of mineralogical composition vs. $\delta^{18}O$ value in suites of related size fractions (usually from a single small oil field) to extrapolate to the $\delta^{18}O$ values of the diagenetic kaolinite and illite/smectite components of each of these suites. The K-Ar ages were used to distinguish and eliminate from consideration those illite/smectites that were predominantly of detrital origin.

The ^{18}O values of the diagenetic components differed markedly from one sample suite to another and were used to constrain the conditions of formation of each of the components on diagrams of temperature vs. $\delta^{18}O_{fluid}$. The temperature of formation of the diagenetic mixed-layers illite/smectite in each field was further constrained using burial history diagrams, assumed geothermal gradients and K-Ar ages. The results indicate that the fluids from which the illite/smectite mixed-layers formed in the more deeply buried samples were warmer and more enriched in ^{18}O than those in the more shallowly buried samples. This is consistent with a mixing model in which cool, ^{18}O-depleted meteoric fluids enter the formation in groundwater recharge zones and mix in varying proportions with warm, ^{18}O-enriched diagenetically modified waters generated at depth. However, the authors suggested an alternative model in which the illite/smectite forming fluids were expelled over a range of burial depths from the thick shales enclosing the relatively thin sandstone. In this model, the chemistry of the fluids would have been fairly similar at all depths, accounting for the relative uniform mineralogy of the diagenetic illite/smectite.

Studies of Arkoses along the Coastal Margin of Angola

Girard et al. (1988, 1989) examined a complex suite of diagenetic phases in Cretaceous arkoses buried along the Angola margin. The times of formation of K-feldspar overgrowths and illitic clay (illite and illite/smectite) were determined using the K-Ar method. Several size fractions of each sample were analyzed to assist in the assessment of whether, and to what extent, the ages are affected by detrital contamination. As discussed earlier, similar K-Ar ages for different size fractions of a sample were interpreted as representing the time of mineral formation. K-Ar ages decreasing with decreasing particle size (Fig. 12) were assumed to be affected by detrital contamination and to represent maximum ages of mineral formation.

The studies of Girard et al. combined the basinal history, petrographic, mineralogic, isotopic and fluid inclusion microthermometry data. The combined data permitted the authors to show that while the depth of burial of the arkoses had increased during most of the post-depositional history of the sequence, temperatures experienced by the samples had decreased during much of that time. They attributed this to the tectonic setting of the sediments, and concluded that the geothermal gradient in the region was anomalously high when the sediments were deposited, as a result of proximity to the mid Atlantic spreading axis. With time, the sampling site moved away from the ridge axis and the geothermal gradient decreased.

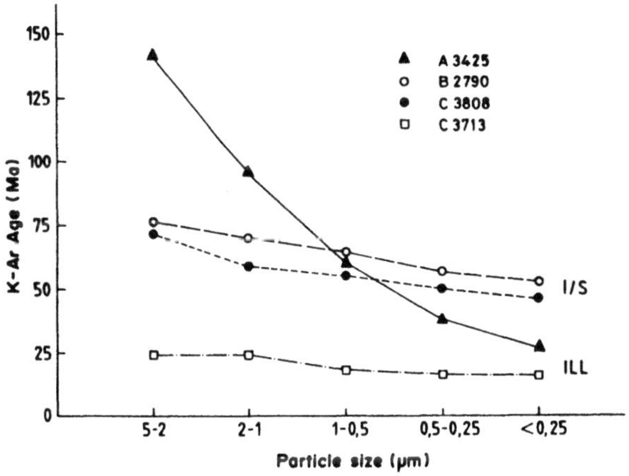

Figure 12: Changes in K-Ar ages with respect to particle sizes for illitic materials from Cretaceous arkoses in the coastal margin of Angola (Girard et al., 1988, 1989)

Studies in the Paris Basin

Mossmann et al.'s (1992) study of diagenetic clay minerals from the Rhaetian (Uppermost Triassic) sandstone of the Paris Basin is another illustration of the advantage of integrating the results of K-Ar analyses of clay minerals with the geologic history of the sedimentary basin and the results of petrographic, mineralogic, and morphologic (TEM and SEM) observations. The clay minerals studied came from a single stratigraphic unit sampled across the basin. Depth of burial ranged from surficial outcrop at the edge of the basin to about 2,700 m deep near the center.

The sandstone samples were gently disaggregated and analysis of the <0.2μm clay fractions by XRD indicated the presence of three

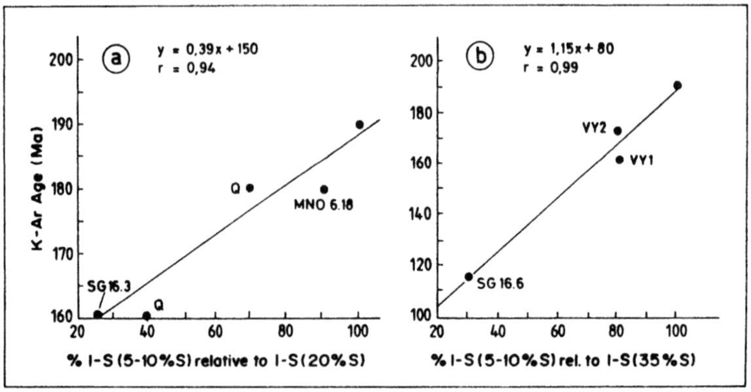

Figure 13: K-Ar ages of different generations of illite-smectite mixed-layer phases in Triassic sediments from Paris Basin (Mossmann et al., in press)

generations of illite/smectite mixed-layer clay: one with 5-10% expandable layers, one with about 20% expandable layers, and one with about 35% expandable layers. The most illitic phase is the dominant clay in most samples. The ages of those clay fractions that consist predominantly (more than 90%) of the illite-rich phase and that are free of any detrital components are between 189 and 192 Ma. Mossman et al. interpreted these ages as being close to the age of crystallization of the illite-rich illite/smectite phase. The times of crystallization of the two other phases were calculated from the mineralogical data and the ages of samples that were two-component mixtures of the highly illitic 190 Ma phase and one of the other illite/smectite components (Fig. 13). The phase with 20% expandable layers formed at about 150 Ma and the phase with about 35% expandable layers formed at about 80 Ma. The calculated temperature of formation of the early formed illite-rich phase is about 150°C or higher and those of the two other illite/smectite components are lower. The temperature of 150°C is higher than would result from burial of these sediments in a region of normal geothermal gradient. Mossman et al. suggested that this temperature may have resulted from upward migration of hydrothermal fluids along fault zones related, as in the previous example of Angola offshore, to a tectonic activity. This interpretation is consistent with inferences by others that contemporaneous tectono-hydrothermal activity had occurred at several other places in western Europe at the time (Bonhomme, 1982, Bonhomme et al., 1983; Bonhomme and Millot, 1987; Liewig et al., 1987b).

Diagenetic Processes in Associated Shales and Sandstones

Rinckenbach (1988) and Rinckenbach et al. (in prep.) studied <0.4μm clay fractions of shales and associated sandstones from a 4,000 m thick Miocene to Pliocene sedimentary succession in the Mahakam delta in eastern Kalimantan (Borneo). The study afforded an examination of the crystallochemical evolution of clay particles in environments of different water-rock ratios at nearly identical temperatures. XRD analyses indicated a classical pattern of evolution of the mixed-layer illite/smectite of both the shales and sandstones with depth, changing from randomly oriented mixed-layer clay in the top 1,000 m, through mixed-layer clays with short-range ordering at intermediate depths, to phases exhibiting long range ordering at depths greater than 3,000 m. The XRD data also emphasized that throughout the sequence, the expandability of the mixed-layer clays in the shales was strictly identical to that of the mixed-layer clays in the adjacent sandstones. Qualitatively, a graph of percentage of illite layers in the mixed-layer clay vs. depth resembled that reported by Hower et al. (1976) for the Gulf Coast shale sequence and Jennings and Thompson (1986) for the delta of the Colorado River.

The clay size-fractions from shales and associated sandstones of the Mahakam delta deposits were separated following a gentle disaggregation technique. In the uppermost part of the sequence, the K–Ar ages of clay size-fractions from the shales are reasonably close to those from adjacent sandstones. In more deeply buried sediments the K-Ar ages of the clay minerals separated from the two lithologies differ considerably from one another (Fig. 14). Ages of the <0.4μm clay fractions from the shales decrease from 80 Ma near the surface to 52 Ma at 3,160 m, whereas those from associated sandstones decrease from 80 Ma near the surface to 25 Ma at 2,595 m. TEM observations showed considerable morphological differences between the clays from the two types of rocks. The $\delta\,^{18}O$ values of the 0.4μm clay fractions from shales decrease from +14.7

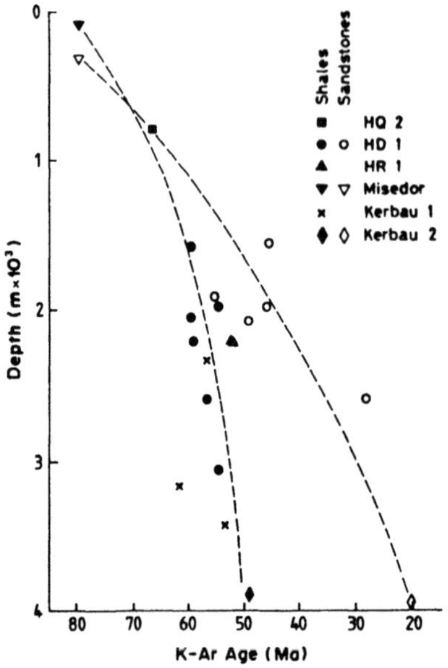

Figure 14: Variations of K-Ar ages of clay fractions (<0.4μm) between shales and associated sandstones from the sedimentary sequence of the Mahakam delta deposits, Indonesia (Rinckenbach, 1988)

per mil to +12.6 per mil, while those from the associated sandstones decrease from +15.6 per mil to +10.5 per mil. Rinckenbach et al. postulated that the isotopic and morphological differences between the clays in the two lithologically distinct rock types require two distinct mechanisms of clay alteration. They proposed that the mixed-layer clays in the shales evolved by a process of transformation within existing clay structures, whereas similar clays in the sandstones grew by a process of dissolution and reprecipitation. Different water/rock ratios, and the resulting differences in chemical environments in the two lithologies seem to have been responsible for the differences in mechanism. Temperature is clearly not the factor responsible for the difference. If the proposal is correct that the alteration of shaly clay minerals during progressive burial does not occur through a mechanism of dissolution-recrystallization of pre-existing clay structures, it might be the reason why complete homogenization of O isotopes among different clay size fractions was not observed, even at great depths, in the Gulf Coast shales studied by Yeh and Savin (1977).

Diagenetic Processes in Bentonites

Mineralogic and K-Ar studies of diagenesis of Gulf Coast shales have shown that the rate and extent of conversion of smectite layers to illite layers in mixed-layer illite/smectite depends upon temperature and time, as well as on the availability of K and probably of Al. But the model of illitization for the Gulf Coast shales may not be applicable for illitization elsewhere. This is evident for the middle Devonian K-rich bentonites in eastern North America. Most of these bentonites were never buried sufficiently deeply to reach the temperatures at which smectite layers were converted to illite layers in the Gulf Coast. Also, the source of K incorporated in the mixed-layer clays of the bentonites was neither feldspar nor mica in the volcaniclastic sediments, as both the feldspar and the mica concentrations remained more or less unchanged as illitization progressed (Huff and Turkmenoglu, 1981).

Elliott and Aronson (1987) analyzed the K-Ar ages of mixed-layer illite/smectite from Middle Ordovician bentonites from Alabama, Tennessee, Virginia, and Kentucky in the southern Appalachian basin. The K-Ar ages of the $<0.125\mu m$ clay fractions, which were free of detrital non-clay minerals, fell in a narrow range between 272 and 303 Ma. This suggested that a widespread illitization episode of short duration had occurred during late Pennsylvanian/early Permian time. Elliott and Aronson proposed that this might have occurred as the result of expulsion of K-rich, hot, saline fluids from deeper parts of the basin to the edges during the Alleghenian orogeny.

ISOTOPIC INDICATORS OF PALEOTEMPERATURES AND PALEOHYDROLOGY IN SEDIMENTARY BASINS

Estimation of Paleotemperatures

Attempts to use O isotopic fractionations between cogenetic

diagenetic minerals in clastic sediments have met with only limited success. This approach is generally not applicable to sandstones because only very rarely does petrographic examination indicate that more than one diagenetic phase forms at a time. Therefore, in general, there is no expectation that the diagenetic minerals in a sandstone formed in isotopic equilibrium with one another.

Yeh and Savin (1977) attempted to use the O isotope fractionation between coexisting fine-grained clay and fine-grained quartz in three sequences of Gulf Coast shales as an isotope geothermometer indicative of the maximum temperature reached during burial. They used a clay fraction $<0.1\mu m$ and a quartz size range of 0.1 to $0.5\mu m$. They reported that isotopic temperatures approached measured borehole temperatures progressively with increasing depth, but the two were never concordant in the Gulf Coast samples, even at temperatures as high as 170°C. It is possible that accurate isotope geothermometry of shales heated to teperatures of 170°C or less might be possible under some circumstances, especially if very fine grained clay and quartz were analyzed, but this has not been well explored. Eslinger and savin (1973) calculated temperatures from $\delta^{18}O$ values of quartz and illite in a series of outcrop samples from the Precambrian Belt Supergroup, Montana. These temperatures, between 225 and 310°C, are consistent with the maximum burial depths of 5.5 to 6 km estimated from stratigraphic and structural considerations.

Characterization of Water-rock Interactions

Eslinger and Savin (1973) calculated the $\delta^{18}O$ value of the fluid with which the clay and quartz of each of several Belt Supergroup samples had equilibrated (at the temperature indicated by the isotopic fractionation between the two phases), and concluded that the fluids were all greatly enriched in ^{18}O (+7 to +11 per mil). Those high $\delta^{18}O$ values indicated that the fluids had evolved through interaction with sediments under conditions of very low water/rock ratio. In other words, each rock volume had been largely closed to the movement of water when isotopic exchange between the quartz and clay occurred. Yeh and Savin (1977) calculated the $\delta^{18}O$ values of the fluids with which the finest size fractions of clay might have equilibrated in their Gulf Coast wells. These calculations were based on the $\delta^{18}O$ values of the clays and measured borehole temperatures. Consistent with the calculation of $\delta^{18}O$ values of diagenetic fluids in the Belt Supergroup (Eslinger and Savin, 1973), the calculated $\delta^{18}O$ values of the fluids which exchanged with the Gulf Coast shales ranged as high as +11 per mil, suggesting very low water/rock ratios. Marked differences between calculated fluid $\delta^{18}O$ values at different depths in the holes suggested a lack of communication between the fluids at different depths. Since that time, Kharaka et al. (1977) and Milliken et al. (1981) have measured $\delta^{18}O$ values of pore waters from the Frio Formation in the Gulf Coast and found values between approximately +3 and +7 per mil, in general agreement with the values calculated by Yeh and Savin (1977).

Suchecki and Land (1983), in a study of the burial diagenesis of the Jurassic and Cretaceous volcanoclastic mudstones and shales of the Great Valley Sequence, California, modeled the change in ^{18}O values of clay and pore water in a hypothetical system of clay-rich sediments undergoing burial and conversion of smectite to illite. The results of the calculation depended upon the parameters used. A reasonable choice of parameters yielded calculated $\delta^{18}O$ values of illite/smectite that were consistent with the measured values. Corresponding calculated $\delta^{18}O_{fluid}$ values were as high as 7 to 10 per mil at depths where temperatures reached 175°C. These values are similar to the values measured in the Gulf Coast shales.

CONCLUSIONS

Clay minerals mostly retain their pre-depositional isotope signature in a sedimentary basin, unless they have undergone major alteration or recrystallization. The isotopic data of clays which may have completely recrystallized can be used as a tracer for syn-sedimentary and burial history of the sediments, which include distinct events of diagenesis, and for reconstructing paleotemperatures and paleohydrological conditions. However, it is often very difficult to determine whether a clay fraction is entirely detrital, authigenic or a mixture of the two.

The mechanisms of epigenetic and diagenetic clay formation have not been discussed here in detail, because these processes have not yet been clearly understood. Presently the debate is between primary precipitation, dissolution-recrystallization, and lately, Oswald ripening advocated by Eberl et al. (1990). It should also be considered at this point that "punctuated diagenesis", as it was suggested by Morton (1985), seems to be only possible under abnormal thermal conditions which cannot be met in sedimentary basins having average thermal gradient.

Among the criteria which should be considered for distinguishing between epigenetic/diagenetic and detrital clays in any study in which time of sedimentation is to be determined by isotopic dating methods, we suggest the following:

-field observations and/or visual examination of the core samples in their stratigraphic and structural context,

-careful petrographic observations to determine the temporal relationships among detrital and authigenic or diagenetic mineral phases,

-careful separation of a range of fine clay fractions following gentle disaggregation of the rocks to avoid size reduction of the coarse detrital components,

-thorough characterization of the clay fractions including determination of their mineralogic and crystallographic properties,

-observations of the clay particle morphology by SEM and TEM,

-chemical analysis and Sr isotopic determinations of paragenetic phases, such as carbonates or sulfates, associated with the clay fractions, or of the formation waters when available.

One of the most important steps in isotopic studies of clay materials is their separation and characterization. Separation procedures are not generally very difficult but are usually very time consuming. To avoid this, attempts are now underway in a number of laboratories to develop microprobe techniques for isotopic analysis of unseparated minerals within rock samples (Shanks and Criss, 1989; Girard and Onstott, 1991; Clauer et al., this volume). These techniques, which involve either secondary ion mass spectrometry (SIMS) or spot-heating with a laser beam, may in the future eliminate some of the experimental problems, and the enormous expenditure of time required for the separation of clay minerals.

ACKNOWLEDGEMENTS

We sincerely express our thanks to Mr. C. Hammel for providing technical assistance in the drafting.

REFERENCES

Amirkhanov Kh.I, Brandt S.B., Bartnitskh E.I., Gurvich V.S. and Gasanov S.A. (1958) On the question of preservation of radiogenic argon in glauconite. Dokl. Acad. Nauk SSSR, 118, 328-330.

Aronson J.L. and Hower J. (1976) Mechanism of burial metamorphism of argillaceous sediment: 2. Radiogenic argon evidence. Geol. Soc. Am. Bull., 87, 738-744.

Ayalon A. and Longstaffe F.J. (1988) Oxygen-isotope studies of diagenesis and porewater evolution in the western Canada sedimentary basin: evidence from the Upper Cretaceous basal Belly River sandstone, Alberta. Jour. Sedim. Petrol., 58, 489-505.

Bailey S.W., Hurley P.M., Fairbairn H.W. and Pinson W.H. (1962) K-Ar dating of sedimentary illite polytypes. Geol. Soc. Am. Bull., 73, 1167-1170.

Biscaye P.E. and Dasch E.J. (1971) The rubidium, strontium, strontium isotope system in deep-sea sediments: Argentine Basin. Jour. Geophys. Res., 76, 5087-5096.

Bird M.I. and Chivas A.R. (1988) Stable-isotope evidence for low temperature kaolinitic weathering and post-formational hydrogen-isotope exchange in Permian kaolinites. Chem. Geol. (Isot. Geosc. Sect.), 72, 249-265.

Bofinger V.M., Compston W. and Vernon M.J. (1968) The application of acid leaching to the Rb-Sr dating of a Middle Ordovician shale. Geochim. Cosmochim. Acta, 32, 823-833.

Boles J.R. and Franks S.G. (1979) Clay diagenesis in Wilcox sandstones of SW Texas: Implication of smectite diagenesis on sandstone cementation. Jour. Sedim. Petrol., 49, 55-70.

Bonhomme M.G. (1982) The use of Rb-Sr and K-Ar methods as stratigraphic tool applied to sedimentary rocks and minerals. Prec. Res., 18, 5-25.

Bonhomme M., Lucas J. and Millot G. (1966) Signification des déterminations isotopiques dans la géochronologie des sédiments.

Actes 151e Coll. Int. CNRS, Nancy, 1965, 541-565.
Bonhomme M. and Prévôt L. (1968) Application de la méthode rubidium-strontium à l'étude radiométrique de quelques dépôts dévono-dinantiens du Massif de la Bruche (Vosges du Nord). Bull. Serv. Carte géol. Als. Lorr., 21, 219-247.
Bonhomme M.G., Bühmann D. and Besnus Y. (1983) Reliability of K-Ar dating of clays and silicifications associated with vein mineralizations in Western Europe. Geol. Rundsch., 72, 105-117.
Bonhomme M.G. and Millot G. (1987) Diagenèse généralisée du Jurassique moyen (170-160 Ma) dans le bassin du Rhône inférieur jusqu'à la bordure des Cévennes (France). Datations K-Ar d'argiles du Trias et du Lias inférieur. C. R. Acad. Sci., Paris, 304 II, 431-434.
Bonhomme M., Clauer N. and Odin G.S. (1970) Résultats préliminaires de datation rubidium-strontium sur des sédiments glauconieux dans le Paléogène d'Angleterre. Bull. Serv. Carte géol. Als. Lorr., 23, 209-214.
Brass G.W. (1975) The effect of weathering on the distribution of strontium isotopes in weathering profiles. Geochim. Cosmochim. Acta, 39, 1647-1654.
Bross R., Stille P., Gauthier-Lafaye F., Weber F. and Clauer N. (1992) Sm-Nd isotopic dating of Proterozoic clay material. Example from Francevillian sedimentary series (Gabon). Earth Plan. Sci. Lett., in press.
Buatier M. (1989) Genèse et evolution des argiles vertes hydrothermales océaniques: Les "monts" du rift des Galapagos (Pacifique Equatorial). Thèse Univ. Strasbourg, 175p.
Burrows S.M. (1985) Oxygen isotope evidence for the conditions of diagenesis of the Muddy Formation, east flank of the Powder River Basin, Montana and Wyoming. M.Sc. Thesis, Case Western Reserve Univ., Cleveland, Ohio.
Burst J.F. (1958a) Mineral heterogeneity in glauconite pellets. Am. Miner., 43, 481-497.
Burst J.F. (1958b) "Glauconite" pellets: their mineral nature and applications to stratigraphic interpretations. Am. Ass. Petrol. Geol. Bull., 42, 310-337.
Chaudhuri S and Faure G. (1967) The geochronology of the Keweenawan rocks, White Pine, Michigan. Econ. Geol., 62, 1011-1033.
Chaudhuri S. and Brookins D.G. (1969) The Rb-Sr age of the Stearn shale, eastern Kansas, before and after leaching experiments. Geol. Soc. Am. Bull., 80, 2605-2610.
Chaudhuri S. and Brookins D.G. (1979) The Rb-Sr systematics in acid-leached clay minerals. Chem. Geol., 24, 231-241.
Clauer N. (1973) Utilisation de la méthode rubidium-strontium pour la datation de niveaux sédimentaires du Précambrien supérieur de l'Adrar mauritanien (Sahara occidental) et la mise en évidence de transformations précoces des minéraux argileux. Geochim. Cosmochim. Acta, 37, 2243-2256.
Clauer N. (1974) Utilisation de la méthode rubidium-strontium pour la datation d'une schistosité de sédiments peu métamorphisés: Application au Précambrien II de la Boutonnière de Bou Azzer-El Graara (Anti-Atlas, Maroc). Earth Plan. Sci. Lett., 22, 404-412.
Clauer N. (1976) Géochimie isotopique du strontium des milieux sédimentaires. Application à la géochronologie de la couverture du craton ouest-africain. Sci. Géol. Mém., Strasbourg, 45, 236p.

Clauer N. (1979) Relationship between the isotopic composition of strontium in newly formed continental clay minerals and their source material. Chem. Geol., 27, 115-124.
Clauer N. (1982) The rubidium-strontium method applied to sediment: certitudes and uncertainties. In: Numerical Dating in Stratigraphy, Odin G.S. (ed.), J. Wiley and Sons Ldt, Chichester, 245-276.
Clauer N., Caby R., Jeannette D. and Trompette R. (1982a) Geochronology of sedimentary and metasedimentary Precambrian rocks of the west African craton. Prec. Res., 18, 53-71.
Clauer N., Hoffert M. and Karpoff A.M. (1982b) The Rb-Sr isotope system as an index of origin and diagenetic evolution of southern Pacific red clays. Geochim. Cosmochim. Acta, 46, 2659-2664.
Clauer N., Giblin P. and Lucas J. (1984) Sr and Ar isotope studies of detrital smectites from the Atlantic Ocean (D.S.D.P., Legs 43, 48 and 50). Chem. Geol. (Isot. Geosc. Sect.), 2, 141-151.
Clauer N., O'Neil J.R., Bonnot-Courtois C. and Holtzapffel T. (1990) Morphological, chemical and isotopic evidence for an early diagenetic evolution of detrital smectite in marine sediments. Clays and Clay Min., 38, 33-46.
Clauer N., Keppens E. and Stille P. (1991) Sr isotope constraints on the process of glauconitization. Geology, in press.
Clauer N., Cocker J.D. and Chaudhuri S. (1992) Isotopic dating of diagenetic illites in reservoir sandstones: Influence of the investigator effect. In: Origin, Diagenesis, and Petrophysics of Clay Minerals in Sandstones, Houseknecht D.W. (ed.), Soc. Econ. Paleontol. Miner., Spec. Pub., 47, in press.
Cocker J.D., Clauer N., Tsui T.F. and Swarbick R.E. (1988) A diagenetic model for the northwest Hutton field. Conf. on Clay Mineral Diagenesis in Hydrocarbon Reservoirs and Shales, Cambridge, UK, 1p.
Compston W. and Pidgeon R.T. (1962) Rb-Sr dating of shales by the total-rock method. Jour. Geophys. Res., 67, 3493-3502.
Cordani U.G., Kawashita K. and Thomaz-Filho A. (1978) Applicability of the Rb-Sr method to shales and related rocks. Am. Ass. Petrol. Geol., Spec. Publ., 6, 93-117.
Cordani U.G., Thomaz-Filho A., Brito-Neves B.B. and Kawashita K. (1985) On the applicability of the Rb-Sr method to argillaceous sedimentary rocks: some examples from Precambrian sequences of Brazil. Giorn. Geol., ser. 3, 47, 253-280.
Cormier R.F. (1956) Rubidium-strontium ages of glauconite and their application to the construction of a post-Precambrian time-scale. PhD Thesis, Massachussetts Institute of Technology, Cambridge, Mass.
Dasch E.J. (1969) Strontium isotopes in weathering profiles, deep sea sediments and sedimentary rocks. Geochim. Cosmochim. Acta, 33, 1521-1552.
Dunoyer de Segonzac G. (1969) Les minéraux argileux dans la diagenèse. Passage au métamorphisme. Mém. Serv. Carte géol. Als. Lorr., 29, 320p.
Eberl D.D., Srodon J., Kralik M., Taylor B.E. and Peterman Z.E. (1990) Ostwald ripening of clays and metamorphic minerals. Science, 248, 474-477.
Elliott W.C. and Aronson J.S. (1987) Alleghanian episode of K-bentonite illitization in the southern Appalachian Basin. Geology, 15, 735-739.
Eslinger E.V. and Savin S.M. (1973) Mineralogy and oxygen isotope

geochemistry of the hydrothermally altered rock of the Ohaki-Broadlands, New Zealand geothermal area. Am. Jour. Sci., 273, 240-267.
Eslinger E.V. and Yeh H.W. (1981) Mineralogy, $^{18}O/^{16}O$ and D/H Ratios of clay rich sediments from Deep Sea Drilling Project Site 180, Aleutian Trench. Clays and Clay Min., 29, 309-315.
Evernden J.F., Curtis G.H., Obradovich J.D. and Kistler R.W. (1961) On the evaluation of glauconite and illite for dating sedimentary rocks by the K-Ar method. Geochim. Cosmochim. Acta, 23, 78-99.
Ghosh P.K. (1972) Use of bentonites and glauconites in potassium 40/argon 40 dating in Gulf Coast stratigraphy. Ph. D. Thesis, Univ. Houston, Texas, 136p.
Girard J.P., Aronson J.L. and Savin S.M. (1988) Separation, K-Ar dating and $^{18}O/^{16}O$ ratio measurements of diagenetic K-feldspar overgrowth: An example from the Lower Cretaceous arkoses of the Angola margin. Geochim. Cosmochim. Acta, 52, 2207-2214.
Girard J.P., Savin S.M. and Aronson J.L. (1989) Diagenesis of the lower Cretaceous arkoses of the Angola margin: Petrologic, K/Ar dating and $^{18}O/^{16}O$ evidence. Jour. Sediment. Petrol., 59, 519-538.
Girard J.P. and Onstott T.C. (1991) Application of $^{40}Ar/^{39}Ar$ laser-probe and step-heating techniques to the dating of diagenetic K-feldspar overgrowths. Geochim. Cosmochim. Acta, 52, 2207-2214.
Glasmann J.R., Clark R.A., Larter S.R., Briedis N.A. and Lundegard P.D. (1989) Diagenesis and hydrocarbon accumulation, Brent Sandstone (Jurassic), Bergen High area, North Sea. Am. Ass. Petrol. Geol. Bull., 73, 1341-1360.
Goldich S.S., Baadsgaard H., Edwards G. and Weaver C.E. (1959) Investigations in radioactivity dating of sediments. Am. Ass. Petrol. Geol. Bull., 43, 654-662.
Hamilton J., Fallick A.E., MacIntyre R.M. and Elliot S. (1987) Isotopic tracing of the provenance and diagenesis of lower Brent Group sands, North Sea. In: Petroleum Geology of North West Europe, Brooks J. and Glennie K. (eds.), Graham and Trotman Ltd, London, 939-949.
Harris W.B. (1976) Rb-Sr glauconite isochron, Maestrichtian unit of Peedee Formation (Upper Cretaceous). Geology, 4, 761-762.
Harris W.B. and Bottino M.L. (1974a) Stratigraphic application of Rb-Sr glauconite ages, Cretaceous Peedee Formation, North Carolina. Geol. Soc. Am., Abstr. with Progr., 6, 4, 361.
Harris W.B. and Bottino M.L. (1974b) Rb-Sr study of Cretaceous Lobate glauconite pellets, North Carolina. Geol. Soc. Am. Bull., 85, 1475-1478.
Harris W.B. and Baum G.R. (1977) Foraminifera and Rb-Sr glauconite ages of a Paleocene Beaufort Formation outcrop in North Carolina. Geol. Soc. Am. Bull., 88, 869-872.
Harris W.B. and Fullagar P.D. (1989) Comparison of Rb-Sr and K-Ar dates of middle Eocene bentonite and glauconite, southeastern Atlantic coastal plain. Geol. Soc. Am. Bull., 101, 573-577.
Hoffert M., Karpoff A.M., Clauer N., Schaaf A., Courtois C. and Pautot G. (1978) Néoformations et altérations dans trois faciès volcanosédimentaires du Pacifique Sud. Oceanolog. Acta, 1, 187-202.
Hofmann A.W., Mahoney J.W. and Giletti B.J. (1974) K-Ar and Rb-Sr data on the detrital and post-depositional history of Pennsylvania clay from Ohio and Pennsylvania. Geol. Soc. Am.

Bull., 85, 639-644.
Hogg A.J.C., Peirson M.J., Fallick A.E., Hamilton P.J. and Mac Intyre R.M. (1987) Clay mineral and isotope evidence for controls on reservoir properties of Brent Group sandstones, British North Sea. Terra cognita, 7, 342.
Hower J., Eslinger E., Hower M.E. and Perry E.A. (1976) Mechanism of burial diagenesis of argillaceous sediments: 1. Mineralogical and chemical evidence. Geol. Soc. Am. Bull., 87, 725-737.
Huff W.D. and Turkmenoglu A.G. (1981) Chemical characteristics and origin of Ordovician K-bentonites along the Cincinnati Arch. Clays and Clay Min., 29, 113-123.
Hunziker J.C., Frey M., Clauer N., Dallmeyer R.D., Friedrichsen H., Flehmig W., Hochstrasser K., Roggwiller P. and Schwander H (1986) The evolution of illite to muscovite: Mineralogical and isotopic data from the Glarus Alps, Switzerland. Contr. Mineral. Petrol., 92, 157-180.
Hunziker J.C., Frey M., Clauer N. and Dallmeyer R.D. (1987) Reply to the comments on the evolution of illite to muscovite by J.R. Glasman. Contr. Miner. Petrol., 96, 75-77.
Huon S. (1985) Clivage ardoisier et réhomogénéisation isotopique dans des schistes paléozoïques du Maroc. Etude microstructurale et isotopique, conséquences régionales. Thèse Univ. Strasbourg, 124p.
Huon S., Clauer N. and Piqué A. (in prep.) Size, orientation and K-Ar data of phyllosilicates from Moroccan Paleozoic slates.
Hurley P.M., Cormier R.F., Hower J., Fairbairn H.W. and Pinson W.H. (1960) Reliability of glauconite for age measurement by K-Ar and Rb-Sr methods. Am. Ass. Petrol. Geol. Bull., 44, 1793-1808.
Hurley P.M., Brookins D.G., Pinson W.H., Hart S.R. and Fairbairn H.W. (1961) K-Ar age studies of Mississippi and other river sediments. Geol. Soc. Am. Bull., 72, 1807-1816.
Hurley P.M., Heezen B.C., Pinson W.H. and Fairbairn H.W. (1963) K-Ar age values in pelagic sediments of the North Atlantic. Geochim. Cosmochim. Acta, 27, 393-399.
James A.T. and Baker D.R. (1976) Oxygen isotope exchange between illite and water at 22°C. Geochim. Cosmochim. Acta, 40, 235-239.
Jennings S. and Thompson G.R. (1986) Diagenesis of Plio-Pleistocene sediments of the Colorado River delta, southern California. Jour. Sediment. Petrol., 56, 89-98.
Jourdan A., Thomas M., Brevart O., Robson P., Sommer F. and Sullivan M. (1987) Diagenesis as the control of the Brent sandstone reservoir properties in the greater Alwyn area (East Shetland basin). In: Petroleum Geology of Northwest Europe, Brooks J. and Glennie K. (eds.), Graham and trotman Ltd, London, 951-961.
Kharaka Y.K., Callender E. and Carothers W.W. (1977) Geochemistry of geopressured waters from the Texas Gulf Coast. In: Proceedings of the third Geopressured-Geothermal Energy Conference, Univ. Southwestern Louisiana, Lafayette, Louisiana, 121-165.
Keppens E., Elewaut E. and Pasteels P. (1978a) Géochronologie rubidium-strontium de glauconies du Crétacé du Nord de la France et du Bassin de Mons, comparaison avec les données potassium-argon correspondantes. Bull. Belg. Ver. Geologie, 87, 153-162.
Keppens E., Elewaut E. and Pasteels P. (1978b) Rubidium-Strontium

and Potassium-Argon radiometric datations on glauconites of the "Bande Noire" (Base of the Asse Clay) from the Belgian Eocene. Ann. Soc. Géol. Belgique, 101, 179-192.

Keppens E. and O'Neil J.R. (1984) Oxygen isotope variations in glauconies. Terra cognita, Spec. Issue, p.42.

Law E. (1983) Petrologic, geochronologic and isotopic investigation of the diagenesis and hydrocarbon emplacement in the Muddy Sandstone, Powder River Basin. Ph. D. Thesis, Case Western Reserve Univ., Cleveland, Ohio, 340 p.

Lawrence J.R. (1979) O^{18}/O^{16} of the silicate fraction of Recent sediments used as provenance indicator in the South Atlantic. Mar. Geol., 33, 1-7.

Lawrence J.R. and Taylor H.P., Jr. (1971) Deuterium and oxygen-18 correlation: Clay minerals and hydroxides in Quaternary soils compared to meteoric waters. Geochim. Cosmochim. Acta, 35, 993-1003.

Lawrence S.R. and Taylor H.P., Jr. (1972) Hydrogen and oxygen isotope systematics in weathering profiles. Geochim. Cosmochim. Acta, 36, 1377-1393.

Lee M. and Aronson J.L. (1986) K/Ar systematics of bentonite and shale in a contact metamorphic zone, Cerillos, New Mexico. Clays and Clay Min., 34, 483-487.

Lee M., Aronson J.L. and Savin S.M. (1985) K-Ar dating of gas emplacement in Rotliegendes sandstones, Netherlands. Am. Ass. Petrol. Geol. Bull., 69, 1381-1385.

Lee M., Aronson J.L. and Savin S.M. (1989) Timing and conditions of Permian Rotliegende sandstone diagenesis, southern North Sea: K/Ar and oxygen isotopic data. Am. Ass. Petrol. Geol. Bull., 73, 195-215.

Liewig N., Clauer N. and Sommer F. (1987a) Rb-Sr and K-Ar dating of clay diagenesis in Jurassic sandstone reservoirs, Noth Sea. Am. Ass. Petrol. Geol. Bull., 71, 1467-1474.

Liewig N., Mossmann J.R. and Clauer N. (1987b) Datation isotopique K-Ar d'argiles diagénétiques de réservoirs gréseux: Mise en évidence d'anomalies thermiques du Lias inférieur en Europe nord-occidental. C. R. Acad. Sci., Paris, 304 II, 707-712.

Lipson J. (1958) Potassium-Argon dating of sedimentary rocks. Geol. Soc. Am. Bull., 69, 137-150.

Longstaffe F.K. (1984) The role of meteoric water in diagenesis of shallow sandstones. In: Clastic Diagenesis, McDonald D.A. and Surdam R.C. (eds.), Am. Ass. Petrol. Geol. Mem., 37, 81-97.

Longstaffe F.K. (1986) Oxygen isotope studies of diagenesis in the basal Belly River sandstone, Pembina I-Pool, Alberta. Jour. Sediment. Petrol., 56, 78-88.

Longstaffe F.J. and Ayalon A. (1987) Oxygen-isotope studies of clastic diagenesis in the Lower Cretaceous Viking Formation, Alberta: implications for the role of meteoric water. In: Diagenesis of Sedimentary Sequences, Marshall J.D. (ed.), Geol. Soc. Amer., Spec. Publ., 36, 277-296.

Longstaffe F.J. and Ayalon A. (1990) Hydrogen-isotope geochemistry of diagenetic clay minerals from Cretaceous sandstones, Alberta: Evidence for exchange. Applied Geochem., in press.

Milliken K.L., Land L.S. and Loucks R.G. (1981) History of burial diagenesis determined from isotope geochemistry, Frio Formation, Brazoria County, Texas. Am. Ass. Petrol. Geol. Mem., 37, 434p.

Moorbath S., Taylor P.N., Orpen J.L., Treloar P. and Wilson J.F. (1987) First direct dating of Archaean stromatolitic

limestone. Nature, 326, 865-867.
Morton J.P. (1985) Rb/Sr evidence for punctuated illite/smectite diagenesis in the Oligocene, Frio Formation, Texas, Gulf Coast. Geol. Soc. Am. Bull., 96, 1043-1049.
Mossmann J.R., Clauer N. and Liewig N. (1992) Dating thermal anomalies in sedimentary basins: The diagenetic history of clay minerals in the triassic sandstones of the Paris Basin (France). Clay Min., 27, in press.
Obradovich J.D. (1965) Problems in the use of glauconite and related minerals for radioactive dating. Ph. D. Thesis, Univ. California at Berkeley.
Odin G.S. (1975) De glauconarium, origine, aetateque. Doc. es-Sciences Thesis, Univ. Paris, 280p.
Odin G.S., Velde B. and Bonhomme M.G. (1977) Radiogenic argon retention in glauconites as a function of mineral recrystallization. Earth Plan. Sci. Lett., 37, 154-158.
Odin G.S. and Matter A. (1981) De glauconarium origine. Sedimentology, 28, 611-641.
Odin G.S., Dodson M.H., Hunziker J.C. and Kreuzer H. (1979) Radiogenic argon in glauconies during their genesis. Bull. Inf. Int. Geol. Corr. Prog., Project 133, 6, 7-8.
Perry E.A. (1974) Diagenesis and K-Ar dating of shales and clay minerals. Geol. Soc. Am. Bull., 85, 827-830.
Perry E. A. and Hower J. (1970) Burial diagenesis in Gulf Coast pelitic sediments. Clays and Clay Min., 18, 165-177.
Perry E.A. and Turekian K.K. (1974) The effects of diagenesis on the redistribution of strontium isotopes in shales. Geochim. Cosmochim. Acta, 38, 929-935.
Powers M.C. (1959) Adjustment of clays to chemical changes and the concept of the equivalue level. Clays and Clay Min., 6, 309-326.
Powers M.C. (1967) Fluid release mechanisms in compacting marine mudrocks and their importance in oil exploration. Am. Ass. Petrol. Geol. Bull., 51, 1240-1254.
Reuter A. (1987) Implications of K-Ar ages of whole-rock and grain-size fractions of metapelites and intercalated metatuffs within an anchizonal terrane. Contr. Mineral. Petrol., 97, 105-115.
Rinckenbach T. (1988) Diagenèse minérale des sédiments pétrolifères du delta fossile de la Mahakam (Indonésie). Evolution minéralogique et isotopique des composants argileux et histoire thermique. Thèse Univ. Strasbourg, 209p.
Rinckenbach T., Clauer N., O'Neil J.R. and Sommer F. (in prep.) Burial diagenesis of oil-bearing Neogene sediments from Mahakam delta (East Kalimantan): 1. Evidence from whole-rock petrography and clay mineralogy and isotope geochemistry.
Salomons W., Hofman P., Boelens R. and Mook W.G. (1975) The oxygen isotopic composition of the fraction less than 2 microns (clay fraction) in Recent sediments from western Europe. Mar. Geol., 18, M23-M28.
Savin S.M. and Epstein S. (1970a) The oxygen and hydrogen isotope geochemistry of clay minerals. Geochim. Cosmochim. Acta, 34, 25-42.
Savin S.M. and Epstein S. (1970b) The oxygen and hydrogen isotope geochemistry of ocean sediments and shales. Geochim. Cosmochim. Acta, 34, 43-63.
Savin S.M., Aronson J.L., Law E. and Burrows S. (in prep.) Diagenesis of the Muddy Sandstone Powder River Basin, Wyoming.

Shanks W.C.III and Criss R.E. (eds.) (1989) New frontiers in stable isotopic research: laser probes, ion probes and small-sample analysis. U.S. Geol. Surv. Bull., 1890, 174p.
Souchez R., Lemmens M., Lorrain R., Tison J.L. and Jouzel J. (1990) Influence of hydroxyl-bearing minerals on the isotopic composition of ice from the basal zone. Nature, 345, 244-246.
Stille P. and Clauer N. (1986) Sm-Nd isochron-age and provenance of the argillites of the Gunflint Iron Formation in Ontario, Canada. Geochim. Cosmochim. Acta, 50, 1141-1146.
Stille P. Clauer N. and Keppens E. (1992) Behavior of Nd-Si isotope systematics during recent glauconitization process. Contr. Mineral. Petrol., in press.
Sucheki R.K. and Land L.S. (1983) Isotopic geochemistry of burial-metamorphosed volcanogenic sediments, Great Valley sequence, Northern California. Geochim. Cosmochim. Acta, 47, 1487-1499.
Thompson G.R. and Hower J. (1973) An explanation for low radiometric ages from glauconite. Geochim. Cosmochim. Acta, 37, 1473-1491.
Tisserant D. and Odin G.S. (1979) Datation isotopique de glauconies miocènes d'Afrique du Nord-Ouest. C.R. Somm. Soc. géol. France, 4, 188-190.
Tsirambides A.E. (1986) Detrital and authigenic minerals in sediments from the western part of the Indian Ocean. Min. Mag., 50, 69-74.
Wasserburg G.J., Hayden R.I. and Jensen K.J. (1956) Ar^{40}-K^{40} dating of igneous rocks and sediments. Geochim. Cosmochim. Acta, 10, 153-165.
Weaver C.E. (1959) The clay petrology of sediments. Clays and Clay Min., 6, 154-184.
Weaver C.E. (1960) Possible uses of clay minerals in search of oil. Am. Ass. Petrol. Geol. Bull., 44, 1505-1518.
Weaver C.E. and Wampler J.M. (1970) K-Ar illite burial. Geol. Soc. Am. Bull., 81, 3423-3430.
Whitney P.R. and Hurley P.M. (1964) The problem of inherited radiogenic strontium in sedimentary age determinations. Geochim. Cosmochim. Acta, 28, 425-436.
Yeh H.W. (1980) D/H ratios and late-stage dehydration of shales during burial. Geochim. Cosmochim. Acta, 44, 341-352.
Yeh H.W. and Epstein S. (1978) Hydrogen isotope exchange between clay minerals and sea water. Geochim. Cosmochim. Acta, 42, 140-143.
Yeh H.W. and Savin S.M. (1976) The extent of oxygen isotope exchange between clay minerals and sea water. Geochim. Cosmochim. Acta, 40, 743-748.
Yeh H.W. and Savin S.M. (1977) The mechanism of burial metamorphism of argillaceous sediments: 3. Oxygen isotopic evidence. Geol. Soc. Am. Bull., 88, 1321-1330.

Sm-Nd ISOTOPES IN FINE-GRAINED CLASTIC SEDIMENTARY MATERIALS: CLUES TO SEDIMENTARY PROCESSES AND RECYCLING GROWTH OF THE CONTINENTAL CRUST

Sam CHAUDHURI[*]
Department of Geology, Kansas State University
Manhattan, Kansas 66506 (USA)

Peter STILLE and **Norbert CLAUER**
Centre de Géochimie de la Surface
1 rue Blessig, 67084 Strasbourg (France)

Fine-grained clastic sedimentary rocks, which make up about 60 to 70 percent of the present-day sedimentary mass, almost always consist of large amounts of detrital minerals derived from multiple source rocks and some small amounts of diagenetic materials. Hence, the tracing of origin of the detrital particles to source rocks has been often extremely difficult. Age determinations by K-Ar and Rb-Sr isotopic methods have shown that many uncertainties attend the determination of source rocks for sedimentary materials by these isotopic methods, because the processes of weathering, erosion, transportation, and diagenesis for sedimentary rocks, and also the processes of metamorphism for metasedimentary rocks, cause unpredictable changes in the relations of parent-daughter isotopes of the clastic materials with respect to that of their sources.

The Sm and Nd elements which belong to a group of rare-earth

---

[*] Presently at the Centre de Géochimie de la Surface, Strasbourg (France)

elements (REE), have close chemical similarity. This behavior allows limited fractionation between them by intracrustal processes. Mantle-crust differentiation is the main process which produces significant differences in the Sm/Nd ratios among large terrestrial reservoirs. Little fractionation seems to occur in the crustal materials by intracrustal sedimentary processes, metamorphism, and even partial melting. As the Sm/Nd ratios seem to remain nearly unaffected against changes by intracrustal processes, the Sm-Nd isotopic systems of clastic sediments and sedimentary and metasedimentary rocks have become subjects of much studies in recent years following the work of McCulloch and Wasserburg (1978) which revealed that such analyses can provide much information about the history of the continental crust (DePaolo et al., 1983; Hamilton et al., 1983; O'Nions et al., 1983; Taylor et al., 1983; Allègre and Rousseau, 1984; Frost and O'Nions, 1984; Goldstein et al., 1984; Miller and O'Nions, 1984; Patchett et al., 1984; Clarke and Halliday, 1985; Davies et al., 1985; Heller et al., 1985; Michard et al., 1985; André et al., 1986; Frost and Winston, 1987; Huhma, 1987; Nelson and DePaolo, 1988; Goldstein and Jacobsen, 1988; McLennan et al., 1990; Nägler, 1990).

The approach to an understanding of the question of continental crustal history based on analysis of sedimentary materials requires that sediments or sedimentary rocks are approximately a good representative of the average composition of a large continental region. Coarse-grained clastic sediments and sedimentary rocks generally have a limited transportation history. The assemblage of detrital particles in these rocks is often biased by the effect of sorting during transportation. Also, the particles in these rocks have often derived by erosion of rocks from a limited continental area. Consequently, coarse-grained sediments and sedimentary rocks are generally not suitable for the Sm-Nd isotopic investigation of crustal evolutionary history. Carbonate sedimentary rocks may contain chemical sedimentary components with a history of differentiation between Sm and Nd and these rocks are generally low in total REE contents as compared to the clastic sedimentary rocks (Haskin et al., 1966; Piper, 1974; Elderfield et al., 1981; Shaw and Wasserburg, 1985). By contrast, fine-grained sediments and sedimentary rocks, in general, consist of large amounts of detrital particles with history of long transportation and well mixing of components derived from a large continental region. Fine-grained clastic sediments and sedimentary rocks generally have a limited range of Sm/Nd ratios (McLennan, 1989). The narrow range of the ratios is probably a reflection of well mixing of the fine particles by their long transport from the site of weathering and erosion in the continent to the site of deposition often in marine sedimentary basins. Because of the very common REE characteristics among vast numbers of fine-grained clastic sediments and sedimentary rocks, and their Sm/Nd ratios being very similar to that of granitoid rocks, the Sm/Nd ratios of the fine-grained materials are taken to be good representatives of average Sm/Nd ratio of large segments of continental crust from which the clastic particles derived. Hence the fine-grained clastic sedimentary materials have been the focus of a large number of Sm-Nd isotopic studies to obtain information about erosion and evolution of the continental crusts. This article summarizes the benefits of Sm-Nd isotopic studies on sediments and sedimentary rocks in understanding sedimentary processes and growth of the continental crust.

The understanding of continental crustal history by the analyses of Sm-Nd isotopes in sedimentary rocks requires an assumption that the two elements were do not significantly fractionated by intracrustal processes. We will briefly review the merits of such an assumption in the analyses of sedimentary materials. Calculations of Sm/Nd model ages are necessary to understand the evolutionary history of sediments and sedimentary rocks. For a non-specialist who is not altogether familiar with the manipulation of Sm-Nd model ages, we will define the meaning of these ages and discuss how different factors could influence the calculated model ages on sedimentary materials. Additional details are given by Faure (1986) and DePaolo (1988b).

MODEL AGES OF SEDIMENTS

The Sm-Nd model age of a sample of clastic sediment or sedimentary rock is calculated based on the comparison between the growth of the $^{143}Nd/^{144}Nd$ ratio of the sample and that of the upper mantle as a function of time. The calculated age relates to the time when the sample and the upper mantle had the same Nd isotopic composition, or in other words, when the sample or its constituent particles appeared for the first time as an entity of the continental crust by differentiation from a depleted mantle source. In a vast majority of studies, the model age is neither the age of deposition of the sediments, nor, in the strict sense of provenance, the age of the source rock, because almost always clastic sedimentary rocks consist of particles of diverse origin. Even in the case of a derivation of a sedimentary rock from a single source rock, the Sm-Nd model age of the sedimentary components will not correspond to the age of formation of the source rock if that source rock itself derived from an older crust. Hence the meaning of a Sm-Nd model age of a sedimentary rock can be best described as an average time of the continental crust forming events of the constituent particles. But it must be clear in mind that the model age does not imply a specific crust forming event. That is why some have preferred to use the term average "crustal residence age," and not a crust-formation age, when referring to the Sm-Nd model age of a sedimentary rock (O'Nions et al., 1983).

The procedure for calculation of a Sm-Nd model age and the the expression of the Nd isotopic composition of a sample in terms of the $_{Nd}$ value can be found in Faure (1986), DePaolo (1988b), and also in the introduction of this volume. The choice of the source mantle for continental crustal rocks will influence the calculated Sm-Nd model age. Most consider a depleted mantle for the derivation of the continental crust, but difference exists as to presumed evolutionary characteristic of the depleted upper mantle and its specific Nd isotopic value, which may lead to a difference in the calculated age by about 200-300 Ma (DePaolo, 1981; Goldstein et al., 1984; Nelson and DePaolo, 1985).

In addition to the choice of appropriate parameters for a depleted mantle source for the evolution of the crustal materials, the Sm-Nd model age of sediments can be influenced by their evolutionary history during the time of their residence in the crust.

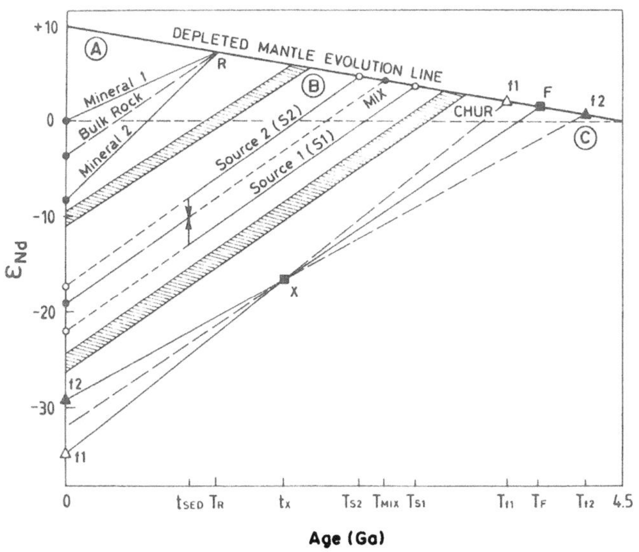

Figure 1: Illustration of the influence of sedimentary or intracrustal processes on the calculated Sm-Nd model age with respect to the crustal forming event. Part (A): transportation related sorting of different components from the same source; T_R is the age of a crust formation event R. Part (B): mechanical mixing of sediments of different sources; t_{sed} is the time of sedimentation, T_{s1} and T_{s2} are periods of crust formation events respectively for sources s1 and s2, T_{mix} is the calculated model age for the mixed sediments. Part (C): t_x is a time of differentiation between Sm and Nd due to diagenesis or metamorphism, T_{f1} is a calculated model age for a component whose Sm/Nd ratio is lower than that of the original source material, T_{f2} is a calculated model age for a component whose Sm/Nd ratio is higher than that of the original source material.

When components formed from a common source are seggregated due to sorting by transportation without having any significant modification of their respective Sm/Nd ratios, each component individually will still have the same model age, although the components will differ in their measured Nd isotopic compositions (Fig. 1 part A). The Nd isotopic evolution and Sm-Nd model age of sediments formed by mixing of materials derived from sources with different continental crustal histories are shown in part B of figure 1. Following the mixing, the Sm-Nd evolution of the mixed sample will occur at a rate intermediate between that of the two components and the resulting model age will be an intermediate value between the model ages of the two components, which is proportional to the relative amounts of each end member, weighted by the rare earth element (REE) concentrations in each end member. The interpretation of a model age can be complicated if the Sm-Nd of a system may be disturbed by an intracrustal process (weathering, diagenesis, and metamorphism) at some time which is much later than the time of formation of the system as a crustal entity, and the error in the calculated age is proportional to the difference of time between the crust formation and the intracrustal Sm-Nd fractionation (Fig. 1 part C). The material whose Sm/Nd ratio has been lowered by any disturbance will yield a lower Sm-Nd model age and the one whose Sm/Nd ratio has been elevated will yield a higher model age than the corresponding undisturbed material.

Sm-Nd FRACTIONATION DURING SEDIMENTARY PROCESSES

Sm-Nd Fractionation during Weathering

Balashov et al. (1964) concluded that subaerial weathering preferentially removes the heavy REE, but the quantity of the REE being removed is extremely small. They attributed this preferential removal of the heavy REE to the increased stability of carbonate complexes of the heavy REE as compared to the light REE. Recently, Goldstein and Jacobsen (1987) have shown that the Sm/Nd ratio of dissolved loads is higher than that of suspended loads in most rivers. Considerable evidence now exists that spring waters and streams selectively remove HREE from source rocks, although the absolute content of the dissolved REE in these waters, having concentrations of tens or few hundred parts per trillion, is extremely small as compared to that in solid source materials having a few hundred parts per million of the REE (Michard et al., 1987; Goldstein and Jacobsen, 1988). The evidence from analyses of dissolved loads of rivers and springs makes evident that in theory at least some difference can be expected in the relative abundances of Sm and Nd between weathered products and their source rocks. But the degree of changes in the Sm/Nd ratios from source materials to weathered products is difficult to document in the evolution of sediments and sedimentary rocks because of the potential for modification of the ratio by sedimentary processes subsequent to weathering. Nevertheless, evidence has been gathered that weathered products could differ substantially from their source materials in their Sm/Nd ratios.

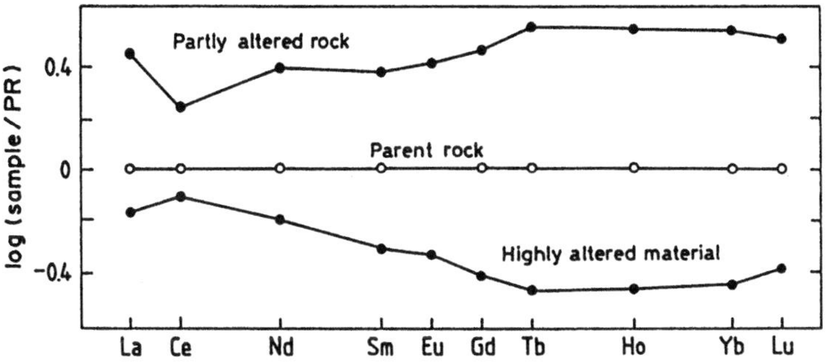

Figure 2: Concentrations of REE in partly and highly weathered material relative to parent unaltered material (after Nesbitt, 1979)

Studies are few which have carefully explored the paths of changes in relative and absolute contents of the REE between source rocks and their associated weathered products. Burkov and Podporina (1967) observed considerable preferential mobilization of the light REE in the formation of kaolinite from weathering of granite and granodiorite in an acidic environment. Roaldset and Rosenqvist (1971b) noted that considerable amounts of REE can be removed from weathered micas by leaching with EDTA, suggesting that the REE can be quite mobile in a weathering environment. Balashov and Girin (1969) also found from experimental work that as much as 95% of the REE in clays can be leached by acid, which suggests that the REE can be mobilized during weathering and subsequent processes.

Nesbitt (1979) compared the absolute and relative abundances of the REE between a parent granodiorite and its associated weathered products. Incipiently to moderately altered parent materials consisted dominatly of vermiculite and Fe-Ti oxyhydroxide with minor amounts of kaolinite and illite as the weathered products, whereas the most altered products consisted essentially of kaolinite and illite with small amounts of Fe-Ti oxyhydroxide and no vermiculite. He observed that while incipiently to moderately altered parent granodiorite became enriched in the absolute REE content and relatively enriched in the heavy REE, the most altered residual product became depleted both in the absolute REE and in the relative abundance of the heavy REE relative to the parent material (Fig. 2). Nesbitt explained that the mobilization of the REE from the most altered zone to partially altered zone was due to increase in pH, but the fractionation of the REE elements was due to mineralogical controls. This study showed that the Sm/Nd ratio of the most altered clayey-residual product was as much as 20% lower than that of the parent rock.

Clauer et al. (1982), who investigated the effects of natural weathering on the REE pattern of biotite, observed general increase in the heavy REE relative to the light REE with increase in the degree of weathering of biotites. The weathered biotites experienced on the average an increase in the Sm/Nd ratio by about 50%, compared to unweathered biotite. Also, the absolute contents of both Sm and Nd, like the other REE, increased in the weathered biotites relative to the fresh biotite.

Middelburg et al. (1988), who investigated the mobilization and redistribution of the REE during weathering of granitic rocks, found that the REE were significantly mobilized and fractionated only during advanced stages of weathering. They concluded that selective leaching of the REE from both stable and unstable minerals and subsequent concentration in secondary phases may account for the observed fractionation.

The preceding discussion makes evident that a measurable degree of remobilization of Sm and Nd and fractionation between the two elements occur in the transformation from parent materials to weathered products. The pattern of these changes is apparently controlled by both minera-logy of the weathered products and chemistry of the transporting fluids being involved.

Sm-Nd Fractionation during Transportation and Deposition

Transportation related differentiation in the Sm/Nd ratio between fine-grained sediments and their source rocks could potentially arise from two effects, one by sorting of weathered and eroded particles having different Sm/Nd ratios and the other by differential amounts of adsorption of Sm and Nd by clay particles from waters due to changes in the pH or alkalinity and changes in the relative abundances of Sm and Nd in the waters. Stordal and Wasserburg (1986) noted a difference of as much as about 23% in the Sm/Nd ratios between two very small clay fractions of a suspended load from a river, but the actual cause of the difference is not known. Study is needed to determine the extent and cause of inter-particle variations in the Sm/Nd ratios among fine suspended loads of rivers.

Cullers et al. (1975) noted that kaolinites can have as much REE as illites and smectites, but, unlike illites and smectites, whose REE patterns approach close to that of average shale, kaolinites have widely varied light enriched REE patterns. Differential transport of clay particles in depositional basin has been known for long times. Kaolinite is often more abundant than other clay minerals in near shore sediments, suggesting that some differences in the Sm/Nd ratio could arise due to differential transport of the clay minerals in a sedimentary basin. Schieber (1986) made a suggestion based on the REE data of sediments from Belt Supergroup in Montana that the length of the transport of clays could have some influence on the relative distribution of the REE in these sediments. Thus, potential preferential sorting of clay minerals in ocean environments could make some degree of difference in the relative abundances of Sm and Nd between fine-grained nearshore sediments and equivalent offshore sediments. The study of Stordel and Wasserburg (1986) has already shown that estuarine entrapment of sediments can have some effect on the Sm-Nd differentiation of sediments that are being delivered by streams from the continents to the oceans.

Sm-Nd Fractionation during Diagenesis

From an investigation of bulk clay-sized fractions in samples of sediments from a deep well in the Gulf Coast, Chaudhuri and Cullers (1979) concluded that diagenesis caused no significant modification in the Sm/Nd ratios of these sediments. Recently, Awwiller and Mack (1991) claimed that diagenesis could have influenced the Sm-Nd model ages of Tertiary sandstones and shales from the Texas Gulf Coast. These ages increased with increased depth of burial. Ohr et al. (1991) found an evidence of diagenetic Nd isotopic equilibrium at depths below the smectite to illite transition zone for very fine particles ($<0.1\mu m$) in sediments from a deep well in the Texas Gulf Coast. They observed a spread in the Sm/Nd ratios between leachates and insoluble residues after acid leaching of the very fine clay particles, although the Nd isotopic compositions between the two phases were nearly identical for about eighty per cent of the cases. They concluded that these isotopic similarities reflected diagenetic isotopic equilibrium and the spread in the Sm/Nd ratio is due to fractionation related to illitization and concurrent precipitation of phosphate minerals. This explanation may not be entirely correct, which has been discussed later in this paper. Even by considering that the isotopic equilibriation occurs during diagenesis of very fine clay particles, the effect of such a process on the model age of fine-grained sediments in general appears to be small, because the amounts of such fine-grained materials in bulk clay sediments or shales are often extremely small.

Potential diagenetic Sm-Nd differentiation has also been suggested by both Stille and Clauer (1986) and Bros et al. (1992) based on Sm-Nd isotopic data defining isochrons. As explained later in this article, these isochrons, although yielding ages which are either the same or less than that of the deposition of the sediments, do not necessarily signify isotopic equilibriation at a post-depositional time.

Integrated Effect of Sedimentary Processes

An individual process may cause some differentiation in the Sm/Nd ratio between sediments and their source rocks, but this effect may be balanced by subsequent processes in the history of sediments. To document competing or complementary influences of weathering, transportation, deposition, and diagenesis in the Sm-

Nd history of sedimentary rocks is extremely difficult or may be impossible, because the effect of these different influences on the Sm/Nd ratios of sediments cannot be easily distinguished from effects due to differences in the provenance or in the mineralogy of what is weathered in a single provenance. Detailed information about the differentiation of the REE are given by Fleet (1984) and McLennan (1989).

What is clearly evident from the REE data on fine-grained sediments and sedimentary rocks and granitoid rocks is that their Sm/Nd ratios are very similar. The Sm/Nd ratios of many different granitoid rocks are commonly between 0.14 and 0.23, an average being about 0.19 (Haskin et al., 1966; Othman et al., 1984; Nelson and DePaolo, 1985; DePaolo, 1988a, DePaolo et al., 1991). Ratios as high as about 0.41 have been found for some granitoids, but they are considered to be few (Cullers and Graf, 1984). Clastic sediments and sedimentary rocks, like the common granitoid rocks, have also a similarly narrow range of Sm/Nd ratios. A large number of studies on clay minerals of sedimentary origin and on fine-grained clastic sedimentary rocks, especially those of continental platforms, have shown that the Sm/Nd ratios very often range between 0.15 and 0.21 with an average of about 0.19 (Haskin et al., 1966; Haskin and Haskin, 1966; Cullers et al., 1975, 1979; Dypvik and Brunfelt, 1976, 1979; Nance and Taylor, 1976; Fleet et al., 1980, Fleet, 1984; O'Nions et al., 1983; Taylor et al., 1983; Goldstein et al., 1984; Miller and O'Nions, 1984; Patchett et al., 1984; Michard et al., 1985; Bhatia, 1985; Nelson and DePaolo, 1988; Goldstein and Jacobsen, 1988; McLennan et al., 1989; Dia et al., 1990; Nägler, 1990). Whereas sediments of the continents and continental shelves have an average Sm/Nd ratio of about 0.19-0.20, those from deep oceans have an average value of about 0.21-0.22 (Patchett et al., 1984). The small difference in the Sm/Nd ratio between the two types of sediments is probably a reflection of the presence of some amounts of products of weathering of basalts or hydrothermal activities in the deep sea sediments. Nevertheless, the close similarity in the Sm/Nd ratios between fine-grained clastic sediments and sedimentary rocks and continental crystalline rocks is a sufficiently justifiable reason to assume that the Sm-Nd ratios are changed little by intracrustal processes of weathering, erosion, deposition, and diagenesis.

Sm-Nd ISOTOPES OF MODERN CONTINENTAL EROSIONAL PRODUCTS

Studies of Goldstein and Jacobsen (1987, 1988) and Goldstein et al. (1984) on river loads have considerably advanced our understanding of the Sm-Nd isotopes of modern erosional products and have given a bedrock knowledge for the interpretation of Sm-Nd isotopic data of sedimentary rocks. The Sm and Nd contents of many present-day river suspended loads range respectively between 3 and 11 $\mu g/g$ and between 12 and 59 $\mu g/g$. The Sm/Nd ratios are between 0.15 and 0.21 and the $^{143}Nd/^{144}Nd$ ratios are between 0.50967, or ϵ_{Nd} value of about -42, and 0.51160, or ϵ_{Nd} value of about -4, for suspended loads of rivers draining primarily Archean, Proterozoic and Paleozoic terrains, whereas the Sm/Nd ratios are as high as 0.28 and the $^{143}Nd/^{144}Nd$ ratios are as high as 0.51221, or ϵ_{Nd} value of about +7, for suspended loads of rivers draining young island or continental arcs, the high values being for the dominant presence of detrital components from a depleted mantle source.

Goldstein and Jacobsen (1988) found a close agreement in the Sm-Nd model ages between suspended loads and their apparent source rocks in the drainage basins underlain by Archean and island or continental arc rocks. But relating the suspended loads of many North American rivers to any clearly definable source rocks is not easy. The authors assumed a close correspondence between basement crystalline rocks in the drainage basin and the suspended loads. Based on this assumption, they found that the Sm-Nd model ages of the suspended loads are on the average 350 Ma younger than the Sm-Nd model ages of the basement rocks in the drainage basin. On the whole, the Sm-Nd model ages agreed reasonably well between the suspended loads and their presumed source rocks (Fig. 3).

The study of Goldstein and Jacobsen (1988) on river suspended loads also revealed an inverse relationship between the Sm/Nd ratios of the river loads and their Sm-Nd model ages (Fig. 4). The

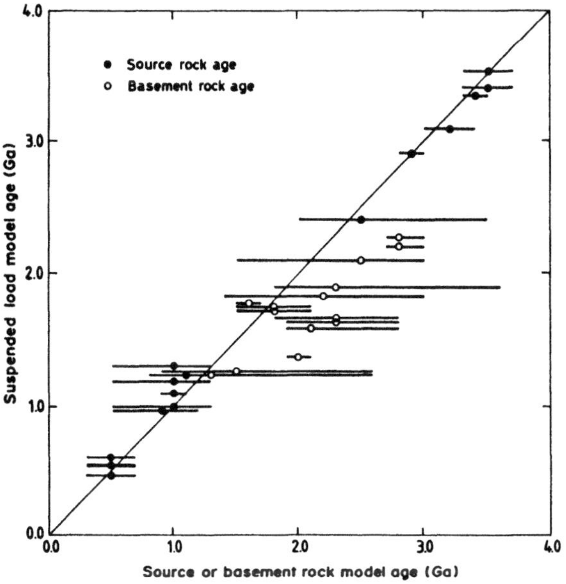

Figure 3: Comparison of depleted mantle based Sm-Nd model ages between river loads and their presumed source rocks (after Goldstein and Jacobsen, 1988)

authors suggested that the Sm/Nd age trend could be due to a progressive change in the Sm/Nd ratios of new additions of crust, possibly reflecting progressive depletion of incompatible elements in the mantle through time. DePaolo (1988a) also inferred a decrease in Sm/Nd ratio with increase in model age for the crust, based on evidence from granitic rocks. The results on modern river suspended loads are not in agreement with the observation made by Taylor and McLennan (1985) who claimed that Archean sedimentary rocks tend to have less fractionated light REE patterns than post-Archean sedimentary rocks. The study by Dia et al. (1990), on the other hand, showed no change in Sm/Nd ratio with depositional age for shales and greywackes ranging in age from 0.2 to 3.3 Ga. Isotopic studies focusing on morphology, mineralogy, and chemistry of the suspended loads are needed to understand the actual cause of the Sm-Nd age trend for modern river loads.

Based on the data on suspended loads of modern rivers, Goldstein and Jacobsen (1988) estimated that the average Sm-Nd model age for the flux of present-day river load from the North

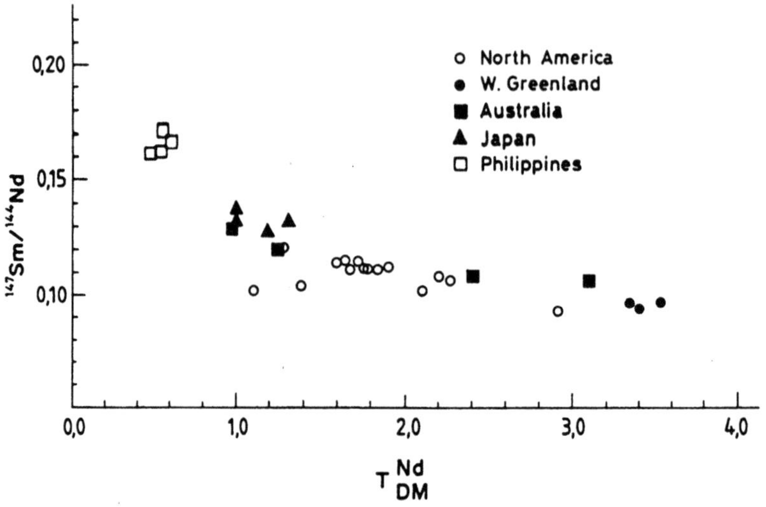

Figure 4: Variation of Sm/Nd ratio relative to the model age for the suspended loads of some major rivers in the world (after Goldstein and Jacobsen, 1988)

American continent to the oceans is about 1.8 Ga. By contrast, the estimated average Sm-Nd model age for the basement rocks of the North American continent is about 2.36 Ga, which makes the model age of the river load to be about 24% lower than that of the basement rocks. The authors explained that the discrepancy in the Sm-Nd model age between the river load and the basement rock reflects a large contribution of suspended material derived from young orogenic areas.

The weighted average Sm-Nd parameters for the loads of all rivers were used by Goldstein and Jacobsen (1988) to calculate the parameters for the upper crust presently exposed to weathering. The average values of different parameters for global river load are: Sm/Nd ratio of about 0.188, ϵ_{Nd} of about -10.6, and Sm-Nd model age of about 1.63 Ga. Based on the age relationship obtained between the river load and the basement rocks for the North American continent, Goldstein and Jacobsen estimated that the present-day upper crust exposed to weathering has an average Sm-Nd model age of about 2.1 Ga. Assuming that the average Sm/Nd of the upper crust exposed to weathering is approximately equal to that of the river load, the authors inferred the average ϵ_{Nd} of the continental upper crust to be about -17.

Sm-Nd isotopic information about the upper continental crust based on data of the present-day river suspended loads should be taken with some caution. Drainage basins of major rivers are mostly in trailing edge and continent-collision tectonic settings, whereas those of many small rivers are in arc-type tectonic settings. But, collectively, the suspended loads from these small rivers can make up a considerable fraction of the total loads removed from continents. Since the Sm-Nd model ages of sediments associated with trailing edge and continent-collision tectonic settings are much higher than those of sediments associated with arc-type settings, as noted by McLennan et al. (1990), a bias in the sampling of the river loads may not allow a reasonably correct estimation of the Sm-Nd isotopic characteristics of the continental upper crust based on the data of the river loads.

Sm-Nd ISOTOPIC VARIATIONS IN A SEDIMENTARY SEQUENCE

The Sm-Nd isotopic data of fine-grained clastic sedimentary rocks and their constituent particles in a sequence of deposition potentially can be of much significance to increased understanding of the depositional and post-depositional history of the sequence. The potential benefits about the latter have been pointed out by Awwiller and Mack (1991) and Ohr et al. (1991) who analyzed Tertiary sediments from the Gulf Coast area.

The data of Ohr et al. (1991) for rocks from a stratigraphic interval of about 3.5 km make evident that the Sm-Nd model ages of untreated bulk samples are higher than that of the corresponding insoluble residues by as much as 250 million years. Also, the dispersion in the model ages for the untreated bulk samples appears to be slightly higher than that for the acid-leached bulk samples. Similar degree of dispersion in the Sm-Nd model ages also exists between untreated and acid-leached clay fractions of $<0.1\mu m$. These disparities in the model ages reflect the presence of varied amounts of Sm and Nd in the acid removing phase or crystal sites which also differ to some extent in the Nd isotopic compositions.

The acid leachate contains Sm and Nd which are mobile, the origin of which could be different from that of the relatively inert REE in the residue. Hence, acid insoluble fraction may be preferable to untreated samples for the Sm-Nd isotopic studies on sediments.

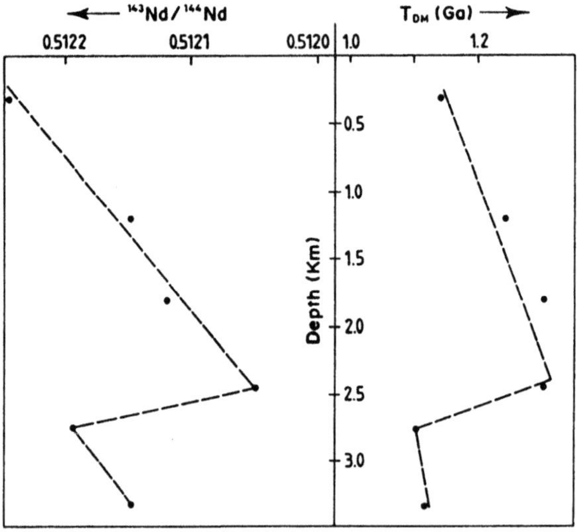

Figure 5: Changes in the Nd isotopic compositions and model ages with depths for HCl insoluble bulk materials of sediments from the Tertiary sequence in the Gulf Coast of Texas (modified from Ohr et al., 1991)

As suggested by the data of Ohr et al. (1991), the Nd isotopic composition decreases and the Sm-Nd model age increases with increase in depth for acid-insoluble bulk materials from the top 2.4 km of the sequence (Fig. 5). Awwiller and Mack (1991) also noted more radiogenic Nd toward the surface. Although some changes in the Sm/Nd ratios occur, no trend with depth could be established. The more radiogenic Nd toward the surface within the top 2.4 km of the sequence naturally reflects increasing input of younger crustal materials as deposition continued. This can be related to several factors such as changes in continental erosion related to sea level changes, slow epirogenic uplift of the land exposing more and more of the young crustal granitoid rocks, erosional order of an older stratigraphic sequence, and possibly some others. Additional studies are needed to determine the cause of such isotopic trends within a sequence of sediments.

The acid-leached bulk samples collectively do not define any discernable trend between the Nd isotopic compositions and the

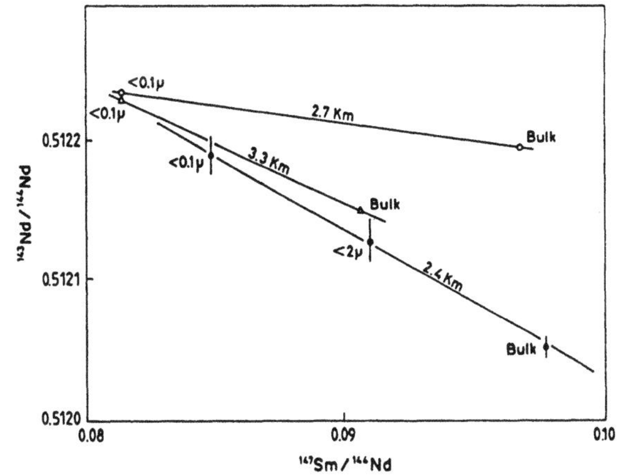

Figure 6: Changes in the Sm/Nd ratios vs. Nd isotopic ratios for acid leached bulk samples and corresponding clay fractions for sediments at three different depths (km) within the Tertiary sequence in the Gulf Coast area of Texas (reconstructed from Ohr et al., 1991)

Sm/Nd ratios. The scatter in this relationship is largely a reflection of varied provenances for the Tertiary sediments in the Gulf Coast area. We have reconstructed some of the data given by Ohr et al. (1991) and come to the conclusion that Sm-Nd isotopic data of each acid-leached bulk sample may be described in terms of mixing of two end-member components (Fig. 6). It is evident from the data on the sample at a depth of about 2.4 km that the Sm/Nd ratio decreases and the $^{143}Nd/^{144}Nd$ ratio increases from the bulk sample to the <2μm clay fraction to the very fine clay fraction. The linearity of the data among the two clay fractions and the bulk material for the sample collected at a depth of about 2.4 km probably represents a mechanical mixing between two end-member components. Similar two-component mixing lines with different slopes potentially exist also for the samples at depths of about 2.7 km and 3.3 km. The compositions of the end-member components may be defined based on some reasonable assumptions. As can be seen from the data presented in figure 6, the three mixing lines appear to have one common end-member component, but with a different second end-member component. The average Sm/Nd ratio of bulk sedimentary rocks is about 0.19, or $^{147}Sm/^{144}Nd$ ratio of about 0.115, which is about the same as that of many granitoid rocks. Assuming that each bulk sample contained a component with a Sm/Nd ratio of about 0.19, the three mixing lines may be extended to this average Sm/Nd ratio to obtain the corresponding Nd isotopic value of the component. The Sm-Nd model ages of the component with the average Sm/Nd ratio of 0.19 in the bulk samples are calculated to be about 2.0 Ga (2.4 km), 1.5 Ga (2.7 km), and 1.8 Ga (3.3 km). The other component which is common to all three samples represents a very fine clay fraction of diagenetic origin. The common end-member component for all three samples appears to have a Sm/Nd ratio about 0.13, or a $^{147}Sm/^{144}Nd$ ratio of about 0.08. By taking the corresponding Nd isotopic value from the extrapolation of the mixing line, the Sm-Nd model age for the common end-member component is calculated to be 1.0-1.1 Ga. The significance of this young model age is not clear, especially in view of the strong diagenetic overprint on the origin of this component.

Sm-Nd ISOTOPIC SIGNATURES AS TRACERS OF DIAGENESIS

The role of diagenesis on the Sm-Nd isotopes of sediments is poorly understood. Analyses of bulk samples of sediments or even <2μm clay fractions have not generally provided any convincing evidence that diagenesis plays any important role in the Sm-Nd chemical fractionation of sediments. Frost and Winston (1987) found in their study of the Middle Proterozoic sedimentary sequence in western North America that the average crustal residence age of fine grained sediments is nearly similar to that of some coarse grained sediments, in spite of wide variations in Sm/Nd ratios of coarse-grained samples. The similarity in the ages suggests that the total effect of weathering, erosion, transportation, and diagenesis is minimal as far as the Sm-Nd model age is concerned. Nelson and DePaolo (1988) observed from a study of Tertiary sediments deposited in a rift basin in north-central New Mexico that the Nd isotopic characteristics of a mudstone are identical to that of sandstones with which the mudstone is interlayered. They suggested that the data on the mudstone and the sandstone argue against any significant fractionation between Sm and Nd due to the combined effects of weathering, erosion, deposition and diagenesis. The results of the studies by Frost and Winston (1987) and by Nelson and DePaolo (1988) are in agreement with that of Chaudhuri and Cullers (1979) who found almost no change in the REE patterns of samples of clays from the deeply buried Tertiary sediments of the Gulf Coast area.

Any variation in the REE contents and Sm-Nd isotopic values among fine-grained sediments or sedimentary rocks can be due to difference in source rocks for the detrital particles, intensity of weathering, depositional environment, burial diagenetic reactions, or a combination of any of these factors. Some studies in about the last five years have either implied or underscored the importance of diagenesis as a major factor in determining the Sm-Nd isotopic characteristics of fine-grained sediments.

Stillc and Clauer obtained a Sm-Nd isochron for four whole rock samples of argillite and a clay fraction from one of the whole rock samples of the Proterozoic Gunflint Formation, which yielded an age of 2.08 $\pm$ 0.59 Ga and initial ϵ_{Nd} value of about -2.1. The authors interpreted the age as being close to the depositional age of the sediments and suggested that the isochron could have resulted from near equilibrium of Nd isotopes during diagenesis shortly after deposition. A Sm-Nd isochron was also observed by Nägler (1990) who analysed whole rock samples of argillites from Spain, but the calculated age of 1524 $\pm$ 173 Ma is obviously geologically meaningless because the age of sedimentation for the argillites is about 490 Ma. As Nägler noted, the isochron for the argillites definitely represents a mixing line.

Recently Bros et al. (1992) analyzed the Sm-Nd isotopic compositions of different size fractions from two different samples about 60 m apart in a stratigraphic column of Lower Proterozoic black shales in southeastern Gabon. The two fine clay fractions of <0.2μm and 0.2-0.4 μm sizes for each of the samples were also leached with dilute HCl for determining the Sm-Nd isotopic compositions of the mobile REE relative to that of the inert REE in the mineral structures. The Sm-Nd isotopic data of the untreated, the acid insoluble and the soluble materials of each of these two fine clay fractions from individual samples independently defined isochron ages of 2099 ± 115 Ma and 2036 ± 79 Ma with very similar initial Nd isotopic compositions. The authors interpreted these dates to be a period of Nd isotopic homogenization related to illitization. The finer clay particles have higher Sm/Nd ratios than the coarser counterparts. Based on mineralogical and chemical evidence, the change in the Sm/Nd ratios suggests that diagenetic mineralization occurred with varied Sm/Nd ratios of the fluids.

The data point of the shale which was richest in the organic content (about 13% kerogen) plotted between the two isochrons defined by the clay fractions of the two other samples which were relatively less enriched in organic C. The agreement between the organic C rich shale, the leached clay particles and the leachates, indicates that it reached a Nd isotopic equilibrium with respect to diagenetic fluids. A fracture-filling bituminous sample showed the highest $^{147}Sm/^{144}Nd$ ratio (about 0.15) and the highest $^{143}Nd/^{144}Nd$ ratio of all materials analyzed. The data of this bituminous material also conform to the isochron trends defined by the silicate minerals. The Sm-Nd isotopic enrichment for the bituminous sample can be assumed to have been related to its migration history, as suggested by Manning et al. (1991) who also observed, as shown in this study, isotopic homogenization between source-rock kerogen and silicate phases of reservoir rocks.

The Sm-Nd isochron from data on clays must be interpreted with caution. The calculated date from an isochron can become much less than, or even approximately close to, that of the age of sedimentation, but this alone is not a good enough reason to assume that the isochron date relates to a diagenetic event. A mixing line with a negative slope at the time of sedimentation could evolve into an isochron with a positive slope, whereby the calculated date becomes less than, or nearly equal to, that of the age of deposition. The Sm-Nd isotopic data of clay fractions and bulk material of a sample of the Gulf Coast sediments of Tertiary age provide clear evidence that such a mixing line with negative slope can exist (Fig. 6).

Geochemical and mineralogical data obtained from analyses of Tertiary sedimentary deposits of the Gulf Coast area in the U.S.A.

have produced a paradigm for clay mineral diagenesis accompanying deep burial of sediments. The most attractive feature of the diagenetic history of these Gulf Coast sediments is the increase in illite content with increase in depth (see companion paper of Clauer et al. in this volume for details). Two recent studies have addressed on the relation of diagenesis to the Sm-Nd isotopic compositions of sediments from the Gulf Coast area (Awwiller and Mack, 1991; Ohr et al., 1991).

The Sm-Nd isotopic analyses by Awwiller and Mack (1991) on bulk samples of sandstones and shales in the Oligocene Frio Formation from south Texas, which consist of detrital particles derived largely from the Tertiary volcanic fields of west Texas and northern Mexico, showed a trend in decrease of the $^{143}Nd/^{144}Nd$ ratio with increase in depth for both types of sedimentary rocks. Their study of sandstones and shales from the Paleocene-Eocene Wilcox Group of south-central Texas indicated a similar trend, but of poorer quality, in the Nd isotopic ratio with depth. The sandstones of the Frio Formation were found to have the Sm-Nd model ages increase with increasing depth, but the shales within the same stratigraphic interval did not have such a record of change. Rather the Sm-Nd model ages of the shales remained nearly similar through much of the stratigraphic interval. Considering the entire stratigraphic interval of about 3.5 km within the Paleocene-Eocene Wilcox Group, from which the bulk samples were taken for the analyses, the Sm-Nd model ages were poorly correlated. Awwiller and Mack's (1991) claim that Sm-Nd model ages increase with increasing depth does not appear to be convincing. The authors suggested that, although the trends in ϵ_{Nd} and the Sm-Nd model ages in sandstones and shales of the Tertiary sequence are possibly due to a detrital control, diagenetic changes in the rocks could have influenced the modifications in the ages. They concluded that the model ages could have been modified by diagenetic reactions involving loss of radiogenic Nd from the sediment and mineralogic fractionation of Sm and Nd during mobilization of the REE. Awwiller and Mack noted that in the Frio sandstones late diagenetic ankerite cement with ϵ_{Nd} value of about -10 contained less radiogenic Nd than early diagenetic calcite cement with a ϵ_{Nd} value of about -6, which suggested early mobilization of Nd from volcanic materials. They also reported that the Sm/Nd ratios of clay fractions of <0.5μm size, which ranged from about 0.15 to 0.17, were lower than that of the equivalent bulk samples whose Sm/Nd ratios were between 0.17 and 0.23.

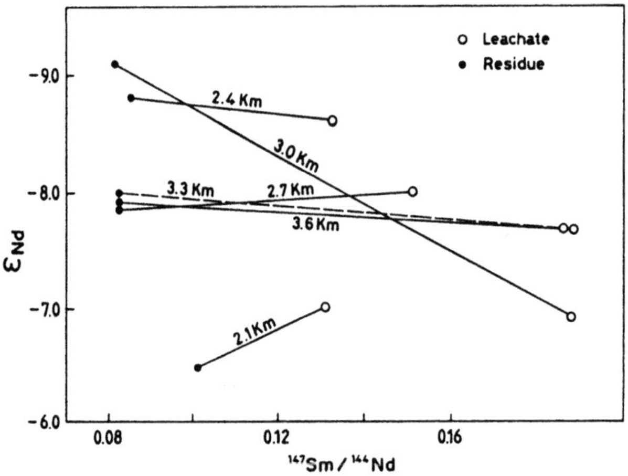

Figure 7: Relationships of Nd isotopic composition and Sm/Nd ratio between acid leachate and corresponding insoluble very fine clay fraction for samples at different depths within the sequence of Tertiary sediments in the Gulf Coast area of Texas. The zone of illitization includes the sequence at 2.4 km depth and lower in the colums (after Ohr et al., 1991)

Ohr et al. (1991) also observed that Sm/Nd ratios of fine clay fractions <0.1μm were lower than that of corresponding bulk samples from the zone of illitization in deeply buried sediments. Since these fine clay fractions consist of large amounts of diagenetic illites, mineralogical fractionation of Sm from Nd might have occurred during diagenetic formation of illites. Analyzing HCl-leached bulk samples and their leachates, and also a coarse clay fraction of one of the samples and its leachate, from the illitic zone, Ohr et al. found considerable disparity in the Nd isotopic composition between an acid insoluble residue and its corresponding leachate. This is expected because such materials consist of particles of diverse origin. But in analyzing the fine clay fractions, they found that, for four out of five fractions from different depths, the Nd isotopic composition of the insoluble residue was identical to that of the corresponding leachate. But for one sample, the leachate and the residue differed in their Nd isotopic compositions. Furthermore, for all five samples they analyzed, the Sm/Nd ratio of the leachate was higher than that of the corresponding insoluble residue (Fig. 7). The similarities in the Nd isotopic compositions between the acid insoluble residue and the corresponding leachate for most of these fine clay fractions suggest isotopic homogeneity during diagenesis. What is also clear from the data is that neither the insoluble residues of the very fine clay fractions themselves nor the leachates themselves are isotopically homogeneous. This indicates that if diagenesis produced isotopic homogeneity, it had to occur only within a limited scale.

We would like to draw attention to the low Sm/Nd ratios of about 0.13-0.14 for the fine clay fractions. These values are lower than the average value of about 0.18 which have been reported in the literature for many illitic, smectitic and glauconitic minerals (Cullers et al., 1975; Fleet et al., 1980). The reason for the difference could be a result of difference in the maturity of evolution between them, but research is needed to determine the real cause. Nevertheless, the Sm/Nd ratio of about 0.13-0.14 for the very fine clay fraction is about the same as that for many K-feldspar minerals in granitoid rocks, which may suggest that these very fine illitic minerals in the Gulf Coast sediments formed from the alteration of K-feldspar minerals. If we assume an approximate conservation of the proportion between Sm and Nd during the transformation of the precursor to the very fine-grained illitic clay minerals, then the average crustal residence age of the precursor for these clay minerals was about 1.0-1.1 Ga.

Undoubtedly, evidence already exists that mineralogical fractionation of Sm from Nd could occur during diagenesis. The question remains about the actual mechanism by which it occurs and the significance of this diagenetic silicate mineralization on the Sm-Nd model ages of the sediments.

Nd ISOTOPES AS POTENTIAL TRACERS FOR OROGENIC EVENTS AND PALEOGEOGRAPHICAL RECONSTRUCTION

The role of the mantle-derived components in the growth of the continent has been a subject of much investigation during the last two decades. Major subduction related orogenic events potentially involve additions of juvenile crust differentiated from a mantle source. The additions of any new mantle component during the orogenic episodes should theoretically be evident in the ϵ_{Nd} values of weathered and erosional products from the orogenic belts. Michard et al. (1985) noted several maxima of ϵ_{Nd} values in a relationship between Nd isotopic compositions and stratigraphic ages for Phanerozoic sediments from France (Fig. 8). The maxima of ϵ_{Nd} values generally coincide with known orogenic events in central Europe, suggesting addition of subduction related crust during these events. The data in this figure also make evident that the effect of the contributions of young mantle Nd is overshadowed in few tens of million years by contributions of relatively non-radiogenic Nd from older crustal rocks, as the Nd-isotope signatures of the sediments are quickly restored to the pre-orogenic background level. In other words, the young crust was uplifted which caused the isotopic signature of the detritus from the uplifted young crust to overwhelm the signature of the detritus from the ancient crust, but the influence of the young crust was soon reduced due to its rapid erosion from uplift. One conclusion which can be made from the study of Michard et al. (1985) is that a prominent maximum of the ϵ_{Nd} values in the spectrum of Nd isotopic compositional variation with the depositional age of sediments may be used for identification and dating of a subductional related, mantle source differentiated crust forming event.

Studies of sedimentary sequence spanning at least several tens of million years of depositional history may reveal a general increase, but without any discernible maximum, in the ϵ_{Nd} values with decreasing ages of the sediments. Such a trend may signify either continuous small additions of new crust produced by a series of short duration subduction or a slow and long duration subduction, or a slow epeirogenic uplift of the crust continuously exposing mantle source derived crystalline rocks to weathering and erosion, or a progressive decrease in the influence of the older tectonic province in the erosion of the continent. Independent evidence is required to explain the cause for a trend of increase in Nd isotope ratio with decrease in stratigraphic age.

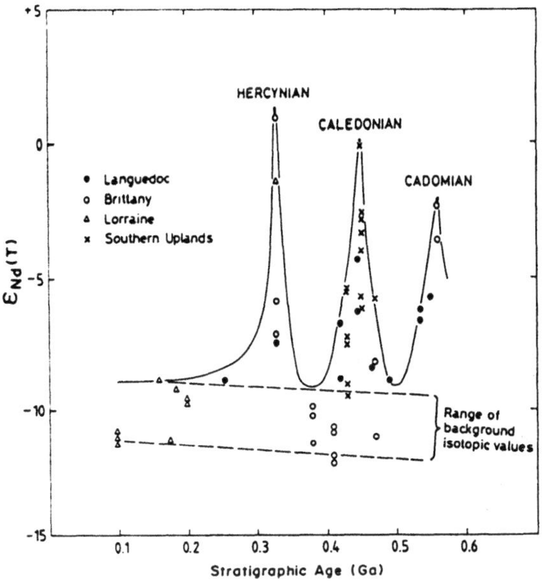

Figure 8: Relationship between stratigraphic age and Nd isotopic composition at the time of sedimentation for samples of Phanerozoic shales and siltstones from central Europe (after Michard et al., 1985)

The Nd isotopic records may reveal regional differences in the responses of sedimentary systems to additions of subduction related crusts. Miller and O'Nions (1984) noted contrasts in Nd isotopic composition among sedimentary rocks between the two sides of the Iapetus suture. They concluded that the Nd isotopic compositions of shales could be a useful tracer in paleogeographical investigations.

Crustal Recycling History

Taking the inventories of all sedimentary and metasedimentary rocks since the Archean time, Garrels and MacKenzie (1971) remarked that the half-life of global sedimentary rocks is about 600 million years. The preponderance of very young sedimentary rocks in the existing continental crust with the long evolutionary history has spurred many different investigations and debates about the crustal erosion and the growth of the continental crust. Different factors relating to intracrustal processes (such as rates of erosion and sedimentation, metamorphism, anatexis, and magmatism) and addition of new mantle differentiated material external to the continental internal cycle are important components in the evolutionary history of the continental crust with its present distribution of rock types exposed to the surface. Contentious views about the crustal growth can be found in the article by Veizer and Janssen (1985). The history of continental crust formation remains a controversial issue. The evolution has been variously explained ranging from a continuous formation (Hurley et al., 1962; Hurley and Rand, 1969; O'Nions and Pankhurst, 1978; Gurnis and Davies, 1986) to an episodic formation (Moorbath, 1977; Patchett et al., 1981; Page et al., 1984; Nelson and DePaolo, 1985; DePaolo et al., 1991), to increased continental mass trough time (Veizer and Janssen, 1979; Dewey and Windley, 1981; McLennan and Taylor, 1982), to steady continental mass (Patterson and Tatsumoto, 1964; Tatsumoto and Patterson, 1964; Armstrong, 1981), and to even decreased continental mass (Fyfe, 1978).

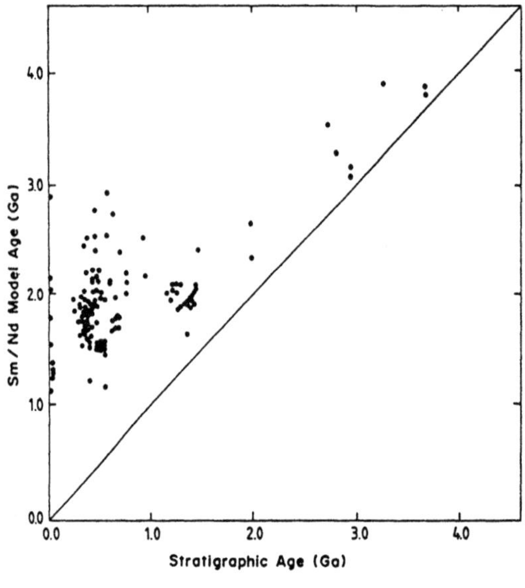

Figure 9: Relationship between stratigraphic age and the Sm-Nd model age for sedimentary and metasedimentary rocks ranging in age from Archean to Recent, including the data for modern river loads

The long term continental crustal growth history has been examined from the Sm-Nd isotope analyses of fine-grained sediments of different geologic ages. But this requires that the continental sedimentary mass should be rapidly recycled so that the records in the sedimentary mass reflect continuous evolution of the average composition of the continental crust. The long-term growth history of the continental crust should be studied only from analyses of well mixed, multi-cycled sedimentary systems, the criteria of which can be based on the REE distribution patterns. Several studies (Allègre and Rousseau, 1984; Frost and O'Nions, 1985; Michard et al., 1985; Frost and Winston, 1987; Dia et al., 1990) have brought attention to the relationship between the Sm-Nd model ages and the stratigraphic ages of sedimentary and metasedimentary rocks from different sedimentary basins in the world and the implication of the relationship concerning the growth of continents through the geologic time. Figure 9 is an illustration of the relationship between the model ages and stratigraphic ages of sedimentary rocks. The progressively greater difference between the crustal residence age and the stratigraphic age for sedimentary rocks younger than about 2.0 Ga suggests an increasing role of intracrustal recycling of older continental crustal materials since about the Proterozoic. The Nd isotopic data can be used to explain the evolution of the continental upper crust, provided the sedimentary rock records are not biased toward specific tectonic settings such as trailing edge and continental collision. Many are inclined to believe in such a bias in sedimentary-rock records (Veizer and Janssen, 1979, 1985). They provide an alternative explanation which suggests that the sedimentary mass essentially consists of reworked sediments and sedimentary rocks and old upper crustal crystalline rocks. They are in agreement with the results of early studies of Patterson and Tatsumoto (1964) and Tatsumoto and Patterson (1964) on Pb isotopes of detrital feldspars, which suggest that a very large segment of the continents was accreted before the Proterozoic, followed by a slow growth during the Proterozoic and the Phanerozoic.

CONCLUSION

The analyses of the Sm-Nd isotopes in sediments and sedimentary rocks have proved useful in reconstructing provenances of sediments, erosional history of the continent, tectonic evolution of the earth, and recycling of sediments. Recent studies have shown that considerable promise lies ahead in understanding diagenesis of sediments by analyses of the Sm-Nd isotopes in very fine-grained clays.

Evidence is sketchy about mineralogical fractionation of Sm from Nd in weathering environment, but considerable degree of fractionation seems possible. Nothing is known yet about the magnitude of change in the Sm-Nd model age which can occur due to weathering. Reports about the effect of sorting of detrital particles on the model ages are limited, but they point to conflicting results between no change and measurable differences in the systematics of Sm-Nd isotopes due to sorting. Influence of depositional environment on the systematics of Sm-Nd isotopes in fine-grained sediments needs investigations. Diagenetic reactions can cause mineralogical fractionation of Sm from Nd, but its real impact on the long-term changes in the Sm-Nd model ages of sedimentary rocks has not been examined. Present evidence shows that the diagenetic records in fine-grained sediments are limited to very fine clay fractions. We have yet to know what happens to the Sm-Nd isotopic systematics of these very fine clays when they mature in their evolutionary history subsequent to the growth in the diagenetic environment.

The comparison between stratigraphic ages of sedimentary rocks and their crustal residence ages estimated from analyses of systematics of the Sm-Nd isotopes has convincingly demonstrated that only small amounts of mantle-differentiated juvenile crust have been added to the continents since about 2.0 Ga ago. Changes in the Nd isotopic compositions of sedimentary rocks through time may provide clues for identification of subduction related crustal forming events and also tectono-eustatic related erosional history of the continent. A future research would be to determine the nature of the relationship between orogeny related Nd isotopic maxima as recorded in the sedimentary rocks and temporal variations in the Nd isotopic compositions of oceans which derived their Nd in large part from the continents.

ACKNOWLEDGEMENTS

This manuscript greatly benefited from review by Bruce K. Nelson (University of Washington at Seattle). We sincerely thank him for this immense help, as well as Mr. C. Hammel who made the draftings.

REFERENCES

Allègre C.J. and Rousseau D. (1984) The growth of the continent through geological time studied by Nd isotope analysis of shales. Earth Plan. Sci. Lett., 67, 19-34.

André L., Deutsch S. and Hertogen J. (1986) Trace element and Nd isotopes in shales as indexes of provenance and crustal growth: the early Paleozoic from the Brabant Massif (Belgium). Chem. Geol., 57, 101-115.

Armstrong R.L. (1981) Radiogenic isotopes: the case for crustal recycling on a near-steady-state no-continental-growth earth. Philos. Trans. Roy. Soc. london, Ser. A 301, 443-472.

Awwiller D.N. and Mack L.E. (1991) Diagenetic modification of Sm-Nd model ages in Tertiary sandstones and shales, Texas Gulf Coast. Geology, 19, 311-314.

Bhatia M.B. (1985) Rare earth element geochemistry of Australian Paleozoic graywackes and mudrocks: Provenance and tectonic. Sediment. Geol., 45, 97-113.

Balashov Y.A., Ronov A.B., Migdisov A.A. and Turanskaya N.V. (1964) The effect of climate and facies environment on the fractionation of the rare earths during sedimentation. Geochem. Intl., 10, 995-1014.

Balashov Y.A. and Girin Y.P. (1969) On the reserve of mobile rare earth elements in sedimentary rocks. Geochem. Intl., 7, 649-659.

Bros R., Stille P., Gauthier-Lafaye F., Weber F. and Clauer N. (1992) Sm-Nd isotopic dating of proterozoic clay materials: an example from the Francevillian sedimentary Series (Gabon). Earth Plan. Sci. Lett., in press.

Burkov V.V. and Podporina Y.K. (1967) Rare earths in granitoid residuum. Doklad. Acad. Nauk Sci., 177, 691-694.

Clarke D.B. and Halliday A.N. (1985) Sm/Nd isotopic investigation of the age of and origin of the Meguma zone metasedimentary rocks. Can. Jour. Earth Sci., 22, 102-107.

Clauer N., O'Neil J.R. and Bonnot-Courtois C. (1982) The effect of natural weathering on the chemical and isotopic compositions of biotites. Geochim. Cosmochim. Acta, 46, 1755-1762.

Chaudhuri S. and Cullers R.L. (1979) The distribution of rare-earth elements in deeply buried Gulf Coast sediments. Chem. Geol., 24, 327-338.

Cullers R.L., Chaudhuri S., Arnold B., Lee M. and Wolf C.W. Jr. (1975) Rare earth distributions in clay minerals and in the clay-sized fraction of the Lower Permian Havensville and Eskridge shales of Kansas and Oklahoma. Geochim. Cosmochim. Acta, 39, 1691-1701.

Cullers R.L., Chaudhuri S., Kilbane N. and Koch R. (1979) Rare-earth in size fractions and sedimentary rocks of Pennsylvanian-Permian age from the mid-continent of the U.S.A. Geochim. Cosmochim. Acta, 43, 1285-1301.
Cullers R.L. and Graf J.L. (1984) Rare earth elements in igneous rocks of the continental crust: Intermediate and silicic rocks - ore petrogenesis. In: Rare Earth Element Geochemistry, P. Henderson (ed.), Elsevier, New York, 275-316.
Davies G., Gledhill A. and Hawkesworth C. (1985) Upper crustal recycling in southern Britain: evidence from Nd and Sr isotopes. Earth Plan. Sci. Lett., 75, 1-12.
DePaolo D.J. (1981) Nd isotopic studies: some new perspectives on earth structure and evolution. EOS, 62, 137-140.
DePaolo D.J. (1988a) Age dependence of the composition of continental crust: evidence from Nd isotopic variations in granitic rocks. Earth Plan. Sci. Lett., 90, 263-271.
DePaolo D.J. (1988b) Neodymium Isotope Geochemistry: An Introduction, Springer-Verlag, New York, 187 p.
DePaolo D.J., Kyte F.J., Marshall B.D., O'Neil J.R. and Smit J. (1983) Rb-Sr, Sm-Nd, K-Ca, O, and H isotopic study of Cretaceous-Tertiary boundary sediments, Caravaca, Spain: Evidence for an oceanic impact site. Earth Plan. Sci. Lett., 64, 356-373.
DePaolo D.J., Linn A.M. and Schubert G. (1991) The continental crustal age distribution: methods of determining mantle separation ages from Sm-Nd isotopic data and application to the southwestern United States. Jour. Geophys. Res., 96, 2071-2088.
Dewey J.F. and Windley B.F. (1981) Growth and differentiation of the continental crust. Philos. Trans. Roy. Soc. London, Ser. A 301, 189-206.
Dia A., Allègre C.J. and Erlank A.J. (1990) The development of continental crust through geological time: the South African case. Earth Plan. Sci. Let., 98, 74-89.
Dypvik H. and Brunfelt A.O. (1976) Rare-earth elements in Lower Paleozoic epicontinental and eugeosynclinal sediments from the the Oslo and Tronheim regions. Sedimentology, 23, 363-378.
Dypvik H. and Brunfelt A.O. (1979) Distribution of rare earth elements in some North Atlantic Kimmeridgian black shales. Nature, 278, 339-341.
Elderfield H., Hawkesworth C.J., Greaves M.J. and Calvert S.E. (1981) Rare earth element geochemistry of oceanic ferromanganese nodules and associated sediments. Geochim. Cosmochim. Acta, 45, 513-528.
Faure G. (1986) Principles of Isotope Geology. J. Wiley and Sons, New York, 589 p.
Fleet A.J. (1984) Aqueous and sedimentary geochemistry of the rare earth elements. In: Rare Earth Element Geochemistry, Henderson P. (ed.), Elsevier, Amsterdam, 343-373.
Fleet A.J., Buckley H.A. and Johnson L.R. (1980) The rare earth element geochemistry of glauconites and celadonites. Jour. Geol. Soc. London, 137, 683-688.
Frost C.D. and O'Nions R.K. (1984) Nd evidence for Proterozoic crustal development of in the Belt-Purcell Supergroup. Nature, 312, 53-56.
Frost C.D. and Winston D. (1987) Nd isotopic systematics of coarse- and fine-grained sediments: examples from the Middle Proterozoic Belt-Purcell Supergroup. Jour. Geol., 95, 309-327.

Fyfe W.S. (1978) The evolution of the Earth's crust: modern plate tectonics to ancient hot spot tectonics? Chem. Geol., 23, 89-114.

Garrels R.M. and Mackenzie F.T. (1971) Evolution of Sedimentary Rock. W.W. Norton Company, Inc., New York, 379 p.

Goldstein S.L., O'Nions R.K. and Hamilton P.J. (1984) A Sm-Nd isotopic study of atmospheric dusts and particulates from major river systems. Earth Plan. Sci. Lett., 70, 221-236.

Goldstein S.J. and Jacobsen S.B. (1987) The Nd and Sr isotopic systematics of river-water dissolved material: implications for the sources of Nd and Sr in seawater. Chem. Geol. (Isot. Geosc. Sect.), 66, 245-272.

Goldstein S.J. and Jacobsen S.B. (1988) Nd and Sr isotopic systematics of river water suspended material: implications for crustal evolution. Earth Plan. Sci. Lett., 87, 249-265.

Gurnis M. and Davies G.F. (1986) Apparent episodic crustal growth arising from a smoothly evolving mantle. Geology, 14, 396-399.

Hamilton P.J., O'Nions R.K., Bridgwater D. and Nutman A. (1983) Sm-Nd studies of Archean metasediments and metavolcanics from West Greenland and their implications for the Earth's early history. Earth Plan. Sci Lett., 62, 263-272.

Haskin M.A. and Haskin L.A. (1966) Rare earth in European shales: a redetermination. Science, 154, 507-509.

Haskin L.A., Wildeman T.R., Frey F.A., Collins K.A., Keedy C.R. and Haskin M.A. (1966) Rare earths in sediments. Jour. Geophys. Res., 71, 6091-6105.

Heller P.L., Peterman Z.E., O'Neil J.R. and Shafiqullah M. (1985) Isotopic provenance of sandstones from the Eocene Tyee Formation, Oregon Coast Range. Geol. Soc. Am. Bull., 96, 770-780.

Huhma H. (1987) Provenance of Early Proterozoic and Archean metasediments in Finland: a Sm-Nd isotopic study. Prec. Res., 35, 127-143.

Hurley P.M., Hughes H., Faure G., Fairbairn H.W. and Pinson W.H. (1962) Radiogenic strontium 87 model of continental formation. Jour. Geophys. Res., 67, 5315-5334.

Hurley P.M. and Rand R.J. (1969) Pre-drift continental nuclei. Science, 164, 1229-1242.

Manning L.K., Frost C.D. and Branthaver J.F. (1991) A neodymium isotopic study of crude oils and source rocks: potential applications for petroleum exploration. Chem. Geol., 91, 125-138.

McCulloch M.T. and Wasserburg G.J. (1978) Sm-Nd and Rb-Sr chronology of continental crust formation. Science, 200, 1003-1011.

McLennan S.M. (1989) Rare earth elements in sedimentary rocks: influence of provenance and sedimentary processes. Rev. Min., 21, 169-200.

McLennan S.M. and Taylor S.R. (1982) Geochemical constraints on the growth of the continental crust. Jour. Geol., 90, 347-361.

McLennan S.M., McCulloch M.T., Taylor S.R. and Maynard J.B. (1989) Effects of sedimentary sorting on neodymium isotopes in deep-sea turbidites. Nature, 337, 547-549.

McLennan S.M., Taylor S.R., McCulloch M.T. and Maynard J.B. (1990) Geochemical and Nd-Sr isotope composition of deep sea turbidites: crustal evolution and plate tectonic associations. Geochim. Cosmochim. Acta, 54, 2015-2050.

Michard A., Gurriet P., Soudant M. and Albarede F. (1985) Nd

isotopes in French Phanerozoic shales: external vs; internal aspects of crustal evolution. Geochim. Cosmochim. Acta, 49, 601-610.
Michard A., Beaucaire C. and Michard G. (1987) Uranium and rare earth elements in CO_2-rich waters from Vals-les-Bains (France). Geochim. Cosmochim. Acta, 51, 901-909.
Middleburg J.J., van der Weijden C.H. and Woittiez J.R.W. (1988) Chemical processes affecting the mobility of major, minor and trace elements during weathering of granitic rocks. Chem. Geol., 68, 252-273.
Miller R.G. and O'Nions R.K. (1984) The provenance and crustal residence ages of British sediments in relation to paleogeographic reconstructions. Earth Plan. Sci. Lett., 68, 459-470.
Miller R.G. and O'Nions R.K. (1985) Source of Precambrian chemical and clastic sediments. Nature, 314, 325-329.
Moorbath S. (1977) Ages, isotopes and evolution of Precambrian continental crust. Chem. Geol., 20, 151-187.
Nägler T. (1990) Sm-Nd, Rb-Sr and common lead isotope geochemistry of fine-grained sediments of the Iberian massif. Ph.D. Thesis, Swiss Feder. Inst. Tech., Zürich, 141 p.
Nance W.B. and Taylor S.R. (1976) Rare earth patterns and crustal evolution. I: Australian post-Archean sedimentary rocks. Geochim. Cosmochim. Acta, 40, 1539-1551.
Nelson B.K. and DePaolo D.J. (1985) Rapid production of continental crust 1.7 to 1.9 b.y. ago: Nd isotopic evidence from the basement of the North American mid-continent. Geol. Soc. Am. Bull., 96, 746-754.
Nelson B.K. and DePaolo D.J. (1988) Comparison of isotopic and petrographic provenance indicators in sediments from Tertiary continental basins of New Mexico. Jour. Sedim. Petrol., 58, 348-357.
Nesbitt H.W. (1979) Mobility and fractionation of rare earth elements during weathering of a granodiorite. Nature, 279, 206-210.
Ohr M., Halliday A.N. and Peacor D.R. (1991) Sr and Nd isotopic evidence for punctuated clay diagenesis, Texas Gulf Coast. Earth Plan. Sci. Lett., 105, 110-126.
O'Nions R.K. and Pankhurst R.J. (1978) Early Archean rocks and geochemical evolution of the Earth's crust. Earth Plan. Sci. Lett., 38, 211-236.
O'Nions R.K., Hamilton P.J. and Hooker P.J. (1983) A Nd isotope investigation of sediments related to crustal development in the British Isles. Earth Plan. Sci. Lett., 63, 229-240.
Othman D.B., Polvé M. and Allègre C.J. (1984) Nd-Sr isotopic composition of granulites and constraints on the evolution of the lower crust. Nature, 307, 510-515.
Page R.W., McCulloch M.T. and Black L.P. (1984) Isotopic record of major Precambrian events in Australia. Prec. Geol., 27th Int. Geol. Congr., 5, 25-72.
Patchett P.J., Kuovo O., Hedge C. and Tatsumoto M. (1981) Evolution of continental crust and mantle heterogeneity: evidence from Hf isotopes. Contrib. Miner. Petrol., 78, 279-287.
Patchett P.J., White W.M., Feldmann H., Kielinczuk S. and Hofmann A.W. (1984) Hafnium/rare earth element fractionation in the sedimentary system and crustal recycling into the Earth's mantle. Earth Plan. Sci. Lett., 69, 365-378.
Patterson C.C. and Tatsumoto P. (1964) The significance of lead isotopes in detrital feldspar with respect to chemical diffe-

rentiation within the earth's mantle. Geochim. Cosmochim. Acta, 28, 1-22.

Piper D.Z. (1974) Rare earth elements in the sedimentary cycle: a summary. Chem. Geol., 14, 285-304.

Roaldset E. and Rosenqvist I.T. (1971a) Unusual lanthanide distribution. Nature, Phys. Sci., 231, 153-154.

Roaldset E. and Rosenqvist I.T. (1971b) Rare earth elements in vivianite from Lake Asrum. Lithos, 4, 417-422.

Schieber J. (1986) Stratigraphic control of rare-earth pattern types in Mid-Proterozoic sediments of the Belt Supergroup, Montana, USA: implications for basin analysis. Chem. Geol., 54, 135-148.

Shaw H.F. and Wasserburg G.J. (1985) Sm-Nd in marine carbonates and phosphates: implications for Nd isotopes in sea water and crustal ages. Geochim. Cosmochim. Acta, 49, 503-518.

Stille P. and Clauer N. (1986) Sm-Nd isochron-age and provenance of the argillites of the Gunflint Iron Formation in Ontario, Canada. Geochim. Cosmochim. Acta, 50, 1141-1146.

Stordal M.C. and Wasserburg G.J. (1986) Neodymium isotopic study of Baffin Bay water: sources of REE from very old terranes. Earth Plan. Sci. Let., 77, 259-272.

Tatsumoto P. and Patterson C.C. (1964) Studies of zircon and K-spar concentrates in Precambrian sands and their significance with respect to continental growth. Jour. Geol., 72, 232-242.

Taylor S.R., McLennan S.M. and McCulloch M.T. (1983) Geochemistry of loess, continental dust composition and crustal model ages. Geochim. Cosmochim. Acta, 47, 1897-1905.

Taylor S.R. and McLennan S.M. (1985) The Continental Crust: its Composition and Evolution. Blackwell, Oxford, 312 p.

Veizer J. and Jansen S.L. (1979) Basement and sedimentary recycling and continental evolution. Jour. Geol., 87, 341-370.

Veizer J. and Jansen S.L. (1985) Basement and sedimentary recycling - 2: Time dimension to global tectonics. Jour. Geol., 93, 625-643.

DEPOSITIONAL HISTORY OF URANIUM ORES: ISOTOPIC CONSTRAINTS

Maurice PAGEL
Centre de Recherches sur la Géologie
des Matières Premières Minérales et Energétiques
and
Groupement Scientifique CNRS-CREGU
BP 23 54501 Vandoeuvre les Nancy Cedex (France)

The elaboration of new and more constrained models for the genesis of uranium deposits has greatly benefited from the application of stable and radiogenic isotope geochemistry during the last two decades. Sulfur isotope and U-Pb geochemistry have been used extensively in the past, but it is only recently that other isotopic systems have been applied (Sm-Nd and Sr isotope geochemistry on pitchblende, K-Ar and $^{39}Ar/^{40}Ar$ on clay minerals, $\delta^{18}O$, δD, $\delta^{13}C$ on minerals and fluid inclusions). An ideal physico-chemical model for the genesis of a U deposit should incorporate the four following stages: (1) leaching from the source-rock, (2) transportation of the elements by the mineralized fluid, (3) some mechanism of elemental deposition, and (4) preservation of the deposit. When studying a U deposit, detailed data acquisition should not be confined to the deposit nor solely to the U; only an integrated approach including the tectonic framework and explanations for the presence of not only the U but also other elements present in the mineralized zones can provide the necessary information to correctly interpret a deposit. Of course, all elements are not necessarily derived from the same source-rock, nor transported by the same fluid, nor deposited at the same time as the U, nor expected to behave similarly during alteration. However, isotopic studies are especially well adapted to recognize the different stages of mineralization in a deposit and to provide valuable insight regarding the source, transport, deposition and preservation of U deposits. The purpose of the present review is to discuss the main avenues of research in isotope geochemistry applied to U deposits in sediments, their applicability and their possible development in the near future.

URANIUM ORES IN SEDIMENTARY ENVIRONMENTS

The main types of U ore deposits in surficial and sedimentary environments and their characteristics are presented in table I. For a more detailed treatment on the subject, the reader is encouraged to consult Nash et al. (1981). From an economic point of view, all these types of U deposits do not have the same value. For example, the calcrete deposits are of low grade and located in areas of difficult access whereas the spatially unconformity related deposits have very high grade up to 10% (Cigar Lake, Saskatchewan, Canada) and high tonnage.

A brief review of U geochemistry is given for those not familiar with U metallogeny. In surficial and sedimentary environments, U could occur in solution in three oxidation states: U(IV), U(V) and U(VI). In natural aqueous solutions, U is transported as the uranyl ion (UO_2^{2+}) which form complexes with carbonate, phosphate, chloride, sulfate,... In many deposits, U is mineralogically present as pitchblende or uraninite where it mainly occurs in the tetravalent state. Therefore, a reduction is necessary for pitchblende or uraninite deposition which could be related to iron, sulfur or organic compounds or hydrogen. In surficial and sedimentary environments, the association with organic matter and the absorption process play an important role in the genesis of a deposit. Furthermore, in some surficial deposits, the U concentration results from deposition of U(VI)-

minerals such as, for example, carnotite in calcrete deposits. In this latter case, no reduction of U is necessary for deposition.

Table 1: Some characteristics of U deposit types in surficial and sedimentary environments, including unconformity type

Deposit type	Host Rocks	U-Mineralogy	Remarks	Example deposits
Quartz-pebble conglomerate	Conglomerate Sandstone	Detrital uraninite, uranothorite, brannerite	Limited to ~ 2.8 to 2.2 b.y.	Witwatersrand (S. Africa) Blind-River Elliot Lake (Ontario, Canada)
Calcrete	Nonpedogenic calcrete	Carnotite	Young (Tertiary) rich in vanadium	Yeelirrie (Australia)
Black shale	Marine carbonaceous shale	U-organic materials	Disseminated	Ranstad (Sweden)
Phosphorite	Marine phosphorite	Uraniferous carbonate apatite	Reworked pebble phosphorite higher grade (~ 0.02 %)	Moroco Phosphoria Fm and Bone Valley Fm (USA)
Sandstone tabular			Rich in organics	Ambrosia Lake district (USA) Grants (USA) Müllenbach (RFA)
Sandstone roll	Fluvial sandstone and arkose	Pitchblende coffinite	Zonality in V, Se, U, Mo	Wyoming (USA) Texas (USA)
Sandstone tectono-lithologic			Control by tectonic features and lithology	Akouta (Niger) Lodève (France) Oklo (Gabon)
Breccia pipe	Fluvial sandstone	Pitchblende Coffinite		Arizona (USA)
Spatially unconformity related	Proterozoic metasediments ± unconformably overlying fluvial sandstone	Pitchblende		Cluff Lake, Rabbit Lake, Key Lake, Collins Bay Cigar Lake, Saskatchewan Ranger Jabiluka, Nabarlek Koongarra (Northern Territory Australia)

SOURCE OF URANIUM

The determination of the source of U that was deposited in many deposits is often extremely difficult to assess other than speculatively. However, the source of U for very young U deposits can be addressed directly by considering the natural $^{234}U/^{238}U$ disequilibrium which exists in most waters and rocks (Cherdyntsev, 1971). In addition, an indirect method has been proposed by using U-Th-Pb isotope systematics (Rosholt and Bartel, 1969). Both these methods are considered below.

Uranium Disequelibrium and the U-Th-Pb Method

$^{234}U/^{238}U$ ratio in recent U-enriched sediments

Compared to other marine sediments, phosphorites always have high U contents, averaging 120 µg/g (Altschuler et al., 1958).

Since phosphorites forming today are known and the $^{234}U/^{238}U$ activity ratio is constant in seawater, if the source of U is seawater it can immediatly be recognized. Phosphorite nodules off the coast of Peru, Chile and Namibia are associated with upwelling of organic-rich waters having $^{234}U/^{238}U$ and $^{230}Th/^{234}U$ activity ratios close to that of modern sea water. However, some $^{234}U/^{238}U$ activity ratios obtained on these nodules are slightly higher than that of modern sea water, indicating that at least one other source is present. It has been suggested that these nodules were formed in organic-rich anoxic sediments, so that U derived from pore water rather than from open sea water directly (Veeh and Burnett, 1982). The slightly higher $^{234}U/^{238}U$ activity ratios result from the back diffusion of ^{234}U lost from previously formed phosphorite nodules (Veeh and Burnett, 1982) or from recoil. Since a marine setting is the only environment where the $^{234}U/^{238}U$ activity ratio is constant, the use of this ratio is very difficult elsewhere.

U-Th-Pb isotope systematics on potential source-rock

There is general agreement that, in the U-Pb system, the two radioactive parents (^{235}U and ^{238}U) and the two radiogenic stable daughters (^{207}Pb and ^{206}Pb) show similar behavior during alteration processes; the equilibrium between ^{232}Th, which is considered immobile, and its radiogenic daughter ^{208}Pb is used as an indicator of U or Pb mobilization. The percentage of U gain or loss is determined by comparing the U content in the sample with the U content calculated from the uranogenic Pb. The age of the disturbance can also be constrained on a concordia diagram based on this technique (Fig. 1). Stuckless (1987) discussed the limitation of this approach. Due to experimental limitations (analytical precision), the use of this approach requires a high initial U to Pb ratio as well as a high radiogenic Pb to common Pb ratio.

A good example for the study of the origin of U is that made by Stuckless and Nkomo (1978) in Wyoming where a volcanic source and/or a granitic source could be considered for the genesis of roll type U deposits. The authors showed that the granite from the Granite Mountains lost large amounts of U during Tertiary times (Fig. 1). As this age is in agreement with the age of ore deposition at 28-35 Ma (Ludwig, 1978, 1979), this event was interpreted as leaching of U in granite and deposition in sediments. However, the isotopic investigation of tuffaceous strata close to the deposits (Nash et al., 1981) was not achieved in spite of the fact that these tuffs could be an U source as well.

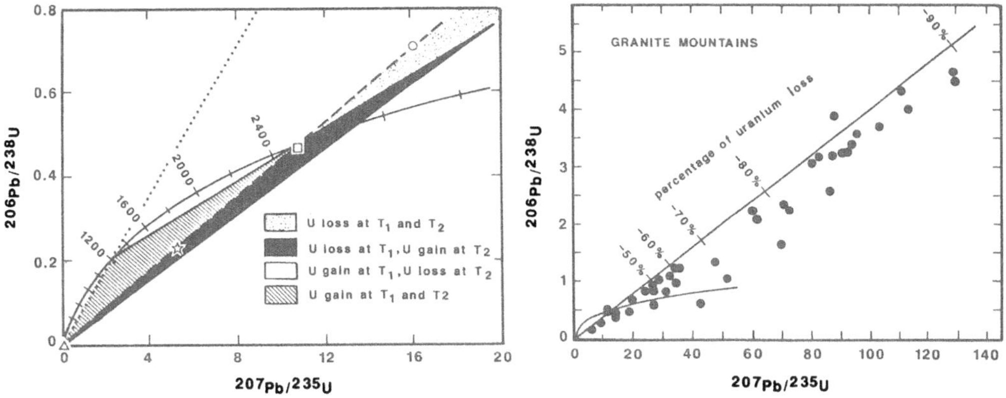

Figure 1: Determination of U loss and gain in a Concordia diagram (after Stuckless, 1987). A: Concordia diagram for a system that formed at 2500 Ma, remained closed until 1250 Ma ago (T1) and was disturbed again very recently (T2). Square = sample showing no disturbation; star = sample which lost 50% of its radiogenic Pb or gained 100% U at T2; triangle = loss of Pb at T2. B: Concordia diagram for samples of the Lankin Dome granite, Granite Mountains, Wyoming. A reference cord from 0 trough the 2600 Ma equilibrium point on the concordia is marked with relative percentage of U loss, assuming closed-system behavior from 2600 Ma until 30 Ma ago

As discussed before, the applicability of U-Th-Pb systematics on potential U source-rocks is favorable when the radiogenic to common Pb ratio is high and the error introduced in the calculation of radiogenic Pb is small. This is especially the case for Precambrian rocks. Although a two event history is the best case, Stuckless and Troeng (1984) have shown that a good set of interpretations could be derived from a three-stage system.

Other Radiogenic Elements (Sr, Nd, Pb, Hf, Os)

The source of other isotopic tracers related to U deposition such as Sr, Nd, Hf, Os or Pb in young deposits or in situations where a precise age correction is possible, is a completely open research field which could have, at least theoretically, great influence on the discrimination of source-rocks. However, during leaching of the source-rock, differences between the whole-rock and the fluid isotopic ratio could be induced by a mineralogical effect. For example if a mineral such as a plagioclase, which contains Sr without Rb, is leached the $^{87}Sr/^{86}Sr$ ratio in the fluid will be lower than the Rb-rich source-rock (Turpin et al., 1990, Franklyn et al., 1991). The reader should also bear in mind that all the elements present in a deposit do not necessarily have to come from the same source-rock. Presently, the only data of this type, available for sedimentary-related U-deposits, are Sr and Nd isotopic ratios (Maas, 1989; Pagel et al., in press) and initial radiogenic Pb (Ludwig et al., 1985; Lancelot and Vella, 1989) on pitchblende. Ludwig et al. (1985) have shown how to decipher the initial radiogenic Pb isotope ratio in order to determine the age and chemical composition of the source of the initial Pb.

The distribution of REE in pitchblende is homogeneous as it is observed in favorable cases by electron microprobe studies. Although, Nd incorporation during its crystallization is my favor hypothesis. Fryer and Taylor (1984, 1987) and Pagel et al. (1987) have analyzed pitchblendes from many deposits and showed that the REE pattern depends on the type of U deposit. For example, the REE chondrite normalized pattern for pitchblende from unconformity-type deposits from Canada and Australia have a bell-shaped distribution with high contents in intermediate REE. On the other hand, the origin of Sr present in pitchblende is not clear while the Rb/Sr is always very low: from 0.003 to 0.105 in the Cigar Lake (Saskatchewan, Canada) pitchblende (Pagel et al., in press). Zirconium is sometimes present in significant amounts in pitchblende from sedimentary U deposits (Forbes et al., 1984) and Hf is also commonly present.

$^{143}Nd/^{144}Nd$ and $^{87}Sr/^{86}Sr$ ratios

Sm-Nd isotopic studies were made on the unconformity-type U deposits of Collins Bay in Saskatchewan (Canada) by Fryer and Taylor (1984) and of Nabarlek and Jabiluka in Northern Territory, Australia by Maas (1989). The important study by Maas (1989) on the Nabarlek, Jabiluka, Koongarra and Ranger deposits (Northern Territory, Australia) shows that, on the basis of the Nd and Sr data, important constraints can be placed on the genesis of these deposits (Fig. 2). Maas (1989) argued that the covariation of U and REE found in the Alligators Rivers U ores suggests simultaneous transportation by the ore-forming solution. This conclusion is strongly supported by the fact that all the pitchblende crystals from the unconformity-type U deposits have the same REE pattern.

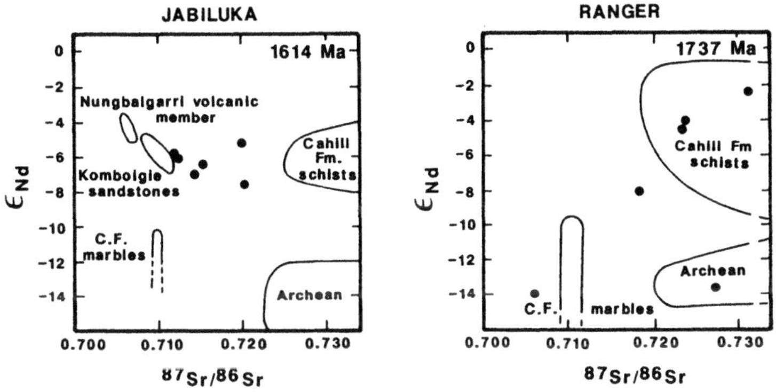

Figure 2: Initial Nd and Sr isotope ratios for primary uraninite from Jabiluka (A) and Ranger (B), Northern Territory, Australia compared with the time-corrected values for potential Nd-Sr source rocks (after Maas, 1989). For Jabiluka, the ore samples lie between the fields of Kombolgie sandstone (and the Nungbalgarri Volcanic Member) and the Cahill Formation schists. For Ranger, five of the six ore samples form a linear array between a carbonate-type and a Cahill schist-type component

A central issue to all these studies on unconformity type U deposits is knowing if the U is derived from the sedimentary cover or if it has been leached from the metamorphic and granitic basement rocks. Late Archean gneisses and the host metamorphic Cahill Formation are the most likely Nd-Sr source-rocks for the Ranger deposit which formed before the deposition of sandstones (1737 ± 20 Ma, Ludwig et al., 1987) if we accept the Rb-Sr age (1648 ± 29 Ma, Page et al, 1980) on syn-Kombolgie Nungbalgarri Volcanics. The Nd and Sr isotope data on the Jabiluka and Nabarlek deposits, formed after the deposition of sandstones, are compatible with the ore fluid acquiring its Nd and Sr isotope signature through mixing of Nd and Sr derived from the Kombolgie sandstones and the metamorphic Cahill Formation (Fig. 2). Despite comparable host-rocks, ore controls and wall-rock alteration from the Jabiluka and Nabarkek deposits, it is not clear why the Ranger deposit could have a different history.

In Saskatchewan, Pagel et al. (in press) showed that $^{87}Sr/^{86}Sr$ in pitchblende is more comparable to that of the Athabasca sandstones, the interbedded volcanics and the upper part of the weathering profile beneath the Athabasca sandstones than to that of the basement at the time of U deposition (Fig. 3).

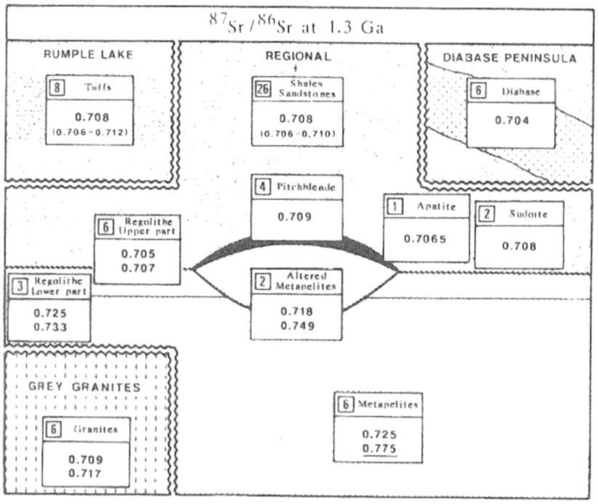

Figure 3: Calculated $^{87}Sr/^{86}Sr$ ratio at 1.3 Ga for different fresh or altered rocks, alteration minerals and pitchblende in the Cigar Lake area and its regional environment (after Pagel et al., submitted)

The possible leaching of U preconcentration in the basement has been excluded by Fryer and Taylor (1987) for Saskatchewan and by Maas (1989) for Australia. They have noted that if recycled Nd from protores is considered, much higher $^{143}Nd/^{144}Nd$ should be expected. This conclusion is ambiguous because it is conceivable that U was leached from the basement rock and since REE have solubilities much more lower than that of U, the REE content of the fluid is dominated by the U-carrying fluid which is derived from sedimentary formation waters (Pagel et al., 1980).

The above approach is difficult to apply to U deposits in sedimentary environments because the U-minerals are generally very fine-grained. The Akouta deposit (Niger) is, however, an interesting exception because centimetric lenses of pitchblende are spatially related to the bottom- and for-sets of the sandstone units. The Sm-Nd isotopic data are scattered (Fig. 4) and no age can be derived from the results (Turpin et al., 1991). If the initial Nd ratio is recalculated for the assumed mineralization age, 200 Ma, and if REE patterns are considered, the REE must have been derived either from different sources or from a heterogeneous source without achievement of REE homogeneity in the ore fluid.

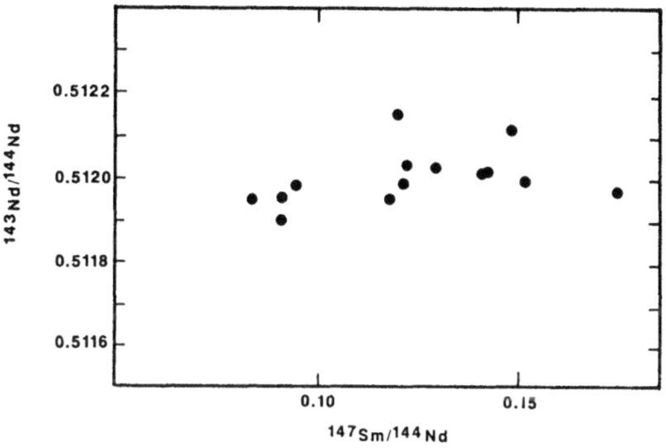

Figure 4: Sm-Nd variation for pitchblende in the Akouta deposit, Niger (after Turpin et al., 1991)

The use of isotopic tracers on minerals associated with pitchblende opens up a wide field of research possibilities; for example, $^{87}Sr/^{86}Sr$ in carbonates, barytes, hydrous aluminous phosphate, apatite, and $^{143}Nd/^{144}Nd$ in fluorite, clays and other minerals. The development of microanalytical techniques such as the ion microprobe, nuclear microprobe and laser ICP-MS show considerable potential in aiding the deciphering of the multistage development of the formation of a U deposit as well as identifying the possible different sources origin of the elements present in the mineralized zone.

Initial $^{207}Pb/^{206}Pb$ in uranium minerals

The existence of a preconcentration event is very difficult to assess in sediment-hosted U deposits, even if the U-Pb data are consistent with the mineralization being epigenetic. If a U preconcentration does exist, the initial $^{207}Pb/^{206}Pb$ would be radiogenic and $^{206}Pb/^{204}Pb$ and $^{207}Pb/^{204}Pb$ would be significantly higher than the common Pb ratio as well. Such a preconcentration has been observed by Lancelot and Vella (1989) in the Rabejac deposit (Lodève Basin, France), this was based on an original interpretation presented on figure 5. This deposit has an age of 183 ± 4 Ma but is emplaced in a Permian formation. The initial $^{207}Pb/^{206}Pb$ is 0.296 whereas the accepted common Pb ratio determined on galena is 0.8573. These authors modelled the evolution of the U-Pb system assuming a preconcentration event during the Permian, its remobilization at 183 Ma and assuming a galena as sample representative of the regional common Pb of the Permian age. They found μ values $(^{238}U/^{204}Pb)_{present}$ around 3000 to 4800. This study is a good illustration of how Pb isotopes can be useful in deciphering the complex evolution of U concentrations where fluid circulation occurred on more than a single occasion.

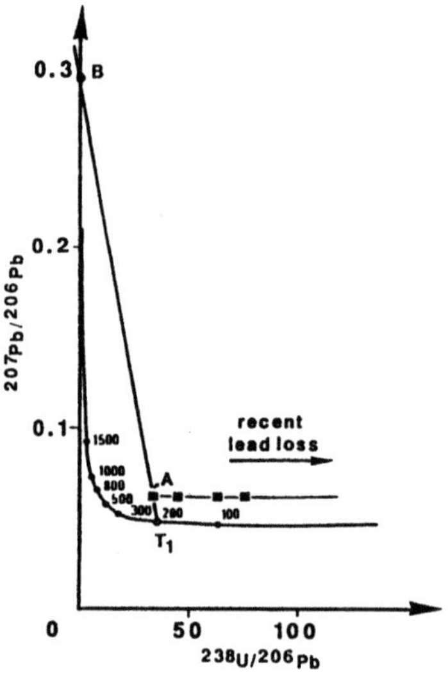

Figure 5: Determination of the initial $^{207}Pb/^{206}Pb$ ratio in pitchblende in a Terra-Wasserburg diagram (after Lancelot and Vella, 1988). Sample A is supposed to have evolved in a closed system since the time of deposition (T1). The initial $^{207}Pb/^{206}Pb$ ratio is determined at point B

Other potential tracers

Among other possible isotopic tracers which could be used are the $^{176}Hf/^{177}Hf$, $^{187}Os/^{186}Os$, and $^{11}B/^{10}B$ ratios, but no data are yet available for U deposits. Despite the presence of significant amounts of Hf in some U minerals, the use of the $^{176}Hf/^{177}Hf$ is limited by (1) the difficulty in analysing Hf by mass spectrometry due to a very low ionization yield, however combination of conventional Hf extraction and SIMS type mass spectrometry has considerably lower the analyzed amount and (2) the small variation of the isotopic ratio. In contrast, the $^{187}O/^{186}Os$ ratio varies widely and with improved techniques for chemically extracting Os from rocks and the use of the ion microprobe very small quantities of Os can be detected. The use of stable B isotopes may provide insight into the origin of tourmaline associated with some U ore deposits, especially those which are spatially unconformity-related in Saskatchewan (Canada) and Northern Territory (Australia). It could result from the remobilization of preexisting tourmaline in the basement, related to evaporites or granitic rocks, or it could be related to some dissolved evaporites previously present in the sedimentary cover.

TRACING THE ORIGIN OF MINERALIZING FLUIDS

Stable isotope geochemistry gives valuable answers to questions about the origin of the mineralizing fluid and the processes of U ore deposition. Excellent reviews by Taylor (1974), Ohmoto (1986) and Sheppard (1986) on the general principles of applying stable isotopes to ore genesis are available. D/H and $^{18}O/^{16}O$ can potentially be used to determine the type of aqueous solutions: meteoric, connate, formation water, etc. The isotopic composition of a fluid can be determined directly from fluid inclusions or indirectly from the mineral, assuming that the mineral-H_2O D/H and $^{18}O/^{16}O$ fractionation factors and temperature are well constrained. Pagel et al. (1980) have directly obtained δD values on fluid inclusions in gangue minerals (quartz and dolomite) from the Rabbit Lake U-deposit. Few data have been published on O isotope ratios in pitchblende and uraninite (eg., Hoekstra and Katz, 1956; Hattori et al., 1978). These minerals are depleted in ^{18}O with $\delta^{18}O$ as low as -30 per mil (SMOW). However, pitchblende and uraninite crystals have been oxidized after deposition. Therefore, the significance of the O isotope data with respect to deposition is unclear due to the probability of isotopic exchange reactions. If the O isotope fractionation factor between U dioxide and water calculated by Hattori and Halas (1982) is correct, the $\delta^{18}O$ of water is negative in most cases. This indicates the involvement of meteoric waters. The determination of C and S isotope compositions of fluids is often difficult because they can more rarely be measured directly on the fluid and isotopic equilibria have not necessarily been established between the fluid species or between the fluid species and the precipitating minerals. Among important processes which can be deciphered by C and S isotopes are the involvement of an organic component in the fluids or the involvement of bacterial or fermentation processes. Under certain conditions, pH and O fugacity can be deduced from isotope geochemistry of the fluid. The major causes for C and S isotopic fractionations in nature are redox reactions involving the isotopic species (e.g. Robinson and Ohmoto, 1973; Ohmoto, 1986), which are very important processes in U ore genesis. Several applications of stable isotope geochemistry are presented below.

Circulation of Sedimentary Formation Waters in Basement Rocks

The formation of U veins in basement rocks is often attributed to the involvement of a fluid derived from sedimentary formations (Turpin, 1984, Turpin et al., 1990; Negga et al., 1986). For the La Lauzière massif (Alps, France), Negga et al. (1986) proposed that U was leached from the basement by meteoric waters and deposition occurred as a response to mixing with more reducing waters from underlying sediments (Fig. 6). The deep sediment-derived waters were confirmed from fluid inclusion data from Mullis and Stalder (1987) who showed that post-Carboniferous sedimentary formation waters can circulate as deep as 1500 m in basement rocks. Some uranium occurrences are known in this context.

In Saskatchewan, the setting is simpler since the relationship between the basement and the cover is nearly unchanged. The δD and $\delta^{18}O$ data on illite and chlorite from the Athabasca sandstones and from altered basement beneath the sedimentary cover (Wallis et al., 1984; Halter et al., 1985, 1987;

Wilson et al., 1987; Wilson and Kyser, 1987; Bray et al., 1988) show that the illite and chlorite minerals were equilibrated with the same fluid at about 200°C. This means in turn that the sheet-minerals associated with the mineralizing event isotopically did equilibrate with the same fluid. Furthermore, all these illites have indistinguishable K-Ar ages (Clauer et al., 1985; Bray et al., 1987), which suggests that the formation waters from the Athabasca sandstones have penetrated the basement and altered it.

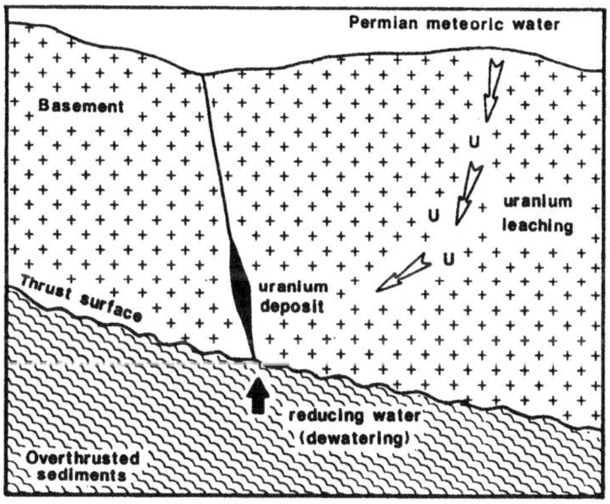

Figure 6: La Lauzière mixing model between Permian oxidized meteoric waters and reducing metamorphic waters (after Negga et al., 1986)

Mineralization Along a Roll Front

Sulfur isotopes have been analyzed in roll-type deposits where negative values become less negative to positive ahead of the roll. Goldhaber et al. (1978) demonstrated a systematic variation of $\delta^{34}S$ values along the Benavides roll in Texas (Fig. 7) which ranges from -29 per mil to +30 per mil. A two-stage process was proposed: (1) isotopically heavy Fe disulfides ahead of the roll were produced by the introduction of fluids containing dissolved H sulfides derived from underlying oil and gas accumulations and were injected via a fault, (2) invasion by oxygenated U-bearing groundwater creates soluble metastable S by oxidation of S and marcasite precipitation was favored in a low pH zone.

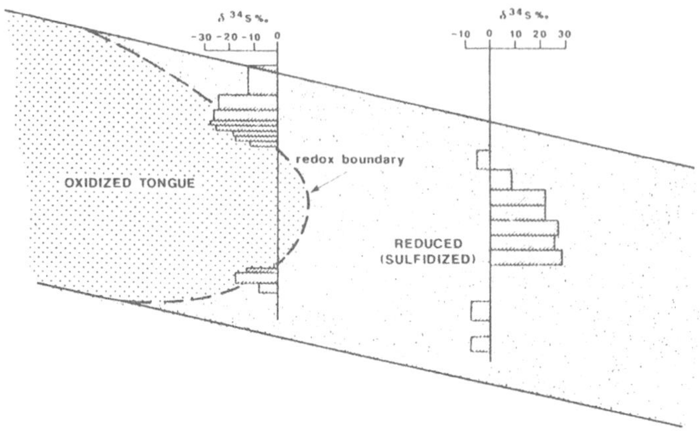

Figure 7: Variation of S isotope ratios along an idealized roll cross-section (after Goldhaber et al., 1978)

Deposition as a Result of Mixtures of Fluids

Goldhaber et al. (1983) have also applied S isotope geochemistry in the Ray Point U-district, south Texas, which is an anomalous area in that it contains FeS_2 minerals in the altered tongue. From S mineralogy and isotope geochemistry, four stages of FeS_2 mineralization are recognized with one pre-ore, another pre-ore to contemporaneous and two post-ore. It is noted that the isotopic composition of the pyrite of stage 3 is similar to that of sour gas in the underlying limestone. The marcasite of stage 4 is related to bacterial activity. Goldhaber et al. (1983) concluded that: "the Ray Point uranium district thus formed and was preserved virtue of its proximity to a zone of interaction of connate brines from depth with shallow meteoric waters".

Involvement of Biogenic Processes

In several deposits of the Grants U Region, isotopically light Fe disulfides suggest that these sulfides were biogenically-derived. During biogenic processes, aqueous sulfide species produced by sulfate-reducing bacteria are enriched in ^{32}S (Thode et al., 1951). For instance, Fishman et al. (1985) have found $\delta^{34}S$ values between -29 to -42 per mil for ore-stage Fe disulfide minerals in the Mariano Lake and Ruby 1 deposits, where U is intimately mixed with organics. This favors the introduction of organic matter in sandstones early in the diagenetic history.

DATING THE DEPOSITION AND REMOBILIZATION OF URANIUM DEPOSITS

The age of U deposition can be obtained from radiogenic systems directly if the isotopes considered are present in the U-minerals, or indirectly if they are located in the gangue minerals, whose textural relationships to the U-minerals indicate that they formed at the same time.

In Uranium Minerals

U-Pb and Pb-Pb isotopic ages

Several ways of exploiting the data have been proposed: Concordia diagram ($^{207}Pb/^{235}U$ vs. $^{206}Pb/^{238}U$ by Wetherill (1956; Fig. 1) or $^{238}U/^{206}Pb$ vs. $^{207}Pb/^{206}Pb$ by Tera and Wasserburg (1972; Fig.5)), U-Pb ($^{235}U/^{204}Pb$ vs. $^{207}Pb/^{204}Pb$ and $^{238}U/^{204}Pb$ vs. $^{236}Pb/^{204}Pb$) and Pb-Pb ($^{206}Pb/^{204}Pb$ vs. $^{207}Pb/^{204}Pb$) isochrons. In U deposits, different disturbing factors generally have to be taken into account: (a) leaching or diffusion of Pb, (b) presence of common Pb in the mineral structure or as a discrete associated mineral, (c) leakage of intermediate radioisotopes such as Ra or Rn by diffusion or recoil, (d) U leaching in oxidizing environments, (e) an initial disequilibrium in the U-Pb system in young deposits, and (f) recent alteration of pitchblende and formation of a U(VI) mineral. It is generally possible to obtain the age for the formation of a hexavalent mineral (low intercept in a Concordia diagram) and the age of the primary pitchblende (upper intercept in a Concordia diagram). A specific methodology and approach is, therefore, required for U-Pb dating of U minerals

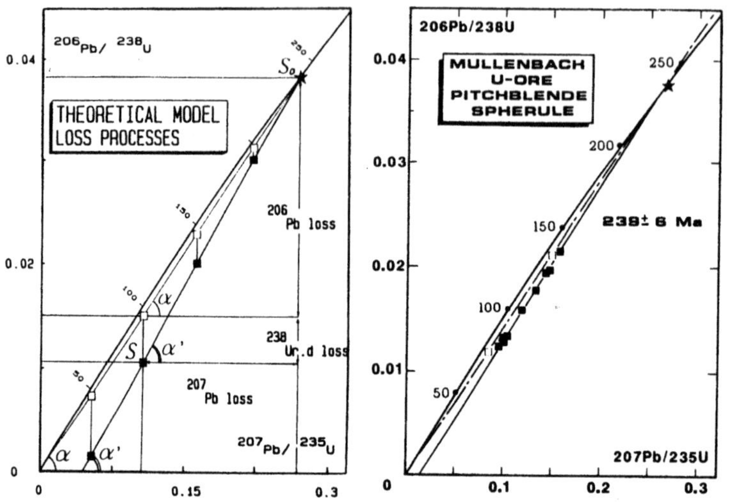

Figure 8: Interpretation of the loss of a ^{238}U radioisotope daughter in a Concordia diagram and its application to the Müllenbach U-deposit (after Holliger et al., 1989). A: Diffusion process in a Concordia diagram for four hypothetical samples. S_0 is the primary age of the U mineralization. The Pb diffusion loss leads to the discordia 0-S_0. The combined Pb diffusion loss and ^{222}Rn leakage Pb to another isochron at the right. S represents a sample where ^{238}U r.d. loss has occurred. B: Concordia plot for the Müllenbach pitchblende. Nine samples define a best-fit line at 239 ± 6 Ma. The two other points are located on the line 0-240 Ma

Many specific guide lines are available for these isotopic determinations. In summary: (1) a preliminary investigation by electron microprobe can distinguish the different generations of U minerals which may exist in the same specimen, and in some favorable cases give an idea of the ages. This approach is especially interesting during prospection. Such chemical ages must, however, always be checked by isotopic analyses. (2) The size of the sample is an important parameter to consider. Ludwig et al. (1981) have shown that for the Midnite Mine, Washington, in samples >10 cm the system is closed for $^{235}U/^{207}Pb$ but not for $^{238}U/^{206}Pb$ (leakage of daughter elements), whereas for samples <1 mm there is migration of Pb and radioactive daughters of ^{238}U. (3) Until now, dating has been mainly done by thermal ionization mass spectrometry (TIMS) after chemical treatment. But recent developement in *in situ* secondary ion mass spectrometry (SIMS) permits U-Pb isotope analysis on a few micron sample (Holliger,

1988). The relative yields of the two elements has been constrained by a well-known, homogeneous and concordant uraninite standard from Zambia. Dating minerals is possible on a single polished section, even if it indicates a multiphase paragenesis. Development of this technique is certainly very important for a better understanding of the mobility of Pb, U and the radioactive daughters of ^{238}U and to select samples with a high $^{206}Pb/^{204}Pb$ ratio. (4) The correction for common Pb is one of the most difficult tasks when $^{206}Pb/^{204}Pb$ is low because of its effect on the calculated age. Several options exist: (a) to calculate the Pb from the model of Stacey and Kramers (1975), (b) to determine the Pb composition in an associated Pb-bearing U-free mineral (galena, pyrite, K-feldspar), (c) to graphically obtain the value using the Pb-Pb isochron. Wendt (1984) has proposed a three-dimensional discordia plane method to yield the concordia intersection ages in the case of a two-stage evolution history without the necessity for assumptions regarding the isotopic composition of the common Pb present in the samples. (5) The leakage of radioactive daughters of ^{238}U is very important in sediment-hosted U deposits (Fig 8) as demonstrated by Ludwig (1978, 1979), Lancelot et al. (1984) and Holliger et al. (1989). The ^{226}Ra and ^{222}Rn are generally the most frequently lost daughters due to recoil and Ra leaching in reduced environment. A Ra emanation from pitchblende can be very significant and variable (Giletti and Kulp, 1955). An example of leakage of radioactive daughter(s) is presented in figure 8 as illustrated by the Müllenbach U ore (Holliger et al., 1989). (6) The effect of initial disequilibrium has to be taken into account for young U mineralization. Wendt and Carl (1985) have shown that the application of a modified Tera-Wasserburg U/Pb concordia plot could overcome this difficulty.

The U-series ages

From $^{230}Th/^{234}U$ and $^{234}U/^{238}U$ activity ratios in recent formed U-minerals (<350 ka), it is possible to derive an age, assuming no ^{230}Th was incorporated during mineral formation. This latter assumption is supported by the absence of ^{232}Th in the mineral. The equation is:

$$^{230}Th/^{234}U = \frac{1-e^{-\lambda_{230}t}}{^{234}U/^{238}U} + \left(1-\frac{1}{^{234}U/^{238}U}\right)\frac{\lambda_{230}}{\lambda_{230}-\lambda_{234}}\left(1-e^{-(\lambda_{230}-\lambda_{234})t}\right)$$

A data set obtained by Kaufman and Ku (1989) on carnotites from Israel suggests that, under some circumstances, an age can be obtained. Wendt and Carl (1985) have shown that the $^{230}Th/^{234}U$ disequilibrium age (132 ± 4 ka) obtained on torbernite from Höhenstein (northern Bavaria) is comparable to the U/Pb age calculated by the modified Tera-Waserburg concordia plot of 137 ± 11 ka. However, if this type of hexavalent U mineral remains in water flow, it acts as a chemically open system, as demonstrated in the Yeeliree deposit (Lively et al., 1979; Airey and Roman, 1981).

Spontaneous fission of ^{238}U

This is an open field of research. Chronologic data can be obtained either by analyzing the fission products, Xe, Kr or Ru for example, or the tracks created in minerals by the fission process. The value of the spontaneous fission decay constant is still under debate, both $8.46.10^{-17}$ y^{-1} (Galliker et al., 1970) and $7.03.10^{-17}$ y^{-1} (Roberts et al., 1968) are commonly used.

U-Xe and U-Kr

Xe_s ($^{136,134,131}Xe$) and Kr_s($^{87,84,83}Kr$) are produced by the spontaneous fission of ^{238}U. The age is given by the following equation:

$$t_{U\text{-}Xe} = \frac{1}{\lambda_\alpha} \ln \left(\frac{Xe_s}{^{238}U} * \frac{\lambda_\alpha}{\lambda_s * \gamma_{Xe_s}} + 1\right)$$

where $t_{U\text{-}Xe}$ is the age of the sample, Xe_s is the atomic concentration of Xe produced by spontaneous fission of ^{238}U, ^{238}U is the atomic concentration of ^{238}U, λ_α is the decay constant of ^{238}U, λ_s is the spontaneous fission decay constant of ^{238}U, and γ_{Xe_s} is the Xe_s fractional yield produced by spontaneous fission.

Recent developments in mass spectrometry have permitted Eikenberg et al. (1989) to date uraniferous minerals, especially pitchblende, by the U-Xe and U-Kr methods. The preliminary results show that if pitchblende has recrystallized, the U-Xe and U-Kr systems date the recrystallization event. On the other hand, it has been shown that the U-Xe and U-Kr systems can survive greenschist facies metamorphism. In the Alps, concordant U-Xe and U-Kr ages on coarse pitchblende have been obtained which correspond also to concordant U-Pb ages of around 250 Ma. Fine grained pitchblende have U-Xe and U-Kr ages which agree with the lower intercept age of about 40 Ma on the U-Pb discordia. This method is very promising and could be applied to other U-rich mineral, such as monazite. However, in many cases U-Xe ages establish only lower limits because of natural Xe losses.

Xe_s-Xe_n spectrum technique

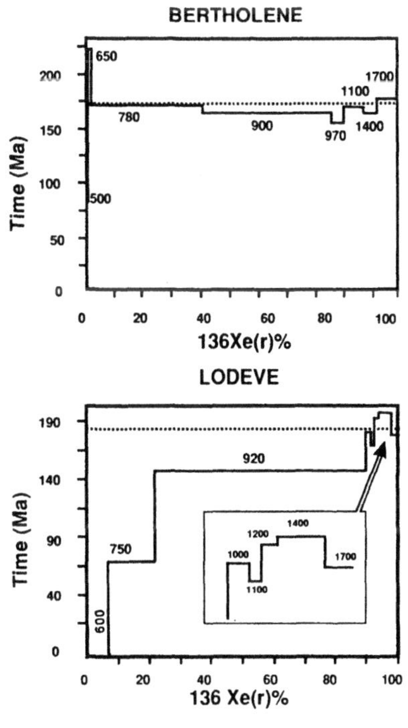

Figure 9: Xe_s-Xe_nspectrum for pitchblende from France (after data from Lévêque and Meshick, 1991). The pattern of apparent age vs. the percentage of radiogenic ^{136}Xe is shown, the temperature (in °C) of gas release is indicated and the Pb-Pb values are given by the dotted line. For Bertholène, a plateau is obtained with six heating steps (167 ± 11 Ma), whereas an anomalous Xe_s-Xe_n ratio could be observed at 650°C. For Lodève, the plateau could be drawn with the flat five heating steps, which represent only 10% of the total gas release

To avoid the problem of natural Xe losses, Shukolyukov and Meshick (1987) and Lévêque and Meshick (1991) have applied the very promising Xe_s-Xe_n spectrum technique (Shukolyukov et al., 1974) to pitchblendes, where Xe_s is the atomic concentration of Xe produced by the spontaneous fission of ^{238}U and Xe_n is the atomic concentration of Xe produced in nuclear reactor by induced fission of ^{235}U. In this technique, only the Xe_s/Xe_n ratio is calculated. To overcome the problem of Xe loss, stepwise heating is performed: the Xe_s/Xe_n ratio will increase because of their different origins. At low temperatures, Xe escapes from the most disturbed zones where Xe_s has already escaped from the pitchblende during its geological history whereas the Xe_n escape depends only on the U distribution in the sample. The ratio will stabilize ultimately giving a plateau. The plateau age corresponds to the initial age of the pitchblende when there are distinct peaks of gas release (Fig. 9).

Compared to U-Pb dating by TIMS, Shukolyukov and Meshick (1987) pointed out that the Xe technique is powerful even if the pitchblende is mixed with a common Pb-bearing mineral, such as a sulfide. In this case, the common-Pb correction introduces a large error. The disadvantage of the Xe technique is that it necessitates large grain-size samples, because the range of displacement of fission fragments is about 15 µm. Furthermore, some problems could arise if the sample is not constituted by one generation of pitchblende. It should also be pointed out that not all aspects of the spectrum can be easily explained and that in several cases the Xe age is lower than the U-Pb age, (e.g. Shukolyukov and Meshick, 1987). More work is needed to determine the relation between the escape of Xe and the structure of pitchblende.

U-Ru method

Several Ru isotopes (^{99}Ru, ^{101}Ru, ^{102}Ru, ^{104}Ru) are derived from the spontaneous fission of ^{238}U. It is thus possible to determine an age using an equation similar to the one presented for U-Xe. If it is a closed system, this age is the age of mineral deposition. Preliminary data obtained on deposits from Africa,

Australia, Brazil and Canada show that the ages are higher than the U-lead ages (Maeck et al, 1978).

On Gangue Minerals

The use of spontaneous fission tracks produced by ^{238}U has a high potential for dating U-ore deposition. A pioneer study published by Cunningham et al. (1982) used quartz in contact with pitchblende as an external detector in the Marysvale, Utah, volcanic field. This technique could be applied to sediment-hosted U deposits. However, it is time-consuming because it requires a two-stage replica of the etched quartz surface by methyl acetate and Pt-C, a correction for the inclination of the surface with reference to the C axis, and an extrapolation of the fission tracks densities from the interior of the quartz grain to the contact with the pitchblende. Furthermore, it is of limited use in attempts to determine age of mineralization in many, especially old, deposits.

The use of K-Ar systematics on clay minerals associated with U ores has been applied extensively in France (Bonhomme, 1982). Illite concentrates were principally analyzed, but some recent data on corrensite-rich fractions were also successful (Turpin et al., 1991). The application of the K-Ar method requires careful preparation of the samples. It is necessary to obtain a representative fraction of the clay. It is therefore essential to characterize properly the studied fractions. An extensive review of these problems is presented in a companion paper by Clauer and Chaudhuri (this volume). In sedimentary rocks, Brockamp et al. (1987) have shown that individual K-Ar ages of sedimentary clay fractions associated with U deposits are size-dependent and that pure authigenic fractions are only present in the smallest size fractions. The problem is apparently not important in veins, as Bonhomme et al. (1987) did not observe a correlation between the age and size of the analyzed fractions. In several cases where good agreement has been obtained between U-Pb ages and K-Ar ages (Table 2). This implies that despite their small size the clay minerals retain their argon. However, we will see later that this is not always the case and that strong disturbances of the K-Ar system have been observed in illites from several U deposits (Wilson et al., 1987, Halter et al., 1987). Furthermore, it should be emphasized that in a multistage ore deposit, it is always difficult to correlate the K-Ar age with the mineralization stage.

Table 2: Comparison between Pb-Pb, K-Ar and Xe-Xe ages obtained on the Rabejac (Lodève) and Bertholène deposits (France)

Deposit	Pb-Pb age on pitchblende	K-Ar age on clays	Xe-Xe age on pitchblende
Rabejac	183±4 Ma[1]	182-174 Ma[2] illite-chlorite	182±13 Ma[3]
Bertholène	173±9 Ma[4]	170-175 Ma[5] illite-smectite	167±11 Ma[3]

1 Lancelot and Vella, 1989
2 Bonhomme et al, 1987 (the location of the studied deposit in the Lodève basin is not specified)
3 Levêque et Meshick, 1991
4 Lévêque et al, 1988
5 Schmitt et al, 1984

Dating K-bearing minerals has also been done by laser probe $^{40}Ar/^{39}Ar$ method. Illites from the unconformity-type deposits from Saskatchewan (Bell, 1985, Bray et al., 1987) were analyzed. This aspect is also discussed in detail by Clauer and Chauduri elsewhere in this volume. The stepwise-heating technique may also be utilized on U-deposits where secondary K-feldspar is present in the mineralized zone or where several generations of minerals are present.

In the alteration halos of U deposits there are often clay minerals (chlorite, illite,...), which may be appropriate candidates for Rb-Sr dating. If a situation is considered where all the U ore has been remobilized, it is possible that the clay minerals could survive this disturbance of the U-Pb system. Lee and Brookins (1978) and Brookins (1980) have presented Rb-Sr data on clays associated with barren and mineralized samples in the Ambrosia Lake and Smith Lake districts (San Juan basin, New Mexico), which contain peneconcordant ore-bodies with abundant migrated organic material. For ore-stage chlorites, Brookins determined an age of 139 ± 24 Ma. This age is in accordance with the results of U-Pb data modelling on primary ore which has a minimum age of about 130 Ma (Ludwig et al., 1984). These combined data show that an early deposition of U has occurred in the Morrison formation.

In the unconformity-type U deposits from Australia, the Rb-Sr age of illites associated with U mineralization is 1.7 Ga (Page et al., 1980), whereas the U-Pb ages are significantly lower. The Sm-Nd data (Maas, 1989) prove that the alteration minerals record the age of U deposition, whereas the U-Pb system has been perturbed (Table 4).

The Mo sulfides (molybdenite or jordisite, MoS_2) present in U deposits are rich in Re and contain only radiogenic Os (^{187}Os). If the system remained closed, it is only necessary to analyze ^{187}Re and ^{187}Os to obtain an age (Luck and Allègre, 1982). However, in sediment-hosted U deposits, the Mo sulfide is poorly crystallized (jordisite) and Re may have been leached out. However, Ravizza and Turekian (1989) have successfully determined the depositional age of black shales despite some post depositional mobilization of Re and/or Os on a small spatial scale.

The La-Ba system (Nakai et al., 1986) may be used to date REE-rich minerals, such as monazite and florencite which are sometimes associated with U-minerals.

Comparison of Ages Obtained by Different Techniques

The ages obtained with the various chronometers are not always the same. This could reflect the contrasted geochemical behavior of different chronometric systems during remobilization processes. A comparison among U-Pb and U-Xe and U-Kr data by Einkenberg et al. (1989) has shown four possibilities, one of which has not been observed in nature (Table 3). They point out that fission gas ages are necessary to interpret ambiguous discordant U-Pb data from pitchblende, especially if it contains excess radiogenic Pb derived from a former mineralization event. Many data have been obtained by U-Pb, Sm-Nd, Rb-Sr and K-Ar on the unconformity type deposits from Saskatchewan (Canada) and the Pine Creek Geosyncline (Australia) (Table 4). The U-Pb ages can be, in some cases, significantly younger than the ages obtained by the other methods. The Sm-Nd ages from Nabarlek, Jabiluka II and Kongarra are similar, 1600-1650 Ma, while the U-Pb ages range from 900 to 1437 Ma. In the Cluff D deposit (Carswell structure, Saskatchewan, Canada), the mean K-Ar age on illite is significantly older, 1293 ± 36 Ma, than U-Pb ages obtained on uraninite, 1150 ± 25 Ma, but the reverse is also known. In the Müllenbach U deposit (FRG), the upper intercept in the Concordia diagram yields an age of 239 ± 6 Ma (Holliger et al, 1989) while the K-Ar age on illite is 170 Ma (Brockamp et al., 1987).

Table 3: Systematics between the U-Xe-Kr-Pb system (from Eikenberg et al., 1989)

Case	U/Xe - U/Kr	U-Pb	Significance
1	concordant	concordant	crystallization age
2	concordant	discordant	recrystallization age or mixing age
3	discordant	discordant	loss of daughter products
4	discordant	concordant	not yet observed

These discrepancies between the different chronometers indicate that it is not sufficient to use only one chronometer. More confidence in the results and interpretations can be obtained by using several different systems. Certain trends during remobilization can be defined: (1) U minerals may or may not contain radiogenic Nd and Pb. This will depend on the distance of migration or on geochemical factors, (2) there could be a complete resetting of the U-Pb system without disturbing the K-Ar system in the gangue minerals, especially illite, and (3) a thermal event could reset the K-Ar system without resetting the U-Pb chronometer and the U-Xe and U-Kr systems. For persons who are not familiar with isotopic geochemistry, it should be underlained that it is important to differentiate the given confidence interval on the age, which integrates the analytical error and the natural variation between the analysed samples, with the "geological" error because it is not straitforward to know what is the geological significance of this age.

Table 4: Ages obtained on the Jabulika, Nabarlek (Australia) and Cluff D (Canada) ore deposits

Deposit	Dating method	Age	Reference
Jabiluka	U-Pb on pitchblende, upper intercept	920 Ma	Hills and Richards, 1976
(Australia)	U-Pb on pitchblende, upper intercept	1280±100 Ma	Gulson and Mizon,1980
	U-Pb on ores, upper intercept	1437±40 Ma	Ludwig et al, 1987
	Pb isotopes on barren sulfides, out of the ore	1380±110 Ma	Gulson and Mizon, 1980
	Pb isotopes on barren sulfides in ores	1600±150 Ma	Gulson and Mizon, 1980
	Rb-Sr model ages on hydrothermally altered chlorite schists	1600±80 Ma	Page et al, 1980
	Five resetted samples	900-1260 Ma	Page et al, 1980
	Sm-Nd on ores and uraninite	1614±132 Ma	Maas, 1989
Nabarlek	U-Pb on pitchblende, upper intercept	≈900 Ma	Hills and Richards, 1976
(Australia)	K-Ar ages on sericite	1580-1670 Ma	Page et al, 1980
	Mean Sr ages on sericite	1610±40 Ma	Page et al, 1980
	Sm-Nd on ores and uraninite	1616±50 Ma	Maas, 1989
Cluff D	U-Pb on pitchblende, upper intercept	1150±25 Ma	Bellon et al, 1976
	K-Ar ages on illite	1293±36 Ma	Clauer et al, 1985

Uranium Deposits and Geodynamic Processes

All age data obtained on sediment-hosted U deposits in Europe have shown that there are optimum periods for U mineralization (Fig. 10), and that mineralization can also occur in the basement rocks at the same time. This observation was first published by Bonhomme (1982) and subsequently detailed by many workers (Lancelot et al., 1984; Bouladon and de Graciansky, 1985; Philippe et al., 1987; Lévêque et al., 1988; Holliger et al., 1989; Cathelineau et al., 1990; Pagel, 1990). Four major periods (240 Ma, 170-180 Ma, 100-110 Ma, 20-40 Ma) have been identified in sediment-hosted U deposits which, except for the 240 Ma stage, coincide with remobilization events for U deposits in the basement rocks. These latter U deposits formed mainly during a Permo-Stephanian (290-260 Ma) stage which is associated with brittle deformation related to major shear zone movements (Cathelineau et al., 1990). A large amount of U reserves are located in Permian formations. However, it should be noted that no U deposits located in Permian sedimentary rocks are known to be of Permian age.

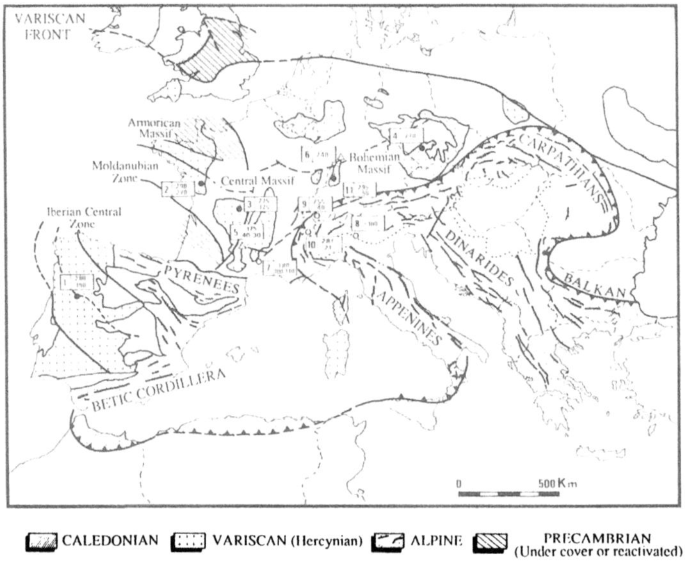

Figure 10: U-Pb and Pb-Pb ages of U deposits in Europe. 1: Urgeiriça, Bica and Carrasca, Portugal (Holliger and Pagel, unpublished); 2: Chardon, Escarpière and Commanderie, France (Holliger and Cathelineau, 1986); 3: Margnac, France (Leroy and Holliger, 1984); 4: Pribam, Czechoslovakia (Kolektiv, 1984); 5: Bertholène, France (Lévêque et al., 1988); 6: Müllenbach, Germany (Holliger et al., 1989); 7: Lodève, France (Lancelot et al., 1984; Lancelot and Vella, 1988); 8: Novazza, Italy (Philippe et al., 1987); 9: Valais, Switzerland (Eikenberg et al., 1989); 10: La Lauzière, France (Negga et al., 1986): 11: Krunkelbach, Germany (Hofmann and Eikenberg, 1991)

A late Permian-early Triassic age has been determined by U-Pb for the Müllenbach U deposit (Germany) (Holliger et al., 1989) which is located in Upper Carboniferous sediments underlying sediments of Permian age. Fluid circulation along a vertical redox front (oxidized Permian-reduced Carboniferous) is responsible for the deposition of Zr-bearing pitchblende and associated Cu, V, Se, Ag, Co, Ni, Mo minerals (Pironon, 1986; Pagel and Pironon, 1986).

Jurassic ages have been determined for the Lodève deposits located in Permian formations (Lancelot and Vella, 1989) and the Bertholène deposit (Lévêque et al., 1988) which is spatially related to an unconformity. The latter is located in an albitized granite underlying the sedimentary cover. Their formation is related to brine circulation during tectonic movements contemporaneous with the opening of the Tethys sea. The Cretaceous age obtained for some Lodève deposits and the Novazza deposit in Italy seems to be a concentrating stage of preexisting U mineralization (Lancelot et al., 1984; Philippe et al., 1987).

The alpine mineralization corresponds to remobilization of pre-existing U mineralization as illustrated by the coffinite in the Bertholène deposit (Lévêque et al., 1988) or small pitchblende spherules admixed to the Permian pichblende in the Swiss Valais (Eikenberg et al., 1989). Uranium deposits are known in Oligocene sediments, as exemplified by the Coutras deposit (Gironde, France) (Meunier et al., 1990) and the Saint Pierre du Cantal deposit, France, although their ages are not known.

RECONSTRUCTION OF THE THERMAL HISTORY IN MINERALIZED AREAS

The reconstruction of the thermal history in sedimentary basins is often based on organic matter indices, kinetic models, fluid inclusions or clay-crystal chemistry. Pertinent data can also be deduced from fission track annealing in apatite between 70°C and 160°C (Meyer et al., 1989) or from stable isotope geochemistry using equilibrium of coexisting minerals.

The basis of fission track analysis is that, depending on time and temperature, there is an annealing of ^{238}U spontaneous fission tracks. For a degree of annealing, it has been experimentally shown that, for a degree of length reduction, the data could plot on an Arrhenius line and that the activation energy varies, the shortest tracks are more resistant to annealing than the longest ones. Various problems have not been totally overcome to provide an entirely confident application of fission tracks (chemical composition of the mineral, leaching conditions, rate of dissolution, observation and counting), but this field is extremely promising and, in my opinion, strong efforts must be devoted to this technique. For the study of sedimentary formations, the range of annealing temperatures for apatite is well adapted, whereas for minerals such as sphene or zircon the annealing temperature is higher and is more probably adapted to anchizonal conditions. For example, Koul et al. (1984) have determined a 1420 Ma fission track age on zircon and deduced that the temperature affecting the Kombolgie Formation was at least 175°C. Meyer et al. (1989) presented a study combining fission track, fluid inclusion and organic matter on breccia pipes mineralized in U from Arizona which tend towards 120-140°C.

In U deposits or in their environment, two cogenetic minerals could be found. This allows the possibility of determining a temperature from the stable isotope data. However, uncertainties in the isotopic fractionation factors are quite large for minerals generally occurring in sedimentary environments (illite, chlorite, kaolinite, smectite). Furthermore, the mineral pair of interest must have crystallized in equilibrium with each other which is not always the case in low temperature environments. Also they must have preserved their isotopic composition during subsequent alteration events. Studying the Akouta U deposit, Forbes (1989) has found that the $\delta^{34}S$ of galena, sphalerite and pyrite have the same values. This is interpreted as resulting from the precipitation of these minerals from a nearly homogeneous fluid without isotopic equilibrium among the minerals at temperatures between 85 and 175°C (from fluid inclusion data) for this deposit.

SPECIFIC BEHAVIOR OF RADIOGENIC AND STABLE ISOTOPES IN AND AROUND URANIUM DEPOSITS

Natural Fission Reactions

Bodu et al. (1972) have identified anomalous isotopic compositions of U in mineralized samples from Oklo (Gabon), and since, concentration of ^{235}U as low as 0.2% have been found in this area. These anomalies have been explained by natural fission reactions, a process which was confirmed by anomalous isotopic compositions of REE and the presence of isotopes related to fission reactions. Gancarz (1978) has dated the primary U mineralization at Oklo at 2.05 ± 0.03 Ga (upper intercept in the Concordia diagram). This age is comparable, within the limits of uncertainty, to the 1.97 ± 0.06 Ga age determined using fission products (Ruffenach,1978). Hagemann et al. (1975) determined that the fission reactions have lasted about 600,000 to 800,000 years based on observations of Pu fission systematics and the ^{100}Ru produced by capture from the fission products, ^{99}Tc and ^{99}Ru. It must be stressed that, if all the Oklo zones are considered, the duration is much longer.

The main factors which are necessary for the occurrences of natural fission reactions are: (a) high U content, (b) high concentration of ^{235}U in the U ore at the time required for the process to occur (in the case of Oklo, this concentration was, at 2 Ga, higher than 3% of the total U), (c) presence of water for moderation of the reactions, and (d) absence or scarcity of elements with high neutron capture, such as B or REE.

From O isotope data obtained on chlorite and illite in and around the reactor zones, Gauthier-Lafaye et al. (1989) deduced that there was an increase of temperature from 175°C far from the reactor to 300-350°C close to the reactor and then a sharp decrease in the reactor to about 125-150°C. They interpret the latter as resulting from the recrystallization of the phyllites during the cooling of the reactor.

Theoretical considerations have shown that such natural fission reactions could have occurred up to 1 Ga (Naudet, 1978) under highly favorable conditions, for example during the formation of the unconformity type U deposits in Saskatchewan (Canada) and Northern Territory (Australia). But search for other natural nuclear fission zones has been unsuccessful. Maas and McCulloch (1990) have shown that there are relatively large Gd and Sm isotopic variations in high-grade ores and uraninites from the

Alligator River Uranium Field. However they conclude that the slow neutron-induced fission of ^{235}U was not adequate for the U ore to become critical. Perhaps this could be due to the high B and/or HREE (high neutron capture cross-sections) contents in these unconformity-type uranium deposits. Some slight $^{235}U/^{238}U$ anomalies have been reported on ore from the Colorado Plateau (Cowan and Adler, 1976).

Behavior of Fission Products in Uranium Ores

A lot of research has been devoted to the Oklo U deposit, because it is a long-time-scale for man-made radioactive waste repositories (Loss et al., 1989). It was recognized that several fission products have not been quantitatively retained in the reactors and that some fission-products, such as Mo, Ru, Pd, Ag and Te, have been retained in the peripheral rocks while others, such as Cd, have been totally removed (Curtis et al., 1989; Loss et al., 1989).

The development of accelerator mass spectrometry, which permits the analysis of very small quantities of fission products, has opened up a new field of research which can be applied especially to the study of weathering rates and environmental subjects. Furthermore, these fission products are excellent natural analogue for high-level radioactive waste. ^{129}I is produced in U ores by the spontaneous fission of ^{238}U and the neutron-induced fission of ^{235}U. Fabryka-Martin et al. (1988) have determined $^{129}I/U$ in the Koongarra ore (Northern Territory, Australia) and have found that ^{129}I is strongly leached from the ore and that the oxic environment around Kongarra is enriched in ^{129}I. The leakage of ^{129}I from U oxide is the general rule (Boaretto et al., 1990). However, in the case of Yeelirrie (Australia), the present value of the $^{129}I/^{127}I$ ratio could be consistent with an age > 1 Ma despite disequilibrium in the ^{238}U series.

^{36}Cl is produced by neutron capture by ^{35}Cl (Kuroda et al., 1957), though $^{36}Cl/^{35}Cl$ may monitor the neutron flux in U ores.

Fabryka-Martin et al. (1988) conclude that "in the weathered primary ore body, these flux estimates exceed those predicted for the present day U content of the ore by factors of 2-3, suggesting that the rate of leaching of U in this part of the deposit is rapid relative to the half-life of ^{36}Cl (3.01×10^5 yr)".

Radiation-Related Retrograde Isotope Exchange in Gangue Minerals

As shown previously, stable isotope studies are widely used to characterize the origin of fluids during ore-forming processes and to determine temperatures of isotopic equilibrium. It has been suggested that radiation could induce retrograde isotope exchange (Leventhal and Threlkeld, 1978 ; Dahl et al., 1988) or could favor isotopic exchange (Halter et al., 1987).

Leventhal and Threlkeld's (1978) study of organic matter from the Grants Uranium District, New Mexico and Dahl et al.'s. (1988) study of the Alum shale kerogen demonstrated that the $^{13}C/^{12}C$ ratio increases when the U concentration increases. In the Grants, Leventhal and Threlkeld (1978) show that the organic C associated with high grade U ore is heavier ($\delta^{13}C$= -16.9 to - 19.6 per mil) than in the adjacent lower grade samples ($\delta^{13}C$= -22.7 and -26.4 per mil). The Alum shale results are more surprising because the U content remains under 200 µg/g. The $\delta^{13}C$ varies from -33.5 to -27.4 per mil. This is explained by the generation of isotopically light hydrocarbon gases by irradiation of the organic matter. Some other chemical changes in the kerogen support this hypothesis. However, the decrease of Tmax with increasing U concentration remains unexplained. The variation of $\delta^{13}C$ with sample age is also puzzling.

In the unconformity-type deposits in Saskatchewan, low δD values have been found chiefly in illite and to a lesser degree in chlorite in the Key Lake, Cluff D and McClean deposits (Wilson et al., 1987 ; Halter et al., 1987 ; Bray et al., 1988). Values as low as -168 per mil have been found while normal values for barren samples are at ~ -50 per mil. Wilson et al. (1987) have proposed that these phyllites exchanged with recent cold meteoric waters.

Halter et al. (1987) obtained data which support this hypothesis, but they proposed that the radiation has facilitated the isotopic exchange and K-Ar exchange of the clays with post-Cretaceous meteoric waters. Halter et al. (1987) and Bray et al. (1988) have found that the δD decreases when the U content increases.

Mineralized bituminens present in U deposits could have low $\delta^{13}C$ values: -43 to -49 per mil at Cluff Lake, Saskatchewan, Canada (Landais and Dereppe, 1985; Leventhal et al., 1987), and -40.8 to -56.5 per mil at Pribram, Czechoslovaquia (Kribek, 1989). According to Leventhal et al. (1987), radiation form isotopically light CH_4. This radiation-induced methane is degraded to solid C which recombine with the solid C and H_2 escape. However, Landais and Dereppe (1985) suggested that these bituminens could be formed by the condensation of light and complex hydrocarbons enriched in ^{12}C isotopes.

CONCLUSION

The study of U deposits in surficial and sedimentary environments has recently opened up a wide field of isotopic research. Isotopic systematics are capable of placing strong constraints on the origin of the U ores. However, the quantification of the chemical processes involved in the deposition of U-ores often requires the use of multiple approaches and techniques involving, for example, fluid inclusions, crystal chemistry, organic matter... However, it seems to me that the lack of data on gangue minerals in ores is a common shortfall of studies. In addition more data on isotopic Sr, Nd, O, H, C are needed to adequately characterize the chemical evolution of the mineralizing fluids. It is hoped that in the near future new technological developments such as the nuclear microprobe, the ion microprobe or the ICP-MS will allow *in-situ* isotopic analyses. More research is needed on fluid inclusions following Changkakoti et al. (1988) who have shown that "radiogenic isotopes in fluid inclusion waters can constitute useful geochronological and genetic tools when applied in conjunction with geological, geochemical and other geochronological studies". However, in U deposits, minerals suitable for fluid inclusion study and isotopic study are rare, small and consequently difficult to use as generally isotope analyses require the collection of the fluid by crushing or heating. This could be overcome in the future by punctual analyses and by using minerals associated with U.

Specific problems related to the presence of high U content have been documented. These include enhancement of H isotopic exchange in phyllites, increase of heavier C content in organic matter and variation in the isotopic composition of REE, especially Nd. This research merits further investigation to determine the confidence limits of stable and radiogenic isotopes in U metallogeny. Unfortunately, due to the actual low price of U, research on U deposits has slowed down. This tendency should be reversed in the near future because U deposits are good natural analogs to waste disposal.

ACKNOWLEDGEMENTS

I wish to thank Norbert Clauer for his encouragements to write this paper. I am indebted to many isotope geochemists and geochronologists with whom I have had the opportunity to work : Norbert Clauer, Philippe Holliger, Martine Juteau, Joël Lancelot, Marie Hélène Lévêque, Annie Michard, Sylvie Philippe, Simon Sheppard, Nelly Toulhoat and Laurent Turpin. I warmfully thank the following, Norbert Clauer, Roland Maas, Simon Sheppard, George Sullivan and Laurent Turpin for their suggestions to improve a preliminary version of this contribution.

REFERENCES

Airey P.L. and Roman D. (1981) Uranium series disequilibria in the sedimentary uranium deposits at Yeeliree, Western Australia. Jour. Geol. Soc. Australia, 28, 357-363.

Altschuler Z.S., Clark R.S.Jr. and Young E.J. (1958) Geochemistry of uranium in apatite and phosphorite. U. S. Geol. Surv., Prof. Pap., 314, 49-90.

Bodu R., Bouzigues H., Morin N. and Pfiffelmann J.P. (1972) Sur l'existence d'anomalies isotopiques rencontrées dans l'uranium du Gabon. C.R. Acad. Sci. Paris, 275, 1731.

Bell K. (1985) Geochronology of the Carswell area, northern Saskatchewan. In: The Carswell Structure Uranium Deposits,

Saskatchewan, Lainé R., Alonso D. and Svab M. (eds.), Geol. Assoc. of Canada, Spec. Pap. 29, 33-46.
Bellon H., Devillers C., Hagemann R. and Touray J.C. (1976) Dater les minéralisations. Mém. h. sér. Soc. Géol. France, 7, 265-268.
Boaretto E., Berkovits D., Hollos G. and Paul M. (1990) Measurements of natural concentrations of 129I in uranium ores by accelerator mass spectrometry. Nucl. Instrum. and Methods in Phys. Res., B 50, 280-285.
Bonhomme M. (1982) Age triasique et jurassique des argiles associées aux minéralisations filoniennes et de phénomènes diagénétiques tardifs en Europe de l'Ouest. Contexte géodynamique et implications génétiques. C. R. Acad. Sci. Paris, 294, 521-524.
Bonhomme M.G., Baubron J.C. and Jebrak M. (1987) Minéralogie, géochimie, terres rares et âge K-Ar des argiles associées aux minéralisations filoniennes. Chem. Geol. (Isot. Geosci. Sect.), 65, 321-339.
Bouladon J. and de Graciansky P.C. (1985) Les minéralisations dites de couverture (plomb, zinc, cuivre, uranium, barytine, fluorine) du Trias au Pliocène, en France.Leurs relations avec les phénomènes connexes de l'ouverture de la Téthys et de l'Atlantique Nord. Chron. Rech. Min., 480, 17-33.
Bray C.J., Spooner E.T.C., Hall C.M., York D., Bills T.M. and Krueger H.W. (1987) Laser probe $^{40}Ar/^{39}Ar$ and conventional K/Ar dating of illites associated with the Mc Clean unconformity related uranium deposits, N. Saskatchewan, Canada. Can. Jour. Earth Sci., 24, 10-23.
Bray C.J., Spooner E.T.C. and Longstaffe F.J. (1988) Unconformity-related uranium mineralization, McClean deposits, north Saskatchewan, Canada: hydrogen and oxygen isotope geochemistry. Canad. Miner., 26, 249-268.
Brockamp O., Zuther M. and Clauer N. (1987) Epigenetic-hydrothermal origin of the sediment-hosted Müllenbach uranium deposit, Baden-Baden, W-Germany. Monogr. Ser. on Miner. Depos., 27, Borntraeger, Berlin, 87-98.
Brookins D.G. (1980) Geochronologic studies in the Grants mineral belt. New Mexico. Bur. Mines Miner. Res., Mem. 38, 52-58.
Cathelineau M., Boiron M.C., Holliger Ph and Poty B. (1990) Metallogenesis of the French part of the Variscan orogen. Part II: Time-space relationship between U, Au and Sn-W ore deposition and geodynamic events - mineralogical and U-Pb data. Tectonophysics, 177, 59-79.
Changkakoti A., Gray J., Krstic D., Cumming G.L. and Morton R.D. (1988) Determination of radiogenic isotopes (Rb/Sr, Sm/Nd and Pb/Pb) in fluid inclusion waters: An example from the Bluebell Pb-Zn deposit, British Columbia, Canada. Geochim. Cosmochim. Acta, 52, 961-967.
Cherdyntsev V.V. (1971) Uranium 234. Israël Program for Scientific Translations, Jerusalem, 244 p.
Clauer N., Ey F. and Gauthier-Lafaye F. (1985) K-Ar dating of different rock types from the Cluff Lake uranium ore deposits (Saskatchewan, Canada). In: The Carswell Structure Uranium Deposits, Saskatchewan, Lainé R., Alonso D. and Svab M. (eds.), Geol. Ass. of Canada, Spec. Pap. 29, 47-53.
Cowan G.A. and Adler H.H. (1976) The variability of the natural abundance of ^{235}U. Geochim. Cosmochim. Acta, 40, 1487-1490.

Cunningham C.G., Ludwig K.R., Naeser C.W., Weiland E.K., Mehnert H.H., Steven T.A. and Rasmussen J.D. (1982) Geochronology of hydrothermal uranium deposits and associated igneous rocks in the eastern source area of the Mount Belknap Volcanics, Marysvale, Utah. Econ. Geol., 77, 453-463.

Curtis D., Benjamin T., Gancarz A., Loss R., Rosman K., De Laeter J., Delmore J.E. and Maeck W.J. (1989) Fission product retention in the Oklo natural fission reactors. Applied Geochem., 4, 49-62.

Dahl J., Hallberg R. and Kaplan I.R. (1988) The effects of radioactive decay of uranium on elemental and isotopic ratios of Alum Shale kerogen. Applied Geochem., 3, 583-589.

Eikenberg J., Köppel V., Labhart T. and Signer P. (1989) U-Pb, U-Xe and U-Kr systematics of a greenschist facies metamorphic uranium mineralization of the Siviez-Mischabel nappe (Valais, Switzerland). Schweiz. Mineral. Petrogr. Mitt., 69, 331-344.

Fabryka-Martin J., Davis S.N., Roman D., Airey P.L., Elmore D. and Kubik P.W. (1988) Iodine-129 and Chlorine-36 in uranium ores. 2 - Discussion of AMS measurements. Chem. Geol. (Isot. Geosci. Sect.), 72, 7-16.

Fishman N.S., Reynolds R.L. and Robertson J.F. (1985) Uranium mineralization in the Smith Lake District of the Grants Uranium Region, New Mexico. Econ. Geol., 80, 1348-1364.

Forbes P., Pacquet A., Chantret F., Oumarou J. and Pagel M. (1984) Marqueurs du volcanisme dans le gisement d'uranium d'Akouta (République du Niger). C. R. Acad. Sc. Paris, 298, 647-650.

Forbes P. (1989) Rôle des structures sédimentaires et tectoniques, du volcanisme alcalin régional et des fluides diagénétiques-hydrothermaux pour la formation des minéralisations à U-Zr-Zn-V-Mo d'Akouta (Niger). Géol. Géochim. Uranium, Mém. Nancy, 17, 376p.

Franklyn M.T., McNutt R.H., Kamineni D.C., Gascoyne M. and Frape S.K. (1991) Groundwater $^{87}Sr/^{86}Sr$ values in the Eye-Dashwa Lakes pluton, Canada : Evidence for plagioclase-water reaction Chem. Geol. (Isot. Geosci. Sect.), 86, 111-122.

Fryer B.J. and Taylor R.P. (1984) Sm-Nd direct dating of the Collins Bay hydrothermal uranium deposit, Saskatchewan. Geology, 12, 479-482.

Fryer B.J. and Taylor R.P. (1987) Rare-earth element distribution in uraninites : implications for ore genesis. Chem. Geol., 63, 101-108.

Galliker D., Hugentobler E. and Hahn B. (1970) Spontane Kernspaltung von ^{238}U und ^{241}Am. Helv. Phys. Acta, 43, 593-606.

Gangarz A.J. (1978) - U-Pb age (2.05 x 10^9 years) of the Oklo uranium deposit. In: Natural fission reactors, IAEA, Vienna, 513-520.

Gauthier-Lafaye F., Weber F. and Ohmoto H. (1989) Natural fission reactors of Oklo. Econ. Geol., 84, 2286-2295.

Giletti B.J. and Kulp J.L. (1955) Radon leakage from radioactive minerals. Am. Min., 40, 481-496.

Goldhaber M.B., Reynolds R.L. and Rye R.O. (1978) Origin of a south Texas roll-type uranium deposit: II. Sulfide petrology and sulfur isotope studies. Econ. Geol., 73, 1690-1705.

Goldhaber M.B., Reynolds R.L. and Rye R.O. (1983) Role of fluid mixing and fault-related sulfide in the origin of the Ray Point uranium district, south Texas. Econ. Geol., 78, 1043-1063.

Gulson B.L. and Mizon K.J. (1980) Lead isotope studies at Jabiluka. In: Uranium in the Pine Creek geosyncline, Ferguson J. and Goleby A.B. (eds.), Vienna, IAEA, 439-456.

Hagemann R., Devillers C., Lucas M., Lecomte T. and Ruffenach J.C. (1975) Estimation de la durée de la réaction - limitations imposées par les données neutroniques. Proc. Int. Symp. on the Oklo Phenomenon, Libreville, IAEA, Vienna, SM-204/28, 415-423.

Halter G., Sheppard S.M.F., Pagel M. and Weber F. (1985) Lithogeochemistry of the alterations correlated with unconformity uranium deposits in the Carswell structure (Saskatchewan, Canada). Terra cognita, 5, 2-3, 153.

Halter G., Sheppard S.M.F., Weber F., Clauer N. and Pagel M. (1987) Radiation-related retrograde hydrogen isotope and K-Ar exchange in clay minerals. Nature, 330, 638-641.

Hattori K. and Halas S. (1982) Calculation of oxygen isotope fractionation between uranium dioxide, uranium trioxide and water. Geochim. Cosmochim. Acta, 46, 1863-1868.

Hattori K., Muehlenbachs K. and Morton R.D. (1978) Oxygen isotope geochemistry of uraninites. Joint Mtg. Geol. Soc. Am. and Geol. Ass. Can., Abstr. with Progr., 10, 417.

Hills J.H. and Richards J.R. (1976) Pitchblendes and galena ages in the Alligator Rivers region, Northern Territory, Australia. Miner. Depos., 11, 133-154.

Hoekstra H.R. and Katz J.J. (1956) The isotope geology of some uranium minerals. In Proc. Intern. Conf. on Peaceful Uses of Atomic Energy, Geneva, 1955, 6, 547-550.

Hofmann B. and Eikenberg J. (1991) The Krunkelbach uranium deposit, Schwarzwald, Germany: correlation of radiometric ages (U-Pb, U-Xe-Kr, K-Ar, ^{230}Th-^{234}U) with mineralogical stages and fluid inclusions. Econ. Geol., 86, 1031-1049.

Holliger Ph. (1988) Ages U-Pb définis in-situ sur oxydes d'uranium à l'analyseur ionique: méthodologie et conséquences géochimiques. C. R. Acad. Sci. Paris, 307, 367-373.

Holliger Ph. and Cathelineau M. (1986) Le chronomètre U-Pb en milieu uranifère: application aux gisements hydrothermaux d'uranium liés spatialement au batholite de Mortagne (Vendée, France). Chron. Rech. Min., 485, 33-43.

Holliger Ph., Pagel M. and Pironon J. (1989) A model for ^{238}U radioactive daughter loss from sediment-hosted pitchblende deposits and the late Permian-Early Triassic depositional U-Pb age of the Müllenbach uranium ore (Baden-Württemberg, F.R.G.). Chem. Geol. (Isot. Geosci. Sect.), 80, 45-53.

Kaufman A. and Ku T.L. (1989) The U-series ages of carnotites and implications regarding their formation. Geochim. Cosmochim. Acta, 53, 2675-2681.

Kolektiv (1984) Ceskoslovenska loziska uranu. Ceskoslovensky uranovy prumysl, Praha, 365 p.

Koul S.L., Wilde A.R. and Tickoo A.K. (1988) A thermal history of the Proterozoic East Alligator River terrain, Northern Territory, Australia: A fission track study. Tectonophysics, 145, 101-111.

Kribek B. (1989) The role of organic matter in the metallogeny of the Bohemian Massif. Econ. Geol., 84, 1525-1540.

Kuroda P.K., Edwards R.R., Robinson B.L., Jonte J.H. and Goolsby C. (1957) Chlorine-36 in pitchblende. Geochim. Cosmochim. Acta, 11, 194-199.

Lancelot J.R., De St André B. and De La Boisse H. (1984) Systématique U/Pb et évolution du gisement d'uranium de Lodève (France). Miner. Depos., 19, 44-53.

Lancelot J. and Vella V. (1989) Datation U-Pb liasique de la pechblende de Rabejac. Mise en évidence d'une préconcentration uranifère permienne dans le bassin de Lodève (Hérault). Bull. Soc. géol. France, 8, V, 309-315.

Landais P. and Dereppe J.M. (1985) A chemical study of the carbonaceous material from the Carswell structure. In: The Carswell structure uranium deposits, Saskatchewan, Lainé R., Alonso D. and Svab M. (eds.), Geol. Ass. of Canada, Spec. Pap., 29, 165-174.

Lee M.J. and Brookins D.G. (1978) Rubidium-strontium minimum ages of sedimentation, uranium mineralization and provenance, Morrison Formation (Upper Jurassic), Grants mineral belt, New Mexico. Am. Ass. Petrol. Geol. Bull., 62, 9, 1673-1683.

Leroy J. and Holliger Ph. (1984) Mineralogical, chemical and isotopic (U-Pb method) studies of Hercynian uraniferous mineralizations (Margnac and Fanay mines, Limousin, France). Chem. Geol., 45, 121-134.

Leventhal J.S., Grauch R.I., Threlkeld C.N., Lichte F.E. and Harper C.T. (1987) Unusual organic matter associated with uranium from the Claude deposit, Cluff Lake, Canada. Econ. Geol., 82, 1169-1176.

Leventhal J.S. and Threlkeld C.N. (1978) Carbon-13/carbon-12 isotope fractionation of organic matter associated with uranium ores induced by alpha irradiation. Science, 202, 430-432.

Lévêque M.H., Lancelot J.R. and George E. (1988) The Bertholène uranium deposit-Mineralogical characteristics and U-Pb dating of the primary U mineralization and its subsequent remobilization : consequences upon the evolution of the U deposits of the Massif Central, France. Chem. Geol., 69, 147-163.

Lévêque M.H. and Meshick A.P. (1991) The Xe_s-Xe_n spectrum technique applied to French uranium deposits and showings. In: Source, Transport and Deposition of Metals, Pagel M. and Leroy J.L.(eds), Balkema, 401-404.

Lively R.S., Harmon R.S., Levinson A.A. and Bland G.J. (1979) Disequilibrium in the 238 uranium series in samples from Yeelirree, Western Australia. Jour. Geochem. Expl., 12, 57-65.

Loss R.D., Rosman K.J.R., De Laeter J.R., Curtis D.B., Benjamin T.M., Gancarz A.J., Maeck W.J. and Delmore J.E. (1989) Fission-product retentivity in peripheral rocks at the Oklo natural fission reactors, Gabon. Chem. Geol., 76, 71-84.

Luck J.M. and Allègre C.J. (1982) The study of molybdenites through the 187Re-187Os chronometer. Earth Plan. Sci. Lett., 61, 291-296.

Ludwig K.R. (1978) Uranium-daughter migration and U/Pb isotope apparent ages of uranium ores, Shirley basin, Wyoming. Econ. Geol., 73, 29-49.

Ludwig K.R. (1979) Age of uranium mineralization in the Gas Hills and Crooks Gap districts, Wyoming as indicated by U-Pb isotope apparent ages. Econ. Geol., 74, 1654-1668.

Ludwig K.R., Grauch R.I., Nutt C.J., Nash J.T., Frishman D. and Simmons K.R. (1987) Age of uranium mineralization at Jabiluka and Ranger deposits, Northern Territory, Australia : new U-Pb isotope evidence. Econ. Geol., v. 82, 857-874.

Ludwig K.R., Nash J.T. and Naeser C.W. (1981) U-Pb isotope systematics and age of uranium mineralization, Midnite Mine, Washington. Econ. Geol., 76, 89-110.
Ludwig K.R., Simmons K.R. and Webster J.D. (1984) U-Pb isotope systematics and apparent ages of uranium ores, Ambrosia Lake and Smith Lake districts, Grants Mineral belt, New Mexico. Econ. Geol., 79, 322-337.
Ludwig K.R., Wallace A.R. and Simmons K.R. (1985) The Schwartzwalder uranium deposit, II: Age of uranium mineralization and lead isotope constraints on genesis. Econ. Geol., 80, 1858-1871.
Maas R. (1989) Nd-Sr isotope constraints on the age and origin of unconformity-type uranium deposits in the Alligator Rivers Uranium Field, Northern Territory, Australia, Econ. Geol., 84, 64-90.
Maas R. and McCulloch M.T. (1990) A search for fossil nuclear reactors in the Alligator River Uranium Field, Australia : Constraints from Sm, Gd and Nd isotopic studies. Chem. Geol., 88, 301-315.
Maeck W. J., Apt K.E. and Cowan G.A. (1978) A possible uranium-ruthenium method for the measurement of ore age. In: Natural Fission Reactors, IAEA, Vienna, 535-540.
Meunier J.D., Trouiller A., Bruhlet J. and Pagel M. (1989) Uranium and organic matter in a paleodeltaic environment: the Coutras deposit (Gironde, France). Econ. Geol., 84, 1541-1556.
Meyer A.J., Landais P., Brosse E., Pagel M., Carisey J.C. and Krewdl D. (1989) Thermal history of the Permian formations from the Breccia Pipes area (Grand Canyon region, Arizona). Geol. Rundsch. 78, 427-438.
Mullis J. and Stalder H.A. (1987) Salt-poor and salt-rich fluid inclusions in quartz from two boreholes in northern Switzerland. Chem. Geol., 61, 263-272.
Nakai S., Shimizu H. and Masuda A. (1986) A new geochronometer using lanthanum-138. Nature, 320, 433-435.
Nash J.T., Granger H.C. and Adams S.S. (1981) Geology and concepts of genesis of important types of uranium deposits. Econ. Geol., 75th Anniv. Vol., 63-116.
Naudet R. (1978) Etude paramétrique de la criticité des réacteurs naturels. In: Les réacteurs de fission naturels, AIEA, Vienne, 589-599.
Negga H.S., Sheppard S.M.F., Rosenbaum J.M. and Cuney M. (1986) Late Hercynian U-vein mineralization in the Alps: fluid inclusion and C, O, H isotopic evidence for mixing between two externally derived fluids. Contrib. Mineral. Petrol., 93, 179-186.
Ohmoto H. (1986) Stable isotope geochemistry of ore deposits. In: Stable isotopes in high temperatures geologic processes, Valley J.W., Taylor H.P.Jr, O'Neil J.R. (eds.), Min. Soc. Am. Rev. Min., 16, 491-559.
Page R.W., Compston W. and Needham R.S. (1980) Geochronology and evolution of the late-Archean basement and Proterozoic rocks in the Alligator Rivers uranium field, Northern Territory, Australia. In: Uranium in the Pine Creek geosyncline, Ferguson J. and Goleby A.B. (eds.), IAEA, Vienna, 39-68.
Pagel M. (1990) Le Permien et la métallogénie de l'uranium. Chron. Rech. Min., 499, 57-68.
Pagel M., Michard A., Juteau M. and Turpin L. Sm-Nd, Pb-Pb and Rb-Sr isotopic constraints on the evolution of the basement in

the Cigar Lake area (Saskatchewan - Canada). Can. Jour. Earth Sci. (submitted).
Pagel M., Pinte G. and Rotach-Toulhoat N. (1987) The rare earth elements in natural uranium oxides. In: Monogr. Ser. on Miner. Depos., 27, 81-85.
Pagel M. and Pironon J. (1986) Un modèle de formation de gisements d'uranium dans les shales noirs continentaux. Sci. Géol., Bull., 39, 277-292.
Pagel M., Poty B. and Sheppard S.M.F. (1980) Contribution to some Saskatchewan uranium deposits mainly from fluid inclusions and isotopic data. In: Uranium in the Pine Creek geosyncline, Ferguson J. and Goleby A. (eds.), IAEA, Vienna, 639-654.
Philippe S., Villemaire C., Lancelot J.R., Girod M. and Mercadier H. (1987) Données minéralogiques et isotopiques sur deux gîtes hydrothermaux uranifères du bassin volcano-sédimentaire permien de Collio Orobico (Alpes Bergamasques): mise en évidence d'une phase de remobilisation crétacée. Bull. Min., 110, 283-303.
Pironon J. (1986) Zonalités géochimiques et minéralogiques dans les bassins continentaux uranifères. Exemples de St Hippolyte (massif Vosgien), Müllenbach (Forêt Noire, RFA), Salamanière (Massif central français). Géol. Géochim. Uranium. Mém., Nancy, 13, 296p.
Ravizza G. and Turekian K.K. (1989) Application of the ^{187}Re-^{187}Os system to black shale geochronometry. Geochim. Cosmochim. Acta, 53, 3257-3262.
Roberts J.H., Gold R. and Armani R.J. (1968) Spontaneous fission decay constant of 238U. Phys. Rev., 174, 1482-1484.
Robinson B.W. and Ohmoto H. (1973) Mineralogy, fluid inclusions and stable isotopes of the Echo Bay U-Ni-Ag-Cu deposits, Northwest Territories, Canada. Econ. Geol., 68, 635-656.
Rosholt J.N. and Bartel A.J. (1969) Uranium, thorium and lead systematics in the Granite Mountains, Wyoming. Earth Planet. Sci. Lett., 7, 141-147.
Ruffenach J.C. (1978) Les réacteurs d'Oklo: Etude des migrations de l'uranium et des terres rares sur une carotte de sondage et application à la détermination de la date des réactions nucléaires. In: Natural fission reactors, IAEA, Vienna, 441-472.
Sheppard S.M.F. (1986) Characterization and isotopic variations in natural waters. In: Stable Isotopes in High Temperature Geological Processes, Valley J.W., Taylor H.P.Jr. and O'Neil J.R. (eds.), Min. Soc. Am. Rev. Miner., 16, 165-184.
Shukolyukov Yu. A., Kirsten T. and Jessberger E.K. (1974) The Xe-Xe spectrum technique, a new dating method. Earth. Planet. Sci. Lett., 24, 271-281.
Shukolyukov Yu. A. and Meshick A.P. (1987) Application of xenon isotopes for dating pitchblendes. Chem. Geol. (Isot. Geosci. Sect.), 66, 123-136.
Schmitt J.M., Baubron J.C. and Bonhomme M.G. (1984) Pétrographie et datations K/Ar des transformations minérales affectant le gîte de Bertholène (Aveyron, France). Miner. Depos., 19, 123-131.
Stacey J.S. and Kramers J.D. (1975) Approximation of terrestrial lead isotope evolution by a two stage model. Earth Planet. Sci. Lett., 26, 207-221.
Stuckless J.S. (1987) A review of applications of U-Th-Pb isotope systematics to investigations of uranium source-rocks. In:

Concentration Mechanisms of Uranium in Geological Environments, Poty B. and Pagel M. (eds.), A Conference Report. Uranium, 3, 235-244.
Stuckless J.S. and Nkomo I.T. (1978) Uranium-lead isotope systematics in uraniferous alkali-rich granites from the Granite Mountains, Wyoming: Implications for uranium source rocks. Econ. Geol., 73, 427-441.
Stuckless J.S. and Troëng B. (1984) Uranium mineralization in response to regional metamorphism at Lilljuthatten, Sweden. Econ. Geol., 79, 509-528.
Taylor H.P.Jr. (1974) The application of oxygen and hydrogen isotope studies to problems of hydrothermal alteration and ore deposition. Econ. Geol., 69, 843-883.
Tera F. and Wasserburg G.J. (1972) U-Th-Pb systematics in the three Apollo 14 basalts and the problem of the initial Pb in lunar rocks. Earth Planet. Sci. Lett., 14, 281-304.
Thode H.G., Kleerboper H. and McElcheran D. (1951) Isotopic fractionation in the bacterial reduction of sulfate. Research, London,v. 4, 581-582.
Turpin L. (1984) Altérations hydrothermales et caractérisation isotopique (O-H-C) des minéraux et des fluides dans le massif uranifère de St Sylvestre. Extension à d'autres gisements intragranitiques français. Géol. Géochim. Uranium Mém., Nancy, 6, 190 p.
Turpin L., Leroy J. and Sheppard S.M.F. (1990) Isotopic systematics (O, H, C, Sr, Nd) of superimposed barren and U-bearing hydrothermal systems in a Hercynian granite, Massif Central, France. Chem. Geol., 88, 85-98.
Turpin L., Clauer N., Forbes P. and Pagel M. (1991) U-Pb, Sm-Nd and K-Ar systematics of the Akouta uranium deposit, Niger. Chem. Geol. (Isot. Geosci. Sect.), 87, 217-230.
Veeh H.H. and Burnett W.C. (1982) Carbonate and phosphate sediments. In: Uranium Series Disequilibrium. Applications to Environmental Problems, Ivanovich M. and Harmon R.S., (eds.) Clarendon Press, Oxford, 459-480.
Wallis R.H., Saracoglu N., Brummer J.J. and Golightly J.P. (1984) The geology of the McClean uranium deposits, northern Saskatchewan. Canad. Min. Metall. Bull., 1-28.
Wendt I. (1984) A three-dimensional U-Pb discordia plane to evaluate samples with common lead of unknown isotopic composition. Isot. Geosc., 2, 1-12.
Wendt I. and Carl C. (1985) U/Pb dating of discordant 0.1 Ma old secondary U minerals. Earth Planet. Sci. Letters, 73, 278-284.
Wetherill G.W. (1956) Discordant uranium-lead ages. Trans. Am. Geophys. Union, 37, 320-326.
Wilson M.R., Kyser T.K., Mehnert H.H. and Hoeve J. (1987) Changes in the H-O-Ar isotope compositions of clays during retrograde alteration. Geochim. Cosmochim. Acta, 51, 869-878.
Wilson M.R. and Kyser T.K. (1987) Stable isotope geochemistry of alteration associated with the Key Lake uranium deposit, Canada. Econ. Geol., 82, 1540-1557.

INDIRECT DATING OF SEDIMENT-HOSTED ORE DEPOSITS: PROMISES AND PROBLEMS

Norbert CLAUER
Centre de Géochimie de la Surface
1, rue Blessig 67084 Strasbourg (France)

and

Sam CHAUDHURI(*)
Department of Geology, Kansas State University
Manhattan, Kansas 66506 (U.S.A.)

Chronological information on sediment-hosted ore deposits is essential for the detailed construction of their genesis. Unfortunately, direct dating of ore minerals, such as sulfides or oxides, is often difficult if not impossible by common isotopic methods, because of both the difficulty of measuring very small amounts of radiogenic isotopes due to the very low parent element to daughter element ratios of these minerals and the lack of precise knowledge of the isotopic compositions of the minerals at their formation. Uranium minerals are among the very few ore minerals known to contain significantly large amounts of radiogenic Pb isotopes, making them amenable to dating by the U-Pb method. However, the time of U deposition measured directly from U minerals is often questioned, because these minerals frequently lose intermediate isotopes of the U decay chain when exposed to near surface condi-

---

(*) Presently at the Centre de Géochimie de la Surface, Strasbourg (France) surface conditions (Miller and Kulp, 1963). Maas et al. (1986) and Turpin et al. (1991) suggested recently that U minerals could contain measurable amounts of radiogenic Nd because of favorable Sm/Nd ratios. As the rare-earth elements (REE) are hardly fractionated by geologic processes in surface or near-surface conditions, the Sm-Nd isotope method might prove useful for direct dating of different types of ore deposits, as shown recently by Chesley et al. (1991) on fluorite. Although this direct evidence for the time of ore mineralization is generally difficult to obtain, isotopic studies, especially with the Pb-Pb method, have been found useful to trace the sources of metal concentrates (e.g. Faure, 1986).

In the recent years, the limited available means for direct dating of ores have prompted studies of minerals from barren zones close to ore concentrations. The minerals which have been sought for such purposes include clay minerals, alkali-feldspars, and other silicate phases, because they are potentially capable of recording chemical disturbance events related to ore concentrations. Some isotopic studies of fluid inclusions in ore minerals or gangue minerals for determining their periods of mineralization have also been considered. As fluid inclusions do not occupy any specific atomic sites in minerals, we have considered the isotopic analyses of their fluids as an indirect means of dating ores, much the same way isotopic dating of minerals outside the ore zones provides information about the period of ore concentration. The present chapter reviews the results of attempts to determine the isotopic ages on ore deposits hosted in sedimentary rocks, and analyzes the merits or potentials of methods which may determine indirectly the time of mineralization of these ores.

DATING OF FLUID INCLUSIONS

Studies of fluid inclusions in ore minerals for dating purposes are few, and they have been carried out exclusively with the Rb-Sr method. Before discussion, some cautionary remarks on the existing analytical approach might be useful. Indeed, during sample preparation, special care should be taken to avoid any contamination, as the Rb and Sr concentrations in fluid inclusions are very low, often only a few tenths of $\mu g/g$ (Shepherd et al., 1982; Lange et al., 1983; Darbyshire and Shepherd, 1985). Sample crushing should be done thoroughly, and the leaching of powder with dilute acid (1N HCl) seems appropriate. Mass balance calculations and petrographic data are desirable to assess the sources of Rb and Sr in the minerals.

Some Case Studies

Shepherd et al. (1982) reported a study on the Pb-Zn-fluorite-barite deposits of the North Pennine orefield in Scotland, providing a Rb-Sr isochron on quartz fluid inclusions from "Great Sulphur" vein. Previous fluid inclusion microthermometry studies established that fluorite was deposited from moderately hot (90 to 160°C) and highly saline brines (Sawkins, 1966; Smith, 1974). The Rb-Sr isochron fitted through 8 data points with $^{87}Rb/^{86}Sr$ ratios from 0.5 to 1.83, and gave an age of 206 $\pm$ 9 Ma with an initial $^{87}Sr/^{86}Sr$ ratio of 0.7106 $\pm$ 0.0002 (Fig. 1). Unfortunately the ana-

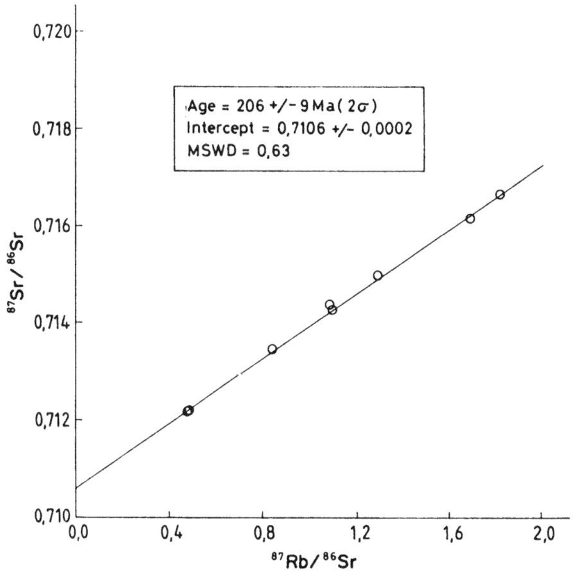

Figure 1: Fluid inclusion Rb-Sr isochron of Great Sulphur vein, North Pennine orefield (Shepherd et al., 1982)

analytical data were not provided in the publication. As this Triassic age agrees with results reported for other fluorite deposits in Western Europe (Baubron et al., 1980), Shepherd et al. claimed that the isochron dates the fluorite deposition, although previous investigation by Bunham et al. (1968) argued for a Permo-Carboniferous depositional age. Shepherd et al. also suggested, on the basis of the $^{87}Sr/^{86}Sr$ ratio of the isochron intercept, that the most likely source rocks for the Sr, and hence possibly for the metals of the deposit, are the Carboniferous shales of the sedimentary sequence subjected to the ore-bearing veins. Based on Sm-Nd data, Halliday et al. (1986) ruled out a plutonic source for the Nd and Sr of the mineralizing fluids, and suggested that most of Nd and Sr derived from Carboniferous sediments.

Lange et al. (1983) studied the Sr isotope chemistry of barites and the Rb-Sr systematics of fluid inclusions contained in cubic galena samples of Mississippi Valley-type ore deposits in southeast Missouri. The temperature of the ore fluids, based again on fluid inclusion microthermometry, could have been as high as 110°C (Leach, 1979). Four samples with $^{87}Rb/^{86}Sr$ spread between 0.13 and 2.16 defined an Rb-Sr isochron, but a fifth sample was significantly off the line. The calculated age was 392 $\pm$ 21 Ma with an initial $^{87}Sr/^{86}Sr$ ratio of 0.7104 $\pm$ 0.0004. Kish and Stein (1979), Posey et al. (1983), Grant et al. (1984), and Stein and Kish (1985) reported Rb-Sr isochron and conventional ages on glauconites separated from ore-bearing strata of the district, generally between 350 and 400 Ma. Hearn and Sutter (1985) published a $^{40}Ar/^{39}Ar$ age of 290 $\pm$ 20 Ma for authigenic feldspars, suggesting ore deposition was possibly connected with the brine migration at that time. In addition, K-Ar data on Upper Cambrian glauconies from the Ozark region suggest two recrystallization ages, one at 385-390 Ma and the other less than 355 Ma (Clauer and Chaudhuri, unpublished). On the basis of all data, it seems that the southeast Missouri region underwent at least two hydrothermal events which could have been related to ore deposition: one at about 390 Ma, and the other possibly as late as at 290 Ma. The existence of an early thermal episode in the region is also suggested by an age of 380 $\pm$ 25 Ma obtained for the Avon diatremes (Zartman et al., 1966). In contrast to all these Devonian or younger ages, York et al. (1981) reported a $^{40}Ar/^{39}Ar$

isochron age of 549 ± 20 Ma for pyrite samples from same deposits, the meaning of which remains questionable; one reason being possible recoil effects.

Discussion

A very basic question shrouds the meaning of Rb-Sr isochrons from fluid inclusions: why are the lines stratigraphically meaningful isochrons and not meaningless mixing lines? The ore fluids in the Viburnum district, for instance, are known to have had varied Sr isotope compositions during the entire period of the mineralization, as suggested by the data obtained on different gangue minerals (e.g. Kessen et al. 1981; Chaudhuri et al., 1983; Baumann and Hofmann, 1988). For example, early dolomite associated with octahedral galena had low initial $^{87}Sr/^{86}Sr$ ratios between 0.7093 and 0.7097, while that of the intermediate dolomite and calcite, and of some quartz associated with cubic galena ranged from 0.7102 to 0.7106; the late calcite displaying $^{87}Sr/^{86}Sr$ ratios of 0.7119 to 0.7123 (Fig. 2). These values are systematically higher than those of the host carbonate rocks, which range from 0.7083 to 0.7089. Thus, considering this variation of the Sr isotope composition of the Viburnum ore fluids, the isochron obtained on fluid inclusions of the cubic galena samples could have been biased by samples characterized by high $^{87}Sr/^{86}Sr$ and high Rb/Sr ratios: in this case the line would be a mixing line. The magnitude of that bias would, of course, depend on the differences in both the initial $^{87}Sr/^{86}Sr$ and the Rb/Sr ratios of the end-members. Hence, the isotopic age for the Viburnum galena would then only set a maximum value for the mineralization event. Lines of evidence in favor of the existence of an isochron, instead of a mixing line, are that the initial $^{87}Sr/^{86}Sr$ ratios of both the fluid inclusions in the galena samples and the glauconitic minerals from the nearby ore-bearing strata are the same within analytical uncertainty, and that their ages are identical at 392 ± 21 Ma and 393 ± 6 Ma, respectively. These similarities favor also the idea of migrating chemically homogeneous ore fluids at that time.

The constancy of the $^{87}Sr/^{86}Sr$ ratio during ore formation is, indeed, an essential requirement for dating of fluid inclusions in ore-minerals or associated gangue minerals by the Rb-Sr method. As the conclusions of the study on the Viburnum deposits suggest, this requirement should be critically tested in any study by analyzing the Sr isotope compositions of associated Rb-free gangue minerals. Sample selection should be critically guided by the study of the paragenetic mineral sequences of the ores to be dated.

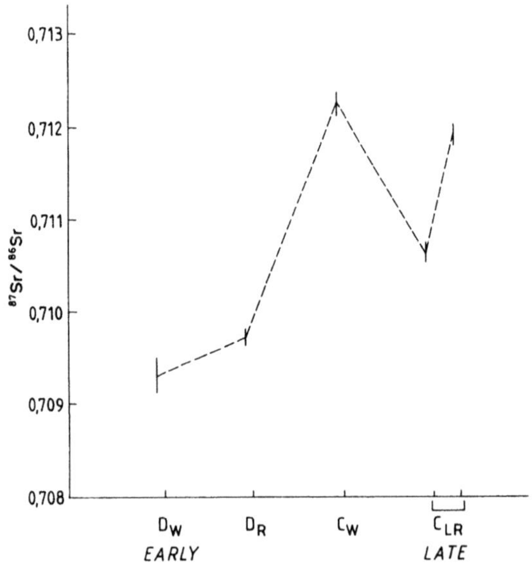

Figure 2: Temporal variation in the $^{87}Sr/^{86}Sr$ ratio of carbonate minerals at Bushy Creek, Viburnum Trend, Southeast Missouri (Chaudhuri et al., 1983). D_w = white rhombic dolomite; D_r = reddish white dolomite; C_w = white calcite; C_{lr} = large yellowish rhombohedral calcite

Aside from the difficulty of ascertaining whether or not the samples had the same $^{87}Sr/^{86}Sr$ ratio during deposition, the Rb-Sr isochron technique suffers basically from the disadvantage of not having sufficient spread in the Rb/Sr ratio among the samples. The ratios range, for example, from 0.05 to 0.75 in the Viburnum galena, which induces a large factor of uncertainty on the calculated ages.

Some potentially new lines of research on isotopic dating of fluid inclusions could be developed. For instance, crushing of ore samples under vacuum in an apparatus equipped with a mobile pestle and connected to a high vacuum gas extraction line should theoretically allow application of the K-Ar method. The amount of K could be measured on the same split after Ar extraction and analysis or on a separate split, by isotope dilution. Although technically feasibile, this sort of investigation should be approached with cautions because a study of Rama et al. (1965) indicated that "excess" ^{40}Ar could be present in the fluid inclusions. Kelley et al. (1986) outlined in great detail a study of Ar isotope chemistry in fluid inclusions generated during mineralization processes. They could relate the "excess" ^{40}Ar observed in their study to the chlorine-Ar component of the brines present in the inclusions. This behavior, if general, would limit the routine application of this method, as liquid phases in fluid inclusions are often found to be Cl-rich.

Another potentially useful alternative to the Rb-Sr and K-Ar methods related to fluid inclusions could be the $^{40}Ar/^{39}Ar$ laser-fusion technique. Application of this technique, which is discussed more extensively in a later section of this chapter, would avoid the tedium of separation of the minerals and of the chemical preparation prior to mass spectrometric analyses. However, indiscriminate application of this method can lead to questionable data, as Kelley et al. (1986) observed ^{39}Ar recoil losses from samples studied in sealed vials. Recoil-related losses were strongly hinted by a study of pyrite by York et al. (1981) who found ages to be higher by about 170 Ma than all other ages obtained with independent means. This recoil effect is not surprising considering that fluids of inclusions do not occupy specific atomic sites but are trapped in vugs generated during mineral growth. Among the causes for these recoil effects are the occurrence of cracks near the surface of the host grains and the decrepitation of some inclusions induced by temperature increase during the necessary irradiation of the samples in a reactor (Kelley et al., 1986). However, the $^{40}Ar/^{39}Ar$ laser-fusion technique has still application possibilities, as fluid inclusions sometimes contain solid phases beside the usual liquid and vapor phases. When the solids are authigenic sericite-type minerals, they can be dated by $^{40}Ar/^{39}Ar$ laser-fusion. A successful application of this technique seems to depend on careful sample selection to avoid the effects of induced artifacts.

DATING OF COMMON GANGUE MINERALS

Gangue minerals of ore deposits are very much devoid of radiogenic isotopes because of the very low parent to daughter isotope ratio. Consequently these minerals are not generally amenable to dating by conventional methods. However, studies may be made of the parts of host rocks which are most intimately associated with ore or gangue minerals to determine the time of mineralization.

A Case Study

Ruiz et al. (1984) suggested a new Rb-Sr approach to determine the period of mineralization of ore bodies by analyzing the Sr isotopic compositions and the Rb/Sr ratios of common gangue minerals and their associated wall rocks. This technique has the potential of overcoming the small spread in the Rb/Sr ratios among the minerals of an ore body, as the spread in the Rb/Sr ratio might be favorably influenced by the wall rocks, which commonly have much higher Rb/Sr ratios than the gangue minerals. Nevertheless, the technique requires, as demanded by the principle of isotopic dating, that the gangue minerals and their wall rocks have the same isotopic composition at the time of mineralization. Failure to comply with this requirement produces a model age derived from average data of the wall rocks and based on an assumed initial $^{87}Sr/^{86}Sr$ ratio, which can be very different from the real age of mineralization. The precision of the calculated mineralization ages of samples from zones where isotopic homogeneity occurred, are less dependent on the exact value of the initial Sr ratios which may be deduced from the values of the gangue minerals. Ruiz et al. elaborated on the conditions which govern the precision of the ages. To demonstrate the feasibility of using the Rb-Sr isotopic method, they reported isotopic results of gangue minerals and associated wall rocks from three separate ore districts. The ages of the mineralizations were previously established by other means. Ruiz et al. found that the age based on assumed isotopic homogeneity between gangue minerals and wall rocks was reasonably in agreement

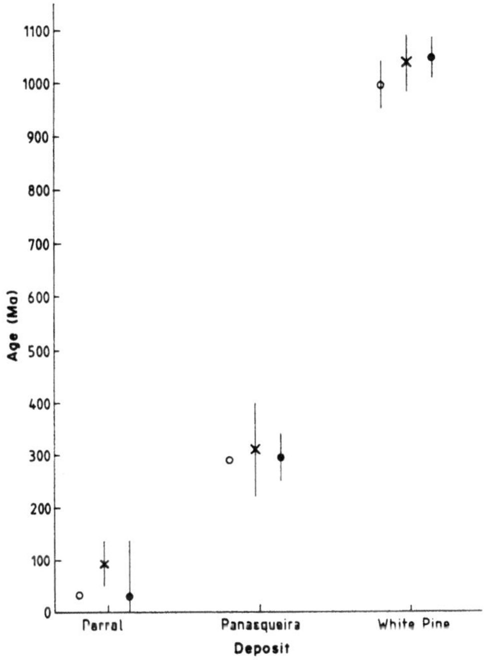

Figure 3: Comparison of mineralization dates for Parral, Panasqueira and White Pine (Ruiz et al., 1984). Squares are ages with errors obtained by other workers, bars with stars and triangles indicate ages and errors obtained by Ruiz et al.'s technique

with previously published data, for the White Pine Cu deposit in the Late Precambrian Nonesuch Shale in Michigan. For the two other examples, the Parral Pb-Zn ores in Mexico and the Panasqueira Sn-W deposit in Portugal, the generated data were greatly lacking in the precision necessary to equate the isotopic ages with the presumed mineralization ages (Fig. 3).

Discussion

The Rb-Sr dating of ore deposits by the method suggested by Ruiz et al. (1984) should be approached with carefully considered supporting information. The failure to do so would add more confusion to the already existing problem of finding the age of ore concentrations. The foremost question of this dating technique is whether or not the wall rocks and the gangue minerals had the same or nearly identical Sr isotopic compositions at the time of mineralization. From above discussions, any claim of isotopic homogeneity between wall rocks and gangue minerals is of extremely dubious nature. It must be demonstrated by complementing chemical or petrographic data, and not simply assumed. In one instance, we have failed to observe an isotopic homogeneity between host rock and vein even on a scale of few millimiters (Clauer et al., 1989). Defining the zone of isotopic homogeneity within wall rocks is critical to the selection of the most appropriate samples, and in this respect analyzing whole rocks is the least desirable option, as they are multi-component systems. A particular group of minerals, but not all minerals, in a wall rock could have isotopically equilibrated with the mineralizing fluids, and analysis of its separates would be far more desirable and accurate than that of the probably more heterogeneous whole rock.

Two potential sources of error in the age of ore deposits from wall-rock analyses are analytical scatter of the Rb/Sr ratios of the wall rocks and of the $^{87}Sr/^{86}Sr$ ratio of the fluids at the time of mineralization. Ore fluids often appear to have varied Sr isotopic compositions, as has been reported in studies on gangue minerals and ore deposits (Chaudhuri et al., 1983; Kesler et al., 1983; Fontbote and Gorzaw-ski, 1988). Ruiz et al. remarked that precise knowledge of the Sr isotopic composition of the mineralizing fluid is less critical when the mean Rb/Sr ratio of the wall rocks is high. Accuracy and precision of the ages obtained by this method are, in fact, not only dependent on the knowledge of the Rb/Sr and initial $^{87}Sr/^{86}Sr$ ratios of the wall rocks, but also on the degree of scatter in the Rb/Sr ratios. The most preferred rocks are those with high but no scatter in the Rb/Sr ratios. In general, shales, sandstones and siltstones among sedimentary rocks are high, but spatially much varied in their Rb/Sr ratios. Selection of samples from the same bed, or from different beds with similar Rb/Sr ratios, could, therefore, improve the precision of the model ages.

DATING OF ASSOCIATED CLAY MINERALS

Clay minerals are common products which precipitate or recrystallize during low-temperature hydrothermal processes. Their isotopic ages may therefore be used to document the time of hydrothermal processes which might have induced ore concentrations. A number of studies with somewhat varied analytical approaches among them have addressed the issue of dating ore deposits through isotopic analyses of the clay components. Most of these studies are reviewed here to point out to what extent this approach provides reliable results, and to suggest some precautionary measures that might avoid failure of getting the intended ages.

Results of Case Studies

In the early 1970's, Ineson and co-workers reported a number of K-Ar data obtained from clay minerals of wall rocks or vein gouges of metallogenic igneous provinces from United Kingdom and Norway (Ineson and Mitchell, 1972; Ineson, 1974, 1975; Ineson et al., 1975). The basis of these datings was that cogenetic minerals, as long as non unambiguous evidence for the age of the ores is available, could be dated to constrain the age of vein emplacements. The clay fraction analyzed consisted mainly of illite-chlorite assemblages, very often mixed with kaolinite and occasionally with zeolites. Unfortunately, in most of the studies mentioned, both the lack of knowledge about the time of mineralization constrained by other means, and the lack of information about the meaning of some minerals like kaolinite and about the morphology of the clay particles, which both will be discussed later, prevent any assessment of the real significance of the ages from these clay fractions. However, the authors claimed that: "clay minerals from host rocks do constitute reliable geochronometers, at least for vein deposits which are post-Carboniferous."

Lee and Brookins (1978), and Brookins (1980) reported on an extensive Rb-Sr study of <2μm bulk clay separates related to U mineralization in the Upper Jurassic Morrison Formation of New Mexico. They defined the earliest mineralization period at 139 $\pm$ 9.5 Ma, based on an isochron of smectite-rich clay fractions from barren rocks and chlorite-rich clay fractions from ore zones. The U-Pb data on the same ore deposit yielded consistently lower and discordant ages from 87 to 113 Ma. These low values and their relatively large scatter might have resulted from loss of intermediate daughter products of the U decay chain by the U minerals near surface. This discrepancy strongly suggests that the U-Pb method applied to altered U minerals systematically underestimates the time of U deposition. Lee and Brookins also obtained low Rb-Sr isochron ages from 110 to 115 Ma for clay minerals from Laguna district, which they interpreted as later periods of mineralization. The latest ore concentration in the district might have occurred about 9 to 10 Ma ago, based on the U-Pb data of an uranophane. In contrast to the Rb-Sr data, the K-Ar values of all the clay minerals were widely scattered between 138 and 49 Ma. This scatter could be due to more or less extensive resets affecting the K-Ar systems during the different mineralization events.

Halliday (1977, 1980a, 1980b), Bonhomme et al. (1980), Pop et al. (1980), Bonhomme (1982), Bonhomme et al. (1983), and Halliday and Mitchell (1983, 1984) reported many K-Ar results obtained on clay fractions associated with ore mineralizations mainly in igneous rocks, but also in sedimentary sequences. They considered these K-Ar ages as being minimum ages for the mineralization, and that the dates probably reflect late hydrothermal activities in the host rocks. The results are widely scattered in most of these investigations, and the values which are believed to be "true ages" are choosen on a basis of "regional consistency" with other previously published results. In fact, most of these data did not provide unambiguous interpretation, because of the lack of information about separation and characterization of the clay fractions.

Clauer et al. (1985) reported K-Ar results of carefully separated and characterized <2μm bulk clay fractions associated with U-ore deposits from Cluff Lake in Saskatchewan, Canada. Eleven clay fractions were selected from mylonitized parts of the sedimentary sequence which were affected by the same tectono-thermal event as the basement. They consisted largely of illite (> 50%) with smaller amounts of chlorite, and their K-Ar apparent ages averaged 1293 $\pm$ 36 Ma. These clay samples are never located farther than 40 m from the massive U ore-deposits and were, therefore, considered to have experienced the tectono-thermal event responsible for the

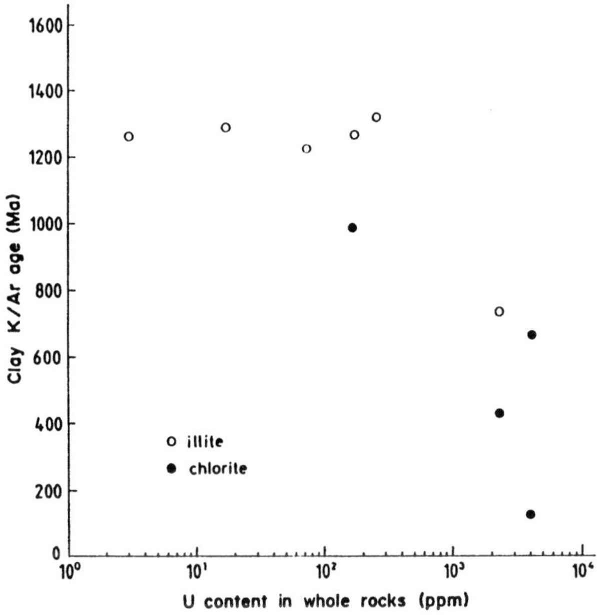

Figure 4: Relationship between K-Ar ages of clay minerals and U content of the corresponding host whole rocks (Halter et al., 1987)

major U concentration in the area. Previously published U-Pb dates on the ore minerals were systematically lower at 1050 to 1150 Ma (Bellon et al., 1976; Gancarz, 1979). Only recently, Philippe (1988) obtained U-Pb ages of 1278 ± 31 Ma on unaltered pitchblende from the same area. To elucidate further on the extent of the thermal event in this region, eight additional <2μm clay fractions were analyzed by Clauer et al. (1985). Five were taken from tectonically undisturbed barren sandstones close to the ore pit, and the three others were collected outside the Carswell structure, several km away from any U show. This second set of clay fractions gave a K-Ar isochron age of 1222 ± 52 Ma, which is similar to the age of the clay fractions associated with the ore-bearing zone. The similarity in ages suggests that the hydrothermal event was widespread at this time, and that the U concentrated only locally at the contact between basement and the sedimentary cover. The temperature of this event was estimated at 200 ± 50°C, based on microthermometry determinations in the fluid inclusions of the quartz overgrowths (Pagel et al., 1980). Wilson et al. (1987), combining

K-Ar data with stable isotope chemistry on clay fractions from fracture zones or highly altered zones, reported K-Ar apparent ages much lower than 1200 Ma. These data seem to have no geological meaning, but they record external effects on the K-Ar system of clay particles. These results are complemented by a study of Halter et al. (1987) on clay fractions directly in contact with the U-concentrates. Relationships between the U content in the whole rocks, the K-Ar data, the D values, and the water content of the clay fractions could be observed (Fig. 4). They emphasized radiation-catalyzed retrograde exchanges of the clay particles, which have been damaged by the irradiations, with recent meteoric waters at low temperatures. These studies emphasize how important it is to characterize thoroughly the clay fractions extracted from well documented samples.

An U chemical age of about 160 Ma was obtained for pitchblende of the sediment-hosted U deposit of Muellenbach, near Baden-Baden in Western Germany (Zuther, 1983), but U-Pb determinations of pitchblende from the same region gave higher values of 240 $\pm$ 6 Ma (Holliger et al., 1989). This age was derived from a model based on continuous Pb and Ra diffusion. Also determined were K-Ar apparent ages of $<2\mu$m bulk clay fractions from different types of sedimentary rocks which have been altered to different degrees during the thermal event responsible for the U deposition (Brockamp et al., 1987). Seven illite-rich clay fractions of arkoses near the U-ore body gave K-Ar values between 150 $\pm$ 3 Ma and 159 $\pm$ 4 Ma, whereas two clay fractions immediately adjacent to the ore were found to be 143 $\pm$ 3 Ma old. As discussed in the preceeding section, this difference between the two ages could have resulted from irradiation damages, but it could also reflect a span of time during which the U-ores were deposited. A noteworthy aspect of this study is that the clay fractions from all arkosic rocks, whether collected from the highly altered ore zone, the moderately altered sericitized zone, or the weakly altered albitized zone, always gave similar K-Ar values averaging 155 $\pm$ 4 Ma. The clay fractions extracted from other lithologic units, such as shales or silts, gave systematically higher K-Ar values, suggesting that only the arkoses, with high porosities and permeabilities tended to have clay mineral assemblages that were in chemical equilibrium with the hydrothermal fluids. Based on coalification measurements, the temperatures of the hydrothermal solutions responsible for the U ores near Baden-Baden, were probably somewhere between 190 and 240°C. This is high enough, according to Hunziker et al. (1986), to erase the Ar memory of any detrital illite-type components of sedimentary occurring rocks. Clearly, in this instance, the clay fractions of the arkoses were able to record a thermal event about 155 Ma ago. The major U concentration could, however, have occurred about 50 Ma earlier, as suggested by the U-Pb data, during another event, the memory of which could have been erased entirely from clay fractions of the highly permeable arkoses and, to a lesser extent, from clay fractions of the other lithologies.

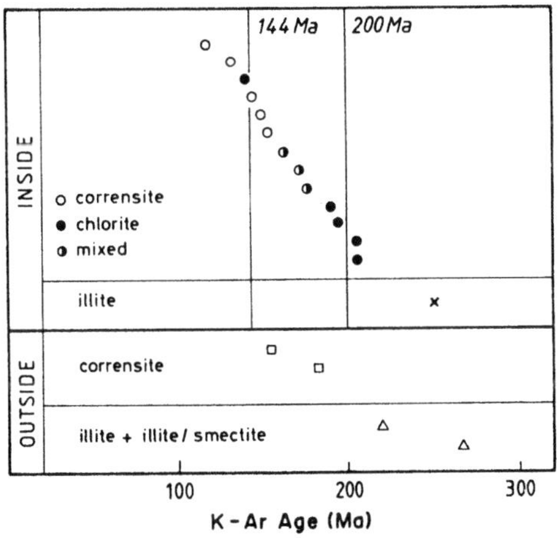

Figure 5: K-Ar ages of the clay fractions associated with the Akouta U deposits (Turpin et al., 1991). The numbers at the top represent the U-Pb ages of the pitchblende

The temperature of the mineralyzing fluids from which the U deposits of Akouta in Central Niger precipitated in sandstones, and the temperature of the mineralizing fluids could have been as high as 160°C (Forbes et al., 1987). Turpin et al. (1991) contended that the deposition occurred between 260 and 150 Ma, based on U-Pb determinations on well characterized pitchblende. The clay fractions in close proximity to the ore gave K-Ar ages of 148 ± 6 Ma for the corrensite-rich ones, while the chlorite-rich assemblages suggested the existence of an earlier hydrothermal event at 199 ± 8 Ma (Fig. 5). Again, slightly lower K-Ar data at 134 ± 4 Ma and

117 $\pm$ 6 Ma were obtained on clay fractions from the ore body itself. Additional age determinations are needed to constrain better the U-Pb age, to define the extent of this early hydrothermal event in the region, and to determine the relationship of this activity to the U deposition.

Figure 6: Scanning and transmission electron microscope photographs. 6a: poorly crystallized illite with ill-defined edges (bar = 2.5μm); 6b: well crystallized illite with well defined edges (bar = 1μm); 6c: authigenic feldspar in a sandstone (bar = 10μm); 6d: detrital altered feldspar in a sandstone (bar = 10μm)

A promising isotopic technique for dating clay minerals associated with ore deposits appears to be the $^{40}Ar/^{39}Ar$ method. Halliday (1978) reported results of $^{40}Ar/^{39}Ar$ stepheating determinations on six clay concentrates from ore bodies in Ireland. The age distribution patterns from stepheating were complex, suggesting that irradiation of the clay particles in the reactor induced ^{39}Ar and radiogenic ^{40}Ar losses. These effects seem, as hinted by Halliday (1978), to be related to the mineralogy of the clay frac-tions, the size of the grains, and the morphology of the particle edges. Reuter and Dallmeyer (1989), who carefully documented the ^{39}Ar recoil effect on clay fractions and the related redistribution of ^{39}Ar from high-K to low-K bearing phases, could show that both, recoil and redistribution seem to occur preferentially in particles with large surface areas and poorly defined grain edges, which relate to poorly crystallized minerals, relative to well crystallized counterparts with straight edges (Fig. 6a and b). Bell (1985) reported some $^{40}Ar/^{39}Ar$ data obtained by laser-fusion technique on "alteration products" of basement crystalline rocks near U shows from southern Carswell structure in Saskatchewan, Canada. As mentioned by the author, the size of the individual clay particles being of 1 to 10μm, laser-fusion required analysis of grain clusters, because the diameter of the laser beam was of 200 to 300 μm. The ages of the nine samples ranged from 500 $\pm$ 50 Ma to 1615 $\pm$ 5 Ma; most of them being between 1175 $\pm$ 15 Ma and 1615 $\pm$ 5 Ma. Bray et al. (1987) presented a study in which they compared laser-

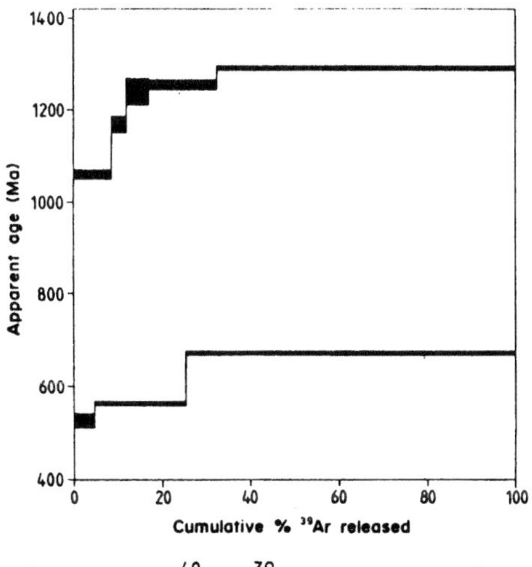

Figure 7: $^{40}Ar/^{39}Ar$ stepheating age spectra for potassic silicates, Mc Clean deposits, Northern Saskatchewan (Bray et al., 1987). Top diagram for sample 1208-458 and bottom diagram for sample 2045-517

probe $^{40}Ar/^{39}Ar$ and conventional K-Ar datings of illite associated with U deposits from northern Saskatchewan, Canada. The illite fractions collected in the alteration halo next to the mineralization, had a larger scatter for the conventional K-Ar ages, from 1321 ± 44 Ma to 1002 ± 33 Ma, than for the laser-probe $^{40}Ar/^{39}Ar$ ages, from 1319 ± 3 Ma to 1178 ± 4 Ma. The larger scatter in the conventional K-Ar values could be due to sample preparation and clay fraction purification. As in the previous example, the scatter in the $^{40}Ar/^{39}Ar$ ages could be partly due to the energy of the beam and its size. Fusion of nearby minerals has to be expected as long as the size of the laser beam is larger by one to two orders of magnitude than the size of the particles. But this effect will have no consequences when the adjacent grains are depleted in Ar, like quartz. The results of Bray et al. (1987) look promising: they obtained some flat $^{40}Ar/^{39}Ar$ step-heating spectra (Fig. 7), compared to the complicated ones presented by Reuter and Dallmeyer (1989; Fig. 8,a and b). The plateau ages, however, remained scattered.

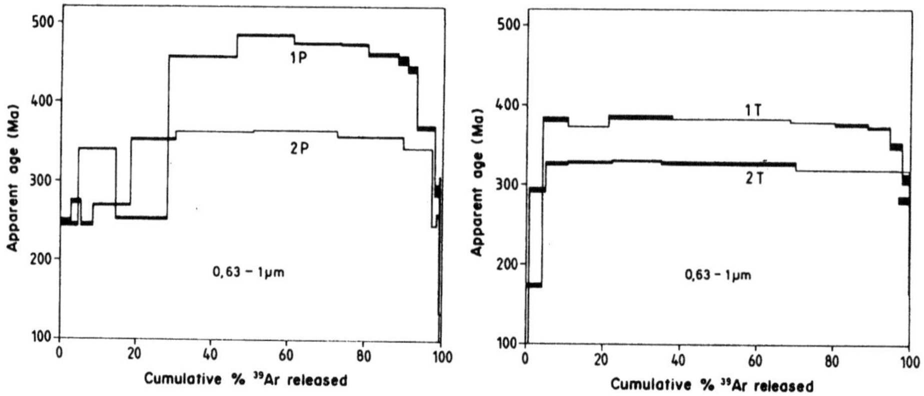

Figure 8: $^{40}Ar/^{39}Ar$ steapheating age spectra for 0.63-1μm clay size fractions (Reuter and Dallmeyer, 1989). 1P: clay fraction extracted from a phyllite of middle anchizone metamorphic intensity. 2P: clay fraction from a phyllite of upper anchizone intensity. 1T: clay fraction extracted from an associated metatuff of middle anchizone intensity; 2T: clay fraction extracted from an associated metatuff of upper anchizone intensity. Note that the clays from metatuff give systematically less complicated patterns than the counterparts from phyllites. The upper anchizone clays have also less complicated patterns than the counterparts from middle anchizone

Discussion

To our experience, many problems of clay isotope dating arise because investigators do not take necessary care in sample preparation and clay separation prior to mass spectrometric determinations. Liewig et al. (1987) and Clauer et al. (1992 and in this volume) have addressed some of these problems in recent studies on K-Ar dating of authigenic clay minerals in oil-bearing sandstone reservoirs. Routine crushing of the whole-rock samples systematically produces overcrushing of detrital components to clay-sized particles, the presence of which automatically invalidates the isotopic data of the intended clay minerals, as discussed earlier. Appropriate gentle disaggregation techniques must be adopted to avoid incorporation of artificially created clay-sized grains in the separated clay fractions.

The analytical procedures are suspect, for instance, when the K-Ar ages of clay fractions which are supposedly related to the same thermal event are widely scattered. Arbitrary selections of numbers from clouds of data can no longer be considered as "true ages" just because they seem consistent on a regional basis. Interpretation of isotope data must counter all possible factors, which fall into both isotopic and non-isotopic categories, contributing to the dispersion of the data. The presence of contaminating minerals in clay samples and the mixtures of clay particles of different generations are very common reasons for data clouds. Depending on their origin, contaminating components, such as feldspars, might influence the isotopic values of the $<2\mu m$ bulk clay fractions. If feldspar peaks can be recognized in X-ray diagrams, their relationship with the clay minerals should be examined under electron microscopy. If these feldspars are authigenic, they display characteristic forms by SEM (Fig. 6c), and they could have formed during the same hydrothermal event as the clay minerals, which in turn implies that these feldspar grains should not influence the isotopic results of the clay fractions. If, on the other hand, the feldspars are detrital in origin, as those on figure 6d, they will definitely "increase" the K-Ar or the Rb-Sr ages of the associated clay assemblages, because they contain an excess of radiogenic ^{40}Ar or ^{87}Sr relative to the clays formed during the hydrothermal event.

As the study of Wilson et al. (1987) demonstrated, clay fractions containing kaolinite should also be avoided for dating purposes, because the conditions favorable to their genesis are often conductive to leaching of interlayer cations from the K-(and Rb) bearing clay minerals, which should adversely affect their isotopic ages. Clay fractions collected within U-ore deposits should also be considered with special care, as Halter et al. (1987) noted that radiation damage tends to promote loss of radiogenic isotopes from the altered lattices of the clay minerals.

However, the overall conclusion of the previous studies is that any lowering in the age of clay fractions can be caused by crystallization of younger generations of clay minerals. Diffusion of radiogenic ^{40}Ar from clays is not a commonly observed behavior, as it occurred only in cases related to irradiation damages of the mineral lattices, or to chemical exchange with environmental fluids at low temperature in deeply weathered zones. This lack of exchange at low temperatures is a basic requirement for dating clay fractions, and it has been recently substantiated by O'Neil (1987), on the basis of the stable isotope chemistry.

To determine isotopic ages by the ^{40}Ar/^{39}Ar method, the samples have to be irradiated with fast neutrons in a reactor to produce ^{39}Ar from ^{39}K. The temperature in the reactor can be as high as 200°C, which, of course, is expected to induce radiogenic ^{40}Ar loss from minerals that have crystallized at lower temperatures. This loss of ^{40}Ar might occur in addition to a ^{39}Ar loss from recoil effect induced by the neutron activation, and detected preferentially in small and poorly crystallized particles. We suspect that in many published accounts of ^{40}Ar/^{39}Ar data obtained on clay minerals, a combined loss of ^{39}Ar and radiogenic ^{40}Ar produced isotopic dates that appear reasonably close to the expected ages, although the final results are technically faulty. Before the ^{40}Ar/^{39}Ar technology can be applied routinely to the dating of clay particles, with or without a laser-probe, much more basic work is needed to determine the effects of irradiation and laser fusion. Studies of clay samples loaded in breakseal quartz vials before irradiation would certainly be helpful to understand more about the effects of combined losses of recoiled ^{39}Ar due to neutron activation and radiogenic ^{40}Ar due to temperature in the reactor.

While paying close attention to the technical aspects and selection of samples for clay age determinations, geological considerations are equally important in the interpretation of the ages. It must be kept in mind that hydrothermal activities affect the evolution of clay minerals in rocks by either inducing partial or complete alteration of preexisting clay fractions, or promoting crystallization of new minerals. These activities, however, do not always relate to ore concentrations. If dates of associated clay minerals do not match the ages of the ore deposition, as might be the case for the U ore-deposit of Muellenbach in Germany, a geologic explanation should be favored, before considering obscure Ar diffusion effects. On the other hand, when the K-Ar ages of very carefully characterized clay fractions are higher than those of the U-minerals, like in the Akouta deposit of Niger, a mineralogic reason might be searched such as the presence of illite in the clay fractions (Fig. 5), which is considered here of detrital origin.

Lithology is also an important consideration in the isotopic dating of ore mineralizations by the means of clay minerals from host-rocks, as suggested by the study on the U-ore deposit of Muellenbach in Germany (Brockamp et al., 1987). Clay minerals from arkoses, and by extension from sandstones, are apparently more suited for dating than those from siltstones and shales. Sandstones with high porosity and permeability facilitate pervasive fluid migration, offering more opportunities than siltstones and shales for the growth of neoformed or recrystallized clay minerals and the alteration of unstable detrital components. Hence, if the occurrence of sandstones and arkoses increase the possibilities of dating properly the time of hydrothermal activity related to ore deposition because their clay minerals are most likely completely recrystallized, they also require special care in sample preparation because the detrital components such as the feldspars are often deeply altered so that routine crushing might reduce these altered detrital components to clay-sized grains and introduce artifacts, as suggested by Clauer et al. (1992).

DATING OF OTHER SILICATE MINERALS

Beside clay minerals, some other alkali-bearing silicates can be occasionally found associated with ore deposits, and therefore represent potential materials for dating ores by isotopic means. Among these minerals are K-feldspars (adularia and orthoclase). While Hearn and Sutter (1985) reported promising results on orthoclases related to the Mississippi-Valley type (MVT) deposits, Halliday and Mitchell (1976) published scattered K-Ar analytical data between 156 $\pm$ 3 and 220 $\pm$ 3 Ma for adularia feldspars from Lizard Complex in England. Halliday and Mitchell could relate the age variations to changes in the triclinicity of the minerals. They also measured $^{40}Ar/^{39}Ar$ on the same adularia splits, which matched with the K-Ar data suggesting that probably no recoil effect occurred. The results were explained in terms of post-crystallization reorganization of the lattices of the minerals, which seem to have induced Ar loss by interaction with low-temperature alkali solutions. These results on adularia feldspars suggest limitations to the use of these minerals for dating ore deposits.

CONCLUSIONS

Presently, the dating of sediment-hosted ore deposits is proceeding along two major lines of approach. One is the isotopic dating of fluid inclusions in ore minerals or gangue minerals, and the other is the isotopic dating of clay minerals associated with ores. Careful selection of samples based on factual information is needed to help avoid ambiguity in the translation of the isotopic data to a meaningful geologic age, which has relevancy to the ore deposition.

Critical to isotopic dating of fluid inclusions in ore or gangue minerals by the Rb-Sr isotopic method is total knowledge of any variation of the isotopic composition of Sr in the fluids during the entire mineralization episode. A Rb-Sr isochron from fluid inclusions will indicate a mineralization age, provided the ore fluids had the same or nearly the same Sr isotopic composition during mineral growths. When the ore fluids varied in their initial Sr isotopic compositions, sample selection should be guided by information on the paragenetic mineral associations, as only minerals of the same generation as the ore fluids can be expected to have the same or nearly the same Sr isotopic composition. Hence, all gangue minerals should be analyzed in an ore district for their $^{87}Sr/^{86}Sr$ ratios to determine both the magnitude and the direction of the Sr isotopic change in the fluids during mineralization.

Clay minerals in host rocks or ore deposits have been viewed as suitable materials for K-Ar and $^{40}Ar/^{39}Ar$ isotopic dating to define the period of ore mineralizations. Ore-related hydrothermal events can be intense enough to either erase previous Ar (or Sr) isotopic memories from detrital clay components or cause authigenesis of new clay particles in the host rocks. Some confusion exists in the literature about the meaning of the ages obtained from clay minerals. Selection of samples, separation and characterization of the clay size-fractions, and, in some situations, analytical problems have greatly contributed to the confusion. Lithology of the rocks is certainly another important consideration. Clay minerals of sandstones, arkoses, and even siltstones, seem to meet the test for age determinations as the porosity in these rocks often facilitates pervasive fluid migration.

Careful mineral separation and detailed characterization of clay minerals are essential preliminary steps toward isotopic dating of ore mineralization. Any clay fraction containing a recognizable contaminant should be discarded for mass spectrometer determinations. Routine characterization of the separated clay fractions by X-ray diffraction should be complemented by transmission electron microscope (TEM) observations.

The use of a laser-probe combined $^{40}Ar/^{39}Ar$ technique, providing in situ isotopic determinations, would solve the delicate aspect of clay separation. Some above mentioned data are promising, but more systematic studies are needed before the technique can be applied on a routine basis on clay particles. It now seems from the available data that the small size of clay particles (<2μm) is not the major reason for the ^{39}Ar loss due to the recoil effect. The recoil loss is insignificant when the particles are well crystallized, which can be verified under TEM by the presence of their sharp and straight edges. It should also be noted that all $^{40}Ar/^{39}Ar$ dating on clay particles should be complemented by conventional K-Ar determinations on the same size fractions, so that the ^{39}Ar recoil effects, the redistribution of ^{39}Ar from K-rich illite into K-poor components, such as chlorite, kaolinite or plagioclases, and the eventual loss of radiogenic ^{40}Ar could be clearly evaluated. Clay fractions containing these K-poor phases should be avoided for $^{40}Ar/^{39}Ar$ determinations, as they tend to obscure the patterns. In addition, irradiation of clay minerals should be performed in sealed quartz vials, which would allow to determine more precisely the combined effects of losses of ^{39}Ar and ^{40}Ar from the minerals. One should also be concerned about the temperature attained during irradiation of clay in a reactor. Radiogenic ^{40}Ar diffusion might be expected out of mineral structures that crystallized at temperatures lower than that reached in the reactor.

ACKNOWLEDGEMENTS

We are thankful to P. Stille (Centre de Géochimie de la Surface, Strasbourg, France), to J.T. Chesley (University of Michigan, Ann Arbor, USA) who carefully reviewed the manuscript and helped with constructive comments, and to Mr. C. Hammel who drafted the figures.

REFERENCES

Baubron J.C., Jebrak M., Joannes C., Lhegu J., Touray J.C. and Ziserman A. (1980) Nouvelles datations K/Ar sur des filons à quartz et fluorine du Massif Central français. C. R. Acad. Sci., Paris, 290D, 951-953.

Baumann A. and Hofmann R. (1988) Strontium isotope systematics of hydrothermal vein minerals in deposits of West Germany. Geol. Rundsch., 77, 747-762.

Bell K. (1985) Geochronology of the Carswell area, Northern Saskatchewan. In: The Carswell Structure Uranium deposits, Saskatchewan. Laine R., Alonso D. and Svab M. (eds.), Geol. Ass.

Can. Sp. Paper, 29, 33-46.
Bellon H., Devillers C., Hagemann R. and Touray J.C. (1976) Dater les minéralisations. Mém. Hors Série de la Soc. géol. Fr., 7, 265-268.
Bonhomme M. (1982) Age triassique et jurassique des argiles associées aux minéralisations filoniennes et aux phénomènes diagénétiques tardifs en Europe de l'Ouest. Contexte géodynamique et implications génétiques. C. R. Acad. Sci., Paris, 294D, 521-524.
Bonhomme M. G., Yerle J.J. and Thiry M. (1980) Datation K-Ar de fractions fines associées aux minéralisations. Le cas du bassin uranifère permo-houiller de Brousse-Broquies (Aveyron). C. R. Acad. Sci., Paris, 291D, 121-124.
Bonhomme M., Buhmann D. and Besnus Y. (1983) Reliability of K-Ar dating of clays and silicifications associated with vein mineralizations in Western Europe. Geol. Rundsch., 72, 105-117.
Bray C.J, Spooner E.T.C, Hall C.M., York D., Bills T.M. and Krueger H.W. (1987) Laser probe ^{40}Ar-^{39}Ar and conventional K/Ar dating of illites associated with the McClean unconformity-related uranium deposits, North Saskatchewan. Canada. Can. Jour. Earth Sci., 24, 10-23.
Brockamp O., Zuther M. and Clauer N. (1987) Epigenetic-hydrothermal origin of the sediment-hosted Muellenbach uranium deposit, Baden-Baden, W-Germany. Mon. Series on Min. Dep., 27, 87-98.
Brookins D.C. (1980) Geochronologic studies in the Grants mineral belt. New Mex. Bur. Mines & Min. Res. Mem., 38, 52-58.
Chaudhuri S., Clauer N. and Ramakrishna S. (1983) Strontium isotopic composition of gangue carbonate minerals in the lead-zinc sulfide deposits at the Bushy Creek Mine, Viburnum trend, Southeast Missouri. In: Intern. Conf. on Mississippi Valley-Type lead-zinc deposits. Kisvarsanyi G., Grant S.k., Pratt W.P. and Koenig J.W. (eds.), Univ. Missouri-Rolla, Rolla MI, 140-144.
Chesley J.T., Halliday A.N. and Scrivener (1991) Samarium-Neodymium direct dating of fluorite mineralization. Science, 252, 949-951.
Clauer N., Chaudhuri S. and Subramanium R. (1989) Strontium isotopes as indicators of diagenetic recrystallization scales within carbonate rocks. Chem. Geol. (Isot. Geosc. Sect.), 80, 27-34.
Clauer N., Ey F. and Gauthier-Lafaye F. (1985) K-Ar dating of different rock types from the Cluff Lake uranium ore deposits (Saskatchewan, Canada). In: The Carswell Structure Uranium Deposits, Saskatchewan. Laine R., Alonso D. and Svab M. (eds.), Geol. Ass. Can. Sp. Pap., 29, 47-53.
Clauer N., Cocker J.D. and Chaudhuri S. (1992) Isotopic dating of diagenetic illites in reservoir sandstones: Influence of the investigator effect. In: Origin, Diagenesis, and Petrophysics of Clay Minerals in Sandstones, Houseknecht D.W. (eds.), Soc. Econ. Paleont. Miner., Spec. Pub., 47,
Darbyshire D.P.F. and Shepherd T.J. (1985) Chronology of granite magmatism and associated mineralization, SW England. Jour. Geol. Soc. London, 142, 1159-1177.
Dunham K.C., Fitch F.J., Ineson P.R., Miller J.A. and Mitchell J.G. (1968) The geochronological significance of argon^{40}/argon^{39} age determinations on White Whin from the northern Pennine orefield. Proc. Roy. Soc. London, A, 307, 251-266.
Faure G. (1986) Principles of Isotope Geology. Second Edition. J.

Wiley, New York, N.Y., 589p.
Fonbote L. and Gorzawski H. (1988) Isotopic trends (Sr, C, O, S) during diagenetic formation of stratabound Pb-Zn-(F-Ba) deposits in carbonate basins. Terra cognita, 8, 31.
Forbes P., Landais P., Pagel M. and Meyer A. (1987) Thermal evolution of the Guezouman Formation in the Akouta uranium deposit (Niger). Terra cognita, 7, 343.
Gancarz A.J. (1979) Chronology of the Cluff Lake uranium deposit, Canada. In: Intern. Uranium Symp. on the Pine Creek Geosyncline, N.T., Australia. Ext. Abstr., 91-94.
Grant N.K., Laskowski T.E. and Foland K.A. (1984) Rb-Sr and K-Ar ages of Paleozoic glauconites from Ohio and Missouri, U.S.A. Isot. Geosc., 2, 217-239.
Halliday A.N. (1977) K-Ar dating of mineralization episodes- A discussion. Econ. Geol., 72, 870-871.
Halliday A.N. (1978) ^{40}Ar-^{39}Ar stepheating studies of clay concentrates from Irish orebodies. Geochim. Cosmochim. Acta, 42, 1851-1858.
Halliday A.N. (1980) The timing of early and main stage ore mineralisation in S.W. Cornwall. Econ. Geol., 75, 752-759.
Halliday A.N. (1980) K-Ar dating of mineral deposits. Jour. Geol. Soc. London, 137, 101-102.
Halliday A.N. and Mitchell J.G. (1976) Structural, K-Ar and ^{40}Ar-39Ar age studies of adularia K-feldspars from the Lizard Complex, England. Earth Plan. Sci. Lett., 29, 227-237.
Halliday A.N. and Mitchell J.G. (1983) K-Ar ages of clay concentrates from Irish orebodies and their bearing on the timing of mineralisation. Trans. Royal Soc. Edinburgh, Earth Sci., 74, 1-14.
Halliday A.N. and Mitchell J.G. (1984) K-Ar ages of clay-size concentrates from the mineralisation of the Pedroches batholith, Spain, and evidence for Mesozoic hydrothermal activity associated with the break up of Pangaea. Earth Plan. Sci. Lett., 68, 229-239.
Halliday A.N., Shepherd T.J., Dickin A.P., MacLaren F. and Darbyshire F. (1986) Sm-Nd dating and fingerprinting of the North Pennine fluorite deposits. Terra cognita, 6, 227.
Halter G., Sheppard S.M.F., Weber F., Clauer N. and Pagel M. (1987) Radiation-related retrograde hydrogen isotope and K-Ar exchange in clay minerals. Nature, 330, 638-640.
Hearn P.P. and Sutter J.F. (1985) Authigenic potassium feldspar in Cambrian carbonates: Evidence of Alleghanian brine migration. Science, 228, 1529-1531.
Holliger P., Pagel M. and Pironon J. (1989) Losses in lead and radon as a mean of estimating the age of sedimentary pitchblende deposits: The Mullenbach uranium ore. Chem. Geol. (Isot. Geosc. Sect.), 80, 45-53.
Hunziker J.C., Frey M., Clauer N., Dallmeyer R.D., Friedrichsen H., Flehmig W., Hochstrasser K., Roggwiller P. and Schwander H. (1986) The evolution of illite to muscovite: mineralogical and isotopic data from the Glarus Alps, Switzerland. Contr. Mineral. Petrol., 92, 157-180.
Ineson P.R. (1974) K-Ar isotopic age determinations from some Lake district mineral localities. Geol. Mag., 111, 521-537.
Ineson P.R. (1975) K-Ar isotopic age determinations from some Welsh mineral localities. Inst. Min. Metall. Trans., B, 84, 7-16.
Ineson P.R. and Mitchell J.C. (1972) Isotopic age determinations on clay minerals from lavas and tuffs of the Derbyshire ore-

field. Geol. Mag., 109, 501-512.
Ineson P.R., Mitchell J.C. and Vokes F.M. (1975) K-Ar dating of epigenetic mineral deposits: An investigation of the Permian metallogenic province of the Oslo region, Southern Norway. Econ. Geol., 70, 1426-1436.
Kelley S., Turner G., Butterfield A.W. and Shepherd T.J. (1986) The source and significance of argon isotopes in fluid inclusions from areas of mineralization. Earth Plan. Sci. Lett., 79, 303-318.
Kesler S.E., Ruiz J. and Jones L.M. (1983) Strontium-isotopic geochemistry of fluorite mineralization (Coahuila, Mexico). Isot. Geosc., 1, 65-75.
Kessen K.M., Woodruff M.S. and Grant N.K. (1981) Gangue mineral $^{87}Sr/^{86}Sr$ ratios and the origin of Mississippi Valley-type mineralization. Econ. Geol., 76, 913-920.
Kish S.A. and Stein H.J. (1979) The timing of ore mineralization, Viburnum Trend, southeast Missouri lead-zinc district: Rb-Sr glauconite dating. Geol. Soc. Amer. Annual Meeting, Abstr. with Progr., 11, 458.
Lange S., Chaudhuri S. and Clauer N. (1983) Strontium isotopic evidence for the origin of barites and sulfides from the Mississippi Valley-Type ore deposits in southeast Missouri. Econ. Geol., 78, 1255-1261.
Leach D.L. (1979) Temperature and salinity of the fluids responsible for minor occurrences of sphalerite in the Ozark region of Missouri. Econ. Geol., 74, 931-937.
Lee M.J. and Brookins D.G. (1978) Rubidium-strontium minimum ages of sedimentation, uranium mineralization, and provenance, Morrison Formation (Upper Jurassic), Grants mineral belt, New Mexico. Am. Ass. Petrol. Geol. Bull., 62, 1673-1683.
Liewig N., Clauer N. and Sommer F. (1987) Rb-Sr and K-Ar dating of clay diagenesis in Jurassic sandstone oil reservoirs, North Sea. Am. Ass. Petrol. Geol. Bull., 71, 1467-1474.
Maas R., McCulloch M.T., Page R.W. and Campbell I.H. (1986) Sm-Nd isotopic systematics of uranium deposits. Terra cognita, 6, 225-226.
Miller D.S. and Kulp J.L. (1963) Isotopic evidence on the origin of the Colorado plateau uranium ores. Geol. Soc. Am. Bull., 74, 609-630.
O'Neil J.R. (1987) Preservation of H, C, and O isotopic ratios in the low temperature environment. In: Stable isotope geochemistry of low temperature. Kyser T.K. (ed.), Short Course Handbook 13, Min. Ass. Canada, 85-128.
Pagel M., Poty B. and Sheppard S.M.F. (1980) Contribution to some Saskatchewan uranium deposits mainly from fluid inclusion and isotopic data. In: Uranium in the Pine Creek Geosyncline, IAEA, Vienna, 639-654.
Philippe S. (1988) Systématique U-Pb et évolution comparée de minéralisations uranifères du bassin d'Athabasca (Saskatchewan, Canada). Cas de gisements de la structure Carswell et de Cigar Lake. Thèse Univ. Montpellier, France, 350p.
Pop G., Baciu F., Gata G., Popescu G. and Soroiu M. (1980) K-Ar dating of clay minerals produced by hydrothermal alteration, a method for determining the age of ore deposits emplacement. Rev. Roum. Géol. Géophys. Géogr., 24, 277-291.
Posey H.H., Stein H.J., Fullagar P.D. and Kish S.A. (1983) Rb-Sr isotopic analyses of Upper Cambrian glauconites, southern Missouri: Implications for movement of Mississippi Valley-

type ore fluids in the Ozark region. In: Intern. Conf. on Mississippi Valley-Type lead-zinc deposits. Kisvarsanyi G., Grant S.K., Pratt W.P. and Koenig J.W. (eds.), Univ. Missouri-Rolla, Rolla MI, 166-172.

Rama S.N.I., Hart S.R. and Roedder E. (1965) Excess radiogenic argon in fluid inclusions. Jour. Geophys. Res., 70, 509-511.

Reuter A. and Dallmeyer R.D. (1987) Significance of $^{40}Ar/^{39}Ar$ age spectra of whole-rock and constituent grain-size fractions from anchizonal slates. Chem. Geol. (Isot. Geosc. Sect.), 66, 73-88.

Reuter A. and Dallmeyer R.D. (1989) K-Ar and $^{40}Ar/^{39}Ar$ dating of cleavage formed during very low-grade metamorphism: A review. In: Evolution of metamorphic belts. Daly J.S., Cliff R.A. and Yardley B.W.D. (eds.), Geol. Soc. Spec. Publ., Blackwell Oxford, 43, 161-172.

Ruiz J., Jones L.M. and Kelly W.C. (1984) Rubidium-strontium dating of ore deposits hosted by Rb-rich rocks, using calcite and other common Sr-bearing minerals. Geology, 12, 259-262.

Sawkins F.J. (1966) Ore genesis in the North Pennine orefield in the light of fluid inclusion studies. Econ. Geol., 61, 385-401.

Shepherd T.J., Darbyshire D.P.F., Moore G.R. and Greenwood D.A. (1982) Rare earth element and isotope geochemistry of the North Pennine ore deposits. Bull. B.R.G.M., (2), sect. II, 4, 371-377.

Smith F.W. (1974) Factors governing the development of fluospar ore bodies in the North Pennine orefield. Ph.D. Thesis, Univ. Durham.

Stein H.J. and Kish S.A. (1985) The timing of ore formation in southeast Missouri: Rb-Sr glauconite dating at the Magmont mine, Viburnum Trend. Econ. Geol., 80, 739-753.

Turpin L., Clauer N., Forbes P. and Pagel M. (1991) U-Pb, Sm-Nd and K-Ar systematics of the Akouta uranium deposit, Niger. Chem. Geol. (Isot. Geosc. Sect.), 87, 217-230.

Wilson M.R., Kyser T.K., Mehnert H.H. and Hoeve J. (1987) Changes in the H-O-Ar isotope composition of clays during retrograde alteration. Geochim. Cosmochim. Acta, 51, 869-878.

York D., Masliwec A., Hall C.M., Kuybida P., Kenyon W.J., Spooner E.T.C., Scott S.D. and Pye E.G. (1981) The direct dating of ore minerals. Ont. Geol. Surv. Misc. Pap., 98, 334-340.

Zartman R.E., Brock M., Heyl A.V. and Thomas H.H. (1966) K-Ar and Rb-Sr ages of some alkalic intrusive rocks from central and eastern United States. Am. Jour. Sci., 265, 848-870.

Zuther M. (1983) Das Uranvorkommen Muellenbach/Baden Baden, eine epigenetisch-hydrothermale Impraegnationslagerstaette in Sedimente des Oberkarbons. Teil 1: Erzmineralbestand. N. Jb. Miner. Abh., 147, 191-216.

NEODYMIUM, STRONTIUM, OXYGEN AND HYDROGEN ISOTOPE COMPOSITIONS OF WATERS IN PRESENT AND PAST OCEANS: A REVIEW

Peter STILLE[1], Sam CHAUDHURI[2*],

Yousif K. KHARAKA[3] and Norbert CLAUER[1]

[1] Centre de Géochimie de la Surface, 1 rue Blessig, 67084 Strasbourg, (France)

[2] Department of Geology, Kansas State University, Manhattan, Kansas 66506, (U.S.A)

[3] United States Geological Survey, 345 Middlefield, Menlo Park, California 94025, (U.S.A.)

Knowledge of the isotopic compositions of trace elements dissolved in present and ancient sea waters is important for the following reasons.

1) Sea water integrates dissolved elements from continents, which are supplied by rivers, ground waters, terrigenous sediment particles, hydrothermal activities in oceans, submarine weathering of the oceanic crust, and diagenesis of sediments. As the isotopic compositions of some of the dissolved elements in these different sources are sufficiently varied, the isotopic signatures of the analyzed elements in the sea water may provide insight to the inventory of the elements present in sea water.

2) Any intra-oceanic variation in the isotopic compositions

(*) Presently at the Centre de Géochimie de la Surface Strasbourg (France)

may provide clues to the presence of major current systems within ocean basins.

3) Erosional history of the continent may be further understood from information on temporal variations in the isotopic compositions.

4) Quantification of isotopic composition of se water through geologic time is important for utilization of the time-trend isotopic change in problems of stratigraphy, origin of subsurface waters, water-rock interactions, and paleothermometry.

In this review article, we discuss only about the isotope compositions of Sr, Nd, O and H of present-day and ancient sea waters. Discussions on other isotopes such as S and C, have been given elsewhere in this book.

RECENT SEA WATER

Strontium Isotopes

Strontium occurs as a trace element in sea water with concentrations of about 7-8 mg/l. The residence time of Sr in the ocean is about 5 x 10^6 yrs, which is much longer than one cycle mixing time (about 10^3 yrs) of ocean waters. The $^{87}Sr/^{86}Sr$ ratio of present-day sea water is considered to be nearly uniform at about 0.70922 ± 0.00002 (Elderfield, 1986). But the uniformity in the Sr isotopic composition should be viewed with caution, as several studies have shown that small but measurable differences in the Sr isotopic values can exist for waters in different parts of oceans (Veizer, this volume). For example, ocean waters close to areas of hydrothermal activities have been found to have much lower values than the one frequently quoted in the literature (Albarède et al., 1981; Clauer and Olafsson, 1981). Submarine thermal spring waters from sea bed of the Florida continental shelf about 45 km south-southwest of Ft. Meyers, Florida, were found to contain 12.9 mg/l Sr with an $^{87}Sr/^{86}Sr$ ratio of 0.70858 ± 0.00004 (Clauer, unpublished data). Similarly, the Sr isotopic value of some present-day shallow coastal marine waters have been found to deviate from the presumed uniform marine Sr isotopic value: the Sr isotopic values of bay waters can be different from the norm, as noted by Aberg and Wickman (1987) in their study of waters from the Bothnian Bay of the Baltic Sea which yielded $^{87}Sr/^{86}Sr$ ratios between 0.70935 and 0.70951. Some studies of ancient marine sulfate deposits have also underscored differences in the Sr isotopic compositions of these minerals from their presumed contemporary marine waters, suggesting potential atypical Sr isotopic compositions for ocean water in restricted circulation basins adjacent to continents (Clauer, 1981; Burke et al., 1982; Chaudhuri and Clauer, this volume).

The balance of the sea water $^{87}Sr/^{86}Sr$ ratio was initially thought to be governed largely by the continental weathering which is dominated by fluvial inputs of Sr derived from silicate and marine carbonate rocks (Brass, 1976). Recognition of the alteration of oceanic crust at mid-oceanic ridge areas emphasized the importance of the oceanic hydrothermal activities as a factor governing the Sr isotopic composition of sea water (Clauer, 1976; Spooner, 1976). Strontium isotopic studies of interstitial waters in marine carbonate sediments also established that diagenetic flux of Sr from the carbonate sediments could have some bearing on the Sr isotopic budget of sea water (Elderfield and Gieskes, 1982). Runout or ground water can be an additional source of Sr which can augment the influence of continental Sr flux in the mass balance of the isotopic composition of Sr in sea water (Chaudhuri and Clauer, 1986). Assuming a steady-state condition for Sr in the ocean, Chaudhuri and Clauer presented the following equation for the Sr isotopic mass balance of the modern sea:

$$Sr_{rw}R_{rw} + Sr_{gw}R_{gw} + Sr_{dc}R_{dc} + Sr_{ob}R_{ob} = (Sr_{rw} + Sr_{gw} + Sr_{dc} + Sr_{ob})\ R_{sw}$$

where Sr_{rw} = flux of river Sr (2.21×10^{12} gy^{-1}), R_{rw} = isotope composition of river water (0.7111), Sr_{gw} = flux of ground water Sr (2.10×10^{12} gy^{-1}), R_{gw} = isotope composition of ground water (0.7110), Sr_{dc} = flux of diagenetic carbonate Sr (0.48×10^{12} gy^{-1}), R_{dc} = isotope composition of diagenetic carbonate (0.7084), Sr_{ob} = flux of oceanic basalt Sr (1.26×10^{12} gy^{-1}), R_{ob} = isotope composition of oceanic basalt (0.7029), and R_{sw} = isotope composition of modern sea water

(0.7092). Opinions are still divided as to the magnitude of influence between the crustal and the mantle sources of Sr on the Sr isotopic composition of sea water, as significant differences exist in the various estimates of the isotopic compositions and fluxes of the sources of Sr to the oceans (Veizer, 1989; Palmer and Edmond, 1989). Estimates of axial hydrothermal flux of Sr have ranged between 0.42×10^{12} and 1.36×10^{12} gy^{-1}.

Neodymium Isotopes

Both Nd and Sm occur as trace constituents in sea water respectively with average concentrations of about 2.6×10^{-6} mg/l and 0.5×10^{-6} mg/l. The Sm/Nd ratio of sea water is about 0.20, which is about the same as that of average crustal rocks or of average river suspended load. The residence time of Nd is still in question, as the estimates have ranged from about 300 to 7,100 years (Faure, 1986; Goldstein and Jacobsen, 1988).

The knowledge of the Nd isotopic compositions of ocean waters has come from measurements made either on the waters or on sediments precipitated from these waters, such as Mn nodules, phosphate and carbonate sediments, and others (Piepgras et al., 1979; Piepgras and Wasserburg, 1980, 1982; O'Nions et al., 1978; Clauer et al., 1984; Shaw and Wasserburg, 1985; Grandjean et al., 1987, Stille and Fischer, 1990; Stille et al., 1991, 1992). These results have shown that the major ocean basins differ in their Nd isotopic compositions. For example, the ϵ_{Nd} isotopic values of Atlantic Ocean waters are about -12 ± 2, that of Pacific Ocean waters are about -3 ± 2, and that of Indian Ocean waters are about -8 ± 2. Even within a single ocean basin, measurable differences exist in the Nd isotopic values of the waters. The Nd isotopic differences among the basins or within a basin reflect the differences in the isotopic compositions of the sources of Nd in the sea waters. The high negative ϵ_{Nd} values in general reflect contributions of Nd largely from continental sources, as the present-day oceanic basalts have an average ϵ_{Nd} value of about +10. The difference in the Nd isotopic compositions between Pacific and Atlantic ocean waters, the former being less negative than the latter, can be attributed to relatively young continental crusts surrounding the Pacific Ocean and its increased oceanic volcanic activities which produce rocks with elevated $_{Nd}$ values and the dominance of old continental rocks surrounding the Atlantic Ocean basin.

Any differences in the ϵ_{Nd} values of waters within a basin or between basins can be used to study the origins and exchange between different water masses, which has been demonstrated by Piepgras and Wasserburg (1982). These authors found that the Nd isotopic composition of sea water at the Drake Passage (South America) was between that of Pacific and Atlantic ocean waters, and they calculated that about 70% of Atlantic sea water is involved in the Antarctic Circumpolar Currents. Piepgras and Wasserburg (1980, 1982, 1983) also observed from investigations of water columns in the Atlantic Ocean, the Pacific Ocean, and the Saragasso Sea that the source of REE in the surface water of each ocean is distinct from that of the respective deep waters (Fig. 1). Their studies suggested that the isotopic data may be used to reconstruct circulation patterns and mixing mechanisms within the oceans.

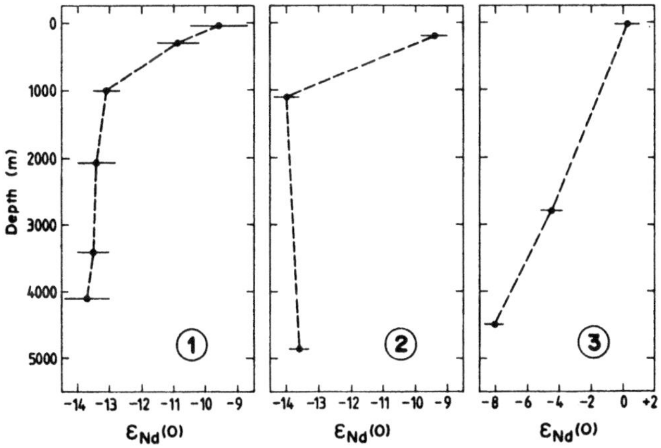

Figure 1: $\epsilon_{Nd}(0)$ as a function of depth in ocean water of Sargasso Sea (1), northwestern Atlantic Ocean (2), and southcentral Pacific Ocean (3) (modified from Piepgras and Wasserburg, 1983)

The Nd isotopic analyses can potentially be useful means of studying basinal flows. This has been demonstrated by Piepgras and Wasserburg (1983) who used the Nd isotopic analyses to study the flows of the water masses into and out of the Mediterranean Sea.

They observed that the Mediterranean outflow causes a shift of more than 2 in the ϵ_{Nd}d value toward more radiogenic Nd at a depth of about 1000 m. The core of the Mediterranean outflow corresponds to an ϵ_{Nd} value of -9.8, whereas the overlying and the underlying waters have $_{Nd}$ values of about -12. The eastern part of the North Atlantic Ocean and the Mediterranean Sea have significantly different Nd isotopic compositions. Based on mass balance calculations, these authors estimated an ϵ_{Nd} value of -6 for the Mediterranean Sea, which agrees well with the value of -7 for biogenic carbonates previously analyzed by Elderfield and Greaves (1982).

The ϵ_{Nd} values of the Mediterranean sea waters are between the more radiogenic values for waters of the Pacific Ocean and the less radiogenic values for waters of the Indian and the Atlantic oceans. Frost et al. (1986) suggested that the Nd isotopic compositions of Mediterranean sea waters probably reflect an influence of the particulate-Nd input rather than of the dissolved-Nd input to the basin. This is in agreement with the results of Elderfield and Greaves (1982) who observed that the surface water layers of the Atlantic Ocean are enriched in REE which led them to suggest the enrichment due to dissolution of the particulate materials delivered to the ocean basin. The idea of dissolution of particles is far from clear, as the enrichment in REE of surface waters is not everywhere evident. For example, Piepgras and Wasserburg (1983) noted Sm and Nd concentrations to increase with increasing nutrient concentrations (PO_4, NO_3, etc.) from surface downward. They suggested that these increases in the elemental contents with depth are induced by progressive increase in the dissolution of particles in the process of settling in deep oceans. Atlantic ocean waters from depths greater than 1000 m have been found with low ϵ_{Nd} values between -13 and -14, as compared to surface waters having ϵ_{Nd} values between -9 and -10. The source of this non-radiogenic Nd component in the bottom water has not yet been clearly identified. Piepgras and Wasserburg (1983) suggested that the low $^{143}Nd/^{144}Nd$ ratios for the deep waters in the Atlantic Ocean are the result of either run-off from Precambrian shields of North America and Asia into the Arctic Ocean by bottom current transport, or REE injection from old components in the deep sea sediments into the overlying water column.

Hydrogen and Oxygen Isotopes

Ocean waters, having their δD and $\delta^{18}O$ values close to zero, are to a small degree spatially varied in their H and O isotope compositions. For example, the $\delta^{18}O$ values of surface waters in large parts of the Atlantic and the Pacific oceans range from about +1.5 to -1.0 per mil. The ^{18}O concentrations also vary with depth, as Craig and Gordon (1965) have shown in their study of Indian Ocean waters pointing to measurable decrease in the concentrations of the heavy isotopes within the upper 1-1.5 km depth, below which the concentration remains unchanged. The average $\delta^{18}O$ contents of deep ocean waters among major basins are varied, ranging from about +0.12 per mil for the North Atlantic, to about -0.17 to -0.21 for the Pacific, to about -0.45 for the Antarctic (Craig and Gordon, 1965). These variations are related to evaporation, precipitation, addition of river water, freezing and melting of ice, and mixing of different water masses.

The H and O isotopic compositions of surface waters of major oceans are approximately linearly related due primarily to the combined effects of evaporation of the surface waters and precipitation of atmospheric moisture. Craig (1965) gave the following relationship for the surface waters of some major oceans:

$$\delta D = m\ \delta^{18}O$$

where m is equal to 7.5 for the North Pacific Ocean, 6.5 for the North Atlantic Ocean, and 6.0 for the Red Sea.

The H and O isotopic variations of waters in restricted or semi-enclosed seas can be significantly different from that in large open oceans. For example, the Gulf of Aqaba is about 180 km long and 14 km wide with average depth of about 800 m and connects to the Red Sea with inflow of surface water and outlow beneath the surface from and into the Red Sea. The surface waters of the Gulf of Aqaba have $\delta^{18}O$ values ranging from +2.0 to +2.5 per mil, as compared to Red Sea surface waters having $\delta^{18}O$ values of generally less than +1.8 (Anati and Gat, 1989). The $\delta^{18}O$ concentration increases to a δ-value of about 2.95 per mil at depth of about 150 m from the surface. Anati and Gat explain such high isotopic values for waters in the Gulf of Aqaba in terms of evaporation exceeding precipitation and runoff in an arid climate.

ISOTOPIC COMPOSITIONS OF OCEAN WATER THROUGH GEOLOGIC TIME

The chemical and isotopic composition of ocean water is mainly controlled by three fluxes, the riverine, the ground water and the hydrothermal input. Mountain building processes, changes in the configuration of the continents, climatic variations, increasing and decreasing subaquatic, hydrothermal activities and changes in the degree of alteration of continental and oceanic crust during the different geologic epochs caused the oscillations of isotopic compositions in sea water with time.

Secular Variation of Strontium Isotopes

The trend of variation in the $^{87}Sr/^{86}Sr$ ratios with time shows a significant shift to a more radiogenic value from Archean to Proterozoic time at about 2.5 Ga, suggesting major influence of run-off influx of radiogenic ^{87}Sr from newly evolved continents with recycling history that is coupled with decreased influx of mantle Sr to the oceans. In a very broad sense, the sea water $^{87}Sr/^{86}Sr$ ratio increased progressively through the Proterozoic and the Phanerozoic. Details about variations in the $^{87}Sr/^{86}Sr$ ratios of sea water during the Proterozoic suffer from lack of data with good stratigraphic control. But the data from sea water during the Phanerozoic show, generally speaking, progressive decrease in the $^{87}Sr/^{86}Sr$ ratios during the Paleozoic which is followed by an increasing trend in the ratio during the Mesozoic and the Cenozoic. Many prominent short-period fluctuations in the $^{87}Sr/^{86}Sr$ ratios have been noted, and these have been explained in terms of the balance between continental Sr input by run-off and run-out and mantle Sr input by oceanic hydrothermal activities.

For more detailed discussions of the secular variation of Sr isotopes, we refer to the companion paper by Veizer in this volume.

Secular Variation of Neodymium Isotopes

The secular variation of Nd isotopes in sea water is poorly known for several reasons.

1) Isotopic homogenization has not been attained between the water masses of different oceans. Therefore, in establishing the secular variation of Nd isotopes, changes in the configuration of the continents and resulting changes of the oceanic basins have to be considered. Even within water columns of present-day oceans large Nd isotopic variations have been observed (see above). Therefore, only samples coming from similar depositional environments (deep sea, shallow water, etc.) might yield a coherent picture of the secular variation. Also, knowledge of the streaming conditions, of the co-circulations, between the different water masses is of immense importance for the sampling and the interpretation of the isotope data. This has been discussed by Stille (1992) showing that during the past 80 Ma at some specific epochs the Nd isotopic composition of the Atlantic ocean water was much more controlled by changing circulation patterns between the North Atlantic and the Tethys oceans, and by the water masses coming from the Indian and Pacific oceans, than by river fluxes from surrounding continental areas.

2) Different materials such as metalliferous sediments, Fe-Mn ores, carbonate fossils, marine phosphates, authigenic argillites, clay fractions and glauconites have been used to estimate the Nd isotopic variation in ancient ocean waters. All of these components might have integrated the Nd and Sr isotopic composition of sea water, but formed in different environments and depths of the ocean and consequently integrated Nd isotopic compositions typical (or peculiar) to the broad environments in which they formed. A common evidence for reliable record of isotopic signatures of ancient sea water is given by phosphate minerals being strongly enriched in Nd and Sr. Conodonts are especially well suited for such isotopic

studies because they can be precisely dated biostratigraphically. Although metalliferous sediments and Fe-Mn ores integrate the Nd isotopic composition of sea water, chemical exchanges over tens of millions of years potentially render them to have typical sea water isotopic compositions (Clauer et al., 1984). Carbonates only contain small amounts of Nd and therefore their Nd isotopic compositions are suspect because they can have been influenced by diagenesis. Authigenic argillites, clay fractions and glauconites can be suitable for the derivation of the sea water Nd isotopic composition, but in many cases it is difficult to exclude contamination by inherited Nd coming from detritus.

These factors are limiting the precise derivation of the Nd secular variation. Nevertheless, for the use of any sedimentary material, one controlling factor has to be systematically considered: to ensure no diagenetic effects or inheritance modified the Nd isotopic compositions of the analyzed samples. One assurance of the choice of appropriate materials is to ensure that the Sr isotopic composition of the materials is identical to that of the contemporaneous sea water.

Nd isotopic variations in the Precambrian sea water

To constrain the secular variation in Precambrian sea water only a very small data base is available, which is founded mainly on information from banded iron formations (BIF) presumably of marine origin from chemical precipitation in the Precambrian oceans (Miller and O'Nions, 1985; Jacobsen and Pimentel-Klose, 1988a, 1988b). Therefore, they are expected to have incorporated the Nd isotopic composition of sea water at the time of their formation. The analyzed samples came from geographically widely different places of the world: Urucum (Bolivia-Brazilia), Sokoman (Labrador), Gunflint (Canada), Hammersly and Cleaverville (Australia), Vermillion and Beartooth (USA), Bjornevann (Norway), Michipicoten (Canada), Fig Tree (South Africa). The data are compiled in figure 2. Also included are data from some presumably authigenic clay materials from the Gunflint BIF, which define a time of diagenesis approximating close to the time of sedimentation, with an isochron age of 2.08 $\pm$ 0.25 Ga (Stille and Clauer, 1986) and some data on phosphorites (Bros, 1992) which formed about 2.04 $\pm$ 0.79 Ga ago within the Francevillian sedimentary basin, Gabon (Bros et al.,

1992).

The data define a huge field with Nd isotopic variations of up to 6 epsilon units for a single BIF. This range of variation is comparable with that of present-day sea water in a single oceanic basin. A trend of increasing $_{Nd}$ values with age is observable. This trend might indicate that the early Precambrian sea water was dominated by a much larger mantle component due to hydrothermal activity than the sea water of the late Precambrian or Phanerozoic. This observation is supported by Sr isotope data (Veizer et al., 1983). Therefore, Jacobsen and Pimentel-Klose (1988a) "argue for a dominantly hydrothermal source (with depleted mantle-like ϵ_{Nd} values) of rare earth elements in BIF's older than 2.0 Ga".

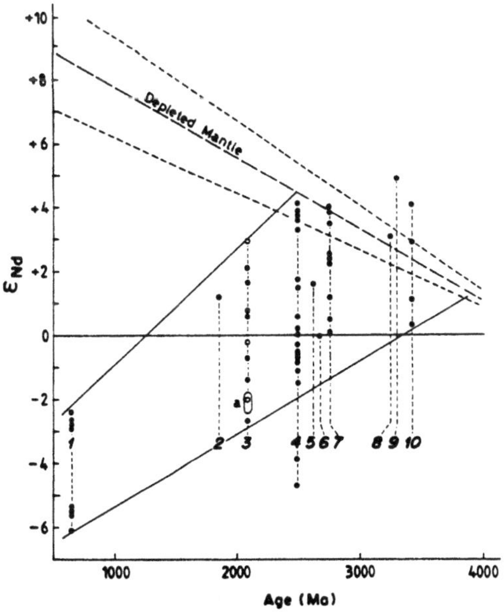

Figure 2: Nd isotopic variations in Precambrian sea water. The initial Nd values are plotted vs. the stratigraphic ages of the samples. Filled circles stand for Banded Iron Formations of Urucum (1), Sokoman (2), Gunflint (3), Hamersley (4), Vermillion (5), Bjornevann (6), Michipicoten (7), Beartooth (8), Cleaverville (9), and Fig Tree (10). Open circles stand for phosphates of Gabon (3). Authigenic clays from Gunflint Banded Formation are in the area (3a). Data are from Miller and O'Nions (1985), Stille and Clauer (1986), Jacobsen and Pimentel-Klose (1988a and b), and Bros (1992)

These authors, however, ignored the partly significantly negative ϵ_{Nd} values of the Hamersley BIF and the Gunflint BIF.

Argillites and a chlorite fraction (<2μm) of the Gunflint BIF define an isochron initial ratio of ϵ_{Nd} = -2.06 at 2.04 Ga, and Stille and Clauer (1986) suggested that the initial ratio of the isochron corresponds to the Nd isotopic composition of sea water some 2.04 Ga ago. Phosphorite sequences from the Francevillian sedimentary series in Gabon support the existence of crustal-like Nd in sea water at that time. These phosphorites show ϵ_{Nd} values ranging between +3 and -2. Therefore, similar to Miller and O'Nions (1985), we suggest that dominantly mantle sources can be discarded for the post-Archean sea water. However, because of the similarity of Nd isotopic composition in the mantle and crustal reservoirs during the Archean we cannot exclude that the Nd budget of the Archean sea water was influenced by mantle sources. The observed large variation of the Nd isotopic composition in the Precambrian sea water may be caused by: (a) post-depositional mobilization of Sm and Nd, (b) presence of mantle and crustal detritus in the investigated samples, (c) existence of different basins which have not been connected with eachother, and (d) different depositional environments and water depths in which the sediments formed.

Nd isotopic variations in the Phanerozoic sea water

<u>Paleozoic</u> - The Paleozoic sea water shows large variations in Nd isotopic compositions with ϵ_{Nd} values ranging from -3 to -20 (Fig. 3). For the time span between 400 Ma and 550 Ma, two different paleo-geography dependent evolutionary trends of ϵ_{Nd} values have been discovered (Keto and Jacobsen, 1987). These authors studied phosphatic brachiopods and conodont fossils from Europe, North America and the Southeastern US and observed that the samples coming from North America are generally much less radiogenic in Nd isotopic composition (ϵ_{Nd} from -10 to -20) than the samples coming from Europe and the southeastern United States (ϵ_{Nd} from -5 to -9). The ϵ_{Nd} values found for the samples from Europe and southeastern United States are identical to those defined earlier by Hooker et al. (1981) for the Iapetus Ocean. Therefore, Keto and Jacobsen (1987) inferred two ocean basins between North America and Baltica separated by an island and/or shoal circulation barrier, the smaller Iapetus Ocean having ϵ_{Nd} values ranging between -5 and -9 and the larger, coeval Pacific-Panthalassa Ocean with ϵ_{Nd} values ranging between -10 to -20. In the Late Ordovician to Silurian

(400-440 Ma) the two evolutionary Nd variation trends join each other. It is the time of the Taconic orogeny in North America when the Iapetus ocean closed again. At the end of the Paleozoic 300 Ma to 250 Ma ago the Nd isotopic composition of sea water strongly increases toward more radiogenic values (ϵ_{Nd} from -3 to -5) pointing to increasingly important volcanic activities and new generation of subduction zones around the Pacific/Panthalassa ocean.

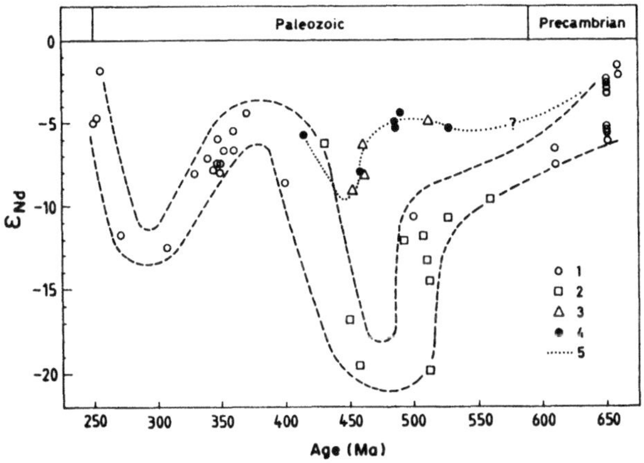

Figure 3: Nd isotopic variation in the Paleozoic sea water. Initial Nd values of mainly phosphates are plotted vs. their stratigraphic ages. Symbols 1 and 2 are for North American samples; 3 is for samples from southeastern United States; 4 is for European samples. The general dashed trend represents the Nd isotopic evolution in the Pacific-Panthalassa Ocean and the dotted line represents the Nd isotopic evolution in the Iapetus Ocean. Data are from Hooker et al. (1981), Shaw and Wasserburg (1985), Keto and Jacobsen (1987), and Banner et al. (1988)

Mesozoic and Cenozoic - This time span comprises the formation of the present-day oceans and therefore is of largest paleoceanographic interest to see the different evolutionary trends of Nd isotopic compositions in the Pacific, the Indian-Tethys-Mediterranean Sea and in the North and South Atlantic (Fig. 4); the variation being to some extent dependent on the communications between these oceans. During the past 220 Ma, the Nd isotopic composition in the Pacific increased with ϵ_{Nd} values ranging from -10 to -7 some 220 Ma ago and present-day values ranging between -1 and -4. The early Tethys sea water had Nd isotopic signatures identical to that of the Pacific, indicating that some 200 to 180 Ma ago Pacific sea water entered in the rift zones of the newly forming Tethys basin. However, continental runoff draining into the rift became more and more important as indicated by the decreasing ϵ_{Nd} values. No data are available for the early Atlantic ocean. At about 100 Ma ago a broad seaway existed between the Tethys and North Atlantic oceans. Therefore, at this time Tethys and Atlantic sea water must have had the same Nd isotopic composition. In the time span between 70 Ma and present-day the Tethys and Atlantic sea water evolved isotopically in different ways. This isotopic decoupling might be caused by the slowly closing of the Tethys seaway.

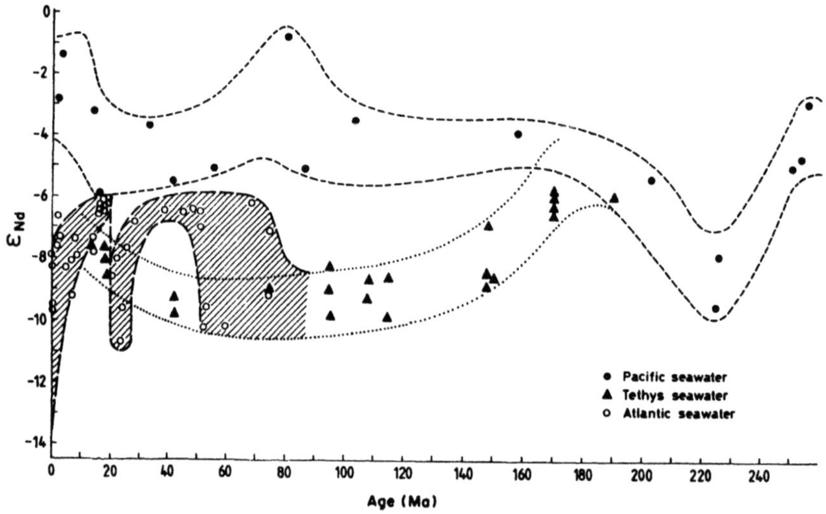

Figure 4: Nd isotopic variation in the Mesozoic and Cenozoic sea water of Pacific, Tethys, and Atlantic oceans. Data are from Piepgras and Wasserburg (1980), Chyi et al. (1984), Shaw and Wasserburg (1985), Staudigel et al. (1985), Grandjean et al. (1987), Keto and Jacobsen (1988), Stille et al. (1989), Stille and Fisher (1990), Steinmann et al. (1991), Stille et al. (1991, 1992), and Stille (1992)

A striking feature in the isotopic variation of the Atlantic sea water is that the Nd isotopic composition changed abruptly twice during the past 70 m.y. but not the Sr isotopic composition (see companion paper by Veizer, this volume) in the time span between 70 Ma and 35 Ma, the Sr isotopic composition of seawater only slightly increased, yet the Nd isotopic composition changed strongly toward more radiogenic values. A similar, even more abrupt change occurred at about 20 Ma. Stille (1992), who described these isotopic variations for the first time, suggests that these changes in Nd isotopic compositions without accompanying changes in Sr isotopic compositions reflect mixing of different water masses rather than changes in runoff due to orogeny. The author relates these variations to major changes in the configuration of the continents, which produced changes in the ocean-water circulation patterns within the Atlantic and changes in the communication of the Atlantic with other oceans.

With the opening of the South Atlantic and the northward drift of India, a broad seaway developped ca. 100 to 80 Ma which allowed westward-traveling currents of the Pacific Ocean to cross into the Indian Ocean and finally enter into the basins of the South Atlantic and later the North Atlantic. Pacific sea water with a more radiogenic Nd isotopic composition (ϵ_{Nd} -5 to -2) began to mix with first the South and then the North Atlantic sea waters. This mixing resulted in an increase of the Nd isotopic composition in the Atlantic seawater. It lasted approximately 40 m.y., reflecting the period of the progressive northward shift of the South-Atlantic-North Atlantic boundary and the continous widening of the passage between the South and North Atlantic. When Australia separated from Antarctica ca. 45 Ma ago, the current pattern in the Indian and Atlantic oceans changed again. The influx of Pacific currents into the South Atlantic must have diminished with the northward drift of Australia and its collision with Indonesia during middle Miocene time. The diminishing influx of Pacific sea water has caused the Nd isotopic composition to evolve back toward more crustal values.

At 25-20 Ma, major changes in the oceanic circulation occurred again, triggered mainly by the opening of the Drake passage and the formation of the deep circum-Antarctic currents. Antarctic Bottom Water (ABW) and Pacific sea water were able to enter the Atlantic Ocean. The new admixture of more radiogenic Pacific sea water (ϵ_{Nd} -6 to -3) and ABW resulted in a very rapid increase of the Nd isotopic composition in the Atlantic at about 20 Ma. The data imply that the Atlantic and Pacific oceans had strong co-circulations during the early Miocene.

Hydrogen and Oxygen Isotopic Secular Variations

There is a great controversy regarding the H and O isotopic composition of ocean water through geologic time. The role of Pleistocene glaciation on the H and O isotopic composition of the ocean is widely recognized, although the magnitude of this effect is uncertain. Craig (1965) calculated that the mean δD and $\delta^{18}O$ values of the ocean could be -0.6 and -5 per mil, respectively, if all the continental glaciers were to melt. The mean δD and $\delta^{18}O$ values of the ocean during the time of maximum glaciation during the Pleistocene epoch would have increased from present values by 1.0 and 7.4 per mil, respectively (see Emiliani and Shackleton, 1974, and Williams et al., 1981 for data and discussions).

There are two contrasting views on the H and O isotopic composition of ocean water through geologic time. The first view is based on the isotopic compositions of marine sediments, including cherts, carbonates, phosphates and glauconites (Weber, 1965; Perry and Tan, 1972; Knauth and Lowe, 1978; Shemesh et al., 1983; Veizer et al., 1986) that suggest that the $\delta^{18}O$ value of ocean water has increased monotonically at a rate of about 5 per mil per billion years. Taylor (1977), on the other hand, concluded that ocean water must have been relatively uniform in δD and $\delta^{18}O$ values during the past 2.5 billion years. Taylor's conclusion is based on the fact that granites affected by meteoric-hydrothermal waters have similar isotopic characteristcs over this time span.

Studies of the $\delta^{18}O$ values of the oceanic crust by Muehlenbachs and Clayton (1976) and Gregory and Taylor (1981) have also led to the conclusion that the $\delta^{18}O$ values of ancient oceans have been relatively constant. Material balance calculations based on the processes that deplete ocean water in ^{18}O (low temperature submarine weathering and sedimentation) and enrich it in ^{18}O (high-temperature hydrothermal alteration of the oceanic crust and the cycling of chemically bound water from the mantle) showed that the $\delta^{18}O$ of the ocean water probably has been held constant at about zero at least since the Cretaceous period. The $\delta^{18}O$-buffering of the ocean at about 0 per mil is also indicated by the patterns of $\delta^{18}O$ distribution in greenschist metabasalts on the modern ocean floor (Muelenbachs and Clayton, 1976), in Cretaceous ophiolitic sections (Gregory and Taylor, 1981) and in the Archean greenstones (Smith et al., 1984).

Assuming an isotopically buffered ocean water, there are two possible explanations for the generalized lowering of $\delta^{18}O$ values of marine sediments with increasing age. The first is based on increased post depositional isotopic exchange with isotopically light meteoric water. The second, used to explain the variations in the $\delta^{18}O$ values of chert (Knauth and Lowe, 1978) and of coexisting chert-phosphate pairs (Karhu and Epstein, 1986), is based on warmer oceanic (>60°C for Precambrian) temperatures. There are objections to these required high temperatures, and the fundamental question of whether the isotopic composition of ocean water has changed significantly over geologic time remains unresolved (for more discussion and references see Veizer et al., 1986; Muelenbachs, 1986; Wadleigh and Veizer, 1992; and Knauth, this volume).

CONCLUSIONS

The sea water $^{87}Sr/^{86}Sr$ ratio increased progressively through the Proterozoic and the Phanerozoic pointing to the increasing influence of run-off influx of radiogenic ^{87}Sr from newly evolved continents with recycling history and coupled with decreased influx of mantle Sr to the oceans.

The many prominent short-period fluctuations in the $^{87}Sr/^{86}Sr$ ratios can be explained in terms of the balance between continental Sr input by run-off and run-out and mantle Sr input by oceanic hydrothermal activities.

The $^{143}Nd/^{144}Nd$ ratio of the Proterozoic and Phanerozoic sea water indicate that dominantly mantle sources can be excluded for post-Archean sea water. In contrast to Sr, however, the secular variation of Nd in sea water is not only controlled by varying proportions of mantle and crustal influx. The residence time of Nd in sea water being much shorter than that of Sr and the time necessary to mix the oceans causes the Nd to be poorly mixed not only among major oceanic basins but also within each basin. Therefore, short periods of fluctuations in the Nd isotopic composition can be induced not only by strong changes of the relative erosion rates of very young and very old crustal material, the input of Nd during times of high hydrothermal activities but also by changes in the paleo-currents or in the communication paths between paleo-oceans.

The phenomenon of significantly different Nd isotopic compositions in the water masses of the recently formed oceans has proved to be useful for potential applications to paleoceanographic reconstructions. This phenomenon also implies that coherent pictures of the Nd secular variation can only be attained by comparing samples comming from similar depositional environments of one specific oceanic basin.

Modern open ocean waters have very limited range of H and O isotopic compositions, mostly close to zero for the δD and $\delta^{18}O$ values. Waters in restricted or semi-closed marginal ocean basins have been found with $\delta^{18}O$ values as high as about +3, which is related to excess evaporation and runoff in arid climate.

Maximum glaciation during the Pleistocene epoch might have caused the increase in mean δD and $\delta^{18}O$ values by about 1.0 and 7.4, respectively, from the present values. Opinions about the stable isotope composition of ocean waters during much of the geologic time are split between monotonous increase in the $\delta^{18}O$ value by about 5 per mil per billion years and relatively uniform during the past 2.5 billion years.

ACKNOWLEDGEMENTS

We would like to express our thanks to Mr. C. Hammel for his assistance in drafting the figures.

REFERENCES

Aberg F. and Wickman F.E. (1987) Variations of 87Sr/86Sr in water from streams discharging into the Bothnian Bay, Baltic Sea. Nordic Hydrol., 18, 33-42.

Albarède F., Michard A., Minster J.F. and Michard G. (1981) 87Sr/86Sr ratios in hydrothermal waters and deposits from the East Pacific rise at 21° N. Earth Plan. Sci. Lett., 55, 229-236.

Anati D.A. and Gat J.R. (1989) Restricted marine basins and marginal sea environments. In: Handbook of Environmental Isotope Geochemistry, Fritz P. and Fontes J.C. (eds.), Elsevier, Amsterdam, 29-73.

Banner J.L., Hanson G.N. and Meyers W.J. (1988) Rare earth element and Nd isotopic variations in regionally extensive dolomites from the Burlington-Keokuk formation (Mississippian): implications for REE mobility during carbonate diagenesis. Jour. Sediment. Petrol., 58, 415-432.

Brass G.W. (1976) The variation of the marine 87Sr/86Sr ratio during Phanerozoic time: interpretation using a flux model. Geochim. Cosmochim. Acta, 40, 721-730.

Brévart O. and Allègre C.J. (1977) Strontium isotopic ratios in limestone through geological time as a memory of geodynamic regimes. Bull. Soc. Géol. France, 19, 1253-1257.

Bros R. (1992) Géochimie isotopique (Sr, Nd, Ar, Pb) appliquée à des processus diagénétiques et hydrothermaux. Evolution du Bassin de Franceville d'âge protérozoique inférieur et influences des réacteurs nucléaires fossiles d'Oklo (Gabon). Thèse Univ. Strasbourg (in prep.).

Burke W.H., Denison R.E., Hetherington E.A., Koepnick R.B., Nelson N.F. and Otto J.B. (1982) Variation of seawater 87Sr/86Sr throughout Phanerozoic time. Geology, 10, 516-519.

Chaudhuri S. and Clauer N. (1986) Fluctuations of isotopic composition of strontium in seawater during the phanerozoic Eon. Chem. Geol. (Isot. Geosci. Sect.), 59, 293-303.

Chyi M.S., Crerar D.A., Carlson R.W. and Stallard R.F. (1984) Hydrothermal Mn-deposits of the Franciscan Assemblage, II. Isotope and trace element geochemistry, and implications for hydrothermal convection at spreading centers. Earth Plan. Sci. Lett., 71, 31-45.

Clauer N. (1976) Géochimie isotopique du strontium des milieux sédimentaires. Application à la géochronologie de la couverture du craton ouest-Africain. Sci. Géol. Mém., Strasbourg, 45, 256p.

Clauer N. (1981) $^{87}Sr/^{86}Sr$ ratios of the Barremian and early Aptian seas. In: Init. Rep. Deep Sea Dril. Proj., Thiede J., Vallier T.L. et al., vol. LXII, 781-783.

Clauer N. and Olafsson J. (1981) Icelandic thermal brines with a mantle Sr isotopic signature. Sci. Géol. Bull., Strasbourg, 34, 243-245.

Clauer N., Keppens E. and Stille P. (1991) Sr isotope constraints on the process of glauconitization. Geology (in press).

Clauer N., Stille P., Bonnot-Courtois C. and Moore W.S. (1984) Nd-Sr isotopic and REE constraints on the genesis of hydrothermal manganese crusts in the Galapagos. Nature, 311, 743-745.

Craig H. (1965) The measurement of oxygen isotope paleotemperatures. In: Stable isotopes in oceanographic studies and paleotemperatures, Tongiorgi E. (ed.), Spoleto, 1965, CNR, Pisa, 161-182.

Craig H. and Gordon L. (1965) Deuterium and oxygen-18 variation in the ocean and the marine atmospheres. In: Stable isotopes in oceanographic studies and paleotemperatures, Tongiogi E. (ed.), Spoleto, CNR, Pisa, 9-13.

Elderfield H. (1986) Strontium isotope stratigraphy. Paleogeo. Paleoclim. Paleoecol., 57, 71-90.

Elderfield H. and Gieskes J.M. (1982) Sr isotopes in interstitial waters of marine sediments from deep sea drilling project cores. Nature, 300, 493-497.

Elderfield H. and Greaves M.J. (1982) The rare earth elements in seawater. Nature, 296, 214-219.

Emiliani C. and Shackelton N.J. (1974) The Brunhes epoch: isotopic paleotemperatures and geochronology. Science, 183, 511-514.

Faure G. (1986) Principles of Isotope Geology. Second Edition, J. Wiley & Sons, New York, 589p.

Faure G. (1982) The marine-strontium geochronometer. In: Numerical Dating in Stratigraphy, Odin G.S. (ed.), J. Wiley & Sons, New York, 1, 73-79.

Faure G., Hurley P.M. and Powell J.L. (1965) The isotopic composition of strontium in surface water from the North Atlantic Ocean. Geochim. Cosmochim. Acta, 29, 209-220.

Frost C.D., O'Nions R.K. and Goldstein S.L. (1986) Mass balance for Nd in the Mediterranean Sea. Chem. Geol., 55, 45-50.

Goldstein S.L. and O'Nions R.K. (1981) Nd and Sr isotopic relationships in pelagic clays and ferromanganese deposits. Nature, 292, 324-327.

Goldstein S.J. and Jacobsen S.B. (1988a) Rare earth elements in river waters. Earth Plan. Sci. Lett., 89, 35-47.

Goldstein S.J. and Jacobsen S.B. (1988b) Nd and Sr isotopic systematics of river water suspended material: implications for custal evolution. Earth Plan. Sci. Lett., 87, 249-265.

Grandjean P., Cappetta H., Michard A. and Albarède F. (1987) The assessment of REE patterns and 143Nd/144Nd ratios in fish remains. Earth Plan. Sci. Lett., 84, 181-196.

Gregory R.T. and Taylor H.P. Jr. (1981) An oxygen isotope profile in a section of Cretaceous oceanic crust, Smail Ophiolite, Oman: Evidence for ^{18}O-buffering of the oceans by deep (>5km) seawater-hydrothermal circulation at mid-ocean ridges. Jour. Geophys. Res., 86, 2737-2755.

Hooker P.J., Hamilton P.J. and O'Nions R.K. (1981) An estimate of the Nd isotopic composition of Iapetus sea water from ca. 490 Ma metalliferous sediments. Earth Plan. Sci. Lett., 56, 180-188.

Jacobsen S.B. and Pimentel-Klose M.R. (1988a) Nd isotopic variations in Precambrian Banded Iron Formations. Geophys. Res. Lett., 15, 393-396.

Jacobsen S.B. and Pimentel-Klose M.R. (1988b) A Nd isotopic study of the Hamersley and Michipicoten banded iron formations: the source of REE and Fe in Archean oceans. Earth Plan. Sci. Lett., 87, 29-44.

Karhu J. and Epstein S. (1986) The implication of the oxygen isotope records in coexisting cherts and phosphates. Geochim. Cosmochim. Acta, 50, 1745-1756.

Keto L.S. and Jacobsen St.B. (1988) Nd isotopic variations of Phanerozoic paleoceans. Earth and Plan. Sci. Lett., 90, 395-410.

Keto L.S. and Jacobsen St.B. (1987) Nd and Sr isotopic variations of early Paleozoic oceans. Earth and Plan. Sci. Lett., 84, 27-41.

Knauth L.P. and Lowe D.R. (1978) Oxygen isotope geochemistry of cherts from Onverwacht Group (3.4 billion years), Transvaal, South Africa, with implications for secular variations in the isotopic composition of cherts. Earth Plan. Sci. Lett., 41, 209-222.

Miller R.G. and O'Nions R.K. (1985) Source of Precambrian chemical and clastic sediments. Nature, 314, 325-330.

Muehlenbachs K (1986) Alteration of the oceanic crust and the ^{18}O history of seawater. In: Stable isotopes in high temperature geological processes, Valley J.W., Taylor H.P. Jr. and O'Neil J.R. (eds.), Rev. Miner., 16, 425-444.

Muehlenbachs K. and Clayton R.N. (1976) Oxygen isotope composition of the oceanic crust and its bearing on seawater. Jour. Geophys. Res., 81, 4365-4369.

O'Nions R.K., Carter S.R., Cohen R.S., Evensen N.M. and Hamilton P.J. (1978) Pb, Nd and Sr isotopes in oceanic ferro-manganese deposits and ocean floor basalts. Nature, 273, 435-438.

Palmer M.R. and Edmond J.M. (1989) The strontium isotope budget of the modern ocean. Earth Plan. Sci. Lett., 92, 11-26.

Perry E.C. and Tan F.C. (1972) Significance of oxygen and carbon isotope variations in early Precambrian cherts and carbonate rocks of South Africa. Geol. Soc. Am. Bull., 83, 683-664.

Piepgras D.J. and Wasserburg G.J. (1983) Influence of the Mediterranean Outflow on the isotopic composition of Nd in waters of the North Atlantic. Jour. Geophys. Res., 88, 5997-6006.

Piepgras D.J. and Wasserburg G.J. (1982) Isotopic composition of neodymium in waters from the Drake Passage. Science, 217, 207-214.

Piepgras D.J. and Wasserburg G.J. (1980) Neodymium isotopic variations in seawater. Earth Plan. Sci. Lett., 50, 128-138.

Piepgras D.J., Wasserburg G.J. and Dasch E.J. (1979) The isotopic composition of Nd in different ocean masses. Earth Plan. Sci. Lett., 45, 223-236.

Shaw H.F. and Wasserburg G.J. (1985) Sm-Nd in marine carbonates and phosphates: implications for Nd isotopes in seawater and crustal ages. Geochim. Cosmochim. Acta, 49, 503-518.

Shemesh A., Kolodny Y. and Luz B. (1983) Oxygen isotope variations in phosphate of biogenic apatites: II. Phosphorite rocks. Earth Plan. Sci. Lett., 64, 405-416.

Smith H.S. O'Neil J.R. and Erlank A.J. (1984) Oxygen isotope composition of minerals and rocks and chemical alteration patterns in pillow lava from the Barberton greenstone belt, South Africa. In: Archean geochemistry: The origin and evolution of Archean continental crust, Kröner A., Hanson G.N. and Goodwin A.M. (eds.), Springer-Verlag, Heidelberg, 115-137.

Spooner E.T.C. (1976) The strontium isotopic composition of seawater, and seawater-oceanic crust interaction. Earth Plan. Sci. Lett., 31, 167-174.

Staudigel H., Doyle P. and Zindler A. (1985) Sr and Nd isotope systematics in fish teeth. Earth and Plan. Sci. Lett., 76, 45-56.

Steinmann M., Stille P. and Winkler W. (1991) Neodynium isotopes applied to stratigraphy of a metamorphic series. Terra nova,

3, 285.
Stille P. (1992) Nd-Sr isotope evidences for dramatic changes of paleo-currents in the Atlantic ocean during the past 80 million years. Geology (in press).
Stille P., Riggs St., Clauer N., Crowson R., Ames D. and Snyder St.W. (1992) Sedimentation through one Miocene depositional cycle based on Sr and Nd isotopic analysis of phosphorite peloids: North Carolina Continental Shelf (Part 1). Jour. Sediment. Petrol. (submitted).
Stille P., Riggs St. and Clauer N. (1991) Nd and Sr isotope stratigraphy of phosphates from the North Carolina Continental Margin. Terra nova, 3, 285.
Stille P., Clauer N. and Keppens E. (1992) Behavior of Nd-Sr isotope systematics during recent glauconitization process. Contr. Mineral. Petrol. (submitted).
Stille P. and Fischer H. (1990) Secular variation in the isotopic composition of Nd in Tethys sea water. Geochim. Cosmochim. Acta, 54, 3139-3145.
Stille P., Clauer N. and Abrecht J. (1989) Nd isotopic composition of Jurassic Tethys sea water and the genesis of Alpine Mn deposits: evidence from Sr-Nd isotope data. Geochim. Cosmochim. Acta, 53, 1095-1099.
Stille P. and Clauer N. (1986) Sm-Nd isochron age and provenance of the argillites of the Gunflint Iron Formation in Ontario, Canada. Geochim. Cosmochim. Acta, 50, 1141-1146.
Veizer J. (1989) Strontium isotopes in seawater through time. Ann. Rev. Earth Plan. Sci., 17, 141-167
Veizer J., Fritz P. and Jones B. (1986) Geochemistry of brachiopods: Oxygen and carbon isotopic records of Paleozoic oceans. Geochim. Cosmochim. Acta, 50, 1679-1696.
Wadleigh M.A. and Veizer J. (1992) $^{18}O/^{16}O$ and $^{13}C/^{12}C$ in Lower Paleozoic articulate brachiopods: implications for the isotopic composition of seawater. Geochim. Cosmochim. Acta, 56, 431-443.
Weber J.N. (1965) Changes in the oxygen isotopic composition of seawater during the Phanerozoic evolution of the oceans. Geol. Soc. Am., Spec. Paper, 82, 218-219.
Williams F.D., Moore W.S. and Fillon R.H. (1981) Role of glacial Arctic Ocean ice sheets in Pleistocene oxygen isotope and sea-level records. Earth Plan. Sci. Lett., 56, 157-166.

STABLE ISOTOPE GEOCHEMISTRY AND ORIGIN OF WATERS IN SEDIMENTARY BASINS

Yousif K. KHARAKA

and

James J. THORDSEN
United States Geological Survey
345, Middlefield, Menlo Park, California 94025 (USA)

Formation waters with salinities ranging from about 5,000 mg/l to more than 350,000 mg/l dissolved solids play a fundamental role in the physical and chemical processes that occur in sedimentary basins. These waters are particularly important in: (a) the generation, transport, accumulation and production of petroleum, (b) the chemical aspects of mineral diagenesis, including dissolution, precipitation and mineral transformations leading to major increases and/or decreases in the porosity and permeability of reservoir rocks, (c) the transport and precipitation of Cu, U, and especially sediment-hosted Mississippi Valley-type Pb-Zn ore deposits, (d) tectonic deformations, (e) the transport of thermal energy for the geothermal and geopressured-geothermal systems, and (f) the interaction, movement and ultimate fate of large quantities of hazardous wastes injected into the subsurface. Despite clear economic and scientific importance, and about a century of investigations, there is still controversy regarding the origin and evolution of these saline waters (Hanor et al., 1988).

Our understanding of the origin of these waters and the

processes that modify their chemical and isotopic compositions, however, has increased considerably in the last 10 years as a result of: (a) improved analytical techniques, especially the use of inductively coupled plasma (ICP) alone or in combination with mass spectrometry (ICP/MS) for the determination of cations and metals, and the use of ion chromatography (IC), gas chromatography (GC) and GC/MS for the analysis of anions and organics (Hull et al., 1985; Hanor et al., 1988), (b) improved sampling methodologies, especially with the use of specially designed downhole samplers (Kharaka et al., 1988a), (c) increased availability and utilization of data on a variety of stable and radioactive isotopes (Fritz and Fontes, 1980, 1986; Faure, 1986; Valley et al., 1986; Hoefs, 1987; Kyser, 1987), (d) major improvements in the chemical thermodynamic data and procedures for applying them to brines and minerals, and (e) development and application of detailed geochemical, hydrological and solute transport codes (Cathles, 1981; Bethke, 1986; Hanor, 1987; Kharaka et al., 1988b; Melchior and Bassett, 1990). We now realize that these fluids are much more mobile and their interactions with rocks much more complex than previously realized. Also, the discovery of high concentrations (up to 10,000 mg/l) of reactive organic species in these waters has led to a new field of organic-inorganic interactions and developed bridges between geochemistries of aqueous fluids, organic matter and petroleum (Kharaka et al., 1985; Gautier, 1986; Surdam et al., 1989; Lundegard and Kharaka, 1990).

In this chapter we review the origin of water and the composition and sources of inorganic and organic solutes from several petroliferous sedimentary basins in North America. Variations in stable isotopes of oxygen and hydrogen in present-day and old meteoric waters are reviewed and used in conjunction with the isotopic and chemical compositions of deep basin waters to indicate their origin and chemical evolution. The isotopic composition of ocean water through geologic time is discussed in a separate contribution of this volume. The emphasis is on the main processes that are responsible for the modification of the chemical and isotopic compositions of these waters including: (a) mixing, (b) dissolution of evaporites, especially halite, (c) incorporation of bitterns, the residual water remaining after the precipitation of evaporites, (d) dissolution and precipitation of minerals other than evaporites, (e) interaction with rocks, principally clays, siltstones and shales that behave as geological membranes, (f) activity of bacteria that can survive in sedimentary rocks to temperatures of up to about 80°C (Davis, 1967; Carothers and Kharaka, 1978), (g) interaction with organic matter, and (h) diffusion, especially in the presence of salt domes.

A number of excellent reviews on stable isotope and elemental geochemistry of waters in sedimentary basins have appeared in the last decade. Books written or edited by Fritz and Fontes (1980; 1986), Faure (1986), Hoefs (1987), Marshall (1987), and Chillingarian and Wolf (1988) contain several excellent chapters with information pertinent to this topic. Recent conference volumes with important contributions relevant to this chapter include Hitchon et al. (1986), Gautier (1986), Naeser and McCullough (1989), Miles (1989), and Melchior and Bassett (1990). Finally, several short-course volumes have recently appeared, including Gautier et al. (1985), Valley et al. (1986), Hanor (1987), Kyser (1987), and Hutcheon (1989). It is clear that water-rock interaction in sedimentary basins is receiving much attention from researchers in several fields of geology, hydrology and isotope geochemistry.

CHEMICAL COMPOSITION OF SUBSURFACE WATERS

Chemical composition of surface and shallow ground waters are highly variable. The waters are generally fresh (<1,000 mg/l dissolved solids) or brackish (1,000 to 10,000 mg/l dissolved solids), but the proportions of the major cations and anions are variable and dependent mainly on the mineral compositions of the rocks with which they come into contact (Freeze and Cherry, 1979; Hem, 1985). Many shallow ground waters are dominated by sulfate, which is replaced by carbonate in deeper ground waters. Dissolved sulfate in these waters is derived from dissolution of gypsum and anhydrite, or from oxidation of pyrite. Under reducing conditions prevalent in deeper waters, dissolved sulfate is reduced by bacteria, and dissolved organic species are oxidized to bicarbonate. The net reaction may be represented by:

$$SO^{2-}_4 + CH_3COO^- \rightleftharpoons HS^- + 2HCO^-_3 \qquad (1)$$

The salinity of oil field waters from different sedimentary basins is highly variable (Table 1, Fig. 1). The variations can be large in waters from different areas of the same basin (Table 2) and even in waters from the same petroleum field (Fig. 2). Water salinities in sedimentary basins range from about 5,000 to more than 350,000 mg/l dissolved solids. In any given basin, salinity generally increases with depth but the rate of increase is obviously variable (Fig. 1). For entire basins (e.g. Central Valley, California) or large areas within a basin (e.g. Corpus Christi area of Texas) that are devoid of evaporites and dominated by shales and siltstones, the salinities of formation waters show a general increase followed by a decrease with increasing depth of the reservoir rocks; salini-

Table 1: Chemical composition (mg/l) of formation waters from Sacramento and San Joaquin Valleys (California), North Slope (Alaska) and central Mississippi Salt Dome Basin (Mississippi)

Area	Sacramento			San Joaquin		North Slope		Central Mississippi	
Field	Grimes	Malton-Black Butte	San Emidio Nose	Wheeler Ridge	Kettleman North Dome	Barrow	Prudhoe Bay	Reedy Creek	West Nancy
Sample #	81-NSV-15	81-NSV-1	74-SEN-3	75-WR-5	912-1	78-AK-52	78-AK-54	84-MS-11	84-MS-1
Well Name	GOU4#2	19-1	21-15	21-28	323-21	S. Barrow 5	Arco 13	W.M. Geiger	W. L. West
Prod. Zone	Forbes	Forbes	Reef Ridge	Tejon	Lower McAdams	Barrow Sandstone	Sadlerochit Group	Rodessa	Smackover
Depth (m)	2,074	1,524	3,337	2,691	3,520	728	2,820	3,486	4,428
Temp. (°C)	65	58	149	117	141	16	94	102	118
TDS	18,600	21,400	10,900	44,300	10,000	22,100	21,900	320,000	275,000
Li	0.32	0.35	1.95	1.95	3.05	2.1	4.0	35	74
Na	6,830	7,510	4,000	7,450	3,760	7,980	7,600	61,700	54,800
K	35.5	28.4	620	135	92.4	3.0	86	990	6,500
Mg	72	148	7.0	27	3.4	67	20	3,050	3,350
Ca	182	331	67	5,550	30.7	119	182	46,600	33,900
Sr	14.3	18.8	8.0	187	4.4	16.1	20.2	1,920	1,670
Ba	6.4	4.6	4.2	12	3.98	175	3.8	60	48
Fe	0.58	54	0.36	2.8	0.31	5.5	63	465	0.47
NH_3	34	30	73	32	8.9	19	17	34	119
F	------	------	3.0	2.0	0.3	1.6	1.5	1.5	11.5
Cl	11,000	12,700	3,460	21,450	4,680	11,800	10,600	198,000	170,000
Br	44	74	57	80	45	62	54	2,020	2,080
I	30	66	14	46	27	28	19	17	80
HCO_3	359	417	2,870	2,210	1.190	1,710	2,930	206	197
SO_4	<0.5	0.9	38	50	0.5	n.d	69	64	161
H_2S	0.07	<0.1	0.02	0.11	3.02	<.1	<.1	<0.02	57.4
SiO_2	31	18	109	46	128	11	62	28	34
B	------	------	92	600	43	42	158	59	341
pH	7.6	7.6	7.7	6.9	7.4	7.2	6.5	5.08	5.48

Note: Depth is depth below ground level of mid-point of perforation. Production zone are those used by oil companies. TDS is calculated total dissolved solids. HCO_3 is field-titrated alkalinity and includes organic and inorganic species.

ties are relatively low reaching values of 5,000 mg/l in the deeper parts of the sections. Finally, a trend of constant water salinities is observed in the North Slope, Alaska, where the salinities are remarkably similar (19,000 to 24,000 mg/l dissolved solids, Table 1), considering the large areal extent of the sampled wells and the great variation in the depth (700 to 2,800 m) and lithologies of the reservoir rocks (Kharaka and Carothers, 1988).

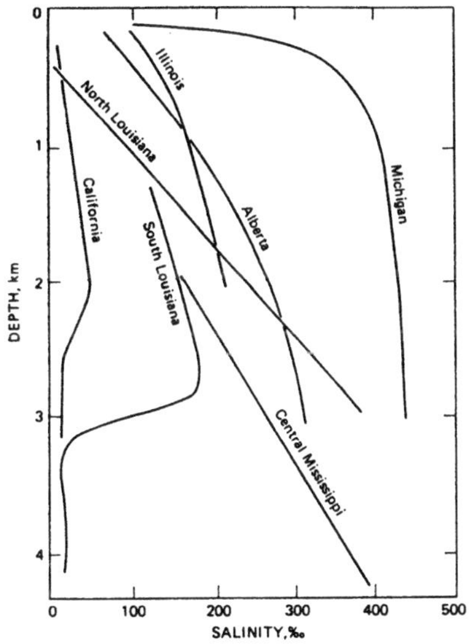

Figure 1: Salinity distribution with depth of reservoir rocks from some basins in North America. Note the different trends and the reversal of salinity in California and South Louisiana

Chloride is the dominant anion in nearly all formation waters associated with petroleum (Tables 1, 2, Fig. 3). The proportion of Cl to total anions (generally 70 to 90%) generally increases as the salinity of water increases reaching values of up to 99%. Acetate may comprise a high portion of total anions, especially in Na-Cl-CH_3COO type waters that are present mainly in Cenozoic reservoir rocks that are at temperatures of 80 to 100°C. In these waters, acetate and other organic acid anions can reach concentra-tions of up to 10,000 mg/l and contribute up to 99% of the measured alkalinities. The concentration of sulfate in oil field waters is low (generally <100 mg/l), even in connate waters of marine origin.

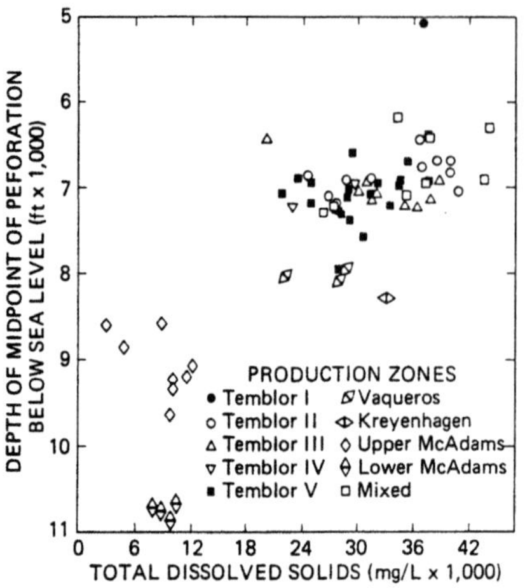

Figure 2: Distribution of salinity of formation waters from Kettleman North Dome oil field (California). Note the much lower salinity of water in the deeper McAdams Formation compared to Temblor Formation (Kharaka and Berry, 1974)

Table 2: Chemical composition (mg/l) and production data of formation waters in geopressured zones from coastal Texas and Louisiana

Area	Lafayette, LA		Houston-Galveston, TX		Corpus Christi, TX		McAllen-Pharr, TX	
Field	Weeks Island	Tigre Lagoon	Chocolate Bayou	Halls Bayou	Portland	East Midway	Pharr	La Blanca
Sample #	77-GG-19	77-GG-55	76-GG-7	76-GG-24	76-GG-63	77-GG-73	77-GG-73	77-GG117
Well Name	St.Un. A#9	Edna Delcambre #1	Angle #3	Houston "FF" #1	Portland A-3	Taylor E-2	Kelly A-1	La Blanca #12
Production Zone	S-Sand	Sand #3	Upper Weiting	Schenck	Morris	Lower Frio	Marks	7150 Sand
Depth (m)	4,275	3,928	3,444	4,161	3,514	3,662	3,018	2,903
Temp. (°C)	117	114	118	150	123	128	127	148
Pressure	43.1	75.8	52.4	80.0	58.0	62.2	52.4	56.6
Fluid Production								
Oil/condensate	21.9	0	0.5	3.8	4.8	2.7	0	0.3
Water	56.0	633	6.7	57.9	7.5	0.2	7.1	51.0
Gas	6.1	7.9	5.1	70.8	25.1	4.9	3.2	17.1
TDS	235,700	112,200	73,300	58,100	17,800	36,000	36,600	7,500
Na	78,000	40,000	26,500	20,500	6,500	13,250	9,420	2,680
Li	16	7.1	9.9	15	3.6	4.2	7.5	1.2
K	1,065	265	400	180	68	72	240	46
Rb	3.4	0.8	0.4	0.9	0.3	0.5	0.8	0.1
Cs	11.8	3.5	----	----	----	----	2.9	0.3
Mg	1,140	270	220	170	15	48	18	3.3
Ca	10,250	1,860	2,000	800	89	330	4,225	150
Sr	920	320	365	170	7.0	23	256	9.6
Ba	185	8.2	290	59	1.4	13	27	1.5
Fe	84	0.4	10.2	22	2.3	1.6	4.1	<0.1
Cl	143,000	67,900	42,700	34,500	9,270	21,000	22,000	3,950
F	0.8	0.8	0.8	----	1.5	7.3	3.9	5.7
Br	419	63	52	32	19	45	78	15
I	18	26	16	11	25	45	22	16
B	44	57	35	91	62	35	105	117
NH_3	100	69	29	13	5.8	13.5	21.5	4.2
H_2S	0.4	0.5	1.2	1.4	<0.1	0.04	<0.1	<0.1
HCO_3	450	1,050	455	409	1,600	1,180	114	400
SO_4	6.4	220	2.7	16	110	42	7.0	57
SiO_2	48	57	87	110	93	132	90	88
pH	6.2	6.3	5.9	6.8	6.8	6.4	6.8	7.3

Note: Depth is depth below ground level of mid-point of perforation. Temp. is measured subsurface temperature. Pressure is original bottom-hole pressure in MPa (1 psi = 6.9 kPa). Liquid production is in m^3/day. Production zones are names used by oil companies. TDS is calculated total dissolved solids. HCO_3 is the field titrated alkalinity and includes organic and inorganic species.

Sodium is the dominant cation in oil field waters, and it generally constitutes 70 to more than 90% of total cations (Tables 1, 2). Calcium generally is the second most abundant cation, and its proportions increase, especially in Na-Ca-Cl type waters in which it may reach concentrations of up to about 50,000 mg/l (Table 1). This increase with salinity, however, is variable in the same basin or in different basins. The concentrations and proportions of Mg are generally much lower than those in ocean water and generally decrease with increasing subsurface temperatures. The concentration and proportions of Sr, Ba, and Fe are generally higher than those in ocean water and generally increase with increasing Ca concentrations. The ratios of Li, K and Ce to Na generally increase with increasing subsurface temperatures, but again the proportions vary from basin to basin.

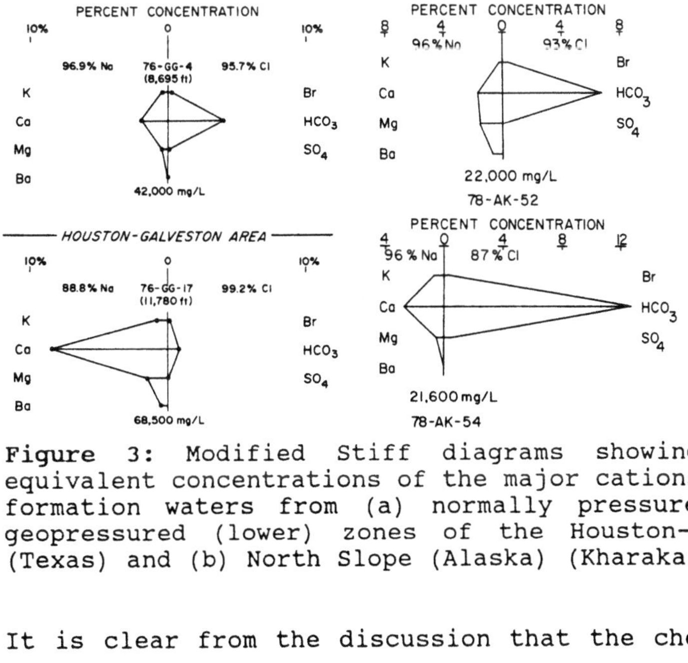

Figure 3: Modified Stiff diagrams showing the percent equivalent concentrations of the major cations and anions in formation waters from (a) normally pressured (upper) and geopressured (lower) zones of the Houston-Galveston area (Texas) and (b) North Slope (Alaska) (Kharaka et al., 1985)

It is clear from the discussion that the chemical composition of subsurface waters, including those of marine origin, is different from that of ocean water. Several investigators (Claypool and Kaplan, 1974; Gieskes, 1981) have noted major changes (with sulfate and Mg depletions and bicarbonate increases) in the chemical composition of sea water trapped in shallow sediments. These early changes become more pronounced with increasing depth of sediments and lead to chemical compositions found in oil field waters.

The chemical composition of formation waters from High Island field, offshore Texas, was investigated by Kharaka et al. (1985) to evaluate the early stages of diagenesis of marine waters (Table 3). The reservoir rocks in this field are present at depths that range from about 1,700 to 3,700 m and are marine sandstones of Pleistocene age. The chemical composition of the waters (Fig. 4), the δD and $\delta^{18}O$ values of the waters (to be discussed later), and the $^{87}Sr/^{86}Sr$ ratios of dissolved Sr show proportional mixing of two end members. The first is a Pleistocene marine connate water represented by samples 83-TX-6 and 83-TX-7 (Table 3); the second is a highly modified marine connate water (83-TX-5) that originated in Miocene sediments at a subsurface temperature of about 130°C (Kharaka et al., 1985).

Table 3: Chemical composition (mg/l) of formation waters from High Island Field (offshore Texas)

Sample #	83-TX-1	83-TX-2	83-TX-3	83-TX-5	83-TX-6	83-TX-7	83-TX-8	83-TX-9	83-TX-10	83-TX-11	83-TX-12
Well Name	A-11A	A-11B	A-10A	A-9A	C-8A	C-14A	B-14A	B-2A	A-14B	B-12A	A-45-1
Depth (m)	2045	1876	2184	2140	1849	1838	1789	1741	2216	2052	3719
Temp. (°C)	70	55	66	67	53	52	50	49	77	64	98
TDS	74,000	80,000	61,000	159,000	44,000	39,000	138,000	92,000	63,000	88,000	91,000
Li	1.79	1.97	1.74	5.03	0.63	0.58	4.89	2.21	1.79	2.25	5.50
Na	26,300	28,400	19,800	53,700	15,400	13,300	47,700	31,800	22,300	30,400	30,900
K	160	172	139	244	110	103	250	167	143	183	271
Rb	0.15	0.16	0.12	0.24	0.08	0.10	0.19	0.14	0.17	0.16	0.39
Cs	0.09	0.07	0.07	0.29	0.03	0.04	0.24	0.15	0.07	0.14	0.21
Mg	788	869	287	1,270	499	349	1,040	1,010	363	873	482
Ca	1,320	1,370	1,030	3,810	679	850	2,840	2,250	1,130	2,010	2,690
Sr	97	119	59	346	36	27	238	147	66	129	146
Ba	77	102	75	316	40	52	256	171	90	113	200
Mn	0.28	0.19	0.16	0.60	0.09	0.23	0.25	0.43	0.29	0.43	2.63
Fe	7.4	5.8	1.7	17.2	4.3	3.4	11.3	4.4	13.9	5.0	6.5
NH_3	140	152	97	206	73	83	182	170	112	182	121
B	23	25	31	33	16	16	40	18	37	24	42
Al-μg/L	----	----	----	<1	15	----	----	----	<1	<1	3
H_2S-Titration	0.25	0.68	0.85	0.08	0.00	0.51	0.68	0.51	0.85	0.93	0.76
H_2S-Electrode	0.04	0.02	0.06	<0.01	0.05	0.04	0.08	0.09	<0.01	0.08	<0.01
Organic(C)	54	60	162	78	73	82	----	36	209	47	933
Inorganic(C)	89	94	71	34	84	76	----	64	44	75	38
F	2.3	2.9	1.9	0.53	2.0	5.9	0.81	0.50	13.1	0.81	0.51
Cl	44,400	48,400	33,400	99,000	26,200	23,000	84,600	56,200	37,700	53,800	53,300
Br	143	88.9	102	177	102	287	90.9	108	185	113	118
I	4.0	4.7	4.5	3.1	5.1	4.3	4.6	4.4	5.6	4.3	6.2
SO_4	12.4	9.9	16.0	7.4	9.9	9.9	0.5	0.4	14.8	15.3	11.5
SiO_2	40.1	38.8	45.4	30.0	40.8	60.0	31.6	32.9	162	41.2	65.8
Alk(HCO_3)	494	546	568	308	480	443	293	350	460	402	2,140
pH	7.51	7.39	7.34	7.25	7.59	7.54	7.09	7.11	7.60	6.97	6.83

Samples from High Island 573 field (Mobil Oil) except sample 83-TX-12, which is from High Island 571 field (Columbia Gas); Depth is that of mid-point of perforation below sea level (depth of water column is ~75m; Temp. is the measured subsurface temperature; TDS - mg/L total dissolved solids; field titrated alkalinity as HCO_3 comprises organic as well as inorganic species.

The concentrations of Na and Cl, total dissolved solids, and δD and $\delta^{18}O$ values of sample 83-TX-6 and 83-TX-7 are about equal to those of ocean water (Table 3). These parameters have behaved relatively conservatively since middle Pleistocene time. The present

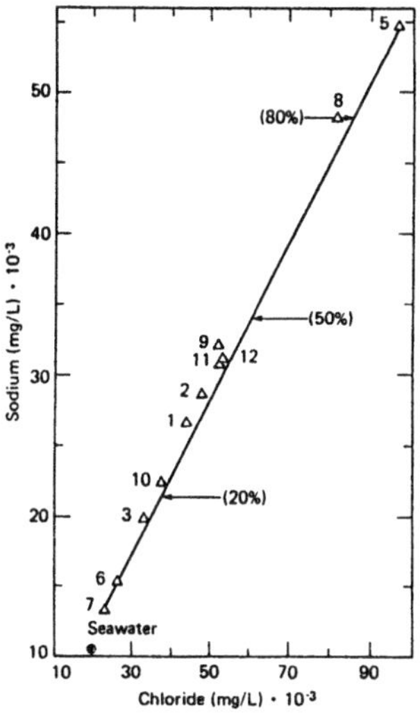

Figure 4: Concentrations of Na and Cl in formation waters from High Island field (offshore Texas) showing proportional mixing between a highly modified water (sample 83-TX-5) and an only slightly modified Pleistocene connate water (sample 83-TX-7). Sample numbers are from table 3 (Kharaka et al., 1985)

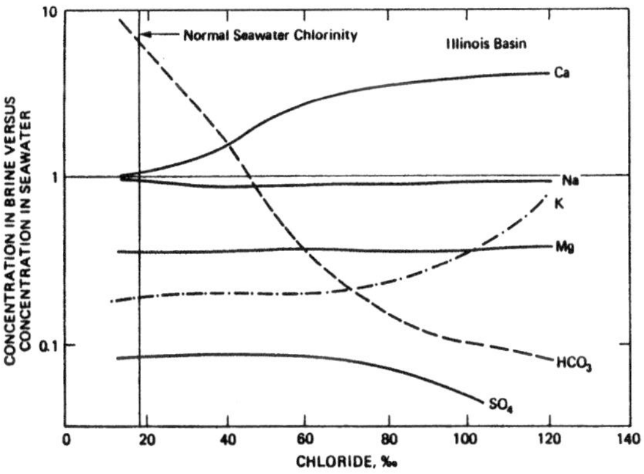

Figure 5: Generalized relative variations in the composition of major ions in formation waters from the Illinois basin. The normalized ratios indicate that the formation waters are depleted in SO_4, Mg and K relative to ocean water, and that they become progressively enriched in Ca and K and depleted in HCO_3 with increasing salinity (modified from Hanor, 1987)

subsurface temperature of these two samples is about 50°C. However, there are major differences between the chemical compositions of these two samples and that of ocean water. The concentrations of sulfate in these waters is lower by about three orders of magnitude from that of ocean water, whereas those of Mg and K are about one third that of ocean water. The concentrations of iodide, organic species, B, Ba, Fe, Si, and Sr are many times higher that those of ocean water. The concentrations of Ca, bromide, and carbonate alkalinity are higher than those of ocean water (Table 3). These differences are even more pronounced when we compare the chemical composition of sample 83-TX-5 with that of ocean water. The concentrations of Cl (99,000 mg/l) and the Br/Cl ratios for this brine indicate dissolution of halite from the Louann Salt structures underlying the field (Kharaka et al., 1985). The variation of major ion concentrations with increasing Cl values (increasing depth) for samples from the Michigan basin continue the trends observed in data from High Island field (Fig. 5).

Metal-rich Brines

The concentrations of metals in oil-field waters, with the exceptions of Fe and Mn, are generally low. In the case of Pb and Zn, the concentrations are less than 100 mg/l, with the exception of fewer than half a dozen localities world-wide (Carpenter et al., 1974; Kharaka et al., 1987). The central Mississippi Salt Dome Basin is one of these metal-rich brine localities that is characterized by extremely high salinities that range to about 350,000 mg/l total dissolved solids (Tables 1, 4). The brines are Na-Ca-Cl type waters with very high concentrations of Ca (up to 50,000 mg/l) and other alkaline-earth metals, but with low concentrations of aliphatic acid anions. The metals are very concentrated in many water samples, reaching values greater than 100 mg/l for Pb, 250 mg/l for Zn, 500 mg/l for Fe and 200 mg/l for Mn. The samples with high metal contents have extremely low concentrations (less than 0.02 mg/l) of H sulfide. Samples obtained from the Smackover Formation (limestone) have low metal contents that are more typical of oil-field waters, but have very high concentrations (up to 85 mg/l) of hydrogen sulfide.

Table 4: Selected metal concentrations (mg/l; *μg/l) of formation waters from central Mississippi Salt Dome Basin (Kharaka et al., 1987)

Sample ID	Fe	Mn	Pb	Zn	Al*	Cd	Cu*
84-MS-1	137	57.5	8.39	49.6	---	0.49	<20
84-MS-2	97.4	38.2	0.07	1.22	59	0.05	<20
84-MS-3	61.9	10.6	0.04	0.53	133	0.02	<20
84-MS-4	346	63.9	53.2	222	267	0.83	<20
84-MS-5	407	70.2	60.5	243	42	0.81	61
84-MS-6	284	21.0	26.8	95.1	67	0.63	21
84-MS-7	261	83.5	34.6	172	79	0.86	<20
84-MS-8	194	69.3	22.8	107	132	0.67	34
84-MS-9	0.54	15.5*	<.5*	12.0*	---	<0.2*	<0.2
84-MS-10	84.9	44.8	0.08	0.31	---	0.02	<20
84-MS-11	465	212	70.2	243	367	0.99	21
84-MS-12	65.3	16.4	0.17	0.28	---	0.03	<20
84-MS-14	0.75	2.78	0.16	0.20	---	<0.02	<20
84-MS-15	223	53.2	2.28	4.10	---	0.02	<20
84-MS-16	0.53	9.98	0.02	0.07	142	<0.02	<20
84-MS-18	0.15	3.2*	<.5*	13*	---	0.08*	0.61
84-MS-19	0.47	1.64	0.04	0.06	---	0.03	<20
84-MS-20	0.07	1.46	0.03	0.16	---	0.05	<20
Field blank*	10.6	1.05	1.04	12.6	---	0.08	<0.20

Organic Species in Subsurface Waters

In the last 15 years there has been considerable interest in the origin and interactions of dissolved organic acids in natural subsurface waters (Willey et al., 1975; Carothers and Kharaka, 1978; Kharaka et al., 1983, 1986; Crossey et al., 1986; Hanor and Workman, 1986; Lundegard and Land, 1986; Lundegard and Kharaka, 1990; MacGowan and Surdam, 1990). Geochemical interest in organic species stems from their important role in mineral diagenesis. In particular, they act as proton donors for a variety of pH-dependent reactions and as pH and Eh buffering agents. They also have the potential to form complexes with metals such as Al, Fe, Pb, and Zn. Finally, organic species have been studied for use as proximity indicators in petroleum exploration (Kartsev, 1974; Carothers and Kharaka, 1978) and as possible precursor for an important portion of natural gas (Kharaka et al., 1983). Kharaka et al. (1986), Lundegard and Kharaka (1990), and MacGowan and Surdam (1990) gave detailed recent reviews of the nature, origin, distribution and interaction of dissolved organics in subsurface waters.

The median concentration of dissolved organic species in shallow ground waters is about 1 mg C/l (Thurman, 1984). An important portion (20 to 40% and up to 90%) of this dissolved C consists of high molecular weight fulvic and humic acids (Thurman, 1984). The proportion of these humic substances in organic species dissolved in the waters of the unsaturated zone is even higher, amounting to 50 to 60%. The median concentrations of dissolved organic C, however, are also higher, ranging from about 20 mg/l in the A soil horizon to 2 mg/l in the C soil horizon.

Concentrations of dissolved organic species in oil-field waters (40 to 200°C) are much higher than in ground waters and may reach values as high as 10,000 mg/l (as acetate) (Kharaka et al., 1986; Lundegard and Land, 1986; Crossey et al., 1986; Hanor and Workman, 1986; MacGowan and Surdam, 1990). Acetate, propionate, butyrate and valerate were identified as the dominant organic species in these waters (Willey et al., 1975; Carothers and Kharaka, 1978). Prior to the identification of these aliphatic acid anions by Willey et al. (1975), these organics were generally lumped and recorded as bicarbonate alkalinity because they are titrated with H_2SO_4 used to measure field alkalinities. Willey et al. (1975) and Carothers and Kharaka (1978) showed that these organic acid anions can contribute up to 99% of the measured alkalinities (Fig. 6). Carothers and Kharaka (1978) and Kharaka et al. (1983) showed that the concentrations of these aliphatic acid anions were controlled mainly by the subsurface temperatures and the ages of the reservoir rocks. The distribution of aliphatic acid anions in oil field waters from several basins shows three distinct

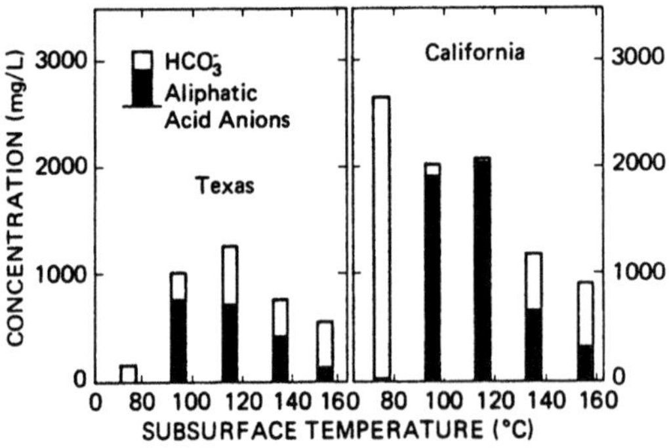

Figure 6: Average concentrations of aliphatic acid anions and bicarbonate alkalinity in oil-field waters from Texas and California (modified from Carothers and Kharaka, 1978). Note that organic acid anions contribute most of the total alkalinity at temperatures of 80 to 140°C

temperature zones (Fig. 7; Kharaka et al., 1988b). Zone 1 is characterized by concentrations of acid anions less than about 500 mg/l and temperatures less than about 800°C. The concentrations of acetate in this zone range from below detection limit to low values, whereas propionate generally predominates. Bacterial degradation is believed to be responsible for the absence of acetate and the low concentrations of organic species in Zone 1 (Carothers and Kharaka, 1978). The highest concentrations of aliphatic acid anions are present in the youngest and shallowest reservoir rocks of Zone 2 (temperature 80 to 100°C) and the concentrations generally decrease with increasing subsurface temperatures and ages of the reservoir rocks (Fig. 7). Acetate constitutes more than 90% of aliphatic acid anions with propionate comprising about 5% of anions (Carothers and Kharaka, 1978). The boundary of Zone 3, where no acid anions are present, is placed at about 220°C and is obtained by extrapolation of the least-squares line through the points of Zonc 2 (Kharaka et al., 1986).

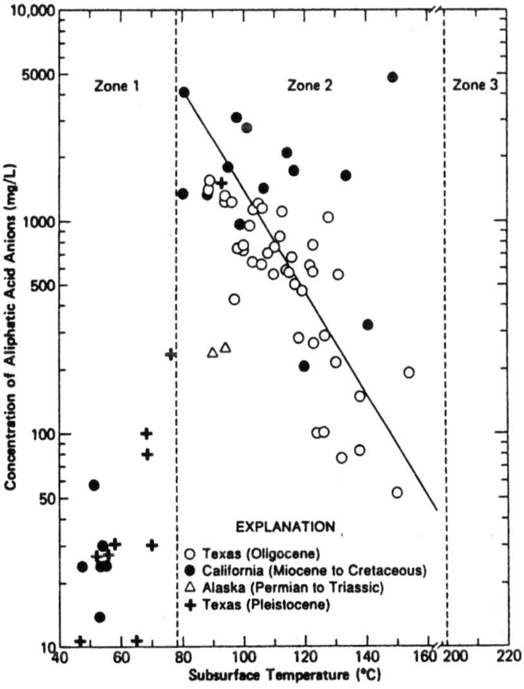

Figure 7: Concentrations of aliphatic acid anions (C125) in formation waters from three sedimentary basins. The highest concentrations are at about 80°C and decrease with increasing temperatures (Kharaka et al., 1988b)

The decrease in the concentrations of acid anions with increasing temperatures and ages of the reservoir rocks (Carothers and Kharaka, 1978) indicates thermal decarboxylation of acid anions to CO_2 and hydrocarbon gases (Fig. 7). Further evidence for thermal decarboxylation of acid anions is obtained from $\delta^{13}C$ values of dissolved bicarbonate, CO_2 and CH_4 in natural gas and diagenetic calcite and ankerite from Gulf Coast and California basins (Boles, 1978; Carothers and Kharaka, 1980; Lundegard and Land, 1986). These $\delta^{13}C$ values indicate that C in CO_2 gas, dissolved carbonate, and carbonate cements was derived largely from an organic source. Experimental studies of acetic acid show that decarboxylation rates are extremely sensitive to temperature and the types of catalytic surfaces available. Rate constants for acetic acid decarboxylation at 100°C differ by more than 14 orders of magnitude between experiments conducted in stainless steel and catalytically less active Ti (Kharaka et al., 1983; Drummond and Palmer, 1986). However, naturally occurring mineral surfaces provide rather weak catalysis effects. Decarboxylation rates calculated from field data provide more realistic half-life values of 10 to 60 million years at 100°C (Kharaka, 1986; Lundegard and Kharaka, 1990).

Table 5: Concentrations of dicarboxylic acids in formation waters from High Island Field, offshore Texas (Kharaka et al., 1985)

Dicarboxylic acid	Carbon No.	Sample number 83-TX-3	83-TX-10	83-TX-12
		Concentration (mg/L)		
Butanedioic acid	4	1.8	0.2	63
Pentanedioic acid	5	1.7	1.2	36
Hexanedioic acid	6	0.5	0.5	---
Heptanedioic acid	7	---	0.6	0.2
Octanedioic acid	8	---	0.7	5.0
Nonanedioic acid	9	---	1.4	6.0
Decanedioic acid	10	---	---	1.3

Dicarboxylic acid anions, which can be expected to form stronger complexes with metals, have also been identified in oil-field waters (Kharaka et al., 1985; Crossey et al., 1986; MacGowan and Surdam, 1990). The concentrations of dicarboxylic acid anions reported in formation waters of High Island Oil field, offshore Texas (Kharaka et al., 1986), are shown in table 5. Methybutanedioic and 2-methylpentanedioic acids were also identified but not quantified in samples 83-TX-1 and 83-TX-12, the only samples tested. The highest concentrations and proportions of dicarboxylic acid anions are present in sample 83-TX-12 where they comprised about 6% by wt of total acid anions. In this sample, butanedioic (C_4) and pentanedioic (C_5) acids are dominant. MacGowan and Surdam (1990) reported values of up to 2,500 mg/l for propane-dioic (malonic) and up to 500 mg/l for ethanedioic (oxalic) acids.

A number of other reactive organic species were identified but not quantified by Kharaka et al. (1986). These include C_6 to C_9 monocarboxylic acid anions, phenol, 2-, 3-, and 4-methylphenol, 2-ethylphenol, 3, 4- and 3, 5-dimethylphenol, cyclohexanone, and 1, 2- and 1, 4-dimethylphenol, cyclohexanone, and 1, 2- and 1, 4-dimethylbenzene. Several amino acids including serine, glycine, alanine and aspartic acid have also been reported in low concentrations (up to 0.3 mg/l) in oil field waters (Rapp, 1976). Additional reactive organic species have been reported by Fisher and Boles (1990) and MacGowan and Surdam (1990).

ORIGIN AND CONTROLS ON SOLUTES IN SUBSURFACE WATERS

The chemical composition of subsurface water in a given basin is determined by the origin of the water and the many chemical, physical and biological processes that modify the original composition of the water. These processes generally act together to increase or decrease the concentrations of the solutes. The ultimate control on the concentration of a given solute, however, generally is the solubility of the least soluble mineral of the solute. For example, the concentration of Ca in water from a given reservoir may increase because of membrane filtration, albitization of plagioclase feldspar and/or dolmitization of limestone. This water, however, will eventually attain saturation with calcite or another carbonate mineral that will be the ultimate control on the Ca (and carbonate) concentration.

The controls on the origins of sulfate and bicarbonate-type

waters, as already stated, are generally well understood. The origins of the great differences in the salinities and ion proportions of Na-Cl, Na-Cl-CH_3COO, or Na-Ca-Cl type waters, however, are not well understood. The main mechanisms proposed to account for their origins are a combination of the following: (a) dissolution of evaporites, especially halite, (b) diffusion, especially where halite and other evaporites are present, (c) incorporation of bitterns that were trapped in the sediments during the precipitation of evaporites, (d) dissolution and precipitation of minerals other than evaporites, (e) interaction of the waters with organic matter present in sedimentary rocks, and (f) interaction with rocks, principally clays, siltstones, and shales, that behave as geological membranes. Other mechanisms including density stratification, and mixing with brines from fluid inclusions were judged to be of minor importance (Mangelsdorf et al., 1970). A detailed discussion of these mechanisms, however, is beyond the scope of this chapter.

A detailed study of the basin that includes its geology, the mineral composition of the rocks, regional hydrology, as well as the chemical and isotopic compositions of the subsurface waters, is generally required to distinguish the different mechanisms responsible for the chemical and isotopic compositions of water. The chemical and other data should be carefully examined by: (a) using plots of various types, (b) subjecting the data to factor analyses (Hitchon et al., 1971), (c) computing the states of reactions of these waters at the subsurface conditions using geochemical models, such as SOLMINEQ.88 (Kharaka et al., 1988b), and (d) comparing the results from (c) with the mineral composition of the reservoir rocks. There have been recent attempts to quantify the results of water-rock interactions from the time of deposition to present, on field and basin-wide scales (Hanor et al., 1988).

ISOTOPIC COMPOSITION OF SURFACE WATERS

Application of stable isotopes has shown that subsurface

waters originate mainly from local meteoric water of different ages and marine connate water deposited with shales and siltstones. The isotopic compositions of present-day and old surface waters are discussed below because an understanding of these is needed in interpreting the origin of deep basin brines. A discussion of the isotopic composition of ocean water has been excluded from this part, but is covered in detail in a separate contribution.

Stable Isotope Compositions of Recent Meteoric Water

The relations governing the distribution of stable isotopes of O and H in present-day surface and shallow ground-waters are reasonably well known (Fig. 8). Craig (1961) showed that surface

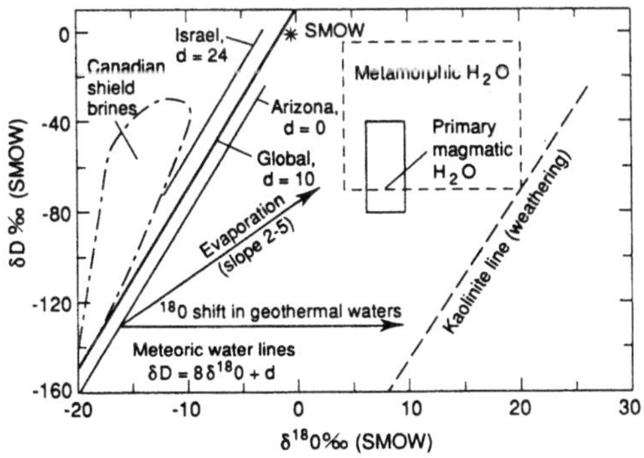

Figure 8: Isotopic compositions of ocean (SMOW), meteoric, metamorphic and primary magmatic waters. The field for the unusual Canadian Shield brines is shown. Also shown are the general trend for a meteoric water exchanging isotopes in a geothermal system and a trend for an evaporating meteoric water

water, other than that from closed basins undergoing extensive non-equilibrium evaporation and water from hot springs, have isotopic compositions that lie on the global meteoric water line (GMWL) that can be described by the equation:

$$\delta D = 8\ \delta^{18}O + 10 \qquad (1)$$

The δ values are given in the per mil (parts per thousand, o/oo) notation. Gat (1980) obtained essentially the same equation (slope

= 8.17, intercept = 10.6) from a plot of the mean δD and $\delta^{18}O$ values of all precipitation samples from the Internatinal Atomic Energy Agency (IAEA) network.

A more general form of the equation describing the relationship between δD and $\delta^{18}O$ values of surface waters is:

$$\delta D = x\, \delta^{18}O + d \qquad (2)$$

where the value of x is generally equal to 8, but is about 6 for precipitation in tropical island stations (Yurtsever, 1975) and is 2-5 in closed basin-lakes undergoing non-equilibrium fractionation during evaporation. Dansgaard (1964) refers to the "d-parameter" in equation (2) as the "deuterium excess". The value of d is generally equal to 10, but may be less than 10 in many areas including wet coastal areas and tropical islands; its value is 22 in the dry eastern Mediterranean area (Gat and Carmi, 1970). The global meteoric water line, however, can be assumed to be applicable to surface and shallow ground waters in any area where no specific line has been established.

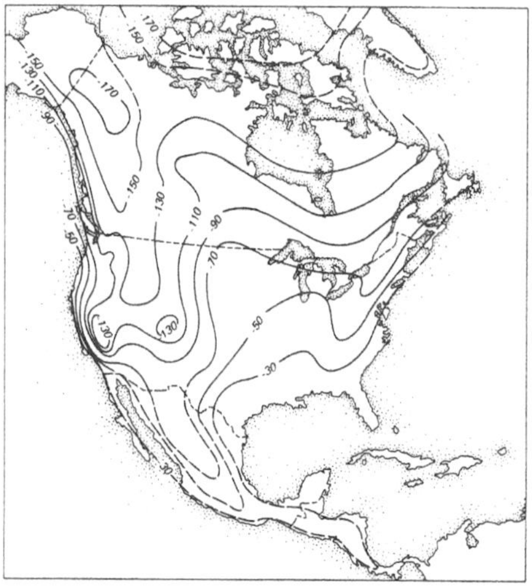

Figure 9: Map of North America showing the approximate distribution of mean annual δ D values of modern meteoric water (modified from Taylor and Margaritz, 1978). Data for Antarctic water from Yurtsever (1975) and Kharaka and Carothers (1988)

The distribution of the mean annual δD values of surface and shallow ground waters (the meteoric water) in North America (Taylor and Margaritz, 1978) is shown in figure 9. The distribution of the $\delta^{18}O$ values for these waters may be approximated, if not already available from equation (1) describing the GMWL. These values cannot be obtained by analyzing water samples from individual storms which, based on season, source of moisture and amount of precipitation, can have large isotopic variations (Fig. 10). Shallow ground waters and cold perennial springs are better representatives of local meteoric water as they give good time integrated samples of precipitation with relatively uniform isotope values (Fig. 10).

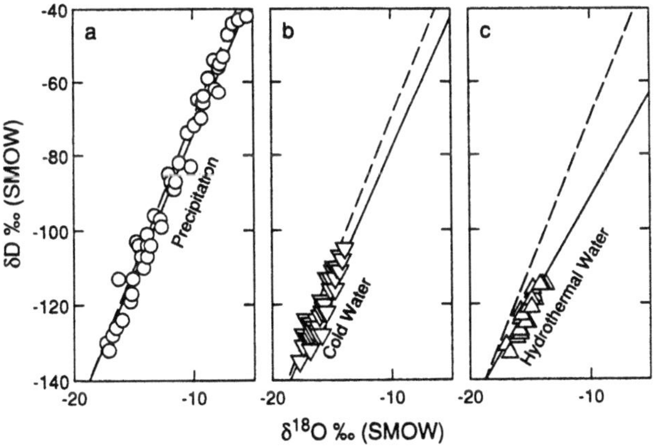

Figure 10: Isotopic composition of (a) precipitation, (b) cold surface and shallow ground waters, and (c) hydrothermal water. Dashed line is the Global Meteoric Water Line (Craig, 1961); solid lines are the least square lines. Note the wide range of isotope values for precipitation and some $\delta^{18}O$ shift for the hydrothermal water (modified from White et al., 1990)

The isotopic composition of mean annual precipitation is mainly controlled by geography (latitude, altitude, distance from the coast, and source of moisture). The most depleted precipitation occurs at high latitudes, high altitudes, and in the centers of large continents (Fig. 9). Yurtsever (1975) demonstrated a decrease in $\delta^{18}O$ of about 0.5 per mil per degree of latitude increase. The altitude effect is variable depending on local topography and climate, but depletions of 12-40 per mil in δD and 1.5-5 permil in $\delta^{18}O$ per km are typical (Gat, 1980). Where the altitude effect can be established, especially in coastal areas and on the windward side of a mountain, it can be used to determine the recharge area for the deep subsurface water in the region (Kharaka et al., 1988a). Generally, the greater the distance from the source of moisture, the more depleted the precipitation. This depletion is caused by colder temperatures in the centers of large land masses during winter, by successive evaporation-condensation cycles as the moisture-laden air mass travels across the continent, and by the reevaporation of isotopically depleted water from continent.

The relationship between mean annual temperature and $\delta^{18}O$ values for northern Atlantic precipitation was given by Dansgaard (1964) as:

$$\delta^{18}O = 0.70t - 13.6 \qquad (3)$$

Combining equations (3) and (1) leads to:

$$\delta D = 5.6t - 98.8 \qquad (4)$$

As the temperature decreases, precipitation becomes more depleted in δD and $\delta^{18}O$ because the fractionation between liquid water and water vapor increases. Unfortunately, the above relationships are not universally applicable (Evans et al., 1978), but they may be used to estimate the mean annual temperatures of recharge during past time periods (Kharaka and Carothers, 1986).

Stable Isotope Compositions of Old Meteoric Water

Knowledge of the variations in the isotopic composition of surface waters through geologic time is necessary in order to study the origin of some deep basin brines which may have been recharged millions of years ago (Kharaka and Carothers, 1986; Knauth, 1988). These variations result from two main causes: changes in the isotopic composition of the oceans through geologic time (see next section), and changes in the climatic and topographic conditions at the area of recharge. These climatic changes in turn result from two factors: continental drift and the phenomena connected with it, including latitude changes, mountain building, and the ensuing changes in oceanic currents and wind circulation patterns, and as-

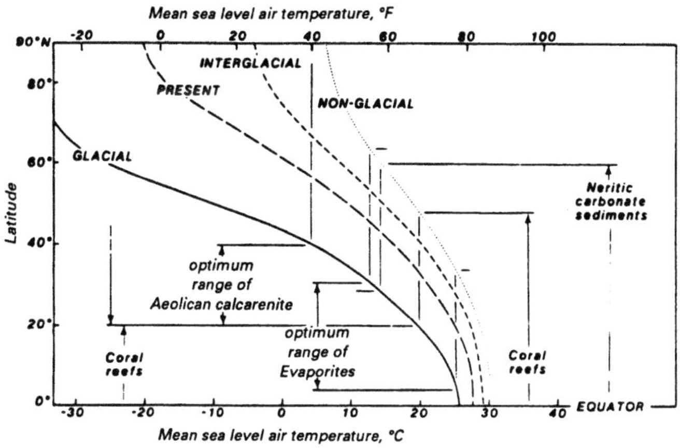

Figure 11: Surface mean temperatures for the present, glacial, non-glacial and interglacial periods at different latitudes (modified from Fairbridge, 1964)

tronomical factors and effects not related to continental drift that might include changes in the sun's radiation and changes in the composition of the earth's atmosphere.

It is generally recognized that "the present is not the key to the past" as far as climate is concerned (Lamb, 1977; Habicht, 1979; and others). This is illustrated by the general distribution of temperatures, as a function of latitude for the present, glacial, interglacial and nonglacial periods (Fig. 11). More detailed studies reported in Lamb (1977) and Gerasimov (1979) show temperature fluctuations of the order of ± 2°C in the middle latitudes within the Holocene Epoch alone. It should be noted that equations (3) and (4) indicate that present day precipitation changes in δD and $\delta^{18}O$ values by 5.6 and 0.7 per mil, respectively, for every 1°C change in mean annual temperature. This shows clearly that paleoclimatic and paleogeographic reconstructions (Habicht, 1979) should be attempted to aid in interpreting the isotopic data.

MODIFICATION OF STABLE ISOTOPES IN SUBSURFACE WATERS

Reactions between water and minerals, dissolved species,

associated gases, and liquids with which they come into contact, modify the isotopic composition of water and the other reactants. In addition to mixing between waters of different isotopic composition, the following are the main processes that modify the isotopic composition of formation waters in sedimentary basins: (1) isotopic exchange between water and minerals; (2) evaporation and condensation; (3) fractionation caused by membrane properties of rocks; (4) isotopic exchange between water and other fluids, especially petroleum.

Isotope Exchange between Water and Minerals

Authigenic minerals generally form in isotopic equilibrium with the associated pore waters at the temperature of their formation. Field data from many investigations show that carbonates, cherts and clay minerals undergo extensive isotopic alterations from reactions with pore waters at sedimentary temperatures (Clayton et al., 1968; Savin, 1980; Milliken et al., 1981; Kharaka and Carothers, 1986; Land and Fisher, 1987; Longstaffe, 1989; and many others). Because rates of diffusion in solid phases are extremely small at low temperatures, it is believed that these isotopic alterations result from chemical reactions involving water that include dissolution, precipitation, mineral transformations (e.g. smectite to illite, Mg-rich calcite to calcite) and recrystallization.

Field and experimental studies have demonstrated that isotopic exchange between water and quartz, feldspars and many other aluminosilicates are negligible at sedimentary temperatures in the absence of chemical reactions. For these minerals and others that have undergone no isotopic exchange with water, their isotope values can be used to calculate the temperature of their formation to study their diagenetic history (Milliken et al., 1981; Kharaka and Carothers, 1986; Longstaffe, 1989; and many others). It should be noted, however, that even in the case of quartz, the most stable and refractory mineral, extensive (generally up to 20% of detrital grains) intergranular pressure solution and reprecipitation is observed in some basins (Houseknecht and Hathon, 1987). The isotopic composition of water and the newly formed quartz, of course, are different from the original values.

Table 6: Oxygen isotope fractionation factors between water and minerals as a function of temperature (T) in °K

Equation	Reference
$1000\ell n\alpha(\text{quartz-}H_2O) = 3.38 \times 10^6 T^{-2} - 2.90$	Friedman and O'Neil (1977)
$1000\ell n\alpha(\text{amorph. silica-}H_2O) = 3.52 \times 10^6 T^{-2} - 4.35$	Kita et al. (1985)
$1000\ell n\alpha(\text{albite-}H_2O) = 2.39 \times 10^6 T^{-2} - 2.51$	Matsuhisa et al. (1979)
$1000\ell n\alpha(\text{K feldspar-}H_2O) = 2.91 \times 10^6 T^{-2} - 3.41$	O'Neil and Taylor (1967)
$1000\ell n\alpha(\text{calcite-}H_2O) = 2.78 \times 10^6 T^{-2} - 2.89$	Friedman and O'Neil (1977)
$1000\ell n\alpha(\text{dolomite-}H_2O) = 3.34 \times 10^6 T^{-2} - 3.34$	Land (1983)
$1000\ell n\alpha(\text{protodolomite-}H_2O) = 2.78 \times 10^6 T^{-2} + 0.11$	Land (1983)
$1000\ell n\alpha(\text{ankerite-}H_2O) = 2.78 \times 10^6 T^{-2} + 0.32$	Dutton and Land (1985)
$1000\ell n\alpha(\text{siderite-}H_2O) = 3.13 \times 10^6 T^{-2} - 3.50$	Carothers et al. (1988)
$1000\ell n\alpha(\text{anhydrite-}H_2O) = 3.21 \times 10^6 T^{-2} - 4.72$	Chiba et al. (1981)
$1000\ell n\alpha(\text{kaolinite-}H_2O) = 10.6 \times 10^3 T + 0.42 \times 10^6 T^{-2} - 15.34$	Savin and Lee (1988)
$1000\ell n\alpha(\text{smectite-}H_2O) = 2.58 \times 10^6 T^{-2} - 4.19$	Savin and Lee (1988)
$1000\ell n\alpha(\text{illite-}H_2O) = 2.39 \times 10^6 T^{-2} - 4.19$	Savin and Lee (1988)
$1000\ell n\alpha(\text{illite/smectite-}H_2O) = (2.58 - 0.19 \times I) \times 10^6 T^{-2}$ -4.19; I = fraction of illite (0.0 - 1.0)	Savin and Lee (1988)
$1000\ell n\alpha(\text{chlorite-}H_2O) = -8.38 \times 10^3 T^{-1} + 4.81 \times 10^6 T^{-2}$	Cole (1985)
$1000\ell n\alpha(\text{pyrophyllite-}H_2O) = 1.08 \times 10^3 T^{-1} + 2.76 \times 10^6 T^{-2} - 5.37$	Savin and Lee (1988)

Formation of authigenic minerals and other mineral reactions discussed above are probably the most important processes that modify the original δD and $\delta^{18}O$ values of water in sedimentary basins. The extent of this modification depends on the following factors.

-The original isotopic compositions of the water and the various minerals with which the water reacts in the sedimentary section.

-The relative number of atoms of O and H in the water and the various minerals (the water/rock ratio) involved. In general, water is the main reservoir for H in sedimentary basins especially at shallow depths where the porosity and permeability of the rocks are high, whereas the minerals are normally the main reservoir for the O atoms. This is the reason why, in the case of water, extensive "oxygen isotope shifts" are common and "hydrogen isotope shifts" are relatively rare.

-The subsurface temperatures at which chemical reactions and isotopic exchange between water and minerals occur. Temperatures are an important variable because the isotopic exchange rates between water and minerals increase with increasing temperature, and the magnitude of fractionation factors between water and minerals are strongly temperature dependent (Figs. 12 and 13, and Tables 6 and 7).

Table 7: Hydrogen isotope fractionation factors between water and minerals as a function of temperature (T) in °K

Equation	Temp. °C	Reference
$1000\ell n\alpha(\text{kaolinite} - H_2O) = -4.53 \times 10^6 T^{-2} + 19.4$	<230	Lambert and Epstein (1980)
$1000\ell n\alpha(\text{illite/smectite} - H_2O) = -19.6 \times 10^3 T^{-1} + 25$	29 - 120	Yeh (1980)
$1000\ell n\alpha(\text{brucite} - H_2O) = 8.72 \times 10^6 T^{-2} - 3.86 \times 10^4 T^{-1} + 14.5$	144 - 510	Satake and Matsuo (1984)
$1000\ell n\alpha(\text{brucite} - H_2O) = -33$	100	Satake and Matsuo (1984)
$1000\ell n\alpha(\text{epidote} - H_2O) = 29.2 \times 10^6 T^{-2} - 139$	150 - 300	Graham et al. (1987)
$1000\ell n\alpha(\text{epidote} - H_2O) = -36$	300 - 650	Graham et al. (1987)
$1000\ell n\alpha(\text{zoisite} - H_2O) = -15.1 \times 10^6 T^{-2} - 28$	280 - 650	Graham et al. (1987)
$1000\ell n\alpha(\text{glauconite} - H_2O) = -70$	6	Savin and Epstein (1970)
$1000\ell n\alpha(\text{gibbsite} - H_2O) = -15 \text{ to } 0$	<25	Savin and Lee (1988)

The isotopic shifts resulting from reactions between water and minerals can be calculated using the law of conservation of mass. In a closed system, and for water reacting with a single mineral only, the equation is:

$$W\delta^i_w + M\delta^i_m = W\delta^f_w + M\delta^f_m \qquad (5)$$

where i and f are the initial and final δ values for water and minerals, respectively, and W and M are the mole fractions of atoms of O or H in water and minerals, respectively. Equation (5) may be rearranged to give:

$$\delta^f_m - \delta^i_w = M/W \; (\delta^i_m - \delta^f_m) \qquad (6)$$

where M/W is the water/rock ratio and

$$M + W = 1 \qquad (7)$$

Assuming isotopic equilibrium between water and this mineral results in:

$$\alpha_{m-w} = \frac{1000 + \delta^f_m}{1000 + \delta^f_w} \qquad (8)$$

where α_{m-w} is the fractionation factor between water and this mineral at the temperature of reaction. Equation (7) can be approximated to:

$$\delta^f_m - \delta^f_w = 1000 \ln\alpha_{m-w} \qquad (9)$$

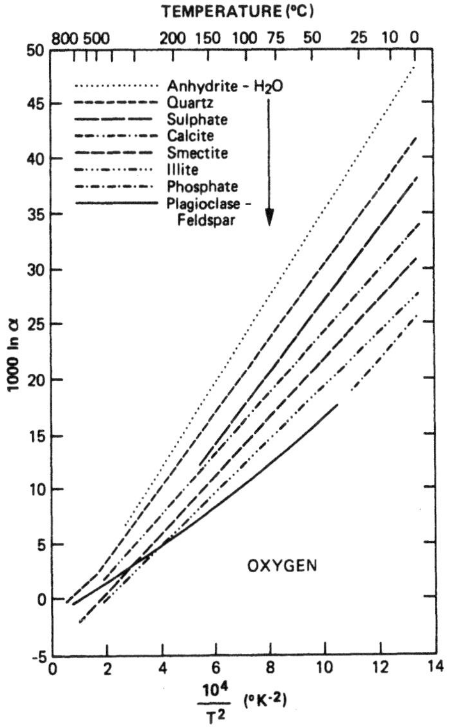

Figure 12: Oxygen isotope fractionations (α = fractionation factor between water and selected minerals plotted as a function of temperature (T in °K) showing the general decreases in the fractionation factors with increasing temperatures. Modified from Savin (1980) and Kharaka and Carothers (1986)

Equations (5) to (9) contain seven independent variables and three primary equations. Thus, all the unknowns can be calculated if values for four of the variables are known. In cases where several minerals are involved in isotopic exchange reactions, where only a portion of a given mineral has exchanged with water or where the section investigated must be treated as an open system, equations (5) to (9) can easily be modified for a more realistic treatment of field situations (Suchecki and Land, 1983; Criss et al., 1987; Dickson, 1987; Longstaffe, 1989).

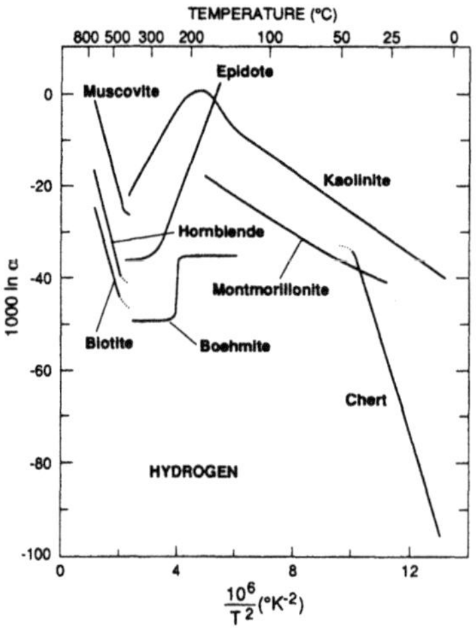

Figure 13: Hydrogen isotope fractionations (α = fractionation factor) between water and selected minerals plotted as a function of temperature (T in °K). Note the pronounced change in the slope of many curves around 200°C. Modified from Savin (1980), Kharaka and Carothers (1986) and Kyser (1987)

In sedimentary basins, reactions with carbonates and clay minerals are the main factors responsible for changes in the isotopic composition of formation waters because of mass balance considerations and the relative ease of isotope exchange. Formation of authigenic quartz, feldspars and other minerals, however, can be quantitatively important in the determination of the final isotopic composition of formation water. Material balance calculations that can be performed with equations (5) to (9) have only recently been attempted (Dickson, 1987; Longstaffe, 1989) because combined and detailed mineral and isotopic analyses simply have not been carried out on basin wide scales. Also, the permeabilities and thus the flow rates in the sedimentary section are often not available, but these determine the effective water/rock ratios and thus the final isotopic composition of water and minerals. Qualitatively, however, the trends observed in the $\delta^{18}O$ values of formation waters in deep basins can be explained by isotopic exchange between water and minerals. Such exchange almost always results in an increase in the $\delta^{18}O$ value of the water and a decrease in the $\delta^{18}O$ values of the minerals involved (Fig. 14). The "oxygen isotope shift" in the case of water almost always increases as the subsurface temperature and depth of the reservoir rocks increase. This increased "oxygen isotope shift" with increasing temperatures result from an enhanced degree of exchange in the low water/rock ratio environment and the fact that the fractionation factors for O between water and all the important minerals with which the water may exchange in sedimentary basins decrease with increasing temperatures (Fig. 12).

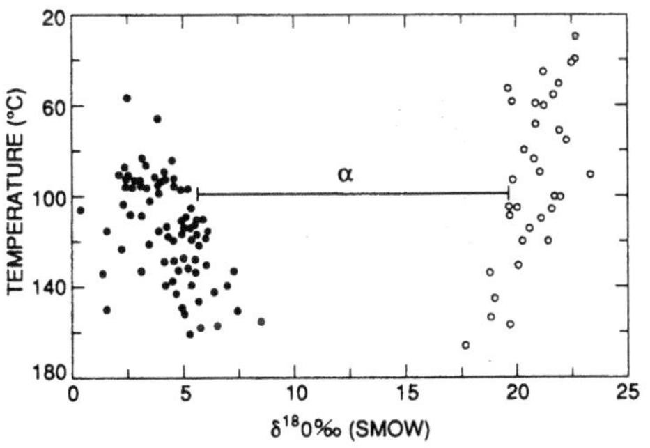

Figure 14: $\delta^{18}O$ values of formation water from Frio sandstones (dots), and the <1μm fraction (essentially smectite-illite) of Frio shale (circles). Line (a) gives the equilibrium fractionation between water and 50% smectite-illite at 100°C using the equation in Table 1. Modified from Land and Fisher (1987)

Hydrogen Isotope Exchange between Water and Clay Minerals

Clay minerals are the most important reservoirs of H in rocks. In many sedimentary basins (northern Gulf of Mexico, Central Valley, California, and others), shales comprise the bulk of the rocks in the basin, and the structural H in clay minerals comprises an important portion of the total H in these basins (Fig. 15). It should be noted that H isotope fractionation factors between clay minerals and water are less than 1.0, indicating that at equilibrium the clay minerals will be depleted in D with respect to the water, and formation of clays from minerals with no structural H (e.g. feldspars) leads to enrichment of water in D (Fig. 13 and Table 6). Also the D values of detrital clay minerals in sedimentary basins generally vary between about -80 to -25 per mil and may be lower than that of the water with which they may exchange. Isotopic exchange between water and clays, thus, can result in increases or decreases in the δD of formation water (Kharaka and Carothers, 1986).

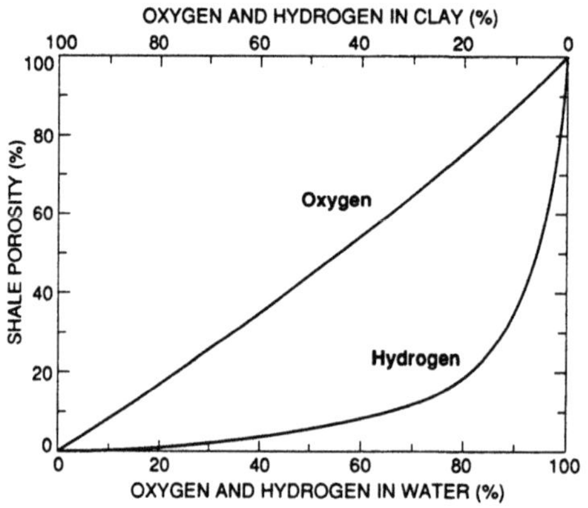

Figure 15: Relative abundance of O and H in pore water and structural illite (close to smectite) as a function of shale porosity. It is assumed that illite comprises 50% by weight of the shale. Note that there is more O and H in shale than pore water at porosities <40% and <6%, respectively

Experimental and field evidence indicate that significant H isotope exchange takes place between water and clay minerals with which they come into contact at all sedimentary temperatures, and that at temperatures higher than about 100°C equilibrium is generally reached in sedimentary basins. Bird and Chivas (1988) reported significant H isotope exchange for Permian kaolinite at temperatures of less than 100°C. Longstaffe and Ayalon (1990) reported H isotope reequilibration between formation water and kaolinite and illite at temperatures of less than 50°C. The clay minerals from three deep wells in the northern Gulf of Mexico basin reported by Yeh (1980) are in apparent isotopic equilibrium with the formation waters in the wells at temperatures higher than about 100°C. Isotopic equilibrium between water and clays in this basin, which results from transformation of smectite to illite, is inferred from the decrease in the range of the δD values of clays in the different size fractions from the same depth at temperatures higher than about 100°C and from the fact that the increase of δD values of clays with increasing temperature are mirrored by, and can be related to the decrease in the δD value of formation waters (Fig. 16 and see Formation Waters of Connate Origin) from the same area (Kharaka et al., 1979; Kharaka and Carothers, 1986).

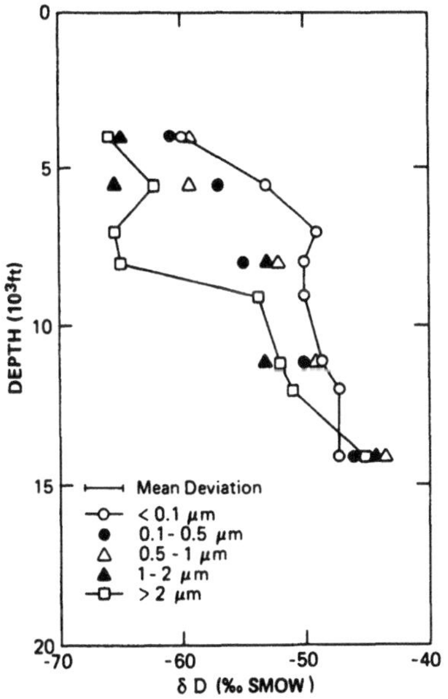

Figure 16: Variations in the δD values of shale with burial depth from CWRU Gulf Coast Well No. 10 (Yeh, 1980). Note the general increase in the δD values of shale with increasing temperatures as a result of exchange with formation waters and that the smaller size particles (<0.1μm) are modified at shallower depths

Experiments on H and O isotopic exchange of illite, smectite, and kaolinite with water conducted by O'Neil and Kharaka (1976) in the temperature range 100-350°C showed significant (up to 28% for smectite) H isotope exchange even at 100°C. The exchange increased with increased temperatures and duration of runs (up to about 2 years) and were much higher for smectite than for the other minerals at all temperatures. At 200°C the maximum H isotope exchange for kaolinite, illite and smectite were 29, 28, and 82%, respectively. These extensive isotope exchanges took place with no detectable changes in the minerals. Oxygen isotope exchange in the same systems was minor to negligible at all temperatures, indicating that H isotope exchange occurred by a mechanism of proton exchange. These experiments showed near complete O isotope exchange at 350°C between kaolinite and water, however, this resulted from the complete transformation of kaolinite to pyrophyllite and diaspore.

Evaporation and Condensation

Evaporation and boiling concentrate the light isotopes of O (at all temperatures) and H (at temperatures less than about 240°C) in the vapor phase, and result in higher δD and $\delta^{18}O$ values of the remaining water with increasing evaporation and boiling. Evaporation from surface waters, especially from lakes in closed basins and in the unsaturated zone, is very important in determining the isotopic composition of these waters. The δD and $\delta^{18}O$ values from these evaporating waters generally plot on straight lines with slopes that vary (depending on humidity and local conditions of evaporation) from 2 to 5, and that intersect the "meteoric water line" at the initial (unevaporated) values characteristic of the area. Boiling and evaporation modify the isotopic composition of water in many geothermal systems (Truesdell and Hulston, 1980), however, these processes probably are not important in sedimentary basins except in the case of gas wells. The isotopic and chemical composition of formation waters produced from many gas fields can be modified significantly by mixing with variable amounts of condensed water which is present as water vapor in the gas phase at subsurface conditions but condenses to liquid water as the gas cools on entering the well. The amount of this condensed water, which can be calculated, increases as the subsurface temperatures increase and pressures and salinities decrease (Kharaka and Carothers, 1986).

Membrane Properties of Rocks

There is field, laboratory, and theoretical evidence suggesting that shales and clays act as membranes fractionating the stable isotopes of water (Graf et al., 1965; Kharaka et al., 1973; Coplen and Hanshaw, 1973; Phillips and Bentley, 1987; Demir, 1988). The concentrations of δD and $\delta^{18}O$ are lower in the effluent water compared to their concentrations in the hyperfiltrated water. Graf et al. (1965) showed isotopic fractionations resulting from the passage of water through shales in the Illinois, Michigan, Gulf Coast and Alberta basins (Fig. 17). The δD and $\delta^{18}O$ values shown in figure 17 are residual values as the data from different basins were normalized to the value for meteoric precipitation at Chicago and to the value at 25°C for equilibrium with pre-Tertiary marine limestone. This normalization was carried out in order to separate the membrane fractionation from other effects. Hitchon et al. (1971) and Kharaka et al. (1973) reported isotopic compositions of waters from the western Canada and California basins, respectively, that indicated isotopic fractionations resulting from the passage of water through shale membranes.

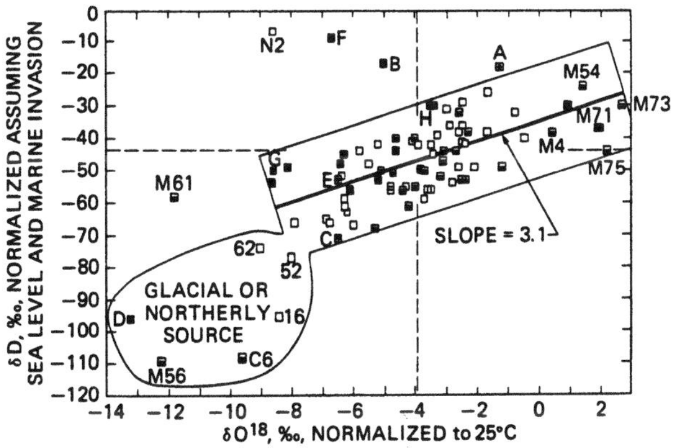

Figure 17: The residual regularity between normalized (see text) δD and $\delta^{18}O$ values of formation waters from Illinois (open squares), Michigan (half full squares), Gulf Coast (full squares) and Alberta (squares with dots) basins showing the membrane effects of shales (Graf et al., 1965)

Results of laboratory experiments (Coplen and Hanshaw, 1973) show that the effluent waters were depleted by 2.5 per mil and 0.8 per mil in δD and $\delta^{18}O$, respectively, relative to the hyperfiltrated (residual) solutions when distilled water and dilute NaCl solution were forced through compacted montmorillonite at ambient temperatu-res. A plot of δD versus $\delta^{18}O$ values from these experiments fall on a line with a slope of 3.1 (Fig. 18). The slope of this line, it should be noted, is exactly the same as that given by Graf et al. (1965) for the residual δD and $\delta^{18}O$ values of formation water that they attributed to membrane effects (Fig. 17). Benzel and Graf (1984) and Demir (1988), however, obtained somewhat different isotopic fractionation factors as functions of temperature, compac-tion and hydraulic pressures and water compositions.

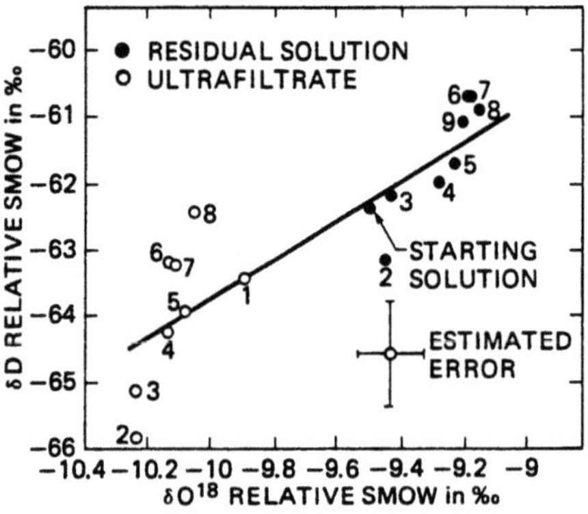

Figure 18: Isotopic compositions of residual (hyperfiltrated) and ultra-filtrate (effluent) solutions of 0.01N NaCl forced through compacted clay (Coplen and Hanshaw, 1973). The slope of the line (3.1) in this figure is equal to that obtained from natural systems (Fig. 17)

Isotope Exchange between Formation Water and other Fluids

Hydrogen isotope exchange between H_2S and water and O isotope exchange between CO_2 and water will occur in nature where water comes in contact with H_2S or CO_2 rich gases. In general, however, isotopic exchange with these gases will not be quantitatively important in significantly modifying the isotopic composition of formation waters from deep sedimentary basins. On a local scale, isotopic exchange between water and these gases may be significant, especially in the case of H_2S in sour gas fields (Hitchon and Friedman, 1969) because of the large fractionation factor between H_2S and water. Isotopic exchange between water and gaseous or liquid hydrocarbons is probably not quantitatively important because exchange rates are extremely low at sedimentary temperatures, but additional investigations are needed to properly evaluate these effects (Kharaka and Carothers, 1986). Finally, isotopic exchange between water and H gas may be very important in sedimentary basins because the fractionation factors are large ($\alpha_{H2O(v)} - H_{2(g)}$ = 1,400-640 at 0 to 200°C; Bottinga, 1969) and H gas may be an important component in basins with their rich organic matter, but the importance of this reaction needs investigation (Sheppard, 1986).

ORIGIN OF WATER IN SEDIMENTARY BASINS

Oxygen and H isotopes of water are the most useful tool in the study of the origin of subsurface waters. Prior to the use of isotopes, it was generally assumed that most of the formation waters in marine sedimentary rocks were connate marine (White, 1965). Clayton et al. (1966) were the first to use stable isotopes to show that waters from several sedimentary basins were predominantly of local meteoric origin, the original connate water being lost by compaction and flushing. The extensive use of water isotopes, solutes and associated minerals coupled with studies of regional geology and paleohydrology have led to the realization that subsurface waters generally have complicated histories and frequently are mixtures of waters with different origins (Kharaka and Carothers, 1986; Knauth and Beeunas, 1986; Sheppard, 1986; Hanor, 1987; Longstaffe, 1989; and many others.)

Formation Waters from Holocene Meteoric Waters

The isotopic compositions of formation waters associated with petroleum were shown to be related principally to recent local meteoric water (Clayton et al. 1966; Hitchon and Friedman 1969; Kharaka et al. 1973). Their δD values show a "hydrogen isotope shift" in addition to the generally observed "oxygen isotope shift" (Fig. 19). The origins of the shifts vary from basin to basin, but the lines fitted to the δD and $\delta^{8}O$ plots always intersect the meteoric water line (Craig, 1961) at values that approximate those of present-day meteoric water in the area of recharge (Fig. 19).

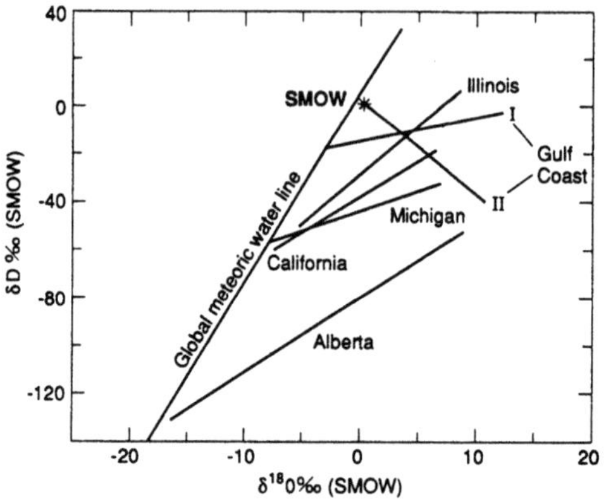

Figure 19: Isotopic compositions of oil-field waters from several basins in North America. Original lines from Clayton et al. (1966), Hitchon and Friedman (1969), Kharaka et al. (1973) and Kharaka et al. (1979). The isotope lines intersect the GML at points with isotope values of present-day meteoric water, except for Gulf Coast II

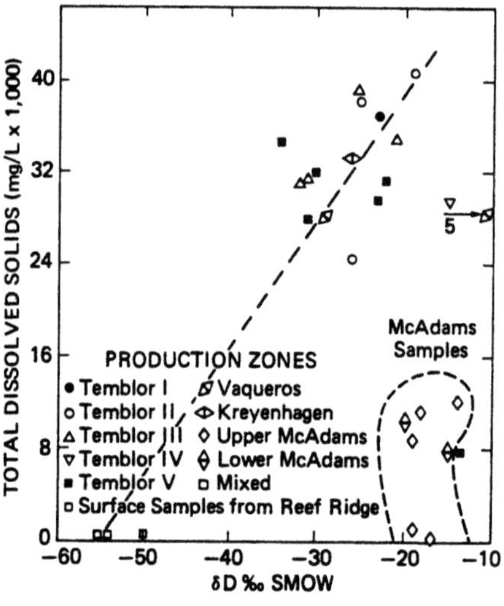

Figure 20: Relationship between δD and salinity of formation and ground waters from the Kettleman North Dome (California). The data from the McAdams samples fall on a separate grouping from the remainder of the samples (Kharaka et al., 1973)

Oxygen and H isotopes of water are not sufficient by themselves to indicate the origin of water. Kharaka et al. (1973) showed that plots of δD values and total dissolved solids indicated a connate marine origin for the waters from McAdams Formation (Eocene) and a meteoric origin for those from overlying formations (Fig. 20); the isotope data alone could indicate a meteoric origin for all the waters as they plot on the same trend (Fig. 21). Also, a recent detailed study (Connolly et al., 1990) of δD and $\delta^{18}O$ values of water and minerals, combined with Cl and Sr concentrations and $^{87}Sr/^{86}Sr$ ratios, suggests that the formation waters in the Alberta basin, Canada, are not all meteoric as may be indicated by figure 19, but that they form three hydrochemical groups. The upper group is dominated by modern meteoric water; the other two groups are a mixture of old (Laramide age) meteoric water and original bittern or residual brine.

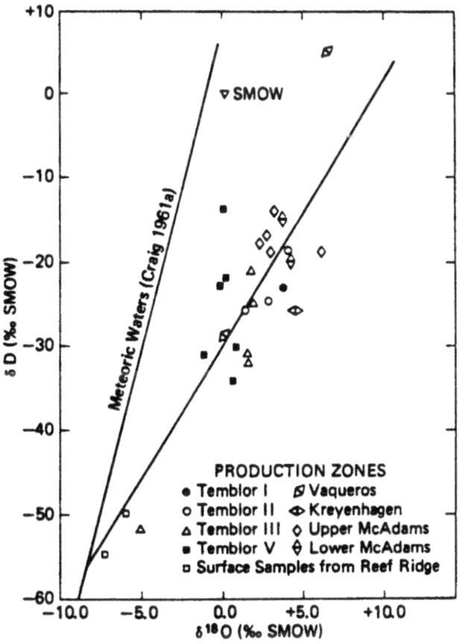

Figure 21: Isotopic composition of formation and ground waters from Kettleman North Dome oil field, California. The solid line is the least-squares line drawn through the data for the formation waters (Kharaka et al., 1973)

Formation Waters from "Old" Meteoric Water

Stable isotopes of water in addition to age determinations using ^{14}C, ^{36}Cl and other isotopes and flow rates, have shown that the waters in many aquifers are "old" meteoric water, i.e., older than Holocene and probably recharged during the Pleistocene Epoch

(Pearson and White, 1967; Bath et al., 1978; Fontes, 1980; Coplen, 1989). Clayton et al. (1966) showed that a number of formation waters from the Michigan and Illinois basins were probably recharged during the Pleistocene Epoch because they had $\delta^{18}O$ values that were much lower than those for present-day meteoric water in the respective areas.

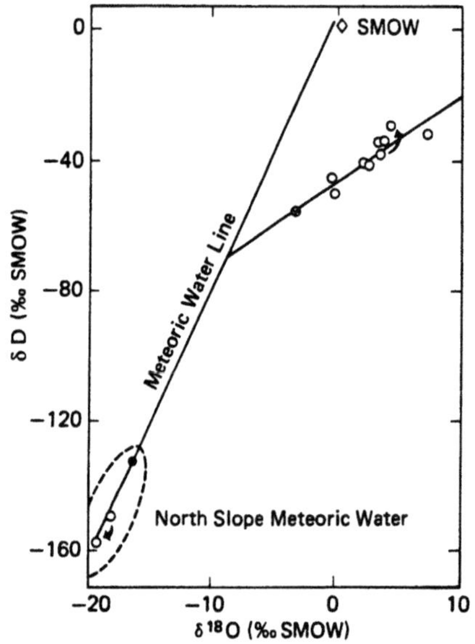

Figure 22: Isotopic composition of formation waters from the North Slope, Alaska. The solid line is the least-squares line drawn through the data. Also shown are values for SMOW, meteoric water in the area and the GMWL. The line through the data does not pass through SMOW or the local meteoric water

Kharaka and his colleagues, (Kharaka et al., 1979; Kharaka and Carothers, 1988) were probably the first to give evidence of meteoric water older than the Pleistocene Epoch in sedimentary basins. These authors presented isotopic and chemical data of formation waters from exploration and producing oil and gas wells from the North Slope of Alaska. The water samples were obtained from reservoir rocks varying in depth from 700 to 2,800 m and in age from Triassic to Mississippian. The chemical data show that the formation waters, regardless of the wells sampled and the age, depth, or lithology of the reservoir rocks, are remarkably similar in total dissolved solids (19,000-24,000 mg/l) and distribution of major cations and anions. The δD and $\delta^{18}O$ values for these waters with the least-squares line drawn through them are shown in figure

22. The least-squares line intersects the meteoric water line at δD and $\delta^{18}O$ values of -65 and -7 per mil, respectively. This line for formation waters, it should be noted, does not pass through the values for standard mean ocean water (SMOW) or the present-day meteoric water of the region.

The conclusion drawn by Kharaka and colleagues was that these formation waters were meteoric in origin, but the recharge occurred when North Slope had an entirely different climate. This climate may be inferred from the data of Dansgaard (1964), who showed that present day meteoric waters change in δD and $\delta^{18}O$ by 5.6 and 0.7 per mil, respectively, for every 1°C change in mean annual temperature. This relationship, which is rather approximate, suggests that the mean annual temperature at the Brooks Range (the most likely recharge area) was 15° to 20°C higher at the time of recharge than today.

Two possibilities were considered by Kharaka and his colleagues: the first assumes topographic conditions at the time of recharge were similar to present, but that the North Slope was at a latitude of 48°N or further south. A more plausible explanation, however, would keep northern Alaska in its present location, but requires changing the climatic conditions so that the mean annual temperatures were significantly higher, perhaps 15 to 20°C higher than their present values. Mean annual temperatures during parts of interglacial periods may have been higher in northern Alaska than their present values, but the increases were far less than those required by waters of this study (Fig. 11; Habicht, 1979). However, paleoclimatic indicators show that North America (including northern Alaska) had warmer climates than that of the present through most of the Mesozoic and Tertiary (Habicht, 1979; Wolfe, 1980). These studies show that the mean annual temperatures in northern Alaska required by the recharge waters of this study (15 to 20°C higher than at present) were reached as recently as the Miocene, but were also present throughout most of early Cenozoic and Mesozoic. The actual time of recharge cannot be determined for formation waters in northern Alaska, but it is clear that extremely old meteoric waters are present in this and other basins (Knauth, 1988; Connolly et al., 1990).

Formation Waters of Connate Origin

There are several sedimentary basins in the world where geological and hydrodynamic considerations indicate that formation waters are connate in origin. The Gulf of Mexico is a prime example of such basins. The extensive drilling for petroleum, especially in the northern half of this basin, has shown a very thick sequence (up to 15,000 m) of terrigenous shales, siltstones and sandstones of Cenozoic age that lack major unconformities. A thick sequence of fine-grained sediments in the basin is important because these sediments contain the largest portion of connate water at the time of deposition (initial porosity up to 80% for clays), and most of this water is squeezed out into the interbedded sandstones by subsequent compaction.

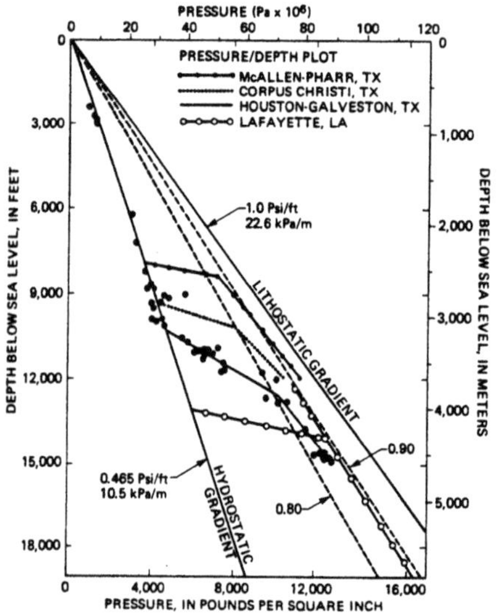

Figure 23: Original bottom-hole pressure from oil and gas wells in coastal Texas and Louisiana plotted against depth below sea level. Note the different depths at which the geopressured zone is encountered in different areas of the Gulf Coast, USA (Kharaka et al., 1978)

Abnormally high fluid pressures (geopressured zones) are another important characteristic of these basins. In the northern Gulf of Mexico basin (Fig. 23), geopressured zones with hydraulic pressure gradients higher than hydrostatic [10.5kPa m^{-1} (0.465psi ft 1)] are encountered at depths that range from about 2,000m in south Texas to more than 4,000m in parts of coastal Louisiana (Kharaka et al., 1985). Fluid potentials calculated from the pressures give hydraulic heads that are higher than any recharge areas in the basin, indicating water flow upward and updip in this section. The geologic and paleogeographic history of the northern Gulf of Mexico basin suggest that the flow has always been updip and toward the recharge areas to the northwest, indicating that the water, at least in the geopressured zones, is connate water squeezed from the marine shales and siltstones in the basin (Kharaka and Carothers, 1986; Land and Fisher, 1987).

The isotopic compositions of formation waters from the geopressured and normally pressured zones and local ground waters from the northern Gulf of Mexico basin are shown in figure 24. The data for formation waters fall on a general trend that passes through SMOW and away from the meteoric water of the area. The isotopic data show that the formation waters in the geopressured and normally pressured zones are not meteoric in origin nor are they the result of mixing of meteoric and ocean waters. These data indicate that the formation waters are connate and represent the original marine water of deposition.

The trend observed in figure 24 is unusual in that the δD values decrease (are lighter) instead of increase with increasing $\delta^{18}O$ values, which in turn increase with increasing subsurface temperatures. Kharaka and Carothers (1986) use equations based on conservation of mass (equ. 5 to 9) to show that isotopic interaction between connate marine water and clay minerals having very light original δD value of -70 per mil (Yeh, 1980) is responsible for this unusual shift. Isotopic exchange with clays appear also to control the $\delta^{18}O$ values of water (Fig. 14) in this basin (Yeh, 1980; Kharaka and Carothers, 1986; Land and Fisher, 1987).

The concentrations and isotopic compositions of noble gases in natural gas produced from geopressured geothermal and natural gas wells in the northern Gulf of Mexico basin also support the conclusion that the formation waters from this basin are marine connate in origin (Mazor and Kharaka, 1981). These authors show

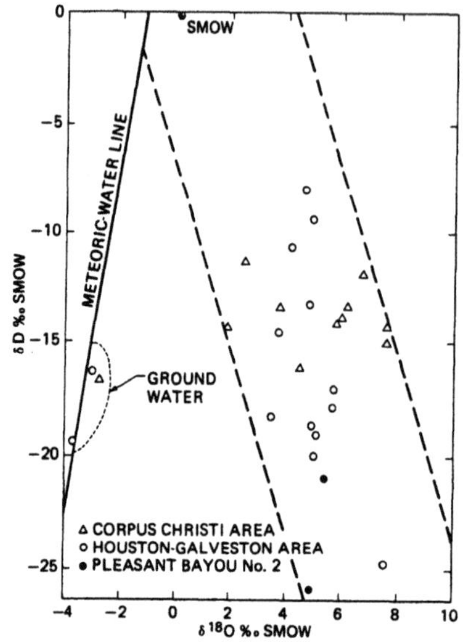

Figure 24: Isotopic composition of formation waters from northern Gulf of Mexico basin. The trend shows decreased δD values with increasing $\delta^{18}O$ values and it goes through SMOW and away from the local meteoric water (Kharaka et al., 1979)

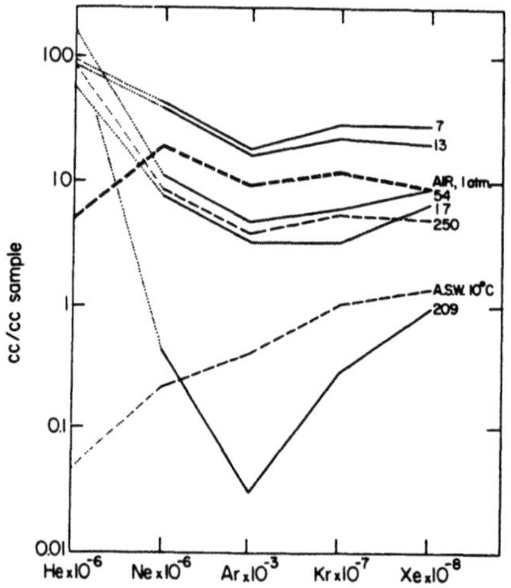

Figure 25: Abundance patterns for noble gases from geo-pressured-geothermal and natural gas wells from coastal Texas and Louisiana, and for air and sea water saturated at 10°C. For Ar, only the atmospheric component is plotted and that the concentration units (cc of gas per cc of water) are at standard conditions (Mazor and Kharaka, 1981)

that the abundance pattern from sample No. 209 (where all the gas produced is solution gas because no separate gas phase is present at depth) is essentially identical to that of air saturated sea water (Fig. 25). The abundance patterns for the other samples where separate gas phases are present, are similar to that of air, indicating degassing of a small fraction (up to about 1%) of dissolved noble gasses from a large volume of connate water of marine origin.

Bittern Connate Water in Evaporites

Evaporated ocean water trapped in salt and associated sediments and adjacent rocks at the time of evaporite precipitation (bittern water) is probably an important component of water in several sedimentary basins that contain thick evaporite sequences (Carpenter et al., 1974; Knauth and Beeunas, 1986; Kharaka et al., 1987; Connolly et al., 1990). The central Mississippi Salt Dome basin represents such a basin and the isotopic composition of its formation water is plotted in figure 26. The isotope values fall into three main groups. Samples from rocks of Jurassic age (Smackover and Norphlet Formations) have δD values (-1 to -3 per mil) that are approximately equal to that of SMOW; they have $\delta^{18}O$ values (5.1 to 7.3 per mil) that are highly enriched in ^{18}O. Samples obtained from rocks of Cretaceous age have δD values (-9 to -13 per mil) that are depleted in D relative to both SMOW and the samples from rocks of Jurassic age. Finally, samples obtained from shallow ground waters in the area have δD and $\delta^{18}O$ values that plot close to the Global Meteoric Water Line (GMWL). The $\delta^{a}D$ values give the D/H measurements in terms of the activities of D in solutions (Fig. 26). These were calculated from the δD values using a correction factor $\Delta\delta$ (Sofer and Gat, 1972) given by:

$$\Delta\delta = 6.1\ m_{Ca} + 5.1\ m_{Mg} + 2.4\ m_{K} + 0.4\ m_{Na} \qquad (10)$$

where m is the molality of the subscripted cation. The correction factors for the formation waters are high, ranging from 2.3 per mil to 10.1 per mil. The $\delta^{a}D$ versus $\delta^{18}O$ (^{18}O values are always reported in terms of activities) plot shows a more linear trend than that obtained using δD values (Fig. 26).

The isotopic compositions of the waters support the conclusion, based on plots of Br versus Cl and combinations of other chemical constituents (Carpenter et al., 1974; Stoessell and

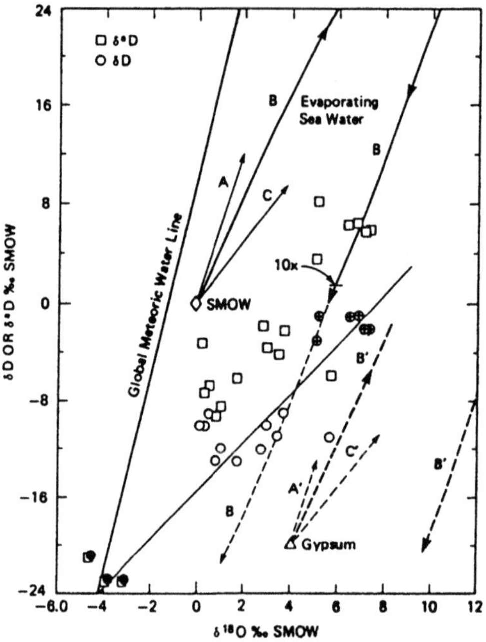

Figure 26: Isotopic composition of ground (solid circles) and formation waters from Jurassic (open circles with +) and Cretaceous (open circles) rocks from central Mississippi Salt Dome basin. $\delta^{a}D$ values were computed from δD values and a correction factor obtained using equation (10) in the text. Lines originating from SMOW and gypsum are the range of trajectories for sea water undergoing evaporation and for hydration water in gypsum precipitated at corresponding points on the evaporation trajectories, respectively. Line B is Holser's (1979) estimate of evaporating sea water in the Gulf Coast. The least-squares line for the δD and $\delta^{18}O$ values is also indicated (from Kharaka et al., 1987)

Carpenter, 1986; Kharaka et al., 1987) that formation waters in rocks of Jurassic age are evaporated seawater. The δD and $\delta^{18}O$ values of evaporating seawater initially increase with increasing evaporation, but then the trend reverses describing a loop on the δ-diagram (Fig. 26; Lloyd, 1966; Holser, 1979). The actual trajectory formed on the -diagram is variable depending on the local conditions of temperature, humidity and wind velocity at the evaporation site. Holser (1979) and Knauth and Beeunas (1986) have shown that line (B) is probably the most likely trajectory for evaporating seawater at the Gulf Coast (Fig. 26). Line (B) passes very close to the point giving the δ-values for the water in the Norphlet Formation. The δ-values for the bittern water at that point are those expected from evaporating seawater at the point of halite precipitation (ratio of remaining water to the original water, R, equal to 11x). The δ-values of the samples from the Smackover Formation were initially approximately equal to those from the Norphlet Formation; the 1 to 2 per mil shifts in $\delta^{18}O$ values are likely related to isotopic interactions with the enclosing carbonate minerals (Heydari and Moore, 1989).

The lower δD and $\delta^{18}O$ values of water from rocks of Cretaceous age can be explained as bittern seawater taken past the point of evaporation for water from rocks of Jurassic age or as a mixture between meteoric water and bittern seawater (Fig. 26). The origin of these waters cannot be explained by mixing between bittern and unevaporated seawater (SMOW), or by water released from the dehydration of gypsum (Fig. 26). The chemical composition of water can be used to show that increased evaporation leading to higher salinities is not a plausible explanation for the origin of waters in rocks of Cretaceous age. The salinities of these waters are considerably lower than those of waters from Jurassic rocks and are likely the result of mixing with meteoric water. A more detailed treatment of the geochemistry of water in this basin (Kharaka et al., 1987) shows that 20 to 80% of the water in Cretaceous rocks is of meteoric origin.

Deep Basin Brines from Mixing of Different Waters

Formation waters in sedimentary basins are highly mobile, resulting in the mixing of waters of the same origin but different age or different origin and age (Kharaka and Carothers, 1986). Hitchon and Friedman (1969) made a very detailed isotopic and chemical study of the surface and formation waters of the Western Canada sedimentary basin. Using mass balance calculations of D and total dissolved solids, they concluded that the observed distribution of D in the formation waters could best be derived through mixing of diagenetically modified seawater with not more than 2.9 times as much fresh meteoric water. They attributed the ^{18}O enrichment of formation waters to extensive exchange with the carbonate rocks (see also the recent study by Connolly et al., 1990).

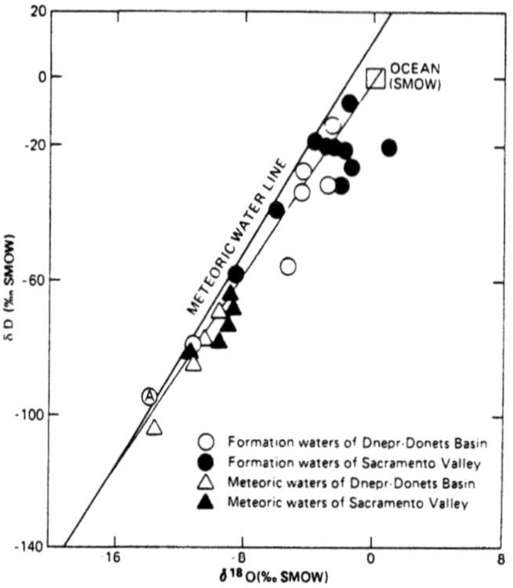

Figure 27: Isotopic composition of meteoric (triangles) and formation waters (circles) from Dnepr-Donets basins, USSR (Vetshteyn et al., 1981), and Sacramento Valley, California, USA. In both basins the formation waters are probably mixtures of meteoric and connate (marine) waters (Kharaka and Carothers, 1986)

Examples of deep-basin waters that can be related to simple proportional mixing between meteoric water and marine connate waters are shown in figure 27. Vetshteyn et al. (1981) showed that the waters from the Dnepr-Donets basin and the Ukrainian Shield are essentially a mixture of meteoric and connate waters (line A-SMOW, Fig. 27). Rozkowski and Przewlocki (1974) reported the isotopic composition of brines from the Upper Silesian Coal basin, Poland, which lie on a simple mixing line between local meteoric water and SMOW. The isotopic data for the Sacramento Valley, California, also shown in figure 27, again indicate mixing of meteoric and marine connate waters, accompanied by some ^{18}O enrichment. Mixing of meteoric and ocean water in the Sacramento Valley basin is supported by geological, hydrologic and isotope evidences. Berry (1973) and Kharaka et al. (1985) used evidence from the chemical composition of formation waters, fluid hydrodynamics, and regional geology to show that connate water was being squeezed from shales in this basin. This water moved upward in the section and mixed with meteoric water that was present in the sandstones from which the samples were obtained.

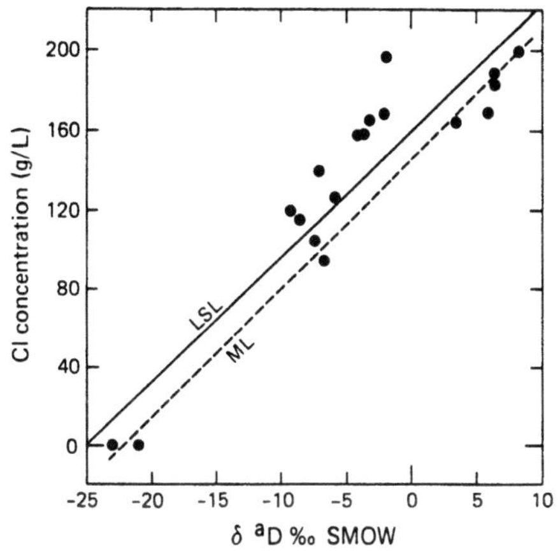

Figure 28: The mixing (ML) and least-squares (LSL) lines based on the concentrations of Cl and δ aD values for formation waters from central Mississippi Salt Dome basin. Samples from Cretaceous rocks plot above the mixing line, indicating halite dissolution (Kharaka et al., 1987)

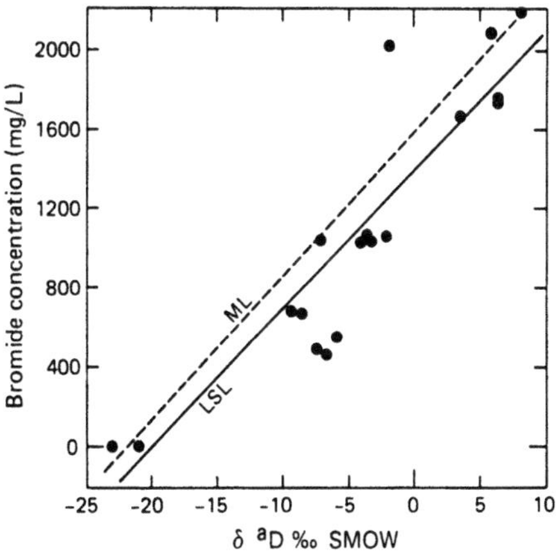

Figure 29: The mixing (ML) and least-squares (LSL) lines based on the concentrations of Br and δ aD values for formation waters from central Mississippi Salt Dome basin. Samples from Cretaceous rocks plot below the mixing line, indicating halite dissolution (Kharaka et al., 1987)

Mixing between meteoric water and bittern connate water was

demonstrated by Kharaka et al. (1987) for formation waters from central Mississippi Salt Dome basin. They used plots of Cl and Br concentrations and $\delta^{a}D$ values to show that meteoric water comprised up to 20% of pore water in Jurassic rocks (Fig. 28, 29); meteoric water comprised 20 to 80% of water in Cretaceous rocks. Knauth (1988) also used water isotopes and chemical data for waters in the Palo Duro Basin, Texas, to indicate extensive mixing between bittern brines and two pulses of meteoric water having different isotopic compositions. Mixing between waters of the same origin but affected by different processes is probably common in sedimentary basins. Kharaka et al. (1985) used δD and $\delta^{18}O$ values and the chemical composition of formation waters from High Island field, offshore Texas, to show that the formation waters in Pleistocene reservoir rocks originated by mixing of two marine connate waters (Fig. 30). One end member is an essentially unmodified marine connate water of Pleistocene age (samples 6 and 7 in Fig. 30) where as the other (sample 5) being highly modified marine connate water possibly of Miocene age. Approximately the same mixing proportions were obtained using stable isotope values (Fig. 30) and concentrations of Na, Cl, Ca, and Li (Fig. 4).

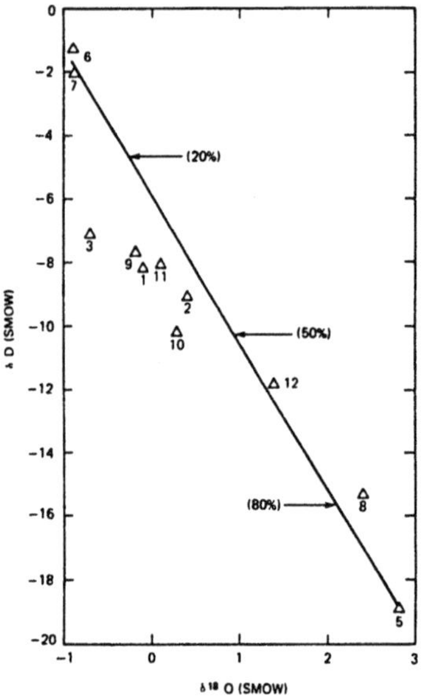

Figure 30: Isotopic composition of formation waters from High Island field, offshore Texas. Percentages representing mixing proportions of the deep brine with Pleistocene connate water (Kharaka et al., 1985). Sample numbers are from Table 3

SUMMARY AND CONCLUSIONS

Formation waters with salinities ranging from about 5,000 mg/l to more than 350,000 mg/l dissolved solids play a major role in the physical and chemical processes that occur in sedimentary basins. Detailed inorganic and organic chemical analyses, together with O and H isotope geochemistry of water and a diverse suite of other stable and radioactive isotope compositions have been reported for formation waters from many sedimentary basins worldwide. Application of these determinations together with local and regional geology and recent and paleohydrology have shown that these fluids are much more mobile and their origins and interactions with rocks and sediments much more complex than previously realized.

The formation waters in the majority of sedimentary basins are dominantly of local meteoric or marine connate origin. However, bittern (residual) water, geologically old meteoric water, and especially waters of mixed origin are important components in many sedimentary basins. The original waters have evolved to Na-Cl, Na-Cl-CH_3COOO, or Na-Ca-Cl type waters by a combination of the following main processes: (a) dissolution of halite, (b) incorporation of bitterns, (c) dissolution and precipitation of minerals other than halite, (d) interaction of the waters with organic matter present in sedimentary rocks, (e) interaction with rocks, principally clays, siltstones, and shales, that behave as geological membranes, and (f) diffusion, which appears more important than previously realized. The important processes responsible for the chemical evolution of water and mineral diagenesis in each basin, however, can be identified using "chemical markers" and isotopes.

The discovery in these waters of high concentrations (up to 10,000 mg/l of mono- and dicarboxylic acid anions, phenols and other reactive organic species has led to numerous field and laboratory studies to determine their distribution and importance in inorganic and organic interactions. The observed concentrations are minimum values because these organics are degraded by bacteria and by thermal decarboxylation. Decarboxylation rates estimated from field data for acetate, the dominant species, give half-life values of 20 to 60 million years at 100°C.

The concentrations of Pb, Zn, Cu, Hg, and several other metals (excluding Fe and Mn) are almost always very low (<1 mg/l) in oil field waters because they are controlled by the very low solubilities of their respective sulfide minerals. Exceptionally rich sources of metals such as redbeds are probably present in the few localities worldwide where metal concentrations are high. Metals leached from these beds first precipitate along with available reduced sulfur; dissolved metal concentrations increase to high values only when the concentrations of H_2S decrease to extremely low values (<0.01 mg/l).

Water-rock interaction in sedimentary basins is and will continue to be the subject of intensive research. The following are several areas that are receiving more attention because of their scientific or economic importance, or the availability of new and applicable equipment.

(1) A plethora of stable and radioactive isotope data are being used to study mineral diagenesis and the origin and age of water and the sources and sinks of solutes in these basins. Determination of accurate fractionation factors and rates of isotopic exchange between waters and minerals at sedimentary conditions are needed for a better interpretation of the isotope data.

(2) Major advances have been made during the last decade documenting the nature, distribution, and interactions of reactive organic species. Additional investigation is needed, especially in determining the stability of organic-inorganic complexes, the rates of decarboxylation in field conditions, and the importance of dissolved organic sulfur and nitrogen compounds.

(3) Geochemical, hydrologic and solute transport codes are being applied to sedimentary basins. To obtain reliable information from this endeavor, accurate, detailed data on aqueous and solid phases must be available and more accurate thermodynamic data are needed for clay minerals, solid-solution minerals, and brines. Also needed are dissolution-precipitation kinetic data that are applica-

ble to field situations.

(4) Geochemistry of Al is poorly understood, especially the transport of Al in formation water.

(5) Membrane behavior of shales in sedimentary basins is also poorly understood and requires additional study.

REFERENCES

Bath A., Edmunds W.M. and Andrews J.N. (1978) Paleoclimatic trends deduced from the hydrochemistry of Triassic sandstone aquifer, United Kingdom. In: Isotope Hydrology, IAEA, Vienna, 2, 545-568.

Benzel W.M. and Graf D.L. (1984) Studies of smectite membrane behavior: Importance of layer thickness and fabric in experiments at 20°C. Geochim. Cosmochim. Acta, 48, 1769-1778.

Berry F.A.F. (1973) High fluid-potentials in the California coast ranges and their tectonic significance. Am. Ass. Petrol. Geol. Bull., 57, 1219-1249.

Bethke C.M. (1986) Hydrologic constraints on the genesis of the upper Mississippi Valley mineral district from Illinois Basin. Econ. Geol., 81, 233-249.

Bird M.I. and Chivas A.R. (1988) Stable-isotope evidence for low-temperature kaolinitic weathering and post-formational hydrogen-isotope exchange in Permian kaolinites. Chem. Geol., (Isot. Geosc. Sect.), 72, 249-265.

Boles J.R. (1978) Active ankerite cementation in the subsurface Eocene of southwest Texas. Contr. Miner. Petrol., 68, 13-22.

Bottinga Y. (1969) Calculated fractionation factors for carbon and hydrogen isotope exchange in the system calcite-CO-graphite-methane-hydrogen and water vapor. Geochim. Cosmochim. Acta, 33, 49-64.

Carothers W.W. and Kharaka Y.K. (1978) Aliphatic acid anions in oil field waters - implications for origin of natural gas. Am. Ass. Petrol. Geol. Bull., 62, 2441-2453.

Carothers W.W. and Kharaka Y.K. (1980) Stable carbon isotopes of HCO^-_3 in oil field waters - implications for the origin of CO_2. Geochim. Cosmochim. Acta, 44, 323-333.

Carothers W.W., Adami, L.H. and Rosenbauer R.J. (1988) Experimental oxygen isotope fractionation between siderite-water and phosphoric acid liberated CO_2-siderite. Geochim. Cosmochim. Acta, 52, 2445-2450.

Carpenter A.B., Trout, M.L. and Pickett E.E. (1974) Preliminary report on the origin and chemical evolution of lead- and zinc-rich brines in central Mississippi. Econ. Geol., 52, 1191-1206.

Cathles L.M. (1981) Fluid flow and genesis of hydrothermal ore deposits. Econ. Geol., 75th Anniversary Volume, 424-457.

Chiba H., Kusakabe M., Hirano S., Matsuo S. and Somiya S. (1981) Oxygen isotope fractionation factors between anhydrite and water from 100°C to 550°C. Earth Plan. Sci. Lett., 53, 55-62.

Chillingarian G.V. and Wolf K.H. (ed.) (1988) Diagenesis, II. Develop. in Sedimentology, 43, 420 p.

Claypool G.E. and Kaplan., I.R. (1974) The origin and distribution of methane in marine sediments. In: Natural Gases in Marine Sediments, Kaplan I.R. (ed.), Plenum Press, New York, N.Y., 99-139.

Clayton R.N., Muffler L.J.P. and White D.E. (1968) Oxygen isotope study of calcite and silicates of the River Ranch No. 1 well, Salton Sea Geothermal field, California. Am. Jour. Sci., 266, 968-979.

Clayton R.N., Friedman I., Graf D.L., Mayeda T.K., Meets W.F. and Shimp N.F. (1966) The origin of saline formation waters, I. Isotopic composition. Jour. Geoph. Res., 71, 3869-3882.

Cole D.R. (1985) A preliminary evaluation of oxygen isotopic exchange between chlorite and water. Geol. Soc. Am., Abstr. with Progr., 17, 550.

Connolly C.A., Walter L.M., Baadsgaard H. and Longstaff F. (1990) Origin and evolution of formation fluids, Alberta Basin, western Canada sedimentary basin: II. Isotope systematics and fluid mixing. Applied Geochem., 5, 397-414.

Coplen T.B. (1989) Environmental isotopes in groundwater studies. In: Handbook on Hydrogeology, Csallany S.C. and Kanivetsky R.A. (eds.), McGraw-Hill, in press.

Coplen T.B. and Hanshaw B.B. (1973) Ultrafiltration by a compacted clay membrane: I. Oxygen and hydrogen fractionation. Geochim. Cosmochim. Acta, 37, 2295-2310.

Craig H. (1961) Isotopic variations in meteoric waters. Science, 133, 1702-1703.

Criss R.E., Gregory R.T. and Taylor H.P. (1987) Kinetic theory of oxygen isotopic exchange between minerals and water. Geochim. Cosmochim. Acta, 51, 1099-1108.

Crossey L.J., Surdam R.C. and Lahann R. (1986) Application of organic/inorganic diagenesis to porosity prediction. In: Relationship of Organic Matter and Mineral Diagenesis, Gautier D.L. (ed.), Soc. Econ. Paleont. Miner., Spec. Publ., 38, 147-156.

Dansgaard W. (1964) Stable isotopes in precipitation. Tellus, 16, 436-468.

Davis J.B. (1967) Petroleum Microbiology. Elsevier, New York, N.Y., 604 p.

Demir I. (1988) Studies of smectite membrane behavior: electrokinetic, osmotic, and isotopic fractionation processes at elevated pressures. Geochim. Cosmochim. Acta, 52, 727-737.

Dickinson W.W. (1987) An oxygen isotope model for interpreting carbonate diagenesis in nonmarine rocks (Green River Basin, Wyoming, U.S.A.). Chem. Geol. (Isot. Geosc. Sect.), 65, 103-116.

Drummond S.E. and Palmer D.A. (1986) Thermal decarboxylation of acetate: II. Boundary conditions for the role of acetate in the primary migration of natural gas. Geochim. Cosmochim. Acta, 50, 813-824.

Dutton S.P. and Land L.S. (1985) Meteoric burial diagenesis of Pennsylvanian arkosic sandstones, southwestern Anadarko Basin, Texas. Am. Ass. Petrol. Geol. Bull., 69, 22-38.

Evans G.V., Otlet R.L., Downing R.A., Monkhouse R.A. and Rae G. (1978) Some problems in the interpretation of isotope measurements in United Kingdom aquifers. In: Isotope Hydrology, 2, Int. Atom. En., Vienna, 679-708.

Fairbridge R.W. (1964) The importance of limestone and its Ca/Mg content to paleoclimatology. In: Problems in Paleoclimatology,

Nairn A.E.M. (ed.), Wiley-Interscience, New York, N.Y., 431-530.
Faure G. (1986) Principles of Isotope Geology, second edition. J. Wiley and Sons, New York, N.Y., 589 p.
Fisher J.B. and Boles J.R. (1990) Water-rock interaction in Tertiary sandstones, San Joaquin basin, California, U.S.A.: Diagenetic controls on water composition. Chem. Geol., 82, 83-101.
Fontes J.Ch. (1980) Environmental Isotopes in Ground-Water Hydrology. In: Handbook of Environmental Isotope Geochemistry, Fritz P. and Fontes J.C. (eds.), vol. 1, Elsevier, New York, N.Y., 75-134.
Freeze R.A. and Cherry J.A. (1979) Ground Water. Prentice Hall, Englewood Cliffs, New Jersey, 604 p.
Friedman I. and O'Neil J.R. 1977. Compilation of stable isotope fractionation factors of geochemical interest. In: Data of Geochemistry, sixth edition. U.S. Geol. Surv. Prof. Paper, 440-KK, 12 p.
Fritz P. and Fontes J.Ch. (eds.) (1980) Handbook of Environmental Isotope Geochemistry, 1. The Terrestrial Environment, Elsevier, Amsterdam, 545 p.
Fritz P. and Fontes J.Ch. (eds.) (1986) Handbook of Environmental Isotope Geochemistry, 2 volumes, Elsevier, Amsterdam, 1, 545 p.
Gat J R. (1980) The isotopes of hydrogen and oxygen in precipitation. In: Handbook of Environmental Isotope Geochemistry, 1. The Terrestrial Environment, Fritz P. and Fontes J.C. (eds.), Elsevier, Amsterdam, 21-48.
Gat J.R. and Carmi I. (1970) Evolution of the isotopic composition of atmospheric waters in the Mediterranean Sea area. Jour. Geoph. Res., 75, 3039-3048.
Gautier D.L. (edit.) (1986) Roles of Organic Matter in Sediment Diagenesis. Soc. Econ. Paleont. and Miner., Spec. Publ., 38, 203 p.
Gautier D.L., Kharaka Y.K. and Surdam R.C. (1985) Relationship of organic matter and mineral diagenesis. Soc. Econ. Paleont. Miner., Short Course, 17, 279 p.
Gerasimov I.P. (1979) Climates of past geological epochs. Proc. of the World Climate Conf., Geneva, WMO, 537, 88-111.
Gieskes J.M. (1981) Deep Sea Drilling Interstitial Waters Studies. Implications For Chemical Alteration of the Ocean Crust, Layers I and II. In: The Deep Sea Drilling Project, Warme J.E., Douglas R.G. and Winterer E.L. (eds.), Washington, D.C., U.S. Government Printing Office, 149-168.
Graf D.L., Friedman I. and Meents W.F. (1965) The Origin of Saline Formation Waters: II. Isotopic fractionation by shale in micropore systems. Illinois State Geol. Surv. Circular, 393, 32 p.
Graham C.M., Viglino J.A. and Harmon R.S. (1987) An experimental study of hydrogen isotope exchange between aluminous chlorite and water. Am. Min., 72, 566-579.
Habicht J.K.A. (1979) Paleoclimate, Paleomagnetism and continental drift. Am. Ass. Petrol. Geol., Studies in Geology, 9, 29 p.
Hanor J.S. (1987) Origin and migration of subsurface sedimentary brines. Soc. Econ. Paleont. Miner., Short Course, 21, 247 p.
Hanor J.S. and Workman A.L. (1986) Distribution of dissolved fatty acids in some Louisiana oil field brines. Applied Geochem., 1, 37-46.

Hanor J.S., Kharaka Y.K. and Land L.S. (1988) Geochemistry of waters in deep sedimentary basins. Geology, 16, 560-561.
Hem J.D. (1985) Study and interpretation of the chemical characteristics of natural water. U.S. Geol. Surv. Water-Supply Paper, 2254, 253 p.
Heydari E. and Moore C.H. (1989) Burial diagenesis and thermochemical sulfate reduction, Smackover Formation, southeastern Mississippi salt basin. Geology, 17, 1080-1084.
Hitchon B. and Friedman I. (1969) Geochemistry and origin of formation waters in the western Canada sedimentary basin: I. Stable isotopes of hydrogen and oxygen. Geochim. Cosmochim. Acta, 33, 1321-1349.
Hitchon B., Bachu S. and Sauveplane C.M. (eds.) (1986) Hydrology of sedimentary basins: Application to exploration and exploitation. Proceed. of Third Canadian/American Conf. on Hydrogeology, National Water Well Association, 275 p.
Hitchon B., Billings G.K. and Klovan J.E. (1971) Geochemistry and origin of formation waters in the western Canada sedimentary basin: III. factors controlling chemical composition. Geochim. Cosmochim. Acta, 35, 567-598.
Hoefs J. (1987) Stable Isotope Geochemistry, third edition. Springer Verlag, New York, N.Y., 241 pp.
Holser W.T. (1979) Trace elements and isotopes in evaporites. In: Marine Minerals, Burns R.G. (ed.), Reviews in Miner., Min. Soc. Am., 295-346.
Houseknecht D.W. and Hathon L.A. (1987) Petrographic constraints on models of intergranular pressure solution in quartzose sandstones. Applied Geochem., 2, 507-522.
Hull R.W., Kharaka Y.K., Maest A.S. and Fries T.L. (1985) Sampling and analysis of subsurface waters: A summary of current methodology. In: Practical Applications of Ground Water Geochemistry, Hitchon B. and Wallick E.I. (eds.), Proceedings First Canadian/American Conference on Hydrology, Banff, Alberta, Canada. National Water Well Association, Worthington, Ohio, 117-126.
Hutcheon I.E. (ed.) (1989) Burial Diagenesis. Miner. Ass. Canada, Short Course Handbook, 15, 409 p.
Kartsev A.A. (1974) Hydrogeology of oil and gas deposits. Natl. Tech. Inf. Service Rept., TT73-58022, 323 p.
Kharaka Y.K. (1986) Origin and evolution of water and solutes in sedimentary basins. In: Hydrology of Sedimentary Basin: Application to Exploration and Exploitation. Proceeding of 3rd Canadian/American Conference on Hydrogeology. National Water Well Association, 173-195.
Kharaka Y.K. and Berry F.A.F. (1974) The influence of geological membranes on the geochemistry of subsurface waters from Miocene sediments at Kettleman North Dome, California. Water Resour. Res., 10, 313-327.
Kharaka Y.K. and Carothers W.W. (1986) Oxygen and hydrogen isotope geochemistry of deep basin brines. Chapter 2. In: Handbook of Environmental Isotope Geochemistry, Fritz P. and Fontes J.C. (eds.), 2, 305-360.
Kharaka Y.K. and Carothers W.W. (1988) Geochemistry of oil-field waters from the North Slope, Alaska, U.S. Geol. Surv. Prof. Paper, 1399, 551-561.
Kharaka Y.K., Berry F.A.F. and Friedman I. (1973) Isotopic composition of oil-field brines from Kettleman North Dome oil field, California, and their geologic implications. Geochim.

Cosmochim. Acta, 37, 1899-1908.
Kharaka Y.K., Brown P.M. and Carothers W.W. (1978) Chemistry of waters in the geopressured zone from coastal Louisiana: Implications for geothermal development. Geoth. Res. Counc. Trans., 2, 371-374.
Kharaka Y.K., Carothers W.W. and Rosenbauer R.J. (1983) Thermal decarboxylation of acetic acid: Implications for origin of natural gas. Geochim. Cosmochim. Acta, 47, 397-402.
Kharaka Y.K., Hull., R.W. and Carothers W.W. (1985) Water-rock interactions in sedimentary basins. In: Relationship of Organic Matter and Mineral Diagenesis, Gautier D.L., Kharaka Y.K. and Surdam R.C. (eds.), Soc. Econ. Paleont. Miner., Short Course, 17, 79-176.
Kharaka Y.K., Lico M.S., Wright V.A. and Carothers W.W. (1979) Geochemistry of formation waters from Pleasant Bayou No. 2 well and adjacent areas in Coastal Texas. Proceed. 4th Geopressured-Geothermal Energy Conf., Austin, Texas, 168-193.
Kharaka Y.K., Law L.M., Carothers W.W. and Goerlitz D.F. (1986) Role of organic species dissolved in formation waters in mineral diagenesis. In: Relationship of Organic Matter and Mineral Diagenesis, Gautier D.L. (eds.), Soc. Econ. Paleont. Miner., Spec. Vol., 38, 111-122.
Kharaka Y.K., White L.D., Ambats, G. and White, A.F. (1988a) Origin of subsurface water at Cajon Pass, California. Geoph. Res. Lett., 15, 1049-1052.
Kharaka Y.K., Gunter W.D., Aggarwal P.K., Perkins E.H. and DeBraal J.D. (1988b) SOLMINEQ.88: A computer program for geochemical modeling of water-rock interactions. U.S. Geol. Surv., Water Resources Investigations Report 88-4227, 420 p.
Kharaka Y.K., Maest A.S., Carothers W.W., Law L.M., Lamothe P.J. and Fries T.L. (1987) Geochemistry of metal-rich brines from central Mississippi Salt Dome basin, U.S.A. Applied Geochem., 2, 543-561.
Kita I., Taguchi S. and Matsubaya O. (1985) Oxygen isotope fractionation between amorphous silica and water at 34-93°C. Nature, 314, 83-84.
Knauth L.P. (1988) Origin and mixing history of brines, Palo Duro Basin Texas, U.S.A.. Applied Geochem., 3, 455-479.
Knauth L.P. and Beeunas M.A. (1986) Isotope geochemistry of fluid inclusions in Permian halite with implications for the isotopic history of ocean water and the origin of saline formation waters. Geochim. Cosmochim. Acta, 50, 419-433.
Kyser T.K. (1987) Equilibrium fractionation factors for stable isotopes. In: Stable Isotopes Geochemistry of Low Temperature Fluids, Kyser T.K. (ed.), Min. Ass. Canada, Short Course, 13, 1-84.
Lamb H.H. (1977) Climate, Present, Past and Future, 2. Methuen, London, 835 p.
Lambert S.J. and Epstein S. (1980) Stable isotope investigations of an active geothermal system in Valles Caldera, Jemez Mountains, New Mexico. Jour. Volcan. Geother. Res., 8, 111-129.
Land L.S. (1983) The application of stable isotopes to studies of the origin of dolomite and to problems of diagenesis of clastic sediments. In: Stable Isotopes in Sedimentary Geology, Soc. Econ. Paleont. Miner., Short Course, 10, 4.1-4.22.
Land L.S. and Fisher R.S. (1987) Wilcox sandstone diagenesis, Texas Gulf Coast: a regional isotopic comparison with the Frio

Formation. In: Diagenesis of Sedimentary Sequences, Marshall J.D. (ed.), Geol. Soc. Am., Spec. Publ., 36, 219-235.

Lloyd R.M. (1966) Oxygen isotope enrichment of seawater by evaporation. Geochim. Cosmochim. Acta, 30, 801-814.

Longstaffe F.J. (1989) Stable isotopes as tracers in clastic diagenesis. In: Burial Diagenesis, Hutcheon I.E. (ed.), Miner. Ass. Canada, Short Course Series, 15, 201-277.

Longstaffe F.J. and Ayalon A. (1990) Hydrogen isotope geochemistry of diagenetic clay minerals from Cretaceous sandstones, Alberta: Evidence for exchange. Applied Geochem., 5, 657-668.

Lundegard P.D. and Kharaka Y.K. (1990) Geochemistry of organic acids in subsurface waters. In: Chemical Modeling of Aqueous Systems II, Melchior D.C. and Bassett R.L. (eds.), Am. Chem. Soc., Symposium Series, 416, 169-189.

Lundegard P.D. and Land L.S. (1986) Carbon dioxide and organic acids: Their origin and role in diagenesis, the Texas Gulf Coast Tertiary. In: Relationship of Organic Matter and Mineral Diagenesis, Gautier D.L. (ed.), Soc. Econ. Paleont. Miner., Spec. Publ., 38, 129-146.

MacGowan D.B. and Surdam R.C. (1990) Importance of organic-inorganic interactions during progressive clastic diagenesis. In: Chemical Modeling of Aqueous Systems II, Melchior D.C. and Bassett R.L. (eds.), Am. Chem. Soc., Symposium Series, 416, 494-507.

Mangelsdorf P.C., Manheim F.T. and Gieskes M.T.M. (1970) Roles of gravity, temperature gradients and ion exchange media in formation of fossil brines. Am. Ass. Petrol. Geol. Bull., 54, 617-626.

Marshall J.D. (ed.) (1987) Diagenesis of Sedimentary Sequences. Geol. Soc. Am., Spec. Publ., 36, 360 p.

Matsuhisa Y., Goldsmith J.R. and Clayton R.N. (1979) Oxygen isotopic fractionation in the system quartz-albite-anorthite-water. Geochim. Cosmochim. Acta, 43, 1131-1140.

Mazor E. and Kharaka Y.K. (1981) Atmospheric and radiogenic noble gases in geopressured-geothermal fluids: northern Gulf of Mexico basin. Proceed. 5th Conf. Geopressured-Geothermal Energy, Louisiana State University, Baton Rouge, LA, 197-200.

Melchior D.C. and Bassett R.L. (eds.) (1990) Chemical modeling of aqueous systems II. Am. Chem. Soc., Symposium Series, 416, 556 p.

Merino E. (1975) Diagenesis in Tertiary Sandstones from Kettleman North Dome, California: I. Diagenetic Mineralogy. Jour. Sediment. Petrol., 5, 320-336.

Miles D.L. (ed.) (1989) Water-rock Interaction WRI-6. A.A. Balkema, Rotterdam, 819 p.

Milliken K.L., Land L.S. and Loucks R.G. (1981) History of burial diagenesis determined from isotopic geochemistry, Frio Formation, Brazoria County, Texas. Am. Ass. Petrol. Geol. Bull., 65, 1397-1413.

Naeser N.D. and McCulloh T.H. (eds.) (1989) Thermal History of Sedimentary Basins: Methods and Case Histories. Springer-Verlag, New York, N. Y., 319 p.

O'Neil J.R. and Kharaka Y.K. (1976) Hydrogen and oxygen isotope exchange reactions between clay minerals and water. Geochim. Cosmochim. Acta, 40, 241-246.

O'Neil J.R. and Taylor H.P., Jr. (1967) The oxygen isotope and cation exchange chemistry of feldspars. Am. Min., 52, 1414-1437.

Pearson F.J., Jr. and White D.E. (1967) Carbon-14 ages and flow rates of water in Carizo Sand, Atascosa County, Texas. Water Resour. Res., 3, 251-261.
Phillips F.M. and Bentley H.W. (1987) Isotopic fractionation during ion filtration: I. Theory. Geochim. Cosmochim. Acta, 51, 683-695.
Rapp J.B. (1976) Amino acids and gases in some springs and an oil field in California. U.S. Geol. Surv. Jour. Res., 4, 227-232.
Rozkowski A. and Przwlocki K. (1974) Application of stable environmental isotopes in mine hydrogeology taking Polish Coal basins as an example. In: Isotope Techniques in Groundwater Hydrology. IAEA, Vienna, 1, 481-488.
Satake H. and Matsuo S. (1984) Hydrogen isotopic fractionation factor between brucite and water in the temperature range from 100°C to 510°C. Contrib. Miner. Petrol., 86, 19-24.
Savin S.M. (1980) Oxygen and hydrogen isotope effects in low-temperature mineral-water interactions. In: Handbook of Environmental Isotope Geochemistry, 1. The Terrestrial Environment. Fritz P., and Fontes J.C. (eds.), Elsevier, Amsterdam, 283-328.
Savin S.M. and Epstein S. (1970) The oxygen and hydrogen isotope geochemistry of clay minerals. Geochim. Cosmochim. Acta, 34, 25-42.
Savin S.M. and Lee M. (1988) Isotopic studies of phyllosilicates. In: Hydrous Phyllosilicates (exclusive of micas). Bailey S.W. (ed.), Rev. in Miner., 19, 189-223.
Sheppard S.M.F. (1986) Characterization and isotopic variations in natural waters. In: Stable Isotopes in High Temperature Geological Processes. Valley J.W., Taylor H.P., Jr. and O'Neil J.R. (eds.), Rev. in Miner., 16, 165-183.
Sofer Z. and Gat J. (1972) Activities and concentrations of 18O in concentrated aqueous salt solutions: analytical and geophysical implications. Earth Plan. Sci. Lett., 15, 232-238.
Stoessell R.K. and Carpenter A.B. (1986) Stoichiometric saturation tests of $NaCl_{1-x}Br_x$ and $KCl_{1-x}Br_x$. Geochim. Cosmochim. Acta, 50, 1465-1474.
Suchecki R.K. and Land L.S. (1983) Isotopic geochemistry of burial-metamorphosed volcanogenic sediments, Great Valley sequence, northern California. Geochim. Cosmochim. Acta, 47, 1487-1499.
Surdam R.C., Crossey L.J., Hagen E.S. and Heasler H.P. (1989) Organic-inorganic interactions and sandstone diagenesis. Am. Ass. Petrol. Geol. Bull., 73, 1-23.
Taylor H.P., Jr. and Magaritz M. (1978). Oxygen and hydrogen isotope studies of the Cordilleran batholiths of western North America. New Zealand Dep. Sc. and Indus. Res. Bull., 220, 151-173.
Thurman E.M. (1984) Humic Substances I. Geochemisty, Characterization and Isolation. J. Wiley and Sons, New York, N. Y., 380 p.
Truesdell A.H. and Hulston J.R. (1980) Isotopic evidence in environments of geothermal systems. In: Handbook of Environmental Isotope Geochemistry, 1, Fritz P., and Fontes J.C. (eds.), Elsevier, New York, N.Y., 179-226.
Valley J.W., Taylor H.P., Jr. and O'Neil J.R. (eds.) (1986) Stable Isotopes in High Temperature Geological Processes. Rev. in Miner., 16, 570 p.
Vetshteyn V.V., Gavish V.K. and Gutsalo L.K. (1981) Hydrogen and oxygen isotope composition of waters in deep seated fault

zones. Intern. Geol. Rev., 23, 302-310
White A.F., Peterson M., Wollenberg H. and Flexser S. (1990) Sources and fractionation processes influencing the distribution of D, 18O and 13C in the Long Valley hydrothermal system, California, USA. Applied Geochem., 5, 571-586.
White D.E. (1965) Saline waters of sedimentary rocks. In: Fluids in Subsurface Environments, Yound A. and Galley G.E. (eds.), Am. Ass. Petrol. Geol. Mem., 4, 342-366.
Willey L.M., Kharaka Y.K., Presser T.S., Rapp., J.B. and Barnes I. (1975) Short-chain aliphatic acids in oil field waters of Kettleman Dome Oil Field, California. Geochim. Cosmochim. Acta, 39, 1707-1710.
Wolfe J.A. (1980) Tertiary climates and floristic relationships at high latitudes in the northern hemisphere. Paleogeogr. Paleoclimat. Paleoecol., 30, 313-323.
Yeh H.W. (1980) D/H ratios and late-stage dehydration of shales during burial. Geochim. Cosmochim. Acta, 44, 341-352.
Yurtsever Y. (1975) Worldwide survey of stable isotopes in precipitation. Isotope Hydrology. International Atomic Energy Agency, Vienna, 40 p.

ISOTOPIC COMPOSITIONS OF DISSOLVED STRONTIUM AND NEODYMIUM IN CONTINENTAL SURFACE AND SHALLOW SUBSURFACE WATERS

Sam CHAUDHURI(*)
Department of Geology, Kansas State University
Manhattan, Kansas 66506 (U.S.A)

and

Norbert CLAUER
Centre de Géochimie de la Surface
1 Rue Blessig, 67084 Strasbourg (France)

Weathering of rocks, denudation of continental surfaces, hydrodynamics of drainage basins, and estimates of fluxes of chemical elements delivered annually to the oceans by the rivers are among central themes in many geochemical studies of continental surface and shallow subsurface waters. Oxygen and H stable isotope geochemistry has long proved very useful in addressing problems in hydrologic studies. But only recently analyses of some commonly occurring natural radiogenic isotopes have become popular means of investigating hydrodynamics of surface and shallow ground waters and modes of present-day continental denudation, although these isotopes had been known as very useful natural tracers for studies of origins of rocks and minerals. Radiogenic isotopic characteristics of rocks and minerals are widely varied due to their varied

---

(*) Presently at the Centre de Géochimie de la Surface Strasbourg (France)

ages and parent to daughter isotope ratios. Natural hydrologic systems evolve through complex mineral-water interactions. Hence common radiogenic isotopes may serve as tracers in defining weathering of rocks, water provenance in drainage basins and average compositions of annually delivered components to the oceans.

Recent literature on radiogenic isotopes in surface and shallow subsurface waters is heavily slanted toward Sr isotopic measurements. This bias is primarily a reflection of the commonly high abundances of Sr in natural waters. Interest in Nd isotopic compositions of waters is growing, but nearly a vacuum exists in the field of Pb isotope geochemistry. The aim of this chapter is to discuss the available isotope data of dissolved Sr and Nd in surface and shallow subsurface waters. As run-off and shallow ground waters are intimately connected, these two water types have been considered as one hydrologic unit in the discussion.

STRONTIUM ISOTOPE COMPOSITIONS OF SURFACE WATERS

Strontium contents in surface waters are widely varied between 0.02 and 2 mg/l and average about 0.06 mg/l (Alexander et al., 1954; Nichols and McNall, 1957; Faure et al., 1967; Mueller, 1969; Vernet et al., 1971; Harmeson et al., 1973; Goldstein and Jacobsen, 1987; Palmer and Edmond, 1989). Values as high as 60 mg/l of Sr are known, but these high values have been related to dissolution of Sr-enriched carbonate and sulfate minerals in the drainage basins (Feulner and Hubble, 1960; Skougstad and Horr, 1963; Carpenter and Miller, 1969). By contrast, Rb contents in surface waters are generally low, ranging between 0.0002 and 0.02 mg/l and averaging about 0.003 mg/l. Rb/Sr ratios commonly vary from 0.003 to 0.06, although values as high as 0.70 were obtained in some instances.

Isotopic Compositions of Strontium in Rivers and Lakes

The study of Faure et al. (1963), reporting on the Sr isotopic compositions of surface waters and associated fresh-water shells from the Kenoran (2.5 Ga), Hudsonian (1.7 Ga), and Grenville (1.1 Ga) age provinces and from the Great Lakes in North America, ushered a new direction of hydrochemical research. Their study showed that the average $^{87}Sr/^{86}Sr$ ratios of surface waters in the Kenoran, the Hudsonian, and the Grenville provinces were similar between 0.718 and 0.719, but that of waters in the Great Lakes were varied, the values being 0.718, 0.711, and 0.708 for Lake Superior, Lake Erie, and Lake Huron, respectively. Faure et al. interpreted these variations as broad reflections of the isotopic differences of the bedrocks in the drainage basins, the low values being related to marine chemical sedimentary rocks and the high values corresponding to very old crystalline rocks with high Rb/Sr ratios. Many subsequent studies used Sr isotopic compositions of surface waters to characterize bedrocks of drainage basins (Hart and Tilton, 1966; Eastin and Faure, 1970; Stueber et al., 1972; Curtis and Stueber, 1973; Steele and Pushkar, 1973; Stueber et al., 1975; Fisher and Stueber, 1976; Albarède and Michard, 1987; Goldstein and Jacobsen, 1987; Palmer and Edmond, 1989).

In some later studies, Faure and his colleagues explored the potential of using Sr isotopes as natural tracers for identification of the source of dissolved constituents in saline lakes in closed valleys (Jones and Faure, 1967; Jones and Faure, 1968; Jones and Faure, 1972). In their studies on the origin of salts in Lake Vanda of the Wright Valley in Antarctica, Jones and Faure (1967) found nearly identical Sr isotopic values of about 0.7146 for waters and sediments of the lake, soils in the valley, and soluble salts leached from the soils. The bedrocks in the valley of the lake consist of marble, dolerite, granite, and granite-gneiss and the $^{87}Sr/^{86}Sr$ ratios of these rocks range from 0.7088 to 0.7285. Based on this evidence, Jones and Faure ruled out trapped sea water and wind transported marine salts as being a major source of salinity of the lake waters. The regolith was probably the major source of the dissolved constituents in the lake, but the authors were unable to defend this assumption for lack of information on the Sr isotopic compositions of the different constituents in the soils.

The collection of studies since the pioneer works of Faure and his associates in the late 1960s and the early 1970s has shown that the isotopic compositions of dissolved Sr in major rivers in the world range from 0.7036 to 0.9430 (Table 1). Nearly 50% of the rivers have $^{87}Sr/^{86}Sr$ ratios between 0.710 and 0.720, about 30% have values less than 0.7100, and the remainder have values higher than 0.720 (Fig. 1). Lithologic diversity in the drainage basins is, as already mentioned, a major control on the Sr isotopic compositions of river waters. Rivers with drainage basins on young volcanic rocks, such as those in Philippine, Japan and Indonesia, have $^{87}Sr/^{86}Sr$ ratios between 0.703 and 0.706, which constitute the low end of the spectrum of the Sr isotopic compositions for global rivers. On the other hand, rivers flowing over very old silicate rocks with high Rb/Sr ratios, as illustrated by three rivers in Greenland, have $^{87}Sr/^{86}Sr$ ratios much higher than 0.730 (Goldstein and Jacobsen, 1987). The $^{87}Sr/^{86}Sr$ ratios of most other rivers in the world reflect derivation of Sr from essentially two sources, one of which is marine chemical sedimentary rocks known to have limited range of $^{87}Sr/^{86}Sr$ values from 0.707 to about 0.709 and the other is silicate crystalline rocks and minerals whose $^{87}Sr/^{86}Sr$ ratios are widely varied depending on age, Rb/Sr ratios, and initial isotopic compositions of the materials.

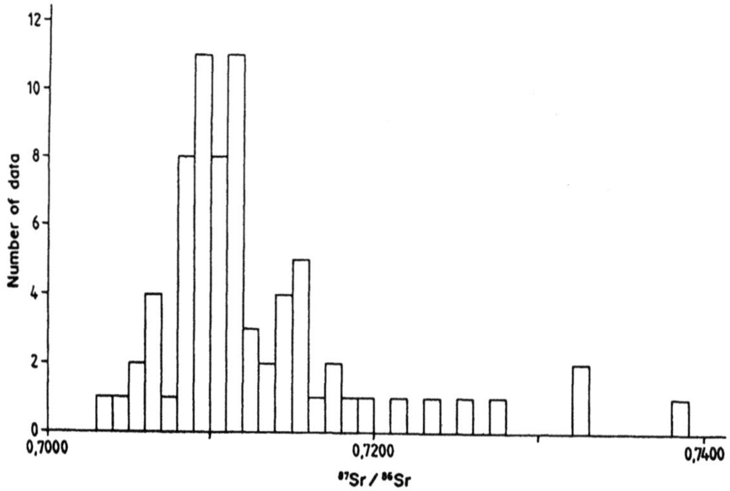

Figure 1: Histogram of Sr isotopic compositions of large rivers in the world

Table 1: Sr isotope data of waters from major rivers and lakes

Sample	Rb(1)	Sr(1)	$^{87}Sr/^{86}Sr$	
North America				
Scioto	--	12.3	0.7085	(1A)
Colorado 1	--	--	0.7086	(2A)
Colorado 2	--	1.16	0.71075	(3B)
Susquehanna 1	--	0.31	0.7090	(4A)
Susquehanna 2	--	0.63	0.7150	(4A)
Chesapeake Bay	--	0.13	0.7125	(4A)
MacKenzie	--	0.17	0.71102	(5A)
Nelson	--	0.08	0.71463	(5A)
Yukon	--	0.11	0.71616	(5A)
St Maurice	--	0.01	0.71121	(5A)
Grande Riviere	--	0.01	0.73844	(5A)
Churchill	--	0.02	0.71996	(5A)
Mississippi 1	--	0.26	0.70953	(3B)
Mississippi 2	--	0.07	0.71530	(3B)
Missouri	--	0.35	0.70998	(3B)
St Lawrence	--	0.18	0.70959	(3B)
Columbia	--	0.09	0.71210	(3B)
Ohio	--	0.20	0.71106	(3B)
Ottawa	--	0.06	0.71134	(3B)
Tennessee	--	0.05	0.71123	(3B)
Hudson	--	0.02	0.71025	(3B)
Pearl	--	0.07	0.7119	(10A)
Rio Grande	--	0.29	0.7092	(10A)
Great Salt Lake	--	--	0.7174	(6A)
Michigan Lake	--	0.13	0.70951	(3B)
Huron Lake	--	0.10	0.71002	(3B)
Superior Lake 1	--	0.02	0.71547	(3B)
Superior Lake 2	--	0.02	0.71512	(3B)
West Greenland				
Isua-B	1.73*	2.38*	0.88907	(3B)
Isua-F	0.22*	3.19*	0.94297	(3B)
Nigsik	3.30*	4.28*	0.82636	(3B)
Australia				
Murray	0.99*	0.21	0.71075	(3B)
Avon	0.02	2.94	0.73255	(3B)
Murchis	5.33*	1.08	0.72796	(3B)
Japan				
Kitikami	1.67*	0.05	0.70633	(3B)
Mogami	0.24*	0.04	0.70711	(3B)
Shinano	2.73*	0.04	0.70825	(3B)
Tone	3.19*	0.09	0.70829	(3B)
Philippines				
Abra	0.42*	0.10	0.70628	(3B)
Agno	4.19*	0.12	0.70452	(3B)
			[illegible]0622	(3B)
Pampanga	2.55*	0.12	0.70540	(3B)
India				
Indus	--	0.30	0.71116	(3B)
Cauveri	--	0.32	0.7131	(2A)
Brahmaputra	--	0.08	0.7210	(10A)
Ganges	--	0.14	0.7257	(10A)
Europe				
Po	--	--	0.7097	(2A)
Rhone	--	0.52	0.70872	(7B)
Garonne	--	0.11	0.71057	(7B)
Rhine	--	0.55	0.7092	(10A)
Danube	--	0.24	0.7089	(10A)
Vistula	--	0.43	0.7094	(10A)
Wesser	--	0.72	0.7089	(10A)
Elbe	--	0.57	0.7097	(10A)
Tisza	--	0.18	0.7096	(10A)
Seine	--	0.40	0.7081	(10A)
Antarctica				
Vanda Lake	--	0.25	0.7145	(8A)
Bonney Lake	--	--	0.7126	(9A)
Africa				
Nile	--	0.23	0.7061	(2A)
Victoria Nil	--	0.10	0.7114	(10A)
Blue Nile	--	0.15	0.7056	(10A)
Zaire	--	0.03	0.7155	(10A)
Niger	--	0.02	0.7140	(10A)
Orange	--	0.15	0.7146	(10A)
South America				
Amazon 1	--	--	0.7111	(2A)
Amazon 2	--	0.03	0.7109	(10A)
Orinoco	--	0.02	0.7183	(10A)
Tocantins	--	0.02	0.7170	(10A)
Xingu	--	0.01	0.7292	(10A)
Tapajos	--	0.01	0.7322	(10A)
Parana	--	0.04	0.7139	(10A)
Hawaii				
Wailuki	--	--	0.7036	(2A)
China				
Yangtze	--	0.18	0.7109	(10A)
Mekong	--	0.30	0.7102	(10A)
Red	--	0.13	0.7114	(10A)
Yellow	--	0.65	0.7111	(10A)

(1) amounts in ug/g except when * in ng/g; (2) Data A are normalized to std E & A = 0.7080, and data B to std NBS = 0.71025 ref. 1= Stueber et al. (1972); 2 = Brass (1976); 3 = Goldstein and Jacobsen (1987); 4 = Fisher and Stueber (1976); 5 = Wadleigh et al. (1985); 6 = Jones and Faure (1972); 7 = Albarede and Michard (1987); 8 = Jones and Faure (1967); 9 = Jones and Faure (1968); 10 = Palmer and Edmond (1989)

Only in limited situations the Sr isotopic compositions of river waters might be very similar to the $^{87}Sr/^{86}Sr$ ratios of the bulk rocks in the drainage basin. Where bedrocks and regoliths in drainage basins consist entirely of either chemical sedimentary rocks or young volcanic rocks with constituent minerals having essentially the same Sr isotopic composition, the waters and the rocks can be isotopically closely similar. Starinsky et al. (1980) noted a close similarity in Sr isotopic values between shallow ground waters and their associated host sedimentary carbonate rocks. They concluded that the isotopic compositions of shallow ground waters may be used to estimate the average Sr isotopic composition of carbonate aquifers in which the waters occur.

The Sr isotopic relations between waters and rocks are complex in drainage basins developed over old immature siliciclastic rocks or crystalline rocks. Minerals in these old rocks are typically heterogeneous in their Sr isotopic compositions due to their varied Rb/Sr ratios, and they have different degrees of susceptibility to weathering. Consequently, surface waters in drainage basins underlain by old siliciclastic or crystalline rocks can have $^{87}Sr/^{86}Sr$ ratios either higher than the whole rocks due to selective dissolution of the Rb-rich minerals such as biotite and K-feldspar, or lower than the rocks due to preferential dissolution of the Rb-poor minerals such as plagioclase.

Reported values in the literature about the Sr isotopic compositions of rivers often relate to a single flow stage of each river. Such a value for a river cannot be considered as its annual mean Sr isotopic value, as many rivers are known to have seasonal variations in dissolved chemical constituents, which are influenced by (a) lithology of the basin, (b) amount and annual distribution of rainfall, (c) temperature, (d) relief, and (e) vegetation. Very few studies have attempted to document the extent of seasonal isotopic variations of dissolved Sr in rivers. The first indication on seasonal isotopic variations came from the study of Eastin and Faure (1970) on dissolved Sr in the Olentangy and the Scioto rivers in Ohio. They observed that during a period of eight months the $^{87}Sr/^{86}Sr$ ratios of the Olentangy River ranged from 0.7086 to 0.7130 and that the isotopic changes did not correlate well with changes either in the discharge, the pH values, or the content of dissolved Sr. By contrast, no isotopic change was apparent for the Scioto River. The drainage basin of the Olentangy River is underlain by till sediments consisting of celestite-rich clasts of sedimentary origin and silicate clasts of igneous and metamorphic origins. Faure et al. attributed the seasonal Sr isotopic variation of the Olentangy River to the relative contribution of Sr from the two types of clasts present in the till sediments in response to changes of hydrodynamic conditions in the drainage basin.

Palmer and Edmond (1989) also provided evidence for the seasonal isotopic variations of dissolved Sr. They analyzed Sr isotopic compositions of dissolved Sr in five different rivers at their low, median and high flow stages (Table 1). Unlike what Eastin and Faure (1970) observed in their study on the Olentangy River in Ohio, Palmer and Edmond found flow stages and Sr contents to be inversely related. Strontium isotopic variations occurred with changes in flow stages, but the direction of isotopic shift was not the same for all streams. Noting these irregular changes in Sr isotopic compositions of rivers, the authors suggested that a single sample may not yield the annual mean $^{87}Sr/^{86}Sr$ ratio.

Additional evidence for temporal Sr isotopic variations was presented recently by Ste Marie (1990) who noted both monthly and annual variations in Sr isotopic compositions of shallow subsurface waters around a bog located in southwestern Ohio (U.S.A.). For example, ground waters collected during April and July of 1986 from a well into ground moraine deposits differed in their $^{87}Sr/^{86}Sr$ ratios by as much as 0.00021. On the other hand, the $^{87}Sr/^{86}Sr$ ratio of waters from a well in outwash deposits changed by only 0.00013 from 0.70837 in 1987 to 0.70850 in 1989. The changes were generally small because Sr derived primarily from carbonate clasts which have a narrow range of Sr isotopic composition and make up bulk (90%) of the sedimentary deposits.

Provenance of Waters in Lakes and Rivers

Faure et al. (1967,a and b) used Sr isotopes as tracers to

investigate the dynamics of waters in Lake Huron, which receives waters from four different drainage systems. Analyzing Sr of waters from the North Channel area, they observed a positive correlation between the $^{87}Sr/^{86}Sr$ and the 1/Sr ratios of the waters with Sr isotopic values ranging from 0.7111 to 0.7181 and Sr contents between 0.084 and 0.022 mg/l (Fig. 2). They related this relationship to mixing in varying amounts of waters from two sources, one of which is the northern drainage system over Precambrian crystalline rocks characterized by high $^{87}Sr/^{86}Sr$ ratios, and the other is the discharge from Lake Michigan which receives Sr primarily from Paleozoic marine chemical sedimentary rocks having low $^{87}Sr/^{86}Sr$ ratios. The correlations between $^{87}Sr/^{86}Sr$ ratios and Sr contents were less distinct for waters from the other areas of Lake Huron.

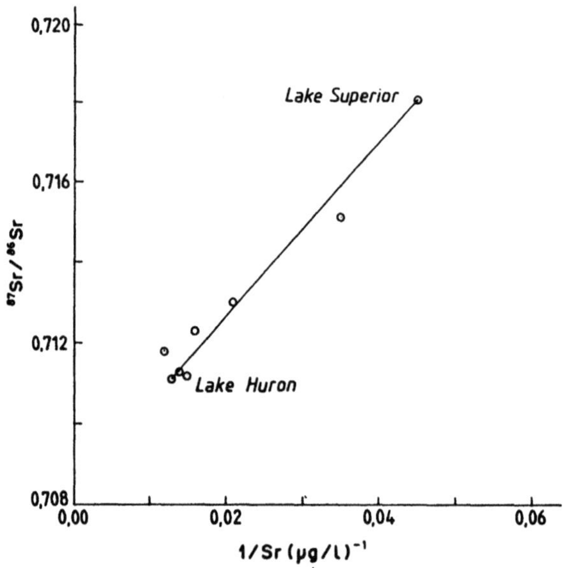

Figure 2: Relationship between $^{87}Sr/^{86}Sr$ ratios and 1/Sr values for waters of the North Channel area of the Lake Huron (Faure et al., 1967)

To assess whether Sr isotopes can be used to characterize hydrologic properties of a major river, Stueber et al. (1975) analyzed the $^{87}Sr/^{86}Sr$ ratios of waters from the Scioto River basin in Ohio, where two contrasting lithologies influence the $^{87}Sr/^{86}Sr$ ratios of ground waters and tributary streams in the drainage basin. In the upper and central reaches of the drainage basin, Pleistocene glacial deposits with abundant locally-derived dolomite clasts overlie Silurian dolomite. Ground waters and tributary streams to the Scioto River in this part of the basin were found to have low $^{87}Sr/^{86}Sr$ ratios between 0.708 and 0.709 with Sr contents from 0.1 to 43 mg/l. The high Sr contents and low Sr isotopic values were believed to be a consequence of dissolution of celestite minerals in the bedrocks and glacial-derived sediments. By contrast, ground waters and streams in the lower part of the Scioto River basin, which is non-glaciated and underlain primarily by Carboniferous shale-sandstone sequences, had high $^{87}Sr/^{86}Sr$ ratios between 0.710 and 0.713 with Sr contents of less than 0.4 mg/l. While ground waters and tributary streams in the drainage basin could be differentiated into two groups based on the Sr isotopic criteria, the $^{87}Sr/^{86}Sr$ ratios were nearly uniform at about 0.7086 for waters along the entire length of the Scioto River within the basin.

The Sr isotopic composition of any system of mixed waters is a function of the amount of water, the Sr content and the Sr isotopic composition of each component in the system. When the input of total Sr from any one component is significantly much higher than the inputs from others and the isotopic compositions of Sr for different components are not greatly different, the Sr isotopic composition of the mixed system is essentially the same as that of the component with relatively very high amounts of Sr. As Stueber et al. (1975) noted, the low input of Sr from tributary streams can explain why the Scioto waters in the lower part of the basin experienced no noticeable change in their Sr isotopic compositions relative to waters in the upper reaches, although ground waters and tributary rivers in the lower part of the basin had measurably higher $^{87}Sr/^{86}Sr$ ratios than ground waters and tributary rivers in upper and central parts of the basin. The study of Stueber et al. has demonstrated that balancing the fluxes of different sources of Sr in a major river might be difficult to achieve by means of Sr isotopic data when the Sr input from the main river far exceeds the Sr inputs from tributaries.

Analyzing waters from a 150 km course of the Susquehanna River and from several of its tributary streams, Fisher and Stueber (1976) noted a scatter in the Sr isotopic compositions of waters from main river and its tributary streams. A wide variety of igneous, metamorphic and sedimentary rocks underlie this part of the Susquehanna drainage basin, but single lithologic types exist in several sub-drainage basins defined by the tributaries. Figure 3 illustrates the distribution of the $^{87}Sr/^{86}Sr$ ratios of waters at the mouths of different tributaries to the Susquehanna River and of the types of bedrocks in the drainage basins. As expected, tributaries with dominance of carbonate bedrocks in their basin have $^{87}Sr/^{86}Sr$ ratios of about 0.709 and relatively high Sr concentrations, whereas those with shale or metamorphic bedrocks have $^{87}Sr/^{86}Sr$ ratios between 0.716 and 0.719 and low Sr concentrations. Tributaries with mixed bedrocks have Sr isotopic values between 0.709 and 0.716. As figure 4 indicates, the $^{87}Sr/^{86}Sr$ and the 1/Sr

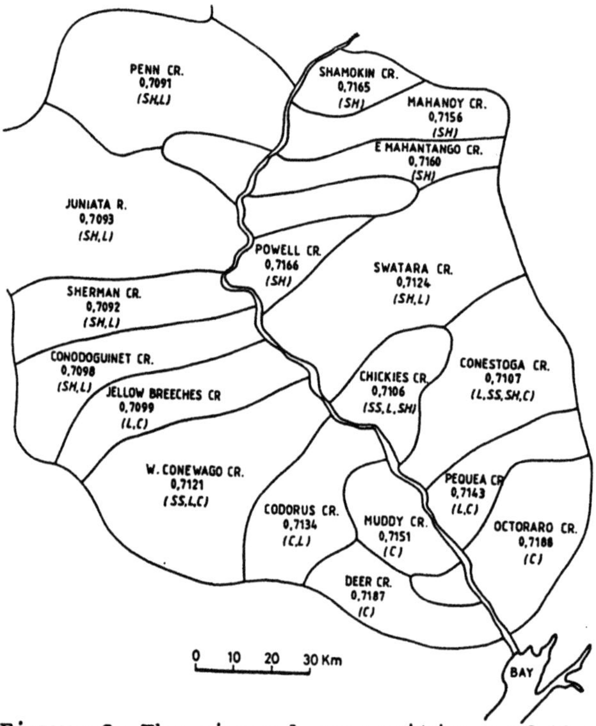

Figure 3: The mineral compositions of the bedrocks (SH=shales; L=limestones; SS=sandstones; C=crystalline rocks) and the $^{87}Sr/^{86}Sr$ ratios of waters in the drainage basins of the tributary streams of the Susquehanna River (Fisher and Stueber, 1976)

ratios for the tributary streams on mixed bedrocks in their basins bear an approximately positive linear relationship, which suggests

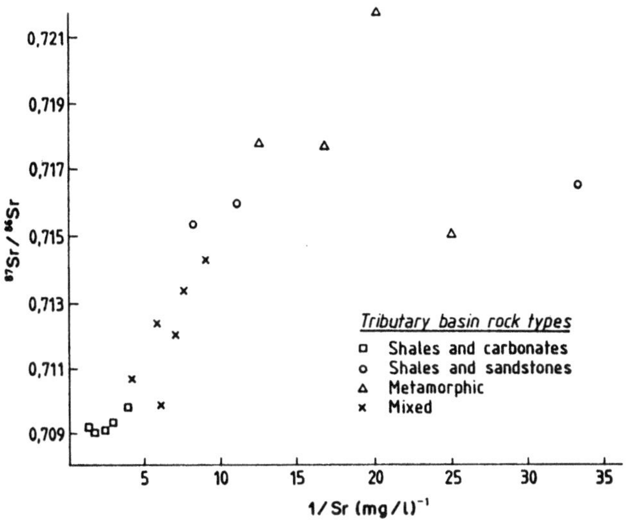

Figure 4: Relationship between $^{87}Sr/^{86}Sr$ ratios and 1/Sr values of waters of tributary streams of the Susquehanna drainage basin, relative to the rock types (Fisher and Stueber, 1976)

that the range of Sr isotopic compositions of the waters could be described in terms of mixing in varying proportions of two isotopically distinct sources of dissolved Sr. But the variations in $^{87}Sr/^{86}Sr$ and 1/Sr values for the waters of the Susquehanna River itself along the whole 150 km course could not be described in a linear trend. Fisher and Stueber suggested that the lack of linearity in the $^{87}Sr/^{86}Sr$ and the 1/Sr values of the waters reflects an incomplete mixing of tributary and main waters due to very small depth to width ratio of the main river channel.

Palmer and Edmond (1989) also examined the relationship between $^{87}Sr/^{86}Sr$ ratios and Sr concentrations of main river and respective tributary waters from Orinoco, Amazon, Ganges rivers and East African rift valleys (Fig. 5). The two variables are well correlated for the waters in the Ganges and, to some degree, for the waters in the Orinoco. The correlation becomes apparent for waters of the East African rift valleys only when the drainage is divided into northern and southern basins. Although the $^{87}Sr/^{86}Sr$ ratios and Sr contents of the waters for the entire Amazon basin are not well-correlated, those of the sub-basins tend to show some good correlation. Palmer and Edmond remarked that the variation in the relationship between $^{87}Sr/^{86}Sr$ and 1/Sr values of waters within a drainage basin or among different drainage basins is related to lithologic, tectonic, and climatic controls in the drainage basins.

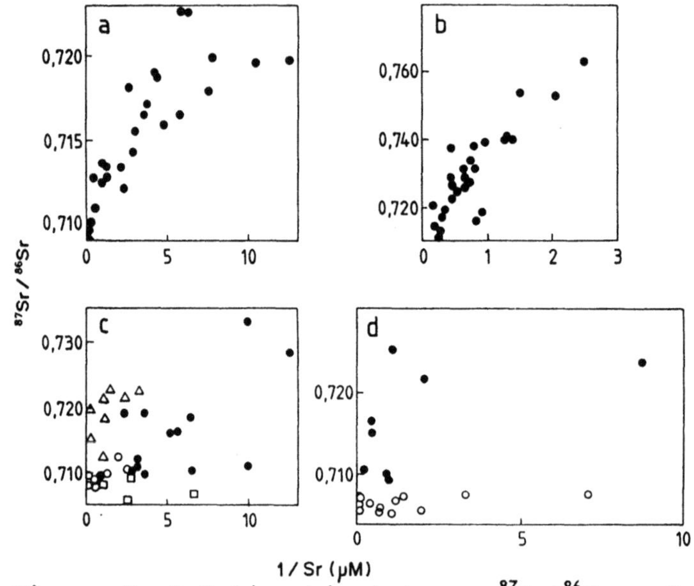

Figure 5: Relationship between $^{87}Sr/^{86}Sr$ and 1/Sr ratios of surface waters from drainage basins of: (a) Orinoco River, (b) Ganges River, (c) Amazon River, open symbols for mainstream at central (circle), northern (squares) and southern Andes (triangles) and closed circles for tributaries at the confluence with mainstream, (d) East African rift (Palmer and Edmond, 1989).

Strontium Isotopic Relations between Suspended and Dissolved Loads

The study of Hart and Tilton (1966) can be considered as the pioneer work on the determination of the relationship between the Sr isotopic compositions of dissolved and suspended loads in fresh water. They noted in Lake Superior in North America, that the $^{87}Sr/^{86}Sr$ ratio of the dissolved Sr was less than that of the Sr in the sediments. The recent study of Goldstein and Jacobsen (1987) is the most informative work to date about the range of isotopic relationships between dissolved and suspended loads in major rivers of the world (Fig. 6). They noted that the $^{87}Sr/^{86}Sr$ ratio of the dissolved load is lower than that of the suspended load for a vast majority of rivers in the world. Nevertheless, a small number of rivers were found with the $^{87}Sr/^{86}Sr$ ratio of their dissolved loads being higher than that of suspended loads, as illustrated by the data on rivers in West Greenland, the Columbia River in North America, and some rivers in Philippines and Japan. Whereas Sr isotopic disequilibrium between suspended and dissolved loads is common among a vast majority of rivers, isotopic equilibrium between the two loads has been observed for a small number of rivers such as the Colorado River in North America and the Tone River in Japan (Goldstein and Jacobsen, 1987). Closely identical Sr isotopic values between dissolved Sr and associated sediments were also reported by Jones and Faure (1967) in their study of Lake Vanda, Antarctica.

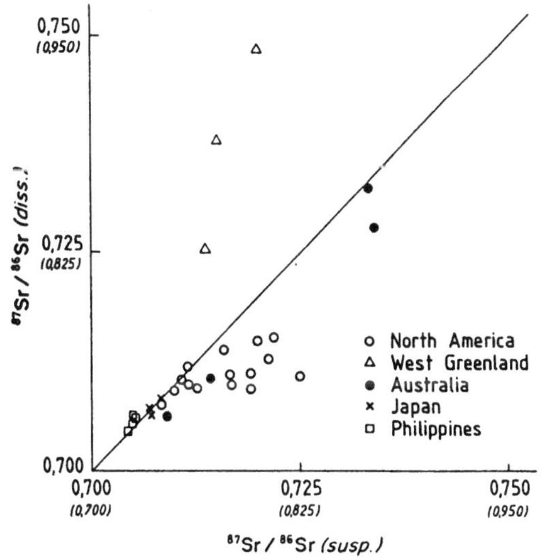

Figure 6: Sr isotopic relationship between suspended and dissolved loads in major rivers of the world. The solid line represents the 1:1 ratio between the two loads (Goldstein and Jacobsen, 1987).

Selective dissolution of minerals and preferential leaching of intracrystalline Sr from isotopically heterogeneous minerals in rocks are a prime cause for the difference in the Sr isotopic composition between dissolved and suspended loads in rivers. Among those rivers which have been found to have the $^{87}Sr/^{86}Sr$ ratio of the dissolved load being lower than that of the suspended load, many drain primarily over both siliciclastic and marine chemical sedimentary rocks. In drainage basins with mixed lithology of siliciclastic and chemical sedimentary rocks, selective dissolutions of calcite, dolomite, gypsum, and celestite, which are typically low in $^{87}Sr/^{86}Sr$, dictate the low $^{87}Sr/^{86}Sr$ ratio for the dissolved loads. Leaching of radiogenic Sr from ^{87}Sr-enriched silicate phases in rocks from drainage basins starved of marine chemical sedimentary rocks will account for the elevated $^{87}Sr/^{86}Sr$ ratios of dissolved loads. Aside from the effects of selective dissolution of minerals and leaching of radiogenic ^{87}Sr from minerals, the isotopic disequilibrium between the two loads in a river could also arise from mixing of the loads in varied amounts from two or more flow systems which are isotopically distinct but each with isotopic equilibrium between its own two loads.

The occurrence of Sr isotopic similarity between the two loads in a river requires specific lithological characteristics of the drainage basin. Isotopic similarity would occur where the drainage basin contains rocks made up of isotopically similar monomineralic phases, such as in limestone, or of isotopically similar polymineralic phases, such as in young volcanic igneous rocks.

Strontium Isotopic Characteristics of Continental Run-Offs

The average isotopic composition and the amount of Sr delivered annually to the oceans by rivers are essential components to the equation for the Sr isotopic mass-balance of the sea water. Knowledge of the global average isotopic composition of the flux of Sr delivered to the oceans by rivers has emerged from two lines of investigation, an indirect approach based on calculations from isotopic mass balance of Sr in sea water and a direct approach based on measurements of Sr contents and the isotopic composition of dissolved Sr in major rivers of the world. Many different suggestions have been made as to the estimate of Sr from ocean ridges in the marine cycle (Albarède et al., 1981; Clauer and Olafsson, 1981). From an estimate of the exchange of Sr between sea water and basalt of the East Pacific Rise, Albarède et al. suggested that the global average isotopic composition of riverine Sr should be between 0.7097 and 0.7113. Taking into account the potential effects of diagenetic reflux of Sr from marine carbonate rocks, in addition to considering the flux of Sr from the mid-oceanic ridge activities, Elderfield and Greaves (1981) concluded that the global average isotopic composition of riverine Sr delivered to the oceans is about 0.7112.

Brass (1976) was the first to suggest a value of 0.7110 for the Sr isotopic composition of global riverine Sr flux based on measurement of the $^{87}Sr/^{86}Sr$ of water from the mouth of the Amazon River, which has the largest drainage basin with nearly 20% of the total river discharge to the ocean. Wadleigh et al. (1985) reported Sr isotopic data for major Canadian rivers which represent about 4% of the total discharge and about 6% of the Sr flux to the oceans. They concluded that the global average isotopic composition of the riverine Sr flux to the oceans is about 0.7111. Goldstein and Jacobsen (1987) published isotopic data of dissolved Sr in some major rivers of the world, which make up about 23% of the total river water being discharged to the oceans. They recommended an average $^{87}Sr/^{86}Sr$ ratio of 0.7101 and the average Sr concentration of 0.062 mg/l for the global annual riverine flux to the oceans. More recently, Palmer and Edmond (1989) suggested, based on data from rivers which constitute about 47% of the total run-off, that the average $^{87}Sr/^{86}Sr$ ratio and Sr content to be 0.7119 and 0.078 mg/l, respectively. The values recommended by Goldstein and Jacobsen are lower than those favored by others because the former included in their estimation a large number of run-offs on young volcanic rocks which have characteristically low $^{87}Sr/^{86}Sr$ ratios. While the disagreement is small among the estimates for average $^{87}Sr/^{86}Sr$ and the flux of continental Sr delivered annually to the oceans, we might recall that $^{87}Sr/^{86}Sr$ ratios for rivers are known to be varied from one season to another or even from one year to another (Eastin and Faure, 1970; Palmer and Edmond, 1989). As recent recommendations concerning the isotopic composition of global riverine Sr flux to the oceans are essentially based on a data set developed from one time measurement for each river, the recommended values can be quite different from the real average value.

The different studies of Sr isotopic compositions of river waters clearly demonstrate that their $^{87}Sr/^{86}Sr$ ratios vary widely and that in many rivers the $^{87}Sr/^{86}Sr$ ratios of dissolved Sr exceed the values of Phanerozoic marine carbonate rocks, which range approximately between 0.7068 and 0.7091. The sources of radiogenic ^{87}Sr in those rivers with $^{87}Sr/^{86}$ ratios higher than that of marine carbonate rocks have not been clearly identified. Clay minerals have been considered by some as a contributor of radiogenic ^{87}Sr to surface waters or shallow subsurface ground-waters because these minerals, especially illite, are abundantly common in shales and soils. While these minerals could have some influence on the Sr isotopic composition of river waters, we do not consider that these minerals would be a major contributor of radiogenic ^{87}Sr to surface waters. The exchangeable Sr in clay minerals is mostly surface adsorbed and not specific to the structural sites and hence the isotopic composition of the exchangeable Sr is very nearly the same as that of Sr in the environment with which the clay minerals are associated (Thellier and Clauer, 1989; Clauer et al., 1990). We believe that weathering of Rb-bearing silicate minerals such as K feldspar and biotite, and even of some plagioclase minerals with high $^{87}Sr/^{86}Sr$ ratios, is the major factor for the high $^{87}Sr/^{86}Sr$ ratios of many river waters.

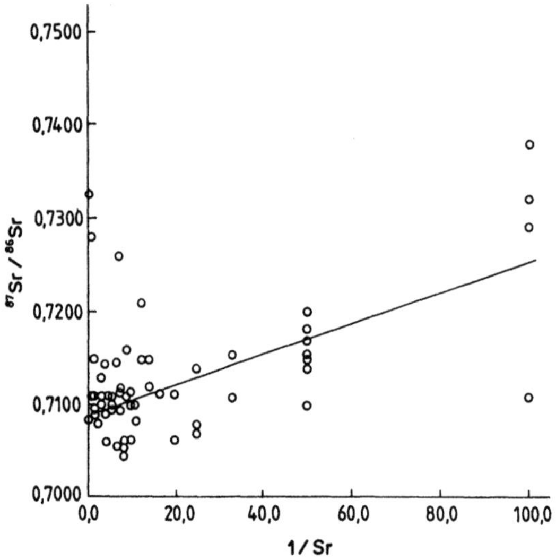

Figure 7a: Relationship between $^{87}Sr/^{86}Sr$ ratios and 1/Sr values of major rivers in the world (Palmer and Edmond, 1989)

The $^{87}Sr/^{86}Sr$ ratio and the Sr concentration of any major river may provide, in some instances, important clues as to the crustal history of rocks in a drainage basin. Palmer and Edmond (1989) observed an approximately linear relationship between Sr isotopic compositions and 1/Sr values among a large number of major rivers in the world (Fig. 7a). This is basically a reflection of two general sources of Sr in the drainage basins. But the Sr isotopic compositions and Sr contents of some rivers, for example those of the Ganges and the Brahmaputra rivers in India and the rivers of Greenland (Fig. 7b), deviate considerably from the global trend. Palmer and Edmond attributed the anomaly found in the Indian rivers to the effect of high weathering rates of the Himalayas. We think that the anomaly is related to the high susceptibilty to weathering of some Sr-enriched silicate minerals, possibly plagioclase minerals, having high $^{87}Sr/^{86}Sr$ ratios. Plagioclase minerals in some leucogranites in the main Higher Himalayan Range have been found with high $^{87}Sr/^{86}Sr$ initial ratios (0.750; Deniel et al., 1987).

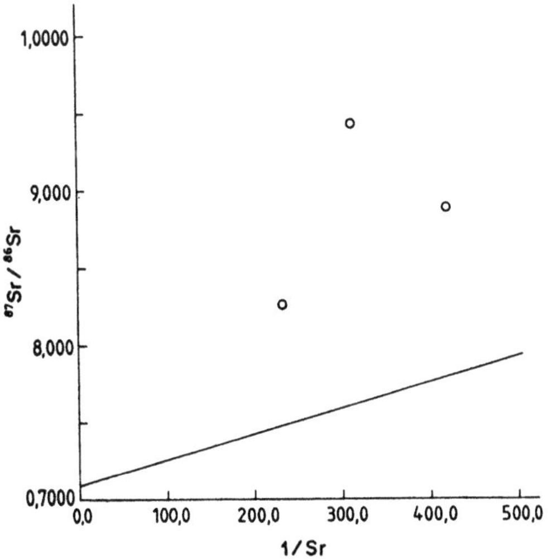

Figure 7b: Relationship between $^{87}Sr/^{86}Sr$ ratios and 1/Sr values of three rivers in Greenland. The linc is that of figure 7a (Palmer and Edmond, 1989)

Similar rocks in the Himalayas could have been a major source of the dissolved Sr in the Ganges and the Brahmaputra rivers. Hence, we believe that the isotopic compositions of dissolved Sr in the Ganges and the Brahmaputra rivers more likely reflect the tectonic evolutionary history of rocks in the Himalayas, rather than the weathering rates as Palmer and Edmund (1989) suggested.

The three major sources for riverine Sr are old continental crystalline rocks, young igneous rocks, and marine carbonate and sulfate rocks. We have yet to determine the relative importance of these sources in the Sr budget for the global river. A simplified view has often sought to find out the relative importance between silicate and carbonate sources for riverine Sr. From a calculation based on Sr/Ca ratios of river waters, Brass (1976) suggested that nearly 80% of the flux of Sr in river waters is from limestone and that the average isotopic composition of the flux of limestone Sr is about 0.708, which would make the average $^{87}Sr/^{86}Sr$ ratio of the flux of silicate Sr in river waters to be about 0.723. From data of river waters in Canada, Wadleigh et al. (1985) found that about 21% of the Sr could have come from a silicate source, a value in agreement with the estimate by Brass. Wadleigh et al., however, also suggested that Sr/Ca and $^{87}Sr/^{86}Sr$ ratios of river waters in Canada can also be generated by mixing of Sr derived from old and young silicate sources, in which case the relative importance of the silicate source may be even higher than the value of Brass (1976).

Goldstein and Jacobsen (1988) gave estimates of 0.7×10^{12} gy^{-1} and 2.52×10^{12} gy^{-1} for the fluxes of Rb and Sr, respectively, in dissolved load of the present-day global runoff, based on the calculation from total runoff of about 4.2×10^{12} gy^{-1}. By comparison, the estimated fluxes of Rb and Sr in the suspended load of the global runoff are 1.47×10^{12} gy^{-1} and 2.98×10^{12} gy^{-1}, respectively. While the fluxes of Sr between the two loads are comparable, more than 95% of the total flux of Rb in the global runoff is transported in the suspended fraction as a reflection of large differentiation between Rb and Sr behaviors during weathering, erosion and transportation. The high degree of differentiation between the two elements is further evident in the unrealistic calculated Rb-Sr isotopic model age for the dissolved load. Goldstein and Jacobsen reported a depleted mantle based Rb-Sr age of about 29 Ga for the dissolved load and about 0.7 Ga for the suspended load.

The discussions earlier emphasized that river waters may or may not be in Sr isotopic equilibrium with bedrocks or soils of the drainage basin, depending on the mineralogy of the bedrocks and the soils, as well as topography, climate, and time which control the degree of weathering in the drainage basin. To date, studies on Sr-isotope systematics in weathering environments are very few (Clauer, 1979), and we have yet to understand fully the various factors which govern the fractionation of ^{87}Sr between waters and bedrocks and soils. Such knowledge is essential to explain clearly the variations in the $^{87}Sr/^{86}Sr$ ratios of run-offs and the significance of the global average $^{87}Sr/^{86}Sr$ ratios of riverine Sr as they relate to continental weathering and denudation.

NEODYMIUM ISOTOPIC COMPOSITIONS OF SURFACE WATERS

From analyses of several run-off samples, Martin et al. (1976) and Keasler and Loveland (1982) reported that Nd contents are less than 9×10^{-5} mg/l. Goldstein and Jacobsen (1987) measured Nd contents of waters from different rivers in Canada, the U.S.A., Australia, West Greenland, Japan, Philippines, and Venezuela and obtained Nd values ranging from 3×10^{-6} to about 3×10^{-3} mg/l. These low Nd contents are thought to be due to the extremely low solubility of the rare-earth elements (REE) in aqueous form.

The paucity of Nd isotopic data of surface waters is largely attributed to the difficulty of analyzing materials with extremely low concentrations. The study of Goldstein and Jacobsen (1987) is at present represents the major source of information (Table 2). It shows that the $^{143}Nd/^{144}Nd$ ratios of the waters range from about 0.509596 to about 0.512192. Like the Sr isotopic compositions, the Nd isotopic compositions of surface waters also broadly reflect the composition of rocks in drainage basins. In general, the more negative are the ϵ_{Nd} values, the greater is the influence of very old continental rocks and less is the influence of young island arc and continental arc volcanic rocks.

Table 2: Nd isotope data of waters from major rivers and lakes

Sample	Sm(1)	Nd(1)	$^{143}Nd/^{144}Nd$
North America			
Mississippi 1	4.86	23.8	0.511412
Mississippi 2	9.59	45.9	0.510854
Missouri	4.98	23.0	0.511451
St Lawrence	2.25	10.7	0.511233
Columbia	4.41	20.1	0.511693
Colorado	1.18	6.0	0.511435
Ohio	1.62	8.2	0.511331
Ottawa	43.5	264	0.510833
Tennessee	2.02	8.5	0.511341
Hudson	20.9	105	0.511429
Michigan Lak	1.40	6.0	0.511196
Huron Lake	1.08	5.3	0.510989
Superior Lake	2.19	9.7	0.511004
West Greenland			
Isua-B	37.3	237	0.509656
Isua-F	40.7	274	0.509645
Nigsik	72.2	478	0.509596
Australia			
Murray	3.49	15.2	0.511539
Japan			
Mogami	--	--	0.511851
Philippines			
Abra	--	--	0.512186
Agno	--	--	0.512190
Cagayan	--	--	0.512192
Pampanga	2.49	8.6	0.512160

(1) amounts in ug/gx10^{-6}; $^{143}Nd/^{144}Nd$ normalized to $^{146}Nd/^{142}Nd$ = 0.636151; the data are from Goldstein and Jacobsen (1987)

Neodymium Isotopic Compositions in Dissolved and Suspended Loads

Goldstein et al. (1984) noted the ϵ_{Nd} values to range from -3.3 to -16.1 for particulate matters of major rivers. But from a study of rivers draining a basin with a varied lithology, Golstein and Jacobsen (1988) observed an even wider range of ϵ_{Nd} values from +7.1 to -44.9. Goldstein and Jacobsen (1987) found that in nearly half of the rivers the $^{143}Nd/^{144}Nd$ ratios of dissolved loads were different from that of the corresponding suspended loads. Dissolved loads of most of the rivers in North America had higher Nd isotopic values and those of the Nigsik River in West Greenland and the Murray River in Australia had lower Nd isotopic values than corresponding suspended loads (Fig. 8). However, the correlation obtained is much better than that for Sr, primarily because little or no fractionation occurs for the Sm-Nd isotope systems, unlike for the Rb-Sr systems, during continental weathering processes. The extent of ϵ_{Nd} differences between the two loads is no more than 4 units, which is significant. The differences can be related to such factors as selective dissolution of minerals with different Nd isoto-

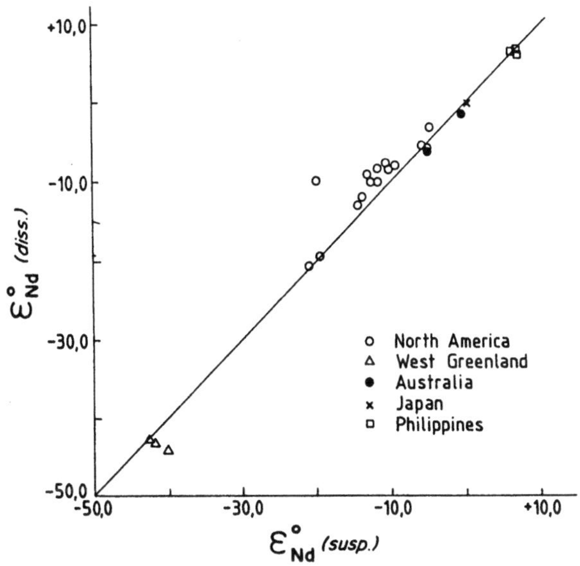

Figure 8: Nd isotopic relationship between dissolved and suspended loads in major rivers of the world (Goldstein et al., 1987)

pic compositions, preferential dissolution of domains in very old minerals having internal isotopic Nd heterogeneities, and varied inputs of loads to a main river by its tributary rivers draining over isotopically different bedrocks in their drainage basins. Mineralogic and isotopic analyses of stream sediments in conjunction with the isotopic analyses of dissolved loads are necessary for a full understanding of the process of rock weathering in drainage basins.

The Sm/Nd ratio of dissolved load is almost always higher than that of suspended load for most rivers (Fig. 9). The Sm/Nd of dissolved load can be as much as 70% higher than that of the suspended load, although on average the ratio is only about 14% higher. Goldstein and Jacobsen (1987, 1988) observed that the increase in the Sm/Nd of the dissolved load is to some extent pH-dependent, and they attributed the high Sm/Nd ratio of the dissolved loads to increased stability of carbonate and hydroxide complexes of the heavy REE. Another factor for the elevated Sm/Nd of dissolved loads could be the higher stability of heavy REE organic complexes as compared to the light REE complexes.

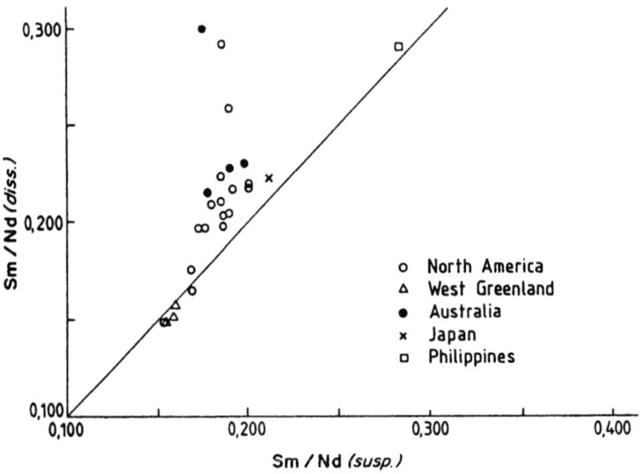

Figure 9: Sm/Nd ratios between dissolved and suspended loads in major rivers of the world (Goldstein and Jacobsen, 1987)

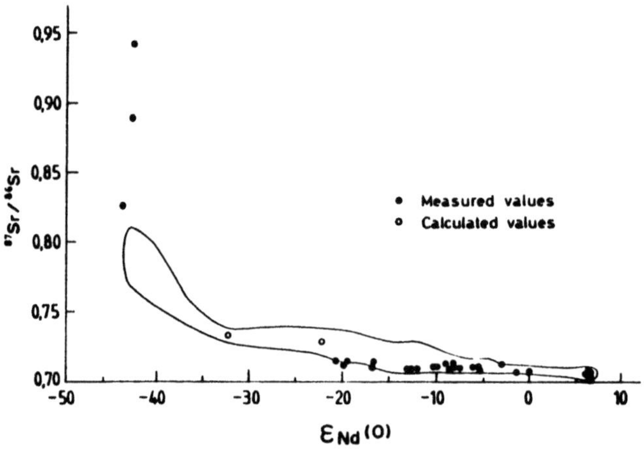

Fig. 10: Relationship between $^{87}Sr/^{86}Sr$ ratio and ϵ_{Nd} values for dissolved loads of rivers. Calculated values are from estimates from data on suspended loads. The encircled area includes the trend between the two parameters (Goldstein and Jacobsen, 1987)

Goldstein and Jacobsen (1987) observed an approximately inverse relationship between $^{87}Sr/^{86}Sr$ ratios and ϵ_{Nd} values for dissolved loads of major rivers of the world (Fig. 10). They also observed a similar relationship for the suspended loads. The relationship relates to the isotopic compositions of the loads being influenced by the ages of the rocks in the drainage basins, older rocks having higher $^{87}Sr/^{86}Sr$ ratios and lower ϵ_{Nd} values than younger rocks.

River waters widely vary in their Nd/Sr ratios from 10^{-1} for those with predominantly crystalline rocks in their drainage basins, to 10^{-5} for those with predominantly marine chemical sedimentary rocks. Mixing of waters from drainage basins with such contrasting lithologies would produce a hyperbolic relationship between the $^{87}Sr/^{86}Sr$ ratios and ϵ_{Nd} values. Goldstein and Jacobsen (1987) observed such a hyperbolic trend among the Lake Michigan, Lake Huron, Mississippi, St Lawrence, Ottawa, and St Louis waters.

Neodymium Isotopic Composition of Global Runoff

Goldstein and Jacobsen (1987) calculated a weighted average ϵ_{Nd} value of the present-day runoffs to the Atlantic, the Pacific, and the Indian oceans. The estimated value for the Atlantic Ocean is -12.6, the Pacific Ocean is -2.9, and the Indian Ocean is -8.7. As these values are close to the average ϵ_{Nd} value of sea water in each of the ocean basins, the authors suggested that the river runoff is a major influence on the ϵ_{Nd} value of sea water. Any definite conclusion about the specific role of the runoff still awaits determination of the importance of various potential roles of the suspended loads in transferring Nd to the oceans.

Goldstein and Jacobsen (1987) also estimated the average ϵ_{Nd} value of global runoff to the oceans to be -10.2 and the average Sm and Nd concentrations to be about 8×10^{-6} mg/l and 4×10^{-5} mg/l, repectively. However, assuming some removal of Nd by estuarine processes, the authors suggested an ϵ_{Nd} value of -8.4 for the effective flux of runoff Nd to the oceans. Goldstein and Jacobsen (1988) calculated a depleted mantle based Sm-Nd model age of 1.63 Ga for the global dissolved load, having the fluxes of Sm and Nd of approximately 2×10^{8} gy^{-1} and 1x10 gy^{-1}, respectively. The Sm-Nd model age for the dissolved load is nearly the same as that for the suspended load. The average model ages reflect the weathering and erosion of rocks on exposed surfaces of the crust. As rocks from young orogenic belts are weathered and eroded to a greater extent than those from old stable cratonic regions with low relief, the calculated Sm-Nd model age of the dissolved or the suspended load can be considered as the minimum average age of the exposed rocks.

CONCLUDING REMARKS

The foregoing discussions have shown that isotopic characteristics of dissolved Sr in waters in a drainage basin are influenced by lithologic composition, tectonic factor, and climatic control. Relations in Sr isotopic compositions between a main river and its tributaries in a drainage basin are often extremely varied. Poor correlation between $^{87}Sr/^{86}Sr$ ratios and 1/Sr values reflects varied inputs of isotopically very different Sr from the tributary rivers to the main river and also incomplete mixing between tributary and main river waters due to low depth to width ratio of the channels. Good correlation between the two parameters suggests a mixing of Sr derived essentially from two sources, which can often be described as being a silicate source with high $^{87}Sr/^{86}Sr$ ratios and low Sr concentrations and a carbonate source with a marine Sr isotopic value and high Sr concentrations.

Seasonal Sr isotopic variations have been noted in some studies on dissolved Sr in river waters. Present estimates of average Sr isotopic composition of global run-offs are based on data from one-time measurements on rivers. As the isotopic compositions of rivers can be significantly varied seasonally and annually, caution should be taken in the use of any of the presently given estimates for the average Sr isotopic compositions of global run-offs.

The Sr isotopic relationships between suspended and dissolved loads of rivers are varied. Isotopic similarities are much less frequent than dissimilarities, but most often these similarities occur with $^{87}Sr/^{86}Sr$ values less than 0.710. This can be attributed to weathering of predominantly marine chemical sedimentary rocks or young volcanic rocks in the drainage basin. Only in very limited cases Sr isotopic similarities were found between dissolved and suspended loads, when the $^{87}Sr/^{86}Sr$ ratio is far in excess of 0.710. But lack of mineralogic and chemical information about the suspended loads presently does not offer the real cause for that isotopic similarity between the loads. Preferential dissolution of phases with low $^{87}Sr/^{86}Sr$ ratios or domains with low $^{87}Sr/^{86}Sr$ ratios in isotopically heterogenous mineral phases and mixing in varied amounts of isotopically different loads from tributary streams may account for the lower $^{87}Sr/^{86}Sr$ ratios of dissolved load relative to corresponding suspended load in a river. Selective removal of radiogenic ^{87}Sr is a major cause for dissolved load to be higher in $^{87}Sr/^{86}Sr$ ratio than suspended load, which may explain why dissolved loads of three rivers in West Greenland were found to be extremely high in $^{87}Sr/^{86}Sr$, from about 0.826 to 0.943.

Like Sr isotopic compositions, the Nd isotopic compositions of dissolved loads can be different from that of suspended loads. The isotopic difference could arise from preferential weathering of minerals which are isotopically heterogenous or from varied mixing of loads with different isotopic compositions from tributary rivers to the main river. River waters are known to have seasonal variations in the flow and the chemical composition, which might induce changes in the inputs by the tributary streams. We, therefore, believe that the differences in the Nd isotopic composition between dissolved and suspended loads of a river is commonly an effect of mixing of loads from tributary streams with different isotopic compositions.

The Sm/Nd ratios range from 0.15 to 0.30 and the ϵ_{Nd} values between +7 and -44 for dissolved loads of major rivers in the world, the more negative ϵ_{Nd} values reflecting the influence of very old continental rocks. In nearly 50% of the rivers, the Nd isotopic composition of dissolved load was found to be different from that of the corresponding suspended load, but the difference in the ϵ_{Nd} values between the two loads is no more than 4 units. The difference can be ascribed to selective dissolutions of minerals with different Nd isotopic compositions or of domains of old minerals with internal chemical heterogeneity, or to varied inputs of loads from isotopically different drainage sub-basins. The Sm/Nd ratios of dissolved loads relative to that of suspended loads tend to increase with increasing pH values, suggesting some degree of differentiation between the two elements as a result of their complexation by carbonate, hydroxide and organic radicals during weathering, erosion and transport.

Both dissolved and suspended loads have an approximately inverse relationship in their $^{87}Sr/^{86}Sr$ and $^{144}Nd/^{143}Nd$ ratios, which apparently relates to the isotopic compositions of the loads being influenced by ages of the rocks in the drainage basins. Calculated weighted average ϵ_{Nd} value of the present-day runoffs to each major ocean basin shows that the Nd isotopic compositions of sea waters are strongly influenced by the runoff inputs. The global runoff ϵ_{Nd} value is about -8.4, and the average Sm-Nd of the dissolved load is about 1.63 Ga, which is about the same as that of the suspended load. This age can be considered as the minimum average age of rocks on the exposed surface of the crust.

Knowledge of the isotopic compositions of dissolved loads in rivers has increased understanding of the weathering of continental crust. Additional understanding would come by combining the chemical and isotopic data of dissolved, suspended, and bed loads with the mineralogic data of the two latter. A cautionary remark must be made here for isotopic studies of all surface waters. Careful filtration of such waters, however difficult it may be, is absolutely necessary to define appropriately the isotopic characteristics of the dissolved constituents. Using $<0.45\mu m$ filter is apparently not enough to trap all the fine suspended materials in waters. This is clealy evident from the data presented by Stordal and Wasserburg (1986). They observed an increase in the Sm/Nd ratio by about 5%, a decrease in the ϵNd value by about 2 units, and a decrease in the ϵSr value by 1 unit for $<0.1\mu m$ filtered water compared to $<0.4\mu m$ filtered water from the Mississippi River. They also reported an increase in the Sm/Nd ratio by about 20% for $<0.1\mu m$ filtered water compared to $<0.4\mu m$ filtered water from the Amazon River.

Recently some investigators, working especially on the effect of atmospheric input of acidity to the waters, have begun to use

Sr isotopes as a tracer to determine the predictability of geochemical models depicting the precise steps of dissolution of minerals in rocks within a drainage basin (Probst et al., 1992).

ACKNOWLEDGEMENTS

We thank P. Stille (Centre de Géochimie de la Surface, Strasbourg, France) and P. Pushkar (Wright State University, Dayton, Ohio) for helpful suggestions during preparation of the text, and Mr. C. Hammel for drafting the figures.

REFERENCES

Albarède F., Michard A., Minster J.F. and Michard G. (1981) $^{87}Sr/^{86}Sr$ ratios in hydrothermal waters and deposits from the East Pacific Rise at 21°N. Earth Plan. Sci. Lett., 55, 229-236.

Albarède F. and Michard A. (1987) Evidence for slowly changing $^{87}Sr/^{86}Sr$ in runoff from freshwater limestones of southern France. Chem. Geol., 64, 55-65.

Alexander G.V., Nusbaum R.E. and Mac Donald N.S. (1954) Strontium and calcium in municipal waters supplies. Am. Works Ass. Journ., 46, 643-654.

Brass C.W. (1976) The variation of the marine $^{87}Sr/^{86}Sr$ during Phanerozoic time: Interpretation using a flux model. Geochim. Cosmochim. Acta, 40, 721-730.

Carpenter A.B. and Miller J.C. (1969) Geochemistry of saline subsurface water, Saline County (Missouri). Chem. Geol., 4, 135-167.

Clauer N. (1979) Relationship between the isotopic composition of strontium in newly formed continental clay minerals and their source material. Chem. Geol., 27, 115-124.

Clauer N. and Olafsson, J., 1981, Icelandic thermal brines with a mantle Sr isotopic signature. Sci. Géol. Bull., 34, 243-245.

Clauer N., O'Neil J.R., Bonnot-Courtois C. and Holtzapffel T. (1990) Morphological, chemical, and isotopic evidence for an early diagenetic evolution of detrital smectite in marine sediments. Clays and Clay Min., 38, 33-46.

Curtis J.B., Jr. and Stueber A.M. (1973) $^{87}Sr/^{86}Sr$ ratios and total strontium concentrations in surface waters of the Scioto River drainage basin, Ohio. Ohio Jour. Sci., 73, 166-175.

Deniel C., Vidal P., Fernandez A., LeFort P. and Peucat P. (1987) Isotopic study of the Manaslu granite (Himalaya, Nepal): inferences on the age and source of Himalayan leucogranites. Contrib. Mineral. Petrol., 96, 78-92.

DePaolo D.J and Wasserburg G.J. (1976) Nd isotopic variations and petrogenetic models. Geophys. Res. Lett., 3, 249-252.

Eastin R. and Faure G. (1970) Seasonal variation of the solute content and the Sr^{87}/Sr^{86} ratio of the Olentangy and Scioto rivers at Columbus, Ohio. Ohio Journ. Sci., 70, 170-179.

Elderfield H., and Greaves M.J. (1981) Strontium isotope geoche-

mistry of Icelandic geothermal systems and implications for sea water chemistry. Geochim. Cosmochim. Acta, 45, 2201-2212.
Faure G., Hurley P.M. and Fairbairn H.W. (1963) An estimate of the isotopic composition of strontium in rocks of the Precambrian shield of North America. Journ. Geophys. Res., 68, 2323-2329.
Faure G., Crocket J.H. and Hurley P.M. (1967) Some aspects of the geochemistry of strontium and calcium in the Hudson Bay and the Great Lakes. Geochim. Cosmochim. Acta, 31, 451-461.
Faure G., Jones L.M., Eastin R. and Christner M. (1967) Strontium isotope composition and trace element concentrations in Lake Huron and its principal tributaries. Ohio State Univ., Dep. Geol. & Water Res. Center, Final Rep. Proj. B-004-OHIO-84, 109p.
Feulner A.J. and Hubble J.H. (1960) Occurrence of strontium in the surface and ground waters of Champaign County, Ohio. Econ. Geol., 55, 176-186.
Fisher R.S. and Stueber A.M. (1976) Strontium isotopes in selected streams within the Susquehanna river basin. Water Resour. Res., 12, 1061-1068.
Goldstein S.J. and Jacobsen S.B. (1987) The Nd and Sr isotopic systematics of river-water dissolved material: Implications for the sources of Nd and Sr in seawater. Chem. Geol. (Isot. Geosc. Sect.), 66, 245-272.
Goldstein S.J. and Jacobsen S.B. (1988) Rare earth elements in river waters. Earth Plan. Sci. Lett., 89, 35-47.
Goldstein S.L., O'Nions R.K. and Hamilton P.J. (1984) A Sm-Nd isotopic study of atmospheric dusts and particulates from major river systems. Earth Plan. Sci. Lett., 70, 221-236.
Harmeson R.H., Larson T.E., Henley L.M., Sinclair R.A. and Neill J.C. (1973) Quality of surface waters in Illinois 1966-1971. Illinois State Water Surv., Urbana, 56, 1-100.
Hart S.R. and Tilton G.R. (1966) The isotope geochemistry of strontium and lead in Lake Superior sediments and water. In: Steinhatt J.S. and Smith T.J. (eds.), Am. Geophys. Union, Geophys. Monogr., 10, 127-137.
Jones L.M. and Faure G. (1967) Origin of the salts in Lake Vanda, Wright Valley, southern Victoria Land, Antarctica. Earth Plan. Sci. Lett., 3, 101-106.
Jones L.M. and Faure G. (1968) Origin of the salts in Taylor Valley. Antarc. Journ. of the U.S., 3, 177-178.
Jones L.M. and Faure G. (1972) Strontium isotope geochemistry of Great Salt Lake, Utah. Geol. Soc. Am. Bull., 83, 1875-1880.
Keasler K.M. and Loveland W.D. (1982) Rare earth elemental concentrations in some Pacific Northwest rivers. Earth Plan. Sci. Lett., 61, 68-72.
Korzum V.I. (ed.) (1978) World water balance and water resources of the earth. Studies and Reports in Hydrology, 25, UNESCO, Rome, 663p.
Martin J.M., Hogdahl O. and Philippot J.C. (1976) Rare earth element supply to the oceans. Jour. Geophys. Res., 81, 3119-3124.
Mueller G. (1969) High strontium content and Sr/Ca ratios in Lake Constance waters and carbonates and their sources in the drainage area of the Rhine River (Alpenrhein). Miner. Depos., 4, 75-84.
Nichols M.S. and Mac Nall D.R. (1957) Strontium content of Wisconsin municipal waters. Am. Water Works Ass. Journ., 49, 1493-1498.

Palmer M.R. and Edmond J.M. (1989) The strontium isotope budget of the modern ocean. Earth Plan. Sci. Lett., 92, 11-26.

Probst A., Fritz B. and Stille P. (1992) Consequence of acid deposition on natural weathering process: field studies and modelling. Seventh Int. Sympos. on Water-Rock Interact., Park City, Utah, 13-18 July, Extended Abstracts (to be published).

Skougstad M.W. and Horr C.A. (1963) Occurrence and distribution of strontium in municipal waters. U.S. Geol. Surv., Water Supply Paper, 1496-D, 1-97.

Starinsky A., Bielski M., Lazar B., Wakshal E. and Steinitz G. (1980) Marine $^{87}Sr/^{86}Sr$ ratios from the Jurassic to Pleistocene: Evidence from groundwaters in Israel. Earth Plan. Sci. Lett., 47, 75-80.

Ste Marie D.J. (1990) $^{87}Sr/^{86}Sr$ ratios as groundwater tracers at Cedar bog, Champaign CO., Ohio. M.S. Thesis, Wright Univ., Ohio.

Steele J.D. and Pushkar P. (1973) Strontium isotope geochemistry of the Scioto river basin and the $^{87}Sr/^{86}Sr$ ratios of the underlying lithologies. Ohio Journ. Sci., 73, 331-338.

Stordal M.C. and Wasserburg G.J. (1986) Neodymium isotopic study of Baffin Bay water: Sources of REE from very old terranes. Earth Plan. Sci. Lett., 77, 259-272.

Stueber A.M., Pushkar P. and Baldwin A.D. (1972) Survey of $^{87}Sr/^{86}Sr$ ratios and total strontium concentrations in Ohio stream and groundwaters. Ohio Journ. Sci., 72, 97-104.

Stueber A.M., Baldwin A.D., Curtis J.R., Pushkar P. and Steele J.D. (1975) Geochemistry of strontium in the Scioto river drainage basin, Ohio. Geol. Soc. Am. Bull., 86, 892-896.

Thellier C. and Clauer N. (1989) Strontium isotopic evidence for soil-solution interactions during evaporation experiments. Chem. Geol. (Isot. Geosc. Sect.), 73, 299-306.

Vernet J.P., Meybeck M., Pachoud A. and Scolari G. (1971) Le Leman: une synthèse bibliographique. Bull. Bur. Rech. Géol. Min., (2), IV, 47-84.

Wadleigh M.A., Veizer J. and Brooks C. (1985) Strontium and its isotopes in Canadian rivers: Fluxes and global implications. Geochim. Cosmochim. Acta, 49, 1727-1736.

SIGNATURES OF RADIOGENIC ISOTOPES IN DEEP SUBSURFACE WATERS IN CONTINENTS

Sam CHAUDHURI(*)
Department of Geology, Kansas State University
Manhattan, Kansas 66506 (U.S.A.)

and

Norbert CLAUER
Centre de Géochimie de la Surface
1, rue Blessig, 67084 Strasbourg (France)

Deep continental waters may be broadly divided into those occurring in sedimentary basins and those found in crystalline rocks of metamorphic and igneous origin. Whereas they have a similar range of salinities, these two water types are chemically dissimilar and have very different flow characteristics. The high salinities common to both waters have been known for more than hundred years due mainly to exploration activities in search of hydrocarbon deposits and to extraction activities of ore deposits from mines in crystalline rocks. In recent years, interest in evaluating the potential of crystalline rocks for storage of wastes in the subsurface and for geothermal energy has enhanced our understanding of the origin of saline waters in these rocks.

The use of radiogenic isotopes is relatively recent in studies

---

(*) Presently at the Centre de Géochimie de la Surface Strasbourg (France)

of waters deep in the continent and only began in the late 1970s with investigations of Sr isotopic compositions of oil-field waters (Chaudhuri, 1978; Sunwall and Pushkar, 1979). Works in the last decade have shown that Sr isotopes can be very useful to study the origin of solutes in these waters and the flow dynamics of subsurface waters in sedimentary basins. By comparison, studies of radiogenic isotopes in waters in crystalline rocks are few, beginning with the work of McNutt et al. (1984). These studies, while providing some important clues to the understanding of origin of the solutes, have offered some fresh perspectives to the intriguing question of any genetic link between the waters in sedimentary basins and those in crystalline rocks in the continent. Although reports on Sr isotopic compositions of deep subsurface waters in the continent are still few, literature on their Nd and Pb isotopic compositions is non-existant. This discussion will therefore focus on the significance of Sr isotopic data.

Deep subsurface waters have been variously termed as brines, formation waters, oil-field waters, or connate waters. The description for each of these terms has been well covered in the publications of White (1965) and Collins (1975). The salinities of deep subsurface waters are varied from a value nearly the salinity of average sea water to as much as ten times that of sea water. Although the term brine has been used loosely by many to include all formation waters, Davis (1964) preferred to apply the term brine to only those waters whose salinities are at least 100,000 mg/l. In this paper we use the term <u>brine</u> or <u>highly saline water</u> interchangeably to include all formation waters whose salinities are higher than that of the average sea water. We also use the terms <u>formation water</u> and <u>oil-field water</u> interchangeably to include those waters which have been produced recently from wells not experiencing any artificial recharge.

DEEP SUBSURFACE WATERS IN SEDIMENTARY STRATA

Oil-field waters are mainly of Na-Cl to Na-Ca-Cl type. Some

perspectives on the ranges of concentrations of the different dissolved constituents in formation waters of sedimentary basins can be obtained from a large collection of data published in articles by White (1965) and Collins (1975).

In early debates the high salinity of waters in deep basins was generally attributed to a connate marine origin. But this idea was radically changed when Clayton et al. (1966) demonstrated from stable isotope compositions that local meteoric water is a major component in many deep basinal waters. Subsequent studies on stable isotope compositions of many formation waters have generally corroborated Clayton et al.'s conclusions. Hitchon and Friedman (1969) showed that the other component of deep basin waters could be either evaporated sea water or diagenetic water. However, as documented by Kharaka et al. (1979) in northern Gulf of Mexico basin, meteoric water can be absent in formation waters such as in geopressured zones. Kharaka and Carothers (1986) provided examples of a wide variety of relationships between D and ^{18}O contents of formation waters and offered genetic implications of these relationships. An elaborate account of stable isotope compositions of formation waters is given in the companion publication of Kharaka and Thordsen (this volume).

Whilst salinities of deep subsurface waters may be compared to that of an average sea water or an evaporated sea water, the two types of water are in many ways different from each other. Collins (1975) provided a convenient way of comparing the chemical compositions of oil-field waters with that of sea water. The concentration of an element in formation water is compared to the expected concentration of the same element in sea water with its chloride content being the same as that in the formation water. In general, deep continental waters are highly enriched in Ca, Sr, and Li contents, slightly to moderately enriched in Br contents, and moderately to highly depleted in Mg, K, and SO_4 contents relative to sea water. The Na concentrations of deep continental waters are either the same as or slighlty different than that of sea water. Relative abundances of some elements between deep continental waters and sea water are illustrated in figure 1.

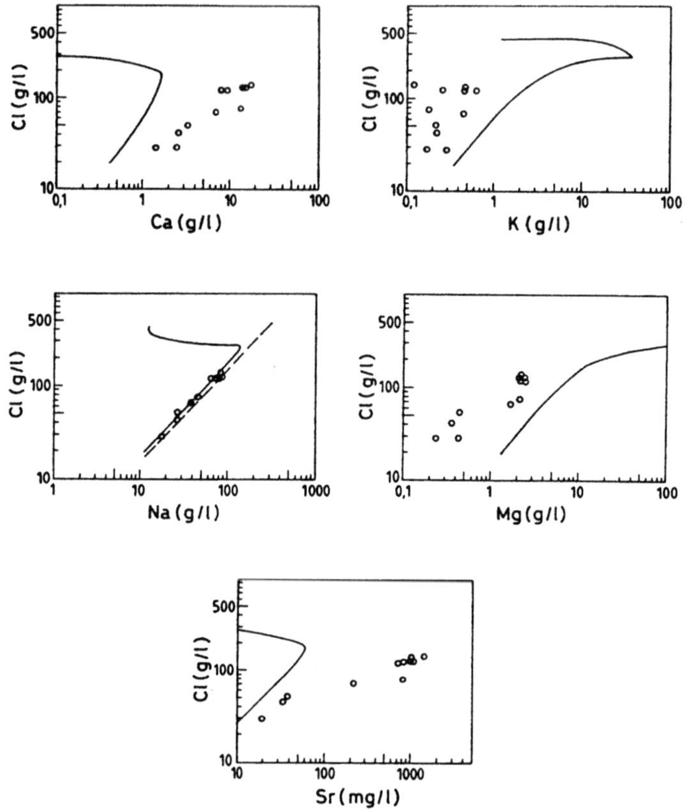

Figure 1: Relative abundances of different elements in deep subsurface waters relative to sea water. In each diagram, the curve represents the evolutionary pathway of the elements in evaporating sea water

Origin of Salinity of Waters in Sedimentary Basins

Notwithstanding the long interest in studies on oil-field waters, interpretation of their salinity or of the origin of their solutes remains controversial. Russell (1933) listed four potentially important mechanisms which can account for the origin of dissolved solids in waters deep in sedimentary basins: (a) infiltration of brines from evaporation of sea water at the surface, (b) exchange and adsorption of anions and cations by clays, (c) dehydration and hydration of minerals, and (d) reverse osmosis. In the current debate, some of Russell's suggestion have been modified or rejected in light of continued emergence of new information about the chemical and isotopic compositions of the waters.

Saline subsurface waters in sedimentary basins have generally been explained in terms of either some sea waters or continental meteoric waters whose chemical compositions have evolved in the subsurface by combined effects of physical and chemical processes including dissolution-precipitation and recrystallization of minerals, oxidation-reduction, ion exchange, and mixing with other waters. The models which have been advanced in recent years to account for the origin of solutes in formation waters in sedimentary basins are: (a) release of interstitial sea water from evaporitic deposits in subsiding basins and subsequent reactions with carbonate and silicate minerals in the overlying sedimentary strata (Carpenter, 1978; Egeberg and Aagaard, 1989), (b) dissolution of salts and subsequent albitization of feldspar, especially plagioclase, and illitization of smectite (Land and Prezbindowski, 1981; Land, 1987), (c) shale-membrane filtration (McKelvey and Milne, 1962; Hanshaw and Coplen, 1973; Kharaka and Berry, 1973), (d) mineral dehydration reactions (Graf and Anderson, 1981), and (e) infiltration of sea water in equilibrium with ice and subsequent reactions with carbonate and clay minerals (Herut et al., 1990).

The influence of membrane filtration by clays and shales, also known as ultrafiltration or reverse osmosis, on the chemical and stable isotope compositions of waters has been documented by experimental studies (McKelvey and Milne, 1962; von Engelhardt and Gaida, 1963; Coplen and Hanshaw, 1973; Hanshaw and Coplen, 1973; Kharaka and Berry, 1973; Fritz and Edy, 1985). These experiments support the importance of this process, but field evidence questions its efficiency. Indeed, many field data have shown that the Ca/Na ratio of subsurface waters often increases with depth, which is contrary to the experimental observations of Kharaka and Berry (1973) who related the discrepancy to the effects of unnaturally high pressure gradients needed to perform the experiments. Ultrafiltration has been viewed skeptically because subsurface brines occur in many areas where sedimentary sequences lack clay-rich beds necessary to generate increased salinity in waters. The difficulty of explaining linear increases in salinity with depth is another reason for questioning the influence of ultrafiltration in the chemical evolution of deep subsurface waters (Dickey, 1969).

Current prevalent ideas on the origin of salinity of formation waters are along the lines of mineral-water interactions. The work of Rittenhouse (1967) on variations in Br contents of oil-field waters provided the basis for developments of some of the more recent chemical models emphasizing the roles of mineral-water interactions in the evolution of high salinity in formation waters. Evaluating Br/Cl of brines in Upper Jurassic sedimentary strata in central Mississippi, Carpenter (1978) contended that many subsurface brines could have evolved from evaporated sea water, possibly carnallite-saturated water, entrapped as interstitial fluids in evaporite deposits in a subsiding basin and subsequently expelled to overlying sedimentary strata where the fluids inherited Na-Ca-Cl characteristics through different diagenetic mineral-water reactions, dolomitization accounting for depletion of Mg and some enrichments of Ca and Sr in the waters. Carpenter also suggested that the study of spatial variations in authigenic minerals and chemical compositions of brines in sedimentary basins could help to establish migration and evolutionary paths of marine brines. Carpenter's model depicts the following reaction for the chemical evolution of brines:

Seawater (carnallite-halite saturated) + Calcite +
(Na-Ca Plagioclase + Kaolinite + Illite) ------->
Na-Ca-Cl Brine + Dolomite + (Albite + Illite + K-Feldspar).

Land and Prezbindowski (1981) presented a different view for the evolution of subsurface brines in carbonate rocks of the Lower Cretaceous Edwards Group in south-central Texas. They questioned Carpenter's model because of lack of evidence of dolomitization in the subsurface in south-central Texas and of the requirement of very high temperature unlikely for the brines to be in thermodynamically equilibrium during calcite-dolomite reaction. They also remarked that Carpenter's model requires an unlikely high amount of carnallite to explain the origin of the brines. Land and Prezbindowski, therefore, suggested that dissolution-recrystallization of halite followed by albitization of plagioclase could be a logical mode of evolution of the brines in south-central Texas. The reaction in Land and Prezbindowski's model may be expressed as:

Halite + Na-Ca Plagioclase + Water -------> Na-Ca-Cl Brine + Albite.

More recently, Land (1987) propounded another model to explain the origin of many Ca-Cl type brines characterized by molar Na/Cl ratio much less than 0.86, a value which is typical for average sea water. According to this model, meteoric water in its involvement with dissolution-recrystallization of salts in the subsurface becomes nearly saturated with respect to halite and subsequently the water evolves through reactions with several different major silicate minerals, which include albitization of detrital feldspar, especially plagioclase, and formation of illite from smectite. The albitization leads to removal of Na from the water and addition of Ca, Sr, K, Ba, etc. to the water and the illitization results in loss of K by the water. The reaction according to this model is expressed as:

Halite + Plagioclase + K-Feldspar + Smectite + Meteoric Water ----------> Na-Ca-Cl Brine + Albite + Illite.

Graf and Anderson (1981) advocated that mineral dehydration reactions are important for the origin of salinity of formation waters in sedimentary basins. They suggested that conversion of gypsum to anhydrite, halloysite to kaolinite, smectite to illite, and analcime to albite could be potentially important reactions which influence the chemical compositions of subsurface waters. Dehydration of both gypsum and halloysite is dependent on the pore-fluid salinity and the reactions can affect the concentrations of some trace elements in the waters. As conversion of analcime to albite and smectite to illite involves other sedimentary minerals, the reactions can lead to significant changes in both major and trace element contents of the waters. Zeolite and halloysite occur in restricted geologic environments, and hence the dehydration reactions involving these minerals appear to have limited roles in influencing the chemical compositions of subsurface waters in sedimentary basins. More recently, Herut et al. (1990) advanced the idea that salinity of formation waters may be linked to mineral-water reactions by residual sea waters after ice formation. The reaction suggested is expressed as:

Sea Water (in equilibrium with ice) + Calcite + (Smectite)
-------> Na-Ca-Cl Brine + Mirabilite + Dolomite + (Chlorite).

Strontium Isotopes and Origin of Salinity in Basinal Waters

Chaudhuri (1978) was the first to suggest that Sr isotopic data could be potentially very useful in the investigation of chemical evolutionary history of subsurface brines. Analyzing Sr isotopic compositions of oil-field waters from Kansas and Colorado, he found that these waters to be significantly enriched in ^{87}Sr relative to their host carbonate rocks, indicating that the waters inherited their isotopic signatures outside the present aquifers and that the Sr isotopic compositions may provide important clues to origin and migration of the fluids and hydraulic relations among different aquifers. Subsequent Sr isotopic studies of other subsurface waters from elsewhere have generally confirmed these observations (Sunwall and Pushkar, 1979; Starinsky et al., 1983; Stueber et al., 1984, 1987; Burtner, 1987; Chaudhuri et al., 1987; McNutt et al., 1987; Lowry et al., 1988; Banner et al., 1989; Egeberg and Aagaard, 1989; Chaudhuri et al., in press).

Table 1 provides a list of Sr isotopic values for formation waters from a wide variety of locations. The Rb/Sr ratios of the waters are very low, generally less than 0.05, due to very low Rb contents of commonly less than 1 mg/l and very high Sr contents of several tens to few hundreds of mg/l. Hence the growth of radiogenic ^{87}Sr from the decay of ^{87}Rb in these waters is negligibly small. The $^{87}Sr/^{86}Sr$ ratios of a large number of oil-field brines and other deep saline waters are between 0.707 and 0.710 (Fig. 2). Illustrations of formation waters with this range of $^{87}Sr/^{86}Sr$ ratios are from the Smackover Formation in southern Arkansas (Stueber et al., 1984) and from Paleozoic strata in the Illinois, the Michigan and the Appalachian basins (Stueber et al., 1987; McNutt et al., 1987). The number of formation waters with $^{87}Sr/^{86}Sr$ higher than 0.710 is relatively small (Chaudhuri, 1978; Burtner, 1987; Chaudhuri et al., 1987; Lowry et al., 1988; Banner et al., 1989; Egeberg and Aagaard, 1989; Chaudhuri et al., in press). The Sr contents of these waters are often limited to only few tens of mg/l or less .

Table 1: Range of Sr contents and $^{87}Sr/^{86}Sr$ ratios in subsurface waters

Samples	Data (1)	Sr (2)	$^{87}Sr/^{86}Sr$	
Israel				
Ss+Ds (U.Mioc.)	2	57-350	0.7087-0.7090	(1)
Ss+Ds (L.Cret.)	14	28-211	0.7075-0.7081	(1)
Kansas				
Ds (Hunton-Sil.)	4	6.9- 7.1	0.7113-0.7118	(2)
Ss (Cherokee-Penn.)	4	45- 115	0.7128-0.7248	(2)
Ls (Warsaw-U.Miss.)	13	40- 46	0.72195-0.72306	(5)
Kansas and Colorado				
Ds (Osage-Miss.)	3	21- 660	0.7129-0.7341	(2)
Ds Spergen-Miss.)	2	22- 42	0.7264-0.7341	(2)
Arkansas				
Ls (Smackover-U.Jur.)	32	169-2960	0.7080-0.7101	(3)
Sh+Ls (Hossten-L.Cret.)	6	239-1540	0.7090-0.7112	(3)
Ontario and Michigan				
Ls (Black River-M.Ord.)	7	418-1319	0.70957-0.71032	(4)
Ls (Trenton-M.Ord.)	28	400- 765	0.70901-0.71041	(4)
Ds (Guelph-M.Sil.)	12	436-1220	0.70849-0.70946	(4)
St (A2 Unit-U.Sil.)	2	1620	0.70853-0.70866	(4)
Ls (Dundee-M.Dev.)	4	12-1120	0.70816-0.70949	(4)
Ss (Berea-L.Miss.)	6	1210-1750	0.70807-0.70915	(4)
Illinois				
Ls (Galena-U.Ord.)	7	34- 873	0.70848-0.71043	(6)
Ds (Kankakee-L.Sil.)	9	6.6- 252	0.70952-0.71084	(6)
Ls (Niagara-U.Sil.)	28	1.7- 908	0.70913-0.71076	(6)
Ds (Geneva-M.Dev.)	2	177- 180	0.70978	(6)
Ls (Lingle-M.Dev.)	3	256- 390	0.70955-0.70996	(6)
Ss (Carper-L.Miss.)	3	133- 745	0.70871-0.70933	(6)
Ls (Ste Genev.-Miss.)	5	131- 641	0.70812-0.70895	(6)
Ss (Aux Vases-Miss.)	3	171- 418	0.70886-0.70967	(6)
Ss (Cypres-Miss.)	2	138- 377	0.70928-0.71041	(6)
Ss (Spoon-Penn.)	2	103- 109	0.71051-0.71052	(6)
Utah and Colorado				
Ss (Weber-Perm.Penn.)	9	4.7- 201	0.7104-0.7195	(7)
Wyoming				
Ss (Tensleep-Perm.Penn.)	5	7.6- 9.5	0.7131-0.7179	(7)
Missouri				
Ls+Ds (Burlingt.-Miss.)	3	8.1-11.6	0.71578-0.71607	(8)

Ls=limestone; Ds=dolostone; Ss=sandstone; Sh=shale; St=salt; (1) number of measurements for each unit; (2) amounts in $\mu g/g$; Ref. 1=Starinsky et al., 1983; 2=Chaudhuri, 1978; 3= Stueber et al., 1984; 4=McNutt et al., 1987; 5=Chaudhuri et al., 1987; 6=Stueber et al. 1987; 7=Burtner, 1987; 8=Banner et al., 1988.

As compared to sea water, many brines are often considerably enriched in Sr. The concentrations of Sr in oil-field brines are highly varied from about 0.1 to more than 2,900 mg/l (Table 1), whereas that of average sea water is about 8 mg/l and that of evaporated sea water could be as much as 20 mg/l. Thus, high Sr contents for many brines suggest that they had to evolve through some processes of Sr enrichment, such as dolomitization, recrystallization of calcite, conversion of calcite to aragonite, conversion of anhydrite to gypsum, albitization of plagioclase, dissolution of celestite, ultrafiltration, and/or ion exchange with clay minerals. Dissolution of gypsum or calcite can also increase the Sr content of the waters, but the amount of Sr which may be released from this process would be generally too small to account for the very high Sr contents of most brines. Depending on the process of Sr enrichment, the marine Sr isotopic signature of the original interstitial water may or may not be obliterated. Dolomitization, recrystallization of marine carbonate minerals, rehydration of anhydrite, and dissolution of celestite are not likely to change significantly the marine isotopic signatures of the waters, because Sr in rocks and minerals involved in these diagenetic processes is also of marine origin. The effect of albitization of plagioclase on the Sr isotopic composition of marine water cannot be accurately predicted, because the Sr isotopic compositions of plagioclase minerals are commonly widely varied, ranging from values which are less than to somewhat more than that of sea water. It can, however, be assumed that plagioclase minerals, unless they are very old, are not very high in $^{87}Sr/^{86}Sr$ ratios due to their very low Rb/Sr ratios, in which case the albitization process may not greatly increase the $^{87}Sr/^{86}Sr$ ratio of the waters. Thus, from the viewpoint of Sr isotopic composition, Sr enrichment processes of subsurface waters cannot be distinguished when the isotopic values of the waters are within the range of values for marine Sr. The data presented in figure 2 indicate that nearly 50% of the samples have Sr isotopic values which are within the Phanerozoic marine Sr isotopic values between 0.7068 and 0.7092. In these instances, as discussed above, while the Sr isotopic data strongly suggest a source for marine Sr in the waters, they do not provide any definitive clue as to the

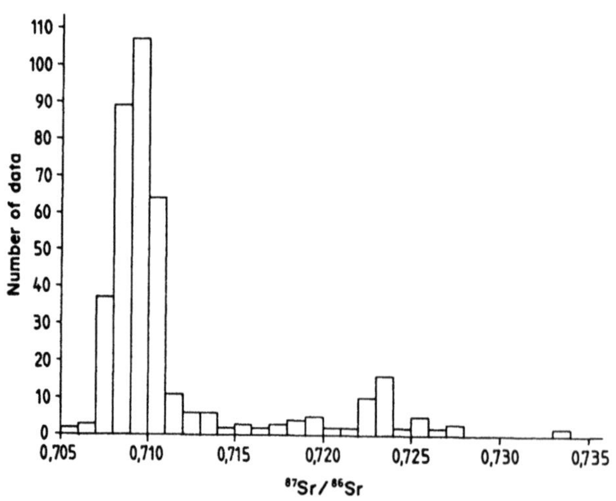

Figure 2: Histogram of Sr isotopic data of formation waters

process that might have caused the enrichment of Sr in these waters.

As noted earlier, many deep subsurface waters have Sr isotopic values which are well above the range of isotopic values for marine Sr. Evidently these waters acquired during their chemical evolution some radiogenic Sr from one or several sources. Different explanations have been offered for these ^{87}Sr enrichments. Ion exchange with clay minerals such as illite (Stueber et al., 1984, 1987; McNutt et al., 1987) and leaching and dissolution of ^{87}Sr-enriched alkali feldspar or even albitization of plagioclase (Burtner, 1987; Chaudhuri et al., 1987) are the common explanations for isotopic signatures above the value of marine Sr. We should add that evaporite deposits could also serve as a source of radiogenic ^{87}Sr in some instances, as Register and Brookins (1980) reported $^{87}Sr/^{86}Sr$ ratios as high as 1.641 for potassium-rich salts, suggesting that dissolution of these salts can contribute to the high $^{87}Sr/^{86}Sr$ ratios for brines.

While illite and illite-smectite mixed-layer minerals do contribute some Sr high in $^{87}Sr/^{86}Sr$ ratios relative to the values for marine Sr, we view the role of ion exchange by clay minerals to be minor in elevating the $^{87}Sr/^{86}Sr$ ratios, at least those of deep subsurface waters whose values are higher than 0.715. The reaction between clay minerals and formation waters can be described as either an ion-exchange phenomenon, or a recrystallization or transformation process of clay minerals. Leaching experiments on illite and illite-smectite mixed-layer minerals have shown that the $^{87}Sr/^{86}Sr$ ratios of the exchangeable Sr, which includes both the surface-adsorbed and interlayered Sr, are seldom higher than 0.714 and often less than 0.712 (Clauer, 1976; Chaudhuri and Brookins, 1978; Clauer et al., 1984; Clauer et al., 1989). These studies also indicate that the amount of exchangeable Sr in illite-rich clay minerals is only about a few micrograms per gram of the materials, while in smectite it is about ten times that in illite. We also measured the total amounts of exchangeable Sr in an illite and a smectite. In leaching 1g of each sample in 25 ml of 0.2 molar HCl at room temperature, the total amount of exchangeable Sr for the illite sample was found to be about 0.1 $\mu g/g$ of material per ml solution, and that for the smectite sample was about 6 times higher. The data on the illite were used to evaluate the possible influence of ion-exchange reactions on the Sr isotopic compositions of brines from Smackover Formation in Arkansas for which Stueber et al. (1984) reported $^{87}Sr/^{86}Sr$ ratios ranging from 0.7100 to 0.7110 and Sr contents between 580 and 2,930 mg/l. If we assume that Sr in these brines derived from two sources, one of which came from a marine source with an $^{87}Sr/^{86}Sr$ ratio of about 0.709 and the other from exchangeable Sr of clay minerals having an $^{87}Sr/^{86}Sr$ ratio of about 0.712, the mass balance calculation shows that about 200 to 980 μg of Sr in each ml of brine had to be derived from clay minerals through ion-exchange process. This would necessitate that each ml of water had to react with 2,000 to 10,000 g of clay minerals. Such a low water/illite ratio renders the ion-exchange with illite as an unlikely cause for the $^{87}Sr/^{86}Sr$ ratio of the brines to be significantly higher than the sea-water value. A similar mass balance calculation with smectite, which has a lower overall $^{87}Sr/^{86}Sr$ ratio than illite, would also require an unreasonably low water/smectite ratio to elevate significantly the $^{87}Sr/^{86}Sr$ of the water above the sea-water value. Ion-exchange with clay minerals is even more unlikely for the waters with Sr isotopic values much above 0.715, such as those found for the oil-field waters in Paleozoic sedimentary rocks in western Kansas and eastern Colorado (Chaudhuri, 1978; Chaudhuri et al., 1987; Chaudhuri et al., in press).

Transformation of smectite to illite is a common process of clay mineral alteration during burial diagenesis. Noting that the $^{87}Sr/^{86}Sr$ ratios of brines in the Illinois basin with values around 0.710 are higher than the isotopic values of marine Sr, Stueber et al. (1987) related the high values for the brines to movements of evaporite-related K-rich brines through shale beds causing transformation of smectite to illite, as Carpenter (1978) previously suggested, with attendant elevation of the $^{87}Sr/^{86}Sr$ ratios of the waters. Many believe that the K necessary for the transformation of smectite to illite in deep diagenetic environment comes from destruction of alkali-feldspar or detrital mica (e.g. Hower et al., 1976). This destruction of alkali-bearing silicate minerals, often highly enriched in ^{87}Sr, would elevate the $^{87}Sr/^{86}Sr$ ratio of the pore waters. Hence, the destruction of alkali-bearing silicate minerals such as feldspar and mica, and not the transformation of smectite to illite, is the most dominant cause for the elevated $^{87}Sr/^{86}Sr$ ratios of brines in the Illinois basin. This argument does not rule out the importance of transformation of smectite to illite in causing elevation of the $^{87}Sr/^{86}Sr$ ratios of pore waters, but the effect by the clay transformation is in most cases overshadowed by the destruction of alkali-bearing silicate minerals. Independent criteria are needed to evaluate the importance of the role of illitization of smectite.

In summary, we believe that reaction with feldspar more than that with clay minerals causes the elevation of the $^{87}Sr/^{86}Sr$ ratio of many deep subsurface waters. The role of clay minerals can become important in the filtration processes, but experimental studies are needed to assess the importance of ultrafiltration in the evolution of Sr isotopic compositions of formation waters.

Strontium Isotopic Relations between Waters and Host Rocks

The Sr isotopic relationship between deep subsurface waters and their present-day host rocks is an important consideration in the reconstruction of the chemical evolution of the waters and in the determination of the role of formation waters in the diagenetic modifications of the host rocks. Many studies on deep subsurface waters have shown that these waters are not in Sr isotopic equilibrium with carbonate minerals of their host rocks. For example, highly saline oil-field waters having $^{87}Sr/^{86}Sr$ ratios of about 0.722 were found in Paleozoic carbonate rocks of west-central Kansas with $^{87}Sr/^{86}Sr$ ratios of about 0.709 (Chaudhuri et al., 1987). Similarly, in southwestern Kansas, the $^{87}Sr/^{86}Sr$ ratios of oil-field waters have been found to be different from $^{87}Sr/^{86}Sr$ ratios of ankerite cements in Pennsylvanian sandstone reservoir rocks (Chaudhuri et al., 1985). McNutt et al. (1987) investigated several reservoir rocks and late-stage diagenetic mineral phases associated with brines in the Michigan and the Appalachian basins. Except in one case where diagenetic calcite appeared to have nearly the same Sr isotopic composition as the brine, secondary mineral deposits and their host rocks were found to be isotopically different from the present-day brines. Stueber et al. (1984), in their studies of brines in the Smackover Formation of Arkansas, also found Sr isotopic dissimilarities between brines and host rocks. Recognizing that oil-field brines are often deficient in Mg, many have been inclined to believe that existing formation waters were responsible for large scale dolomitization of limestones in the sedimentary basins. But the numerous examples of the lack of isotopic equilibrium between deep subsurface waters and the associated reservoir rocks suggest that much of recrystallization-precipitation history of reservoir rocks pre-dates the migration and entrapment of the existing waters. This observation of the lack of isotopic equilibrium between waters and host rocks also strengthens the argument that much of mineralogic stabilization of carbonate rocks in sedimentary basins occurs during early diagenesis as suggested by Clauer et al. (1989).

Aquifer in deep subsurface often contains records of different episodes of influxes of chemically different waters that entered the aquifer. Secondary vug-filling minerals in the same reservoir rock can differ from each other in their Sr isotopic compositions (Cassan et al., 1983; Chaudhuri et al., 1987; McNutt et al., 1987; Stueber et al., 1987). The chemical relationship between the present formation waters and the previous ones, the clue of which resides in the secondary minerals, has not been seriously examined. Studies of fluid inclusions in secondary minerals of reservoir rocks are needed along with the studies of the formation waters to describe time-dependent changes in the chemistry of waters which have infiltrated the aquifer. The work of Rickenbach (1988) has shown considerable promise in the value of combining fluid-inclusion data with the geochemical data on formation waters to understand the hydrochemical evolution of a reservoir. He described two generations of fluid inclusions in quartz overgrowths in sandstones of the Mahakam delta in Indonesia and noted that both the temperature and the salinity of the late included fluids are apparently the same as that of the saline waters presently occurring in the rocks, the temperature and salinity of the early fluid being higher.

Variations in Strontium Isotopic Composition of Formation Waters

Knowledge of Sr isotopic variations of formation waters in different stratigraphic units of a sedimentary basin can give insight to the configurations of fluid paths. Such knowledge can be used to understand movement of ore fluids, migration and accumulation of hydrocarbons, and hydraulic relations among different stratigraphic units in a basin.

A wide variety of Sr isotopic relationships exists among formation waters in different stratigraphic units of a sedimentary basin. Stueber et al. (1987) examined the Sr isotopic compositions of formation waters from stratigraphic units ranging in age from Ordovician to Pennsylvanian in the Illinois basin (U.S.A.). They noted that the $^{87}Sr/^{86}Sr$ ratios of the waters from nearly all major stratigraphic units, with the exception of waters from a Silurian unit, ranged from 0.7087 to 0.7108 and that the Sr contents of these waters differed by as much as 906 mg/l. The isotopic values for waters from the Silurian unit were between 0.7091 and 0.7108, but their Sr contents were widely varied between about 2 and 908 mg/l. The $^{87}Sr/^{86}Sr$ ratios of formation waters in the Silurian unit,

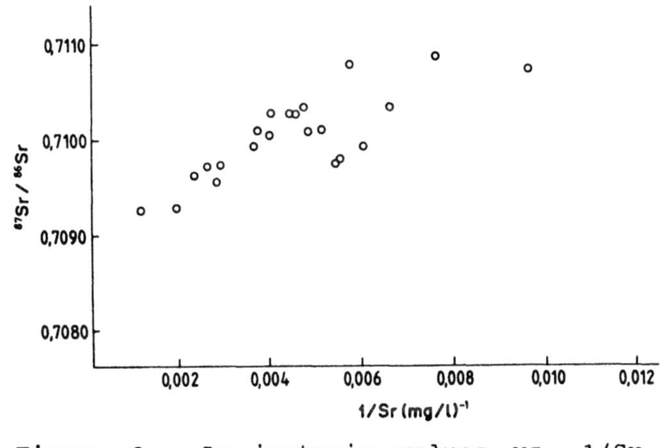

Figure 3a: Sr isotopic values vs. 1/Sr for Silurian and Devonian formation waters in the Illinois Basin (After Stueber et al., 1987)

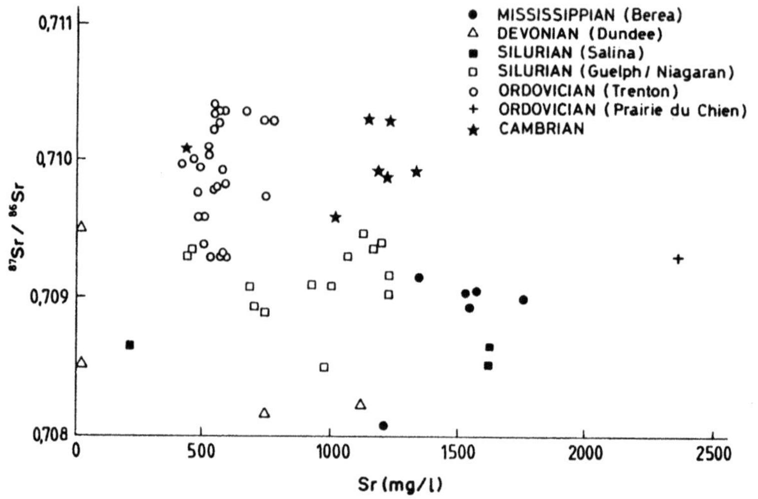

Figure 3b: $^{87}Sr/^{86}Sr$ ratios vs. Sr contents for Paleozoic units other then the Silurian and Devonian strata (After Stueber et al., 1987)

and also of some formation waters in Devonian strata, were found to bear an approximately linear relationship to the 1/Sr ratios of the waters (Fig. 3a). Stueber et al. concluded that the waters from these Silurian and Devonian units were forming a single hydrologic system which evolved through mixing of two isotopically different formation waters. They also suggested that the limited variations in the $^{87}Sr/^{86}Sr$ ratios of the waters in the Silurian strata reflect a period of regional ground-water flow. In contrast to the data from Silurian-Devonian units, the relationship between the $^{87}Sr/^{86}Sr$ ratios and the Sr contents of the waters from other Paleozoic units is highly scattered (Fig. 3b). The high degree of scatter suggests that the evolution of Sr in formation waters within an entire basin is more complex than a simple mixing of two components. The study of Clayton et al. (1966) previously indicated that the stable isotopic compositions of water molecules in formation waters from an entire basin might be explained in terms of mixing of two components, one component of which having the ^{18}O and D contents about the same as the local meteoric water. McNutt et al. (1987) also observed such differences between the Sr isotopic data and the stable isotope data for brines in the Michigan and the Appalachian basins. Varied mixing of two components can explain the stable isotope data of these brines, but not the Sr isotopic data. A possibility for the failure of the Sr isotopic data to reveal the two component mixing is that in addition to two-component mixing, varied rock-water interactions occurred during migration and residence of the waters in the reservoir rock. But no study yet has clearly identified different sources of Sr for brines in sedimentary basins.

Analyzing the Sr isotopic compositions of formation waters in rocks from Cambrian to Mississippian ages in the Michigan and the Appalachian basins in southwest Ontario, Canada and the U.S.A., McNutt et al. (1987) also noted that the $^{87}Sr/^{86}Sr$ ratios of dissolved Sr were limited to a narrow range between 0.7081 and 0.7104 with a wide range of Sr concentrations between 12 and 2,350 mg/l, like the waters in the Illinois basin analyzed by Stueber et al. (1987). Despite limited variations in the Sr isotopic compositions, the data proved useful to differentiate brines produced from different major stratigraphic units (Fig. 4). For example, brines from Cambrian and Ordovician units with $^{87}Sr/^{86}Sr$ ratios between 0.7090 and 0.7104 and Sr contents between 418 and 2,350 mg/l were found to be somewhat enriched in ^{87}Sr compared to brines produced from Mississippian and Silurian units with $^{87}Sr/^{86}Sr$ ratios between 0.7081 and 0.7095 and Sr contents between 214 and 2,060 mg/l. None

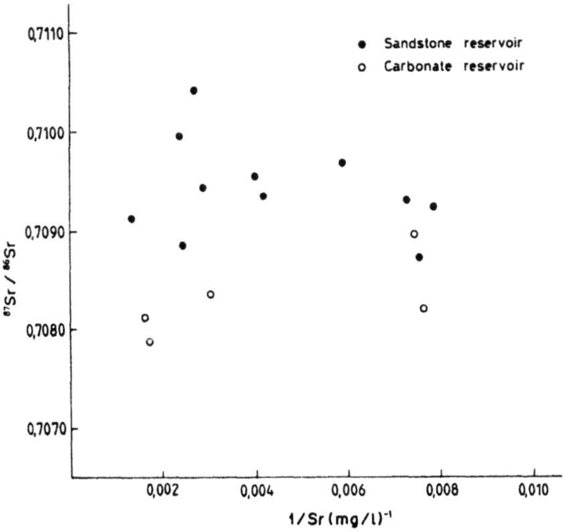

Figure 4: Sr isotopic data of oil-field waters from the Michigan and the Appalachian basins (McNutt et al., 1987)

of the Sr isotopic data for the waters of any of the major stratigraphic units can be described in terms of mixing of two components with distinct Sr isotopic compositions. Nevertheless, McNutt et al. noted that in the Ordovician unit, where the formation waters had a very narrow $^{87}Sr/^{86}Sr$ range (0.7092-0.7105), brines from the edge of the basin in Ontario had nearly the same Sr isotopic values as those from the deep parts of the basin in Michigan. They alluded the isotopic uniformity to migra-tion of brines over hundreds of kilometers, supporting a popular concept of long distance migration for fluids in sedimentary basins.

Unlike formation waters in the Illinois, the Michigan, and the Appalachian basins, those from Paleozoic beds in the northern shelf of the Anadarko basin in Kansas and Colorado are widely varied in $^{87}Sr/^{86}Sr$ ratios between 0.708 and 0.724. For example, waters from carbonate rocks of Upper Pennsylvanian Lansing-Kansas City Groups were found to have $^{87}Sr/^{86}Sr$ ratios between 0.7088 and 0.7102 and Sr contents ranging from 21 to 1,400 mg/l (Stell et al., 1988), but formation waters in carbonate and siliciclastic rocks of Middle to Lower Pennsylvanian, Mississippian, and Cambro-Ordovician were found to have $^{87}Sr/^{86}Sr$ ratios as high as 0.724 with Sr contents often less than 80 mg/l.

The lateral variations in the $^{87}Sr/^{86}Sr$ ratio of formation waters from a single stratigraphic unit could be as much as the vertical variations in the $^{87}Sr/^{86}Sr$ ratios of waters across several stratigraphic intervals. For example, formation waters in Lower Pennsylvanian Morrow sandstones in western Kansas and eastern Colorado have $^{87}Sr/^{86}Sr$ ratios between 0.718 and 0.722 and Sr contents ranging from over 50 mg/l to less than 150 mg/l, but those in Morrow sandstones in southwestern and southern Kansas had $^{87}Sr/^{86}Sr$ ratios between 0.709 and 0.711 with Sr contents as high as 900 mg/l (Fig. 5). The precise cause for these isotopic variations has yet to be determined, although the isotopic data appeared to correspond broadly with the depositional environments of the sandstones. The high $^{87}Sr/^{86}Sr$ ratios were found for waters occurring in fluvial facies in western Kansas and eastern Colorado, whereas the low isotopic values were observed for waters associated with equivalent rocks of marine facies in southern and southwestern Kansas.

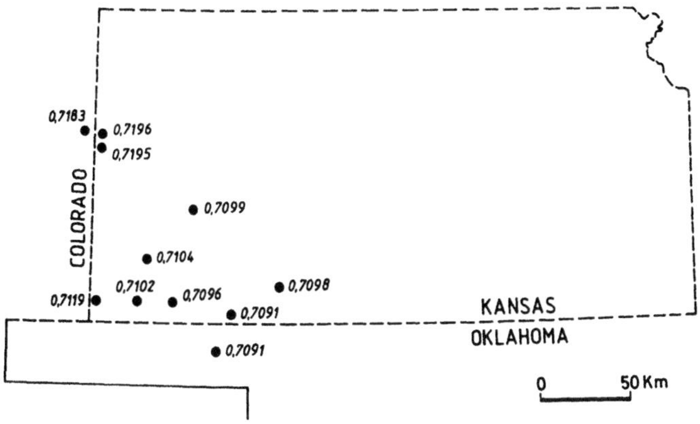

Figure 5: Sr isotopic data of Morrowan formation waters in the Hugoton Basin (Chaudhuri et al., in press)

The Sr isotopic compositions of formation waters have also been found to be varied within a single oil field. Stueber et al. (1984) reported a small range of $^{87}Sr/^{86}Sr$ ratios from 0.70802 to 0.70867 with Sr contents ranging from 169 to about 2,800 mg/l for waters of the Smackover carbonates in Walker Creek field in south Arkansas. Chaudhuri et al. (1987) reported a somewhat larger spread in the $^{87}Sr/^{86}Sr$ ratios from 0.72195 to 0.72300 but with a limited variation in the Sr contents between 40 and 46 mg/l for waters in Mississippian carbonate rocks of Bindley field in west-central Kansas. Isotopic variations of similar magnitude were found in other individual oil-fields in the Illinois, the Michigan, and the Appalachian basins (Stueber et al. ,1987; McNutt et al., 1987). Chaudhuri et al. (1987) suggested that the variations in the $^{87}Sr/^{86}Sr$ ratios among the waters of Bindley field in Kansas reflect isolated pools and that the small isotopic variations obtained are related to limited local dissolutions of the host carbonate rocks. Stueber et al. (1987) and McNutt et al. (1987) similarly explained the intra-field variations in the $^{87}Sr/^{86}Sr$ ratios of formation waters from the Illinois and the Michigan basins.

FORMATION WATERS IN CRYSTALLINE ROCKS

A number of studies in recent years have addressed the origin of saline waters in crystalline rocks (Frape and Fritz, 1981, 1982; Fritz and Frape, 1982; Frape et al.,1984; McNutt et al., 1984; Edmunds et al., 1984; Kelly et al., 1986; Edmunds et al., 1987; McNutt, 1987; McNutt et al., 1987; Fritz and Frape, 1987; Fritz et al., 1987; Nordstrom and Olsson, 1987; Pearson, 1987; Smalley, et al., 1988; Nurmi et al., 1988; Clauer et al., 1989; Moser et al., 1989; Nordstrom et al., 1989a, 1989b, 1989c; McNutt et al., 1990). While reconfirming the results of early studies concerning the broad chemical characteristics of these waters, the recent studies have revealed many different complexities in the source of these waters and also in the evolution of their solutes. But the results have yet to develop a consensus of opinion on the origin of saline waters in crystalline rocks.

Chemical Characteristics of Deep Ground Waters in Crystalline Rocks

Total dissolved solids in waters of crystalline rocks are widely varied ranging from about 50 mg/l at shallow depths of few tens of meters to more than 320,000 mg/l at depths of several hundred meters. Whereas the shallow waters with low salinity are typically of $Ca\text{-}HCO_3$ type, the highly saline deep waters are frequently of Ca-Na-Cl type and less commonly of Na-Ca-Cl type. Fritz and Frape (1982) provided an overview of the composition of dissolved constituents in saline waters in crystalline rocks of the Canadian Shield, and a recent summary of the composition is given by Fritz and Frape (1987). Table 2 is a compilation of major chemical data, following the works of Fritz and Frape (1982), Kelly et al. (1986), Nordstrom et al. (1989b), and McNutt et al. (1990), for saline waters from different localities in Precambrian shields in Canada and the U.S.A. and from a granitic pluton at Stripa in Sweden. The content of total dissolved solids in deep ground waters in Stripa granite is only a small fraction as compared to the total dissolved solids of deep ground waters in the Precambrian shields in Canada and the U. S. A. The Stripa waters are of Na-Ca-Cl type, whereas the waters in the Precambrian shields in Canada and the U.S.A. are primarily of Ca-Na-Cl type. Fritz and Frape (1982) considered that Ca-Na-Cl waters evolve from Na-Ca-Cl waters. While deep ground waters in crystalline rocks range from Na-Ca-Cl type to Ca-Na-Cl type, the common chemical traits between these two types of waters are the very low abundances of K relative to Na and the low contents of Mg relative to Ca. The Na/Cl molar ratios in highly saline waters in crystalline rocks are often less than 0.6.

Table 2: Range of total dissolved solids and contents of some constituents in saline waters from crystalline rocks

Components	Contents in mg/kg
TDS	20,340 - 324,500
Ca	3,740 - 65,000
Na	2,200 - 45,000
Mg	12 - 5,100
K	32 - 495
Sr	65 - 1,640
Cl	11,520 - 207,000
Br	111 - 1,760
SO_4	1 - 405
HCO_3	2 - 141

compiled from Frape and Fritz (1987)

The origin of the dissolved constituents in saline waters in crystalline rocks is complex and a contentious issue. Nordstrom et al (1989) have recently summarized different processes which have been suggested for the source of salinity and the origin of these waters. Taking into consideration of whether or not the major dissolved constituents derived from outside or inside the rock mass, Nordstrom et al. categorized the sources of solutes in the deep saline waters into three groups: allochthonous, autochthonous, or both. Advocates of allochthonous source for the solutes advance several different processes which include involvement of basinal brine or sea water whose composition could have been modified by evaporation or dissolution of marine evaporites (Lindewald, 1981; Lahermo and Lampen, 1987; Michelot et al., 1984; Spencer, 1987), or by long-term low temperature water-rock interactions (Kelly et al., 1986; Guha and Kanwar, 1987). Proponents of autochthonous source emphasize the importance of indigenous water-rock interactions which involve reactions with silicate minerals (Edmunds et al., 1984, 1985, 1987; Kamineni, 1987; Pearson, 1987; Fritz et al., 1987; Clauer et al; 1989), radiolysis of water (Vovk, 1981, 1987), and leakage of highly saline fluid inclusions (Grigsby et al., 1983; Nordstrom et al., 1985; Nordstrom and Olsson ,1987; Peters, 1986; Gascoyne et al., 1987). Both autochthonous and allochthonous sources for the solutes were proposed by Fritz and Frape (1982),

Frape et al. (1984), and Fritz and Frape (1987).

Strontium Isotopes and Origin of the Dissolved Components

Fritz and Frape (1982) and others noted that the stable isotope compositions of deep subsurface waters in crystalline rocks are systematically varied both within a single area and from one area to another. In studies of subsurface waters from several different districts in the Canadian Shield, Fritz and Frape (1987) recognized that in many instances the δ D and $\delta^{18}O$ values of the waters lie above the global meteoric water line and define a linear trend with a slope greater than that of the meteoric water line. The isotopic compositions of the waters from each district may be described by mixing of local meteoric water with a highly saline water source (Fig. 6). Despite wide spatial variations in the occurrences of subsurface waters, the highly saline fluid component of ground water in each of the areas studied appears to have very similar stable isotope compositions. The stable isotope composition of this highly saline component may be roughly described by a $\delta^{18}O$ value above -11 per mil and a δD value above -30 per mil. Although

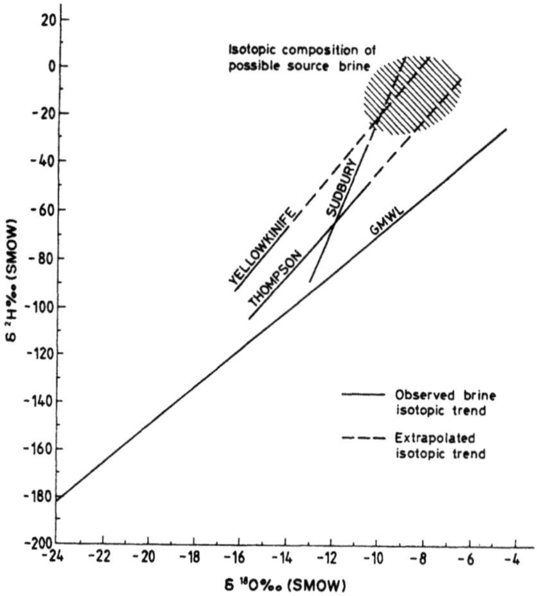

Figure 6: The stable isotope compositions of waters in crystalline rocks (Frape and Fritz, 1987)

the chemical composition of the highly saline component may be approximately described by extrapolations from relationships between the $\delta^{18}O$ values and different chemical parameters, the origin of the saline component, that is the source of the water and the evolution of the chemical components in the water, remains largely unknown.

To assess how Sr and Ca of highly saline waters in crystalline rocks have evolved from water-rock interactions, McNutt et al. (1984) made some Sr isotopic studies on both waters and associated minerals and rocks. This early study showed strong similarity in the range of $^{87}Sr/^{86}Sr$ ratios between the water samples and their associated Precambrian crystalline rocks in the Sudbury district of Canada. The results indicated that the Sr in the water is the result of extensive water-rock interactions within the locality. The authors also observed two isolated pockets of water in the North Range having $^{87}Sr/^{86}Sr$ ratios of 0.711 and 0.716, which they attributed to different modal compositions of the noritic host rocks consisting primarily in plagioclase and pyroxene minerals with different Sr isotopic compositions. McNutt et al. suggested that isotopic exchange at all sites probably continued till the present time. In contrast to the waters in the North Range area, those in the South Range were found to be highly enriched in ^{87}Sr with Sr isotopic values between 0.725 and 0.740. The high content of ^{87}Sr-enriched minerals, such as biotite, and to some extent hornblende, in quartz diorite rocks in the South Range, has been considered to be the cause for the high Sr isotopic values of the waters. In the Sudbury district of Canada, McNutt et al. (1984) noted that fracture-filling gypsum and brines had the same Sr isotopic values. In a later study of highly saline waters in the Canadian Shield, McNutt et al. (1987) noted that many secondary carbonate and sulfate minerals in the crystalline rocks, such as calcite and gypsum, had nearly the same Sr isotopic values as those of the associated brines. The isotopic similarity suggests that either the minerals precipitated from the brines or complete isotopic exchange occurred through interactions between the minerals and the brines.

McNutt et al. (1987) observed Sr isotopic similarity between brines and fracture filling materials, but isotopic dissimilarity between brines and host rocks. This isotopic relationship among host rocks, fracture filling-materials, and brines raised a prospect that ages of the highly saline waters may be determined, assuming that: (1) the water remained in a closed system and did not react with the wall-rock since the formation of the secondary mineral, (2) the reaction was a whole-rock reaction so that the secondary mineral, the brine and the whole rock had the same Sr isotopic composition, and (3) the water at the time of the reaction carried no isotopically different extraneous Sr (McNutt et al., 1990). Because many of the secondary minerals had very low Rb/Sr ratios, their initial isotopic compositions may be taken as the Sr isotopic composition of the rock at their time of formation. And hence from knowledge of the present-day $^{87}Sr/^{86}Sr$ and Rb/Sr ratios of the rock, an age may be obtained which will correspond to the time of the rock-water interaction and the age of the water. But such an approach to date ground waters must be viewed with extreme caution, because the requirements for the dating are too specific to be met by most natural systems.

The works of Fritz et al. (1987) and Clauer et al. (1989) on ground waters and secondary minerals of the Stripa granite in central Sweden have shown that the evolution of secondary minerals might be complex. Fritz et al. observed that the $^{87}Sr/^{86}Sr$ ratios of deep ground waters with values between 0.7475 and 0.7536 are very close to that of fracture coating minerals, which consist mainly of chlorite. They concluded that the chemistry of the waters resulted from long term interactions with the fracture coating minerals and not with the major minerals of the host rock. In a more recent study on fracture minerals of the same granite, Clauer et al. recognized at least three generations of calcite, each having a distinct range of Sr isotopic values and distinctive O and C isotope compositions. The group of calcite with the lowest $^{87}Sr/^{86}Sr$ ratios between 0.7264 and 0.7406 appeared to be the youngest among the different generations. The O and C isotopic data on this group of calcite suggest that the vein calcites formed during low temperature reactions involving methane oxidation or sulfate reduction. Another group of calcite, which was mainly associated with chlorite, had $^{87}Sr/^{87}Sr$ ratios between 0.7406 and

0.7536, which were nearly the same as that of ground waters in the same general locality. The third group of calcite, associated with epidote and chlorite, had very high $^{87}Sr/^{86}Sr$ ratios between 0.7814 and 1.0696. This group of calcite is believed to have formed at temperatures above 200°C during some metamorphic events between 1.4 and 0.8 billion years ago.

In a recent study of deep gound waters, generally brackish to saline and of Ca-Na-Cl type, from Late Archean felsic plutonic rocks located in NW Ontario in the Superior Province of the Canadian Shield, Franklyn et al. (1991) observed that the $^{87}Sr/^{86}Sr$ ratios of the waters were controlled primarily by rapid plagioclase-water interaction relative to other rock forming minerals (K-feldspar and biotite) and water interactions. The $^{87}Sr/^{86}Sr$ ratios of the waters ranged from 0.7057 to 0.7070, which are similar to the values for unaltered plagioclase minerals. The $^{87}Sr/^{86}Sr$ ratios of calcite and gypsum, occurring as fracture filling minerals in the host rocks, were also similar to that of the present deep ground waters.

The studies of Clauer et al. (1989), McNutt et al. (1990), and Franklyn et al. (1991) have clearly shown that very often the Sr of the brines is not in isotopic equilibrium with that of the present-day host rocks. Many explanations may be given for these isotopic differences, but the identification of the cause or causes in any situation remains a formidable task.

CONCLUSION

Studies on Sr isotopic compositions of formation waters in sedimentary rocks can provide critical constraints on a model which may be developed to explain the evolution of chemical compositions of these waters. The dissolved Sr in formation waters are often enriched in ^{87}Sr relative to marine Sr. Most of the enrichment in ^{87}Sr of deep subsurface waters in sedimentary basins cannot be attributed to ion-exchange reaction between clay minerals and water or reaction between plagioclase and water. Illitization of smectite and dissolutions of ^{87}Sr-enriched evaporite minerals, alkali feldspar, and also of detrital mica in the deep subsurface are potential sources for radiogenic ^{87}Sr in formation waters in sedimentary basins.

Investigation of isotopic compositions of dissolved Sr in formation waters is important because such data can greatly expand our knowledge on the hydrology in sedimentary basins. The results of studies made thus far suggest that while pervasive hydraulic intercommunications may exist among some adjacent major stratigraphic units in a sedimentary basin, hydraulic isolations is common among major stratigraphic units on a basin scale. If cross-formational flows in sedimentary basins are of limited nature, as suggested by the Sr isotopic data, then hydrologic models based on major cross-formational flow to drive the process of membrane filtration (e.g. Bredehoeft et al., 1963), appear inadequate for the process of ultrafiltration needed to account for the high salinity of the waters.

In some instances, the array of Sr isotopic data of formation waters can be described in terms of mixing of two distinct components of Sr. A linear relationship between $^{87}Sr/^{86}Sr$ ratios and the reciprocal of Sr contents of the waters is indicative of mixing of Sr from two sources. The apparent linearity suggests either that the dissolved Sr in the waters results from mixing of two isotopically and chemically distinct waters, or that waters with narrow ranges of Sr isotopic compositions and Sr contents derived varying amounts of Sr from one or several types of minerals with narrow limits in $^{87}Sr/^{86}Sr$ ratios and Sr contents. However, non-linear relationships, rather than linear relationships between $^{87}Sr/^{86}Sr$ and 1/Sr ratios, are the most common phenomena for formation waters in sedimentary basins. Any rational solution to the presence of both phenomena of linearity and non-linearity in the $^{87}Sr/^{86}Sr$ and 1/Sr relations in the same sedimentary basin would require integration of inputs of both chemical and physical information of the hydrostratigraphic units.

More studies are needed to derive full benefits of the isotopic data of formation waters relating to basinal hydrology and diagenetic history of sedimentary rocks. Geochemical and isotopic studies on basinal rocks should be conducted simultaneously with those on formation waters. Because of many complexities involved in water-rock interactions, the results from radiogenic isotopes should be integrated with those from the analyses on stable isotope compositions and major and trace elemental compositions of waters to determine paths of chemical evolution of formation waters and hydraulic relationships of different waters in a sedimentary basin.

The deep formation waters in crystalline rocks from the Canadian Shield have total dissolved solids of more than 200,000 mg/l and they are of Ca-Na-Cl type with molar Na/Cl often much less than 0.6. Similar high salinity and Ca-Na-Cl type waters have been found in very deep drillholes in Kola Peninsula. The deep groundwaters of the Stripa granite in central Sweden, on the other hand, generally have TDS of less than 1,300 mg/l and they are of Na-Ca-Cl type with molar Na/Cl of about 0.70. By contrast, formation waters in sedimentary basins often range from nearly Na-Cl type with molar Na/Cl of about 1 to Na-Ca-Cl type with molar Na/Cl around 0.6 and seldom less than 0.5. The comparison between the formation waters in the Stripa granite and those in the Canadian Shield suggests that the progressive development of brines in crystalline rocks takes place along the line from Na-Ca-Cl type to Ca-Na-Cl type. Speculation may be made that very deep brines in crystalline rocks may evolve toward a Ca-Cl end member by appropriate dissolution-precipitation reactions in the subsurface. The overlap in the chemical evolutionary paths between the brines in sedimentary basins and those in crystalline rocks has led to two lines of speculation about the hydrodynamic linkage between the two distinct reservoirs deep in the subsurface (McNutt et al., 1987). One interpretation is that brines formed in crystalline rocks migrate to sedimentary basins where Na-enriched rocks are common to influence the composition of the brines toward a Na-Cl end member, the other interpretation is that the brines formed in sedimentary basins move deep into crystalline rocks where water-rock reactions lead to an evolution of the brines toward a Ca-Cl end member. The debate may be somewhat resolved if the brines in crystalline rocks are proved to be primarily of autochthonous origin, as the evidence is overwhelming that brines in sedimentary basins often appear to have migration histories of several tens or few hundred kilometers suggesting an allochthonous origin with respect to the reservoir rocks.

ACKNOWLEDGEMENTS

We thank P. Stille (Centre de Géochimie de la Surface, Strasbourg, France) and R.H. McNutt (Mac Master University, Hamilton, Canada) for constructive discussions and review of the manuscript, and Mr. C. Hammel for drafting the figures.

REFERENCES

Banner J.L., Wasserburg G.J., Dobson P.F., Carpenter A.B. and Moore C.H. (1989) Isotopic and trace element constraints on the origin and evolution of saline groundwaters from central Missouri. Geochim. Cosmochim. Acta, 53, 383-398.

Bredehoeft J.D., Blyth C.R., White W.A. and Maxey G.B. (1963) Possible mechanism for concentration of brines in subsurface formations. Am. Ass. Petrol. Geol. Bull., 47, 257-263.

Burtner R.L. (1987) Origin and evolution of Weber and Tensleep formation waters in the green River and Uinta-Piceance basins, northern Rocky Mountain area. Chem. Geol., 65, 255-282.

Carpenter A. B. (1978) Origin and chemical evolution of brines in sedimentary basins. Oklahoma Geol. Surv. Circ., 79, 60-77.

Cassan J.P., Clauer N., Fritz B. and Tardy Y. (1983) Water circulations and mass transfers during the diagenesis of oil-bearing reservoirs. In: Extended abstracts, Seki I. (ed.), 4th Int. Symp. Water-Rock Interaction, Misasa, Japan, 75-78.

Chaudhuri S. (1978) Strontium isotopic composition of several oilfield brines from Kansas and Colorado. Geochim. Cosmochim. Acta, 42, 329-332.

Chaudhuri S. and Brookins D. G. (1978) The Rb-Sr systematics in acid-leached clay minerals. Chem. Geol., 24, 327-338.

Chaudhuri S., Clauer N. and O'Neil J.R. (1985) Hydrochemical evolution of some oil-field waters from central Kansas, U.S.A.. Terra cognita, 3, 177-178.

Chaudhuri S., Broedel V. and Clauer N. (1987) Strontium isotopic evolution of oil-field waters from carbonate reservoir rocks in Bindley field, central Kansas, U.S.A. Geochim. Cosmochim. Acta, 51, 45-54.

Chaudhuri S., Robinson R., Clauer. N. and Jones L. M. (in press) Evolution of formation waters in early Pennsylvanian Morrowan sandstones in the Hugoton embayment of the Anadarko basin in southern midcontinent of the U.S.A. Oklahoma Geol. Surv. Circ. (in press).

Clauer N. (1976) Géochimie isotopique du strontium des milieux sédimentaires. Application à la géochronologie de la couver-

ture du craton ouest-Africain. Sci. Géol. Mém. Strasbourg, 45, 256p.
Clauer N., Giblin P. and Lucas J. (1984) Sr and Ar isotope studies of detrital smectites from the Atlantic Ocean (DSDP, Legs 43, 48 and 50). Isot. Geosc., 2, 141-151.
Clauer N., Frape S.K. and Fritz B. (1989) Calcite veins of the Stripa granite (Sweden) as records of the origin of the grounwaters and their interactions with the granitic body. Geochim. Cosmochim. Acta, 53, 1777-1781.
Clayton R.N., Friedman I., Graf D.L., Mayeda T.K., Meents W.F. and Shimp F. (1966) The origin of saline formation waters, I: Isotopic composition. Jour. Geophys. Res., 71, 3869-3882.
Collins A. G. (1975) Geochemistry of oilfield waters. Elsevier, New York, 496p.
Coplen T.B. and Hanshaw B.B. (1973) Ultrafiltration by a compacted clay membrane, I. Oxygen and hydrogen fractionation. Geochim. Cosmochim. Acta, 37, 2295-2310.
Davis S.N. (1964) The chemistry of saline waters - Discussion. Groundwaters, 2, 51.
Dickey P.A. (1969) Increasing concentration of subsurface brines with depth. Chem. Geol., 4, 361-370.
Edmunds W.M., Andrews J.N., Burgess W.G., Kay R.L.F. and Lee D.J. (1984) The evolution of saline and thermal grounwaters in the Carnmenellis granite. Min. Mag., 48, 407-424.
Edmunds W.M., Kay R.L.F. and MacCartney R.A. (1985) Origin of saline groundwaters in the Carnmenellis granite: Natural processes and reaction during Hot Dry Rock reservoir circulation. Chem. Geol., 49, 287-302.
Edmunds W.M., Kay R.L.F., Miles D.L. and Cook J.M. (1987) The origin of saline grounwaters in the Carnmenellis granite, Cornwall (U.K.): Further evidence from minor and trace elements. In: Saline water and gases in crystalline rocks, Fritz P. and Frape S.K. (eds.), Geol. Ass. Canada, Spec. Paper, 33, 127-144.
Egeberg P. K. and Aagaard P. (1989) Origin and evolution of formation waters from oil fields on the Norwegian shelf. Applied Geochem., 4, 131-142.
Engelhardt W. von and Gaida K.H. (1963) Concentration changes of pore solutions during the compaction of clay sediments. Jour. Sediment. Petrol., 33, 919-930.
Franklyn M.T., McNutt R.H., Kamineni D.C., Gascoyne M. and Frape S.K. (1991) Grounwater $^{87}Sr/^{86}Sr$ values in the Eye-Dashwa Lakes pluton, Canada: Evidence for plagioclase-water reaction. Chem. Geol. (Isot. Geosc. Sect.), 86, 111-122.
Frape S.K. and Fritz P. (1981) A preliminary report on the occurrence and geochemistry of saline grounwaters on the Canadian Shield. Atom. Energ. Canada Ltd, Technical Record, 136, 71 pp.
Frape S.K. and Fritz P. (1982) The chemistry and isotopic composition of saline groundwaters from the Sudbury Basin, Ontario. Can. Jour. Earth Sci., 19, 645-661.
Frape S.K., Fritz P. and McNutt R.H. (1984) The role of water-rock interaction in the chemical evolution of groundwaters from the Canadian Shield. Geochim. Cosmochim. Acta, 48, 1617-1627.
Fritz B., Clauer N. and Kam M. (1987) Strontium isotopic data and geochemical calculations as indicators for the origin of saline waters in crystalline rocks. In: Saline water and gases in crystalline rocks, Fritz P. and Frape S.K. (eds.), Geol.

Ass. Canada, Spec. Paper, 33, 121-126.
Fritz P. and Frape S.K. (1982) Saline groundwaters in the Canadian Shield - a first overview. Chem. Geol., 36, 179-190.
Fritz P. and Frape S.K. (eds.) (1987) Saline water and gases in crystalline rocks, Geol. Ass. Canada, Spec. Paper, 33, 253 pp.
Fritz S.J. and Edy C.G. (1985) Hyperfiltration-induced precipitation of calcite. Geochim. Cosmochim. Acta, 49, 761-768.
Gascoyne M., Davison C.C., Ross J.D. and Pearson R. (1987) Saline groundwaters and brines in the Canadian Shield. In: Saline water and gases in crystalline rocks, Fritz P. and Frape S.K. (eds.), Geol. Ass. Canada, Spec. Paper, 33, 53-68.
Graf D. L. and Anderson D. E. (1981) Geochemical inputs for hydrologic models of deep-lying sedimentary units. Jour. Hydrol., 54, 297-314.
Grigsby C.O., Tester J.W., Trujillo P.E., Counce D.A., Abbott J., Holley C.E. and Blatz L.A. (1983) Rock-water interactions in hot dry rock geothermal systems: field investigations of in situ geochemical behavior. Jour. Volcan. Geoth. Res., 15, 101-136.
Guha J. and Kanwar R. (1987) Vug brines-fluid inclusions: a key to the understanding of secondary gold enrichment processes and the evolution of deep brines in the Canadian Shield. In: Saline water and gases in crystalline rocks, Fritz P. and Frape S.K. (eds.), Geol. Ass. Canada, Spec. Paper, 33, 95-102.
Hanshaw B.B. and Coplen T.B. (1973) Ultrafiltration by a compacted clay membrane, II. Sodium ion exclusion at various ionic strengths. Geochim. Cosmochim. Acta, 37, 2311-2327.
Herut B., Starinsky, A., Katz, A. and Bein, A. (1990) The role of sea water freezing in the formation of subsurface brines. Geochim. Cosmochim. Acta, 54, 13-21.
Hitchon B. and Friedman I. (1969) Geochemistry and origin of formation waters in the western Canada sedimentary basin, I. Stable isotopes of hydrogen and oxygen. Geochim. Cosmochim. Acta, 33, 1321-1349.
Hower J., Eslinger E.V., Hower J. and Perry E.A. (1976) Mechanism of burial metamorphism, I. Mineralogical and chemical evidence. Geol. Soc. Am. Bull., 87, 725-737.
Kamineni D.C. (1987) Halogen-bearing minerals in plutonic rocks: a possible source of chlorine in saline groundwater in the Canadian Shield. In: Saline water and gases in crystalline rocks, Fritz P. and Frape S.K. (eds.), Geol. Ass. Canada, Spec. Paper, 33, 69-80.
Kelly W.C., Rye R.O. and Livnat A. (1986) Saline mine waters of the Keweenaw Peninsula, northern Michigan: Their nature, origin and relation to similar deep waters in Precambrian crystalline rocks of the Canadian Shield. Am. Jour. Sci., 286, 281-308.
Kharaka Y.K. and Berry F.A.F. (1973) Simultaneous flow of water and solute through geological membranes: I. Experimental investigations. Geochim. Cosmochim. Acta, 37, 2577-2604.
Kharaka Y.K., Lico M.S., Wright V.A. and Carothers W.W. (1979) Geochemistry of formation waters from Pleasant Bayou N° 2 well and adjacent areas in Coastal Texas. Proc. 4th Conf. Geopressured-Geothermal Resources, Austin, Texas, 168-193.
Kharaka Y.K. and Carothers W.W. (1986) Oxygen and hydrogen isotope geochemistry of deep basin brines. In: Handbook of Environmental Isotope Geochemistry, Fritz P. and Fontes J.C. (eds.),

2, 8, 305-360.

Lahermo P.W. and Lampen P.H. (1987) Brackish and saline groundwaters in Finland. In: Saline water and gases in crystalline rocks, Fritz P. and Frape S.K. (eds.), Geol. Ass. Canada, Spec. Paper, 33, 103-110.

Land L. S. (1987) The major ion chemistry of saline brines in sedimentary basins. In: Physics and Chemistry of Porous Media II, Banavar J.R., Koplik J. and Winkler K.W. (eds.), Am. Inst. of Phys. Conf. Proc., 154, 160-179.

Land L. S. and Prezbindowski D. R. (1981) The origin and evolution of saline formation water, Lower Cretaceous carbonates, south-central Texas, U.S.A. Jour. Hydrol., 54, 51-74.

Lindewald H. (1981) Saline groundwater in Sweden. In: Intruded and relict groundwater of marine origin, 7th Salt-Water Intrusion Meeting, 24-32.

Lowry R.M., Faure G., Mullet D.I. and Jones L.M. (1988) Interpretation of chemical and isotopic compositions of brines based on mixing and dilution, "Clinton" sandstones, eastern Ohio, U.S.A. Applied Geochem., 3, 177-184.

McKelvey J.G. and Milne I.H. (1962) The flow of salt through compacted clay. Clays and Clay Min., 9, 248-259.

McNutt R.H. (1987) $^{87}Sr/^{86}Sr$ ratios as indicators of water/rock interactions: Application to brines found in Precambrian age rocks from Canada. In: Saline water and gases in crystalline rocks, Fritz P. and Frape S.K. (eds.), Geol. Ass. Canada, Spec. Paper, 33, 81-88.

McNutt R.H., Frape S.K. and Fritz P. (1984) Strontium isotopic composition of some Precambrian Shield brines. Isot. Geosc., 2, 205-215.

McNutt R. H., Frape, S. K. and Dollar P. (1987) A strontium, oxygen and hydrogen isotopic composition of brines, Michigan and Appalachian basins, Ontario and Michigan. Applied Geochem., 2, 495-505.

McNutt R.H., Frape S.K., Fritz P., Jones M.G. and MacDonald I.M. (1990) The $^{87}Sr/^{86}Sr$ values of Canadian Shield brines and fracture minerals with applications to groundwater mixing, fracture history, and geochronology. Geochim. Cosmochim. Acta, 54, 205-215.

Michelot J.L., Bentley H.W., Brissaud I., Elmore D. and Fontes J.C. (1984) Progress in environmental isotope studies (^{36}Cl, ^{34}S, ^{18}O) at the Stripa site. Isotope Hydrology, 1983, I.A.E.A., Vienna, 207-229.

Moser H., Wolf M., Fritz P., Fontes J.C., Florkowski T. and Payne B.R. (1989) Deuterium, oxygen-18, and tritium in Stripa groundwater. Geochim. Cosmochim. Acta, 53, 1757-1763.

Nordstrom D.K., Andrews J.N., Carlsson L., Fontes J.C., Fritz P., Moser H. and Olsson T. (1985) Hydrogeological and hydrogeochemical investigations in boreholes: Final report of the phase I geochemical investigations of the Stripa groundwaters. S.K.B. Stripa Project Report, Stockholm, 85-06, 250 p.

Nordstrom D.K. and Olsson T. (1987) Fluid inclusions as a source of dissolved salts in deep granitic groundwaters. In: Saline water and gases in crystalline rocks. Fritz P. and Frape S.K. (edit.), Geol. Ass. Canada, Spec. Paper, 33, 111-120.

Nordstrom D.K., Olsson T., Carlsson L. and Fritz P. (1989a) Introduction to the hydrogeochemical investigations within the International Stripa Project. Geochim. Cosmochim. Acta, 53, 1717-1726.

Nordstrom D.K., Lindblom S., Donahoe R.J. and Barton C.C. (1989b) Fluid inclusions in the Stripa granite and their possible influence on the groundwater chemistry. Geochim. Cosmochim. Acta, 53, 1741-1755.

Nordstrom D.K., Ball J.W., Donahoe R.J. and Whittemore D. (1989c) Groundwater geochemistry and water-rock interactions at Stripa. Geochim. Cosmochim. Acta, 53, 1727-1740.

Nurmi P.A., Ilmo T.K. and Lahermo W. (1988) Geochemistry and origin of saline groundwaters in the Fennoscandian Shield. Applied Geochem., 3, 185-203.

Pearson F.J., Jr. (1987) Models of mineral controls on the composition of saline groundwaters of the Canadian Shield. In: Saline water and gases in crystalline rocks, Fritz P. and Frape S.K. (eds.), Geol. Ass. Canada, Spec. Paper, 33, 39-52.

Peters T. (1986) Structurally incorporated and water extractable chlorine in the Boettstein granite (N. Switzerland). Contrib. Miner. Petrol., 94, 272-273.

Register J.K. and Brookins D.G. (1980) Rb-Sr isochron age of evaporite minerals from the Salado Formation (Late Permian), Southeastern New Mexico. Isochron/West, 29, 39-42.

Rinckenbach T. (1988) Diagenèse minérale des sédiments pétrolifères du delta fossile de la Mahakam (Indonésie). Evolution minéralogique et isotopique des composants argileux et histoire thermique. Thèse Univ. Strasbourg.

Rittenhouse G. (1967) Bromine in oil field waters and its use in determining possibilities of origin of these waters. Am. Ass. Petrol. Geol. Bull., 51, 2430-2440.

Russell W. L. (1933) Subsurface concentration of chloride brines. Am. Ass. Petrol. Geol. Bull., 17, 1213-1228.

Saigal G. C., Morad S., Bjorlykke K, Egeberg P. K. and Aagaard P. (1988) Diagenetic albitization of detrital K-feldspars in Jurassic, Lower Cretaceous and Tertiary clastic reservoir rocks from offshore Norway. I. Textures and origin. Jour. Sedimen. Petrol., 58, 1003-1013.

Smalley P.C., Blomqvist R. and Raheim A. (1988) Sr isotopic evidence for discrete saline components in stratified ground waters from crystalline bedrock, Outokumpu, Finland. Geology, 16, 354-357.

Spencer R.J. (1987) Origin of Ca-Cl brines in Devonian formations, western Canada sedimentary basin. Applied Geochem., 2, 373-384.

Starinsky A., Bielski M., Lazar B., Steinitz G. and Raab M. (1983) Strontium isotope evidence on the history of oilfield brines, Mediterranean coastal plain, Israel. Geochim. Cosmochim. Acta, 47, 687-695.

Stell M. J. (1988) Isotopic and chemical compositions of oil-field waters in Upper Pennsylvanian Lansing-Kansas City rocks in western Kansas. M.S. Thesis, Kansas State University.

Stell M.J., Chaudhuri S., Jones L.M. and Clauer N. (1987) Origin and evolution of oil-field brines in Upper Pennsylvanian Lansing-Kansas City Groups in western Kansas. Trans. Am. Geophys. Union, 68, 1275.

Stueber A. M., Pushkar P. and Hetherington E. A. (1984) A strontium isotopic study of Smackover brines and associated solids, southern Arkansas. Geochim. Cosmochim. Acta, 48, 1637- 1649.

Stueber A. M., Pushkar P. and Hetherington E. A. (1987) A strontium isotopic study of formation waters from the Illinois basin.

Applied Geochem., 2, 477-494.
Sunwall M. T. and Pushkar P. (1979) The isotopic composition of strontium in brines from petroleum fields of southeastern Ohio. Chem. Geol., 24, 189-197.
Vovk I.F. (1981) Radiolytic model of formation of brine compositions in the crystalline basement of shields. Geokhimiya, 4, 467-480.
Vovk I.F. (1987) Radiolytic salt enrichment and brines in the crystalline basement of the East European Plateform. In: Saline water and gases in crystalline rocks, Fritz P. and Frape S.K. (eds.), Geol. Ass. Canada, Spec. Paper, 33, 197-210.
White D. E. (1965) Saline waters of sedimentary rocks. In: Fluids in Subsurface Environments. Young Y. and Galley G.E. (eds.), Am. Ass. Petrol. Geol. Mem., 4, 342-366.

Lecture Notes in Earth Sciences

Vol. 1: Sedimentary and Evolutionary Cycles. Edited by U. Bayer and A. Seilacher. VI, 465 pages. 1985. (out of print).

Vol. 2: U. Bayer, Pattern Recognition Problems in Geology and Paleontology. VII, 229 pages. 1985.

Vol. 3: Th. Aigner, Storm Depositional Systems. VIII, 174 pages. 1985.

Vol. 4: Aspects of Fluvial Sedimentation in the Lower Triassic Buntsandstein of Europe. Edited by D. Mader. VIII, 626 pages. 1985.

Vol. 5: Paleogeothermics. Edited by G. Buntebarth and L. Stegena. II, 234 pages. 1986.

Vol. 6: W. Ricken, Diagenetic Bedding. X, 210 pages. 1986.

Vol. 7: Mathematical and Numerical Techniques in Physical Geodesy. Edited by H. Sünkel. IX, 548 pages. 1986.

Vol. 8: Global Bio-Events. Edited by O. H. Walliser. IX, 442 pages. 1986.

Vol. 9: G. Gerdes, W. E. Krumbein, Biolaminated Deposits. IX, 183 pages. 1987.

Vol. 10: T.M. Peryt (Ed.), The Zechstein Facies in Europe. V, 272 pages. 1987.

Vol. 11: L. Landner (Ed.), Contamination of the Environment. Proceedings, 1986. VII, 190 pages.1987.

Vol. 12: S. Turner (Ed.), Applied Geodesy. VIII, 393 pages. 1987.

Vol. 13: T. M. Peryt (Ed.), Evaporite Basins. V, 188 pages. 1987.

Vol. 14: N. Cristescu, H. I. Ene (Eds.), Rock and Soil Rheology. VIII, 289 pages. 1988.

Vol. 15: V. H. Jacobshagen (Ed.), The Atlas System of Morocco. VI, 499 pages. 1988.

Vol. 16: H. Wanner, U. Siegenthaler (Eds.), Long and Short Term Variability of Climate. VII, 175 pages. 1988.

Vol. 17: H. Bahlburg, Ch. Breitkreuz, P. Giese (Eds.), The Southern Central Andes. VIII, 261 pages. 1988.

Vol. 18: N.M.S. Rock, Numerical Geology. XI, 427 pages. 1988.

Vol. 19: E. Groten, R. Strauß (Eds.), GPS-Techniques Applied to Geodesy and Surveying. XVII, 532 pages. 1988.

Vol. 20: P. Baccini (Ed.), The Landfill. IX, 439 pages. 1989.

Vol. 21: U. Förstner, Contaminated Sediments. V, 157 pages. 1989.

Vol. 22: I. I. Mueller, S. Zerbini (Eds.), The Interdisciplinary Role of Space Geodesy. XV, 300 pages. 1989.

Vol. 23: K. B. Föllmi, Evolution of the Mid-Cretaceous Triad. VII, 153 pages. 1989.

Vol. 24: B. Knipping, Basalt Intrusions in Evaporites. VI, 132 pages. 1989.

Vol. 25: F. Sansò, R. Rummel (Eds.), Theory of Satellite Geodesy and Gravity Field Theory. XII, 491 pages. 1989.

Vol. 26: R. D. Stoll, Sediment Acoustics. V, 155 pages. 1989.

Vol. 27: G.-P. Merkler, H. Militzer, H. Hötzl, H. Armbruster, J. Brauns (Eds.), Detection of Subsurface Flow Phenomena. IX, 514 pages. 1989.

Vol. 28: V. Mosbrugger, The Tree Habit in Land Plants. V, 161 pages. 1990.

Vol. 29: F. K. Brunner, C. Rizos (Eds.), Developments in Four-Dimensional Geodesy. X, 264 pages. 1990.

Vol. 30: E. G. Kauffman, O.H. Walliser (Eds.), Extinction Events in Earth History. VI, 432 pages. 1990.

Vol. 31: K.-R. Koch, Bayesian Inference with Geodetic Applications. IX, 198 pages. 1990.

Vol. 32: B. Lehmann, Metallogeny of Tin. VIII, 211 pages. 1990.

Vol. 33: B. Allard, H. Borén, A. Grimvall (Eds.), Humic Substances in the Aquatic and Terrestrial Environment. VIII, 514 pages. 1991.

Vol. 34: R. Stein, Accumulation of Organic Carbon in Marine Sediments. XIII, 217 pages. 1991.

Vol. 35: L. Håkanson, Ecometric and Dynamic Modelling. VI, 158 pages. 1991.

Vol. 36: D. Shangguan, Cellular Growth of Crystals. XV, 209 pages. 1991.

Vol. 37: A. Armanini, G. Di Silvio (Eds.), Fluvial Hydraulics of Mountain Regions. X, 468 pages. 1991.

Vol. 38: W. Smykatz-Kloss, S. St. J. Warne, Thermal Analysis in the Geosciences. XII, 379 pages. 1991.

Vol. 39: S. E. Hjelt, Pragmatic Inversion of Geophysical Data. Approx. 240 pages (in prep.). 1992.

Vol. 40: S. W. Petters, Regional Geology of Africa. XXIII, 722 pages. 1991.

Vol. 41: R. Pflug, J. W. Harbaugh (Eds.), Computer Graphics in Geology. XVII, 298 pages. 1992.

Vol. 42: A. Cendrero, G. Lüttig, F. Chr. Wolff (Eds.), Planning the Use of the Earth's Surface. IX, 556 pages. 1992.

Vol. 43: N. Clauer, S. Chaudhuri (Eds.), Isotopic Signatures and Sedimentary Records. VIII, 529 pages. 1992.

GPSR Compliance

The European Union's (EU) General Product Safety Regulation (GPSR) is a set of rules that requires consumer products to be safe and our obligations to ensure this.

If you have any concerns about our products, you can contact us on ProductSafety@springernature.com

In case Publisher is established outside the EU, the EU authorized representative is:

Springer Nature Customer Service Center GmbH
Europaplatz 3
69115 Heidelberg, Germany

Batch number: 10838062

Printed by Printforce, the Netherlands